The Atlantic Salmon
Genetics, Conservation and Management

Edited by

Eric Verspoor
FRS Freshwater Laboratory, Scotland

Lee Stradmeyer
FRS Freshwater Laboratory, Scotland

Jennifer Nielsen
United States Geological Survey

© 2007 by Blackwell Publishing Ltd

Blackwell Publishing editorial offices:
Blackwell Publishing Ltd, 9600 Garsington Road, Oxford OX4 2DQ, UK
Tel: +44 (0)1865 776868
Blackwell Publishing Professional, 2121 State Avenue, Ames, Iowa 50014-8300, USA
Tel: +1 515 292 0140
Blackwell Publishing Asia Pty Ltd, 550 Swanston Street, Carlton, Victoria 3053, Australia
Tel: +61 (0)3 8359 1011

The right of the Author to be identified as the Author of this Work has been asserted in accordance with the Copyright, Designs and Patents Act 1988.

All rights reserved. No part of this publication may be reproduced, stored in a retrieval system, or transmitted, in any form or by any means, electronic, mechanical, photocopying, recording or otherwise, except as permitted by the UK Copyright, Designs and Patents Act 1988, without the prior permission of the publisher.

First published 2007 by Blackwell Publishing Ltd

ISBN 978-1-4051-1582-7

Library of Congress Cataloging-in-Publication Data
Verspoor, Eric.
 The Atlantic salmon : genetics, conservation, and management / Eric
Verspoor, Lee Stradmeyer, Jennifer Nielsen.
   p. cm.
 Includes bibliographical references and index.
 ISBN-13: 978-1-4051-1582-7 (hardback : alk. paper)
 1. Salmon stock management.   2. Atlantic salmon–Genetics.
  I. Stradmeyer, L.   II. Nielsen, Jennifer L.   III. Title.
 SH346.V47 2007
 639.9′7756–dc22
                      2006025811

A catalogue record for this title is available from the British Library

Set in 10/12.5pt Sabon
by Graphicraft Limited, Hong Kong
Printed and bound in Singapore
by Fabulous Printers Pte Ltd, Singapore

The publisher's policy is to use permanent paper from mills that operate a sustainable forestry policy, and which has been manufactured from pulp processed using acid-free and elementary chlorine-free practices. Furthermore, the publisher ensures that the text paper and cover board used have met acceptable environmental accreditation standards.

For further information on Blackwell Publishing, visit our website: www.blackwellpublishing.com

Front cover photograph of salmon in a pool in the Cascapedia River, Quebec, courtesy of G. van Ryckevorsel.

This book is dedicated to the memory of Henry Morrice and his long service to salmon conservation and management in Scotland, as Water Bailiff for the Dee District Fisheries Board 1947–1966 and as Superintendent of the Kyle of Sutherland District Salmon Fisheries Board 1966–1990.

Contents

Foreword xiii
Preface xiv
Acknowledgements xvi
Contributors xvii

1 Introduction 1
E. Verspoor

   1.1 Background 2
   1.2 Genetics, management and conservation 6
   1.3 Purpose of this book 8
   1.4 Organisation of this book 9
   1.5 Summary and conclusions 12

Part I Background 15

2 The Atlantic Salmon 17
J. Webb, E. Verspoor, N. Aubin-Horth, A. Romakkaniemi and P. Amiro

   2.1 Introduction 18
   2.2 Taxonomy and geographic range 21
   2.3 Life-history variation 22
   2.4 Biology of anadromous populations 25
         2.4.1 Distribution and life in fresh water 25
         2.4.2 Reproduction 28
         2.4.3 Egg size, development and survival 30
         2.4.4 Emergence and dispersal of fry 32
         2.4.5 Free-swimming juvenile life and production 33
         2.4.6 Sexual maturation of parr 34
         2.4.7 Movements of parr 35
         2.4.8 Smolt migration 36
         2.4.9 Marine life and distribution 36
         2.4.10 Homing and return marine migration 38
   2.5 Biology of non-anadromous populations 40
         2.5.1 Geographic distribution 41

		2.5.2 Life history and behaviour	42
	2.6	2.5.3 Maturation and reproduction	43
		Summary and conclusions	45

3 The Atlantic Salmon Genome   57
P. Morán, E. Verspoor and W. S. Davidson

	3.1	DNA		58
	3.2	Chromatin and chromosomes		60
		3.2.1	Nature and structure	60
		3.2.2	Replication, cell division and growth	60
		3.2.3	Number and ploidy level	63
	3.3	Genes and genome organisation		66
		3.3.1	Molecular nature and structure	66
		3.3.2	Number and molecular distribution	66
		3.3.3	Extragenic DNA	68
	3.4	Genes and development		69
		3.4.1	Genotypes, alleles and loci	69
		3.4.2	Genes and traits	70
		3.4.3	Gene expression	70
	3.5	Variation among individuals		71
		3.5.1	Origin	71
		3.5.2	Scope	72
		3.5.3	Detection	74
	3.6	Summary and conclusions		83

4 Investigating the Genetics of Populations   86
M. M. Hansen, B. Villanueva, E. E. Nielsen and D. Bekkevold

	4.1	Overview		87
	4.2	Population genetics		89
		4.2.1	Basic concepts	89
		4.2.2	Models of population structure	90
		4.2.3	Population differentiation	92
	4.3	Quantitative genetics		93
		4.3.1	How it differs from population genetics	93
		4.3.2	Quantitative genetic variation	94
		4.3.3	Genotype by environment interaction	95
		4.3.4	Integration of molecular and quantitative genetics	95
	4.4	The genetic characterisation of wild populations		95
		4.4.1	Allozyme electrophoresis	96
		4.4.2	Mitochondrial DNA	98
		4.4.3	Microsatellite DNA	99
		4.4.4	Other types of molecular marker	101
	4.5	Studying populations: issues and limitations		102
		4.5.1	Types of study and their limitations	102
		4.5.2	Mixed-stock analysis and assignment tests	103

		4.5.3	Estimating effective population size and detecting population declines	104
		4.5.4	Parentage assignment	104
		4.5.5	Relatedness estimation	106
	4.6	Future perspectives: going beyond quantifying genetic differentiation and understanding local adaptation		107
	4.7	Summary and conclusions		109

Part II Population Genetics — 115

5 Biodiversity and Population Structure — 117
T. L. King, E. Verspoor, A. P. Spidle, R. Gross, R. B. Phillips, M-L. Koljonen, J. A. Sanchez and C. L. Morrison

5.1	Introduction		118
5.2	Evolutionary relatedness to other salmonids		120
5.3	Phylogeographic diversity		121
	5.3.1	Range-wide	123
	5.3.2	Eastern Atlantic	131
	5.3.3	Western Atlantic	137
	5.3.4	Resident (non-anadromous) salmon	143
	5.3.5	Historical origins	147
5.4	Regional and local population structure		153
	5.4.1	Spatial scale and boundaries	154
	5.4.2	Metapopulation structure and gene flow	156
5.5	Overview		158
5.6	Summary and conclusions		159
5.7	Management recommendations		160

6 Mating System and Social Structure — 167
W. C. Jordan, I. A. Fleming and D. Garant

6.1	Introduction		168
	6.1.1	Definitions, approach and organisation	168
	6.1.2	Genetic markers in the analysis of mating system and social structure	169
6.2	Mating system		170
	6.2.1	Effective population size	170
	6.2.2	Factors affecting the variance in reproductive success of male alternative reproductive tactics	171
	6.2.3	Reproductive success estimates and mate choice under natural conditions	177
	6.2.4	Hybridisation	180
6.3	Social structure		184
	6.3.1	Kin recognition and kin-biased behaviour	185
	6.3.2	Patterns of relatedness in nature and fitness	185
6.4	Summary and conclusions		186
6.5	Management recommendations		187

7 Local Adaptation — 195
C. García de Leániz, I. A. Fleming, S. Einum, E. Verspoor, S. Consuegra, W. C. Jordan, N. Aubin-Horth, D. L. Lajus, B. Villanueva, A. Ferguson, A. F. Youngson and T. P. Quinn

7.1	Introduction	196
	7.1.1 Phenotypic diversity and fitness in a changing world	197
7.2	Scope for local adaptations in Atlantic salmon	198
	7.2.1 Genetic variation in fitness-related traits	199
	7.2.2 Environmental variation and differential selective pressures	201
	7.2.3 Reproductive isolation	203
7.3	Evidence for the existence of local adaptations in Atlantic salmon	204
	7.3.1 Indirect, circumstantial evidence for local adaptations	204
	7.3.2 Direct evidence for local adaptations	210
	7.3.3 Challenges to the local adaptation hypothesis	211
7.4	Summary and conclusions	218
7.5	Management recommendations	219

Part III   Management Issues — 237

8 Population Size Reductions — 239
S. Consuegra and E. E. Nielsen

8.1	Introduction	240
8.2	Loss of genetic variation in small populations	240
	8.2.1 Importance of genetic diversity in natural populations	240
	8.2.2 Measuring loss of genetic variation in small populations: heterozygosity and allelic diversity	242
8.3	Effective population size	243
	8.3.1 Minimum effective population size	245
	8.3.2 Relationship between census and effective population sizes	247
	8.3.3 Factors influencing genetically effective population size in Atlantic salmon	247
	8.3.4 Calculating effective population size	249
8.4	The effects of genetic drift and selection in small populations	251
8.5	The effects of inbreeding in small populations: inbreeding depression	256
8.6	Population reductions, gene flow and local adaptation	259
	8.6.1 Small populations of Atlantic salmon and the metapopulation models	263
8.7	Summary and conclusion	264
8.8	Management recommendations	264

9 Genetic Identification of Individuals and Populations — 270
M-L. Koljonen, T. L. King and E. E. Nielsen

9.1	Introduction	271
9.2	Assignment of individuals	273
	9.2.1 Application to Atlantic salmon	274
	9.2.2 Background to methodology	277

	9.3	Identification of population contributions	280
		9.3.1 Application to Pacific salmon fisheries	280
		9.3.2 Application to Atlantic salmon fisheries	282
		9.3.3 Background to mixed-stock analysis	284
	9.4	Resolving power of different markers	289
	9.5	Summary and conclusions	291
	9.6	Management recommendations	292
10	**Fisheries Exploitation**		**299**
	K. Hindar, C. García de Leániz, M-L. Koljonen, J. Tufto and A. F. Youngson		
	10.1	Introduction	300
	10.2	A historical perspective on fisheries exploitation	301
		10.2.1 Catch statistics	303
		10.2.2 Exploitation rates	304
		10.2.3 Potential for selection	304
	10.3	Fisheries exploitation as an ecological and evolutionary force	309
		10.3.1 Undirected genetic erosion	309
		10.3.2 Directed genetic change	312
	10.4	Fishing and effective population size: the evidence	313
	10.5	Phenotypic and evolutionary changes in exploited populations	314
	10.6	Future management of salmon fisheries	317
	10.7	Summary and conclusions	318
	10.8	Management recommendations	318
11	**Stocking and Ranching**		**325**
	T. F. Cross, P. McGinnity, J. Coughlan, E. Dillane, A. Ferguson, *M-L. Koljonen, N. Milner, P. O'Reilly and A. Vasemägi*		
	11.1	Introduction	327
	11.2	Genetic characteristics of wild salmon populations	327
	11.3	Nature of strains reared for stocking and ranching	327
	11.4	Approach based on numbers of salmon present	328
	11.5	Scenario 1: Where salmon are extinct in a river (reintroduction)	330
	11.6	Scenario 2: Where small to near optimal numbers of local population(s) remain (rehabilitation)	339
	11.7	Scenario 3: Attempting to achieve productivity in excess of naturally constrained production (enhancement)	343
	11.8	Scenario 4: Mitigation programmes and conservation hatcheries to counter irreversible loss of natural production (mitigation)	345
	11.9	Summary and conclusions	352
	11.10	Management recommendations	352
12	**Farm Escapes**		**357**
	A. Ferguson, I. A. Fleming, K. Hindar, Ø. Skaala, P. McGinnity, T. F. Cross *and P. Prodöhl*		
	12.1	Introduction	358
	12.2	Magnitude of farm salmon escapes	360

		12.2.1	Identifying escaped farm salmon	360
		12.2.2	Escapes from sea cages	361
		12.2.3	Juvenile escapes	362
	12.3	Genetic differences between wild and farm salmon	362
		12.3.1	Founder effects	362
		12.3.2	Differences due to domestication	363
		12.3.3	Genetic marker differences between wild and farm salmon	364
		12.3.4	Phenotypic differences between wild and farm salmon	365
	12.4	Potential impact of farm escapes on wild populations	368
		12.4.1	Fate of adult escapes	368
		12.4.2	Juvenile escapes	369
		12.4.3	Indirect genetic effects of farm escapes	369
		12.4.4	Direct genetic effects of farm escapes	371
	12.5	Breeding of escaped farm salmon in the wild	372
		12.5.1	Evidence for breeding of escaped farm salmon in the wild	372
		12.5.2	Differences in breeding behaviour of farm and wild salmon	373
		12.5.3	Increased hybridisation with brown trout as a result of farm escapes	374
	12.6	Experimental studies of the impact of farm escapes	375
		12.6.1	Imsa experiment	375
		12.6.2	Burrishoole experiment	377
	12.7	Discussion of genetic implications of farm escapes	382
	12.8	How can the genetic impact of farm escapes be reduced?	385
	12.9	Summary and conclusions	387
	12.10	Management recommendations	388

13 Genetics and Habitat Management
E. Verspoor, C. García de Leániz and P. McGinnity
399

	13.1	Introduction	400
	13.2	Genetic issues	404
		13.2.1	Habitat reduction	407
		13.2.2	Habitat fragmentation	408
		13.2.3	Habitat expansion	409
		13.2.4	Habitat degradation	410
		13.2.5	Loss of biodiversity	411
		13.2.6	Global climate change	412
	13.3	Summary and conclusions	418
	13.4	Management recommendations	419

14 Live Gene Banking of Endangered Populations of Atlantic Salmon
P. O'Reilly and R. Doyle
425

	14.1	Introduction	426
		14.1.1	Genetic concerns associated with the long-term captive rearing of salmonids	426
		14.1.2	Impact of long-term genetic changes on captive populations	431

	14.2	Live gene banking of inner Bay of Fundy Atlantic salmon: a case study	432
		14.2.1 Collection of founder broodstock	434
		14.2.2 Captive rearing of broodstock	435
		14.2.3 Spawning	437
		14.2.4 Captive rearing and river release of offspring	443
		14.2.5 Ongoing founder broodstock collection and recovery of wild-exposed live gene bank salmon	445
	14.3	Conservation and management of small remnant populations of Atlantic salmon	446
		14.3.1 Prioritising rivers for conservation measures	446
		14.3.2 Should very small populations be combined or managed separately?	447
	14.4	Use of cryopreserved sperm in the conservation of Atlantic salmon	451
		14.4.1 Methods for the cryopreservation of milt	451
		14.4.2 Use of cryopreserved milt in the restoration of wild salmon populations	452
		14.4.3 Addition of genetic variation to impoverished future populations	457
		14.4.4 Minimising genetic change between founder and subsequent generations of live gene bank populations	457
	14.5	Research	458
		14.5.1 Monitoring the loss of genetic variation and accumulation of inbreeding	458
		14.5.2 Identification of individuals, and evaluation of the relative efficacy of alternate management strategies	459
	14.6	Summary and conclusions	463
	14.7	Management recommendations	463
15	**Atlantic Salmon Genetics: Past, Present and What's in the Future?** *J. L. Nielsen*		**470**
	15.1	Past	471
	15.2	Present	472
	15.3	Future	473

Glossary of terms   481

Index   490

Foreword

In January 2003 I attended a SALGEN symposium entitled 'Genetics and the Conservation of Atlantic Salmon' held in Westport, Ireland. I did so with a certain amount of trepidation since, as a mere physical chemist, I have to admit that on several occasions during my 20 years as Secretary of NASCO I have left meetings confused and somewhat intimidated by genetic terminology and theory. Moreover, on a number of occasions, particularly when discussing the genetic impacts of cultured salmon on the wild stocks, geneticists appeared to be diametrically opposed in their views and I could not always grasp why!

The SALGEN symposium, however, was different. First, the symposium convenors had given clear guidance to all contributors that their presentations should be comprehensible to resource managers and others who may have no genetic knowledge. This they achieved to a very large extent. Second, this was the first time, to the best of my knowledge, that a three-day international symposium had been held which was devoted to understanding of the genetics of the Atlantic salmon and the implications for the conservation and management of the species. Third, it was clear from the presentations that this understanding has increased dramatically in recent years and that there is now more common ground between geneticists.

The contributions from the symposium have now been collated in this book which reviews the biology of the Atlantic salmon and its genome, considers our understanding of the population genetics of the species and, finally, examines the major management issues from a genetic perspective. These issues include fisheries exploitation, salmon farming, stocking and ranching and habitat management. There is also a very valuable chapter on gene banking of endangered populations.

Given the depressed status of many stocks in both Europe and North America, the pressure for further reductions in exploitation and to undertake stock rebuilding programmes and the continuing concerns about the impact of escapees from fish farms, the publication is timely indeed and an invaluable reference for salmon managers and policy makers. It is clear that rational management of the resource can only be achieved if genetic aspects are given the increasing attention they deserve and are now receiving. Although there is, inevitably, still much genetic terminology here, a real attempt has been made to make this comprehensible to those of us who are not specialists. I commend the editors and authors for the quality and comprehensive coverage of the subjects in the various chapters and recommend it wholeheartedly to those with an interest in the genetics of this remarkable and highly prized resource.

Malcolm Windsor
Secretary of NASCO

Preface

The aim of this book is to provide an accessible, up-to-date synthesis of what is a rapidly advancing field of knowledge, that of genetics and its relevance to Atlantic salmon conservation and management. It has been a long endeavour achieving this aim. Had we known how long it would take and how difficult it would be, we might not have started the journey. It has required bringing together a very large body of information and a broad spectrum of scientists, with diverse expertise and opinions. More challenging, however, has been finding a way to effectively communicate genetic concepts and issues to make them readily accessible to the intended audience of the book – resource managers, biologists, policy makers and others involved with the conservation and management of the Atlantic salmon, most of whom are non-geneticists. Though it has not been an easy endeavour, we have been kept to our task by continuous affirmation, from the intended primary audience, of the need for such a book. Whether we have succeeded in our aim is for them to judge.

As it provides a rare species specific treatment of the implications of genetic variation for conservation and management, it is also hoped the book will be of value and interest to a wider audience, including those involved in the development of conservation policies and programmes for other species, as well as senior undergraduate and postgraduate-level students studying resource management. For them, it may serve as a useful practical example of the application of genetics to conservation.

The book is the product of SALGEN, a project sponsored by the Atlantic Salmon Trust and funded by the European Commission (Contract No. Q5AM-2001-00200). SALGEN, which began in 2001, undertook to comprehensively review existing genetic data on population structuring and adaptive population differentiation in the Atlantic salmon and to critically assess its implications for stock conservation and restoration. Four data analysis and review workshops (one each on protein variation, mitochondrial DNA, nuclear DNA, and quantitative and experimental studies) were convened and four focused scientific reviews produced. These were used as the background for the SALGEN 'Genetics and the Conservation of Atlantic Salmon' symposium held in Westport, Ireland, during January 2003 which brought together researchers from the workshops with resource managers and biologists to discuss the implications of current information for Atlantic salmon conservation and management programmes. The presentations made at the symposium form the basis of the book's chapters.

The chapters, with the exception of the introduction and the summation, are multi-authored and written by leading researchers working on the topics covered. A multi-author approach was taken to ensure as much as possible that the prevailing scientific view of the research community on the topics covered are presented. Providing a consensus view on the topics considered was important, given many of the genetic issues in Atlantic salmon conservation are controversial, in part due to an apparent divergence of views among geneticists.

In the writing of the book, considerable effort has been made to organise the material into chapters which reflect a management rather than genetic perspective, and to minimise technical jargon, so as to maximise accessibility to non-geneticists. Feedback from resource managers at the SALGEN symposium in Westport, Ireland, on the level and content of the presentations was used to improve the content and presentation of the chapters. At the same time, the book has also been reviewed by members of the intended audience of non-geneticists, as well as by experts in the field. This has added to the time taken to write the chapters, but we believe has resulted in a volume useful to geneticists and non-geneticists alike. We hope reading this book proves enjoyable as well as informative.

Eric Verspoor
Lee Stradmeyer
Jennifer Nielsen

Acknowledgements

This book would not have been possible without the help of the EU and input of many people. Production of the book was supported by the EU under Accompanying Measures Contract Q5AM-2001-00200 under the Framework V programme. We are indebted to Phil McGinnity and Gillian Rodger for organising the SALGEN symposium on the 'Genetics and the Conservation of Atlantic Salmon', held in January 2003 in Westport, Ireland, out of which the book's chapters emerged. We are also indebted to the authors for the many hours of their spare time they put in, for the quality of their contributions, and for indulging the editor's requests in our attempts to produce a cohesive tome for non-geneticists. We thank them for their patience in relation to the time it has taken to produce the book. Behind the scenes, the support of the Atlantic Salmon Trust, Jeremy Read, Jenny Sample and Gillian Rodger, with the SALGEN project as a whole, was central to making this book possible. The input of the reviewers, John Thorpe, John Beardmore, Peter Hutchinson and Malcolm Windsor, was invaluable and did much towards helping achieve the book's aims. Thank you to Malcolm Windsor and the North Atlantic Salmon Conservation Organization (NASCO) for moral support and the kind foreword. Gilbert van Ryckevorsel, John Webb and David Hay kindly provided their photographs to bring the book to life and bed it in the reality of our everyday experience, as did Richard Bond and Anders Haukvik. Throughout, the SALGEN project has also been supported by the Scottish Fisheries Research Services. Last but not least, we thank all those who gave moral support and those who helped us with the other things that needed doing as well while the book was in progress.

Contributors

P. Amiro
Fisheries and Oceans Canada, Diadromous Fish Division, Dartmouth, Nova Scotia, Canada

N. Aubin-Horth
Département de Sciences Biologique, Université de Montréal, Pavillon Marie-Victorinqo, Ave. Vincent-D'Indy, Montréal (Quebec), Canada HZV, ZSQ

D. Bekkevold
Danish Institute for Fisheries Research, Department of Inland Fisheries, Silkeborg, Denmark

S. Consuegra
Gatty Marine Institute, University of St Andrews, St Andrews KY16 8LB, Scotland, UK

J. Coughlan
Department of Zoology, Ecology and Plant Science/Aquaculture & Fisheries Development Centre, Environmental Research Institute, National University of Ireland Cork, Ireland

T. F. Cross
Department of Zoology and Animal Ecology, National University of Ireland Cork, Ireland

W. S. Davidson
Department of Molecular Biology and Biochemistry, Simon Fraser University, 8888 University Drive, Burnaby, BC, V5A 1S6, Canada

E. Dillane
Department of Zoology, Ecology and Plant Science/Aquaculture & Fisheries Development Centre, Environmental Research Institute, National University of Ireland Cork, Ireland

R. Doyle
Genetic Computation Ltd, 1031 Beaufort Avenue, Halifax, Nova Scotia, Canada B3H 3Y1

S. Einum
Norwegian Institute for Nature Research, Tungasletta 2, NO-7485 Trondheim, Norway

A. Ferguson
School of Biology & Biochemistry, Queen's University, Belfast BT9 7BL, Northern Ireland, UK

I. A. Fleming
Ocean Sciences Centre, Memorial University of Newfoundland, St John's, Newfoundland A1C 5S7, Canada

D. Garant
Département de Biologie, Faculté des Sciences, Université de Sherbrooke, Sherbrooke, Quebec, J1K ZR1, Canada

C. García de Leániz
University of Wales Swansea, Department of Biological Sciences, Swansea SA2 8PP, UK

R. Gross
Estonian Agricultural University, Tartu, Estonia

M. M. Hansen
Danish Institute for Fisheries Research, Department of Inland Fisheries, Silkeborg, Denmark

K. Hindar
Norwegian Institute for Nature Research (NINA), NO-7485 Trondheim, Norway

W. C. Jordan
Zoological Society of London, Institute of Zoology, Regent's Park, London NW1 4RY, UK

T. L. King
US Geological Survey, Biological Resources Division, Leetown Science Center, Kearneysville WV, USA

M-L. Koljonen
Finnish Game and Fisheries Research Institute, Helsinki, Finland

D. L. Lajus
Zoological Institute, Russian Academy of Sciences, 199034 St Petersburg, Russia

P. McGinnity
Aquaculture and Catchment Management Services, Marine Institute, Newport, Co Mayo, Ireland

N. Milner
Environment Agency, National Fisheries Technical Team, Cardiff, UK

P. Morán
Department of Biochemistry, Genetics and Immunology, University of Vigo, 36200 Vigo, Spain

C. L. Morrison
USGS-BRD Leetown Science Center, Kearneysville WV, USA

E. E. Nielsen
Danish Institute for Fisheries Research, Department of Inland Fisheries, Silkeborg, Denmark

J. L. Nielsen
United States Geological Survey, Alaska Science Center, 1011 East Tudor Road, Anchorage AK, USA

P. O'Reilly
Department of Fisheries and Oceans, Diadromous Fish Division, 1 Challenger Drive, Bedford Institute of Oceanography Dartmouth, Nova Scotia B2Y 4A2, Canada

R. B. Phillips
Department of Biological Sciences, Washington State University, Vancouver WA, USA

P. Prodöhl
School of Biology & Biochemistry, Queen's University, Belfast BT7 1NN, Northern Ireland, UK

T. P. Quinn
School of Aquatic & Fishery Sciences, University of Washington, Seattle, WA 98195, USA

A. Romakkaniemi
Finnish Game & Fisheries Research Institute, Oulu, Finland

J. A. Sanchez
Universidad de Oviedo, Oviedo, Spain

Ø. Skaala
Institute of Marine Research, PO Box 1870, Nordnes, NO-5024 Bergen, Norway

A. P. Spidle
USGS-BRD Leetown Science Center, Kearneysville WV, USA

J. Tufto
Norwegian University of Science and Technology (NTNU), NO-7491 Trondheim, Norway

A. Vasemägi
Swedish University of Agricultural Sciences, Department of Agriculture, Umea, Sweden

E. Verspoor
Freshwater Fisheries Programme, FRS Freshwater Laboratory, Pitlochry PH16 5LB, Scotland, UK

B. Villanueva
Sustainable Livestock Systems Group, SAC, Bush Estate, Penicuik, Scotland, UK

J. Webb
Atlantic Salmon Trust, Moulin, Pitlochry, Scotland, UK

A. F. Youngson
FRS Freshwater Laboratory, Faskally, Pitlochry PH16 5LB, Scotland, UK

1 Introduction

E. Verspoor

A salmon image incised on a stone by ancient Pictish peoples of Scotland more than 1000 years ago. The stone is located in the grounds of the manse in the village of Glamis in the catchment of the River Tay, one of Scotland's largest and most famous salmon rivers, and illustrates that the species has long been a cultural icon in this part of its geographical range. (Crown copyright FRS, reproduced with the permission of FRS Freshwater Laboratory, Pitlochry; photograph by Edward Johnson.)

'. . . consider as true either that which appears most probable on the basis of the available evidence, or that which is consistent with more, or more compelling, facts than competing hypotheses.' (E. Mayr 1982 *The Growth of Biological Thought*)

1.1 Background

The Atlantic salmon, *Salmo salar* L., is an icon and a king in its underwater world (Fig. 1.1). Yet, like many aquatic species in today's world, its survival in many parts of its native range is threatened (Parrish *et al.* 1998; WWF 2001). Indeed, in many rivers, and particularly those at the species' southern distributional limit, it is already extinct. Also, in common with many species, its demise can to a large degree be attributed to human activities, yet ironically because of humans the species has never been more abundant. This paradox arises because the species is one of the world's pre-eminent aquaculture species and, unfortunately, more than 99% of the Atlantic salmon now in existence are of farmed origin.

This demise, and the cultural and economic importance of the Atlantic salmon in Western Europe and Eastern North America (Netboy 1968, 1980), has led to vast sums of money being spent on researching the species' biology and on restoration programmes; the literature contains thousands of scientific papers and books detailing its biology and management (e.g. Saunders 1988). Not surprisingly, understanding of the species has advanced dramatically over the years. Yet, even where the political and social conditions have been favourable, most restoration programmes have failed in their attempts to bring back healthy, self-sustaining runs of salmon in the rivers targeted. The limited success, at least in part, reflects the fact that the scientific basis for restoration and conservation remains inadequate. This is arguably for two reasons. One reason is that aspects of the species biology central to the development of successful conservation and restoration programmes remain to be fully understood. The other reason is that aspects which *are* understood have still to be properly taken into account. One aspect to which both concerns apply is the genetic character of the species which lies at the heart of its biological character, survival and reproduction. Genetics, when used in this

Fig. 1.1 An anadromous Atlantic salmon caught in a coast net fishery in Scotland on its return after one winter at sea (grilse). (Photo credit: J. Webb.)

> **Box 1.1** Genetic controversy: a centuries-old phenomenon.
>
> Genetic issues have been integral to debates on Atlantic salmon management since before the turn of the last century when genetics emerged as a distinct field of study. Andrew Young (1854) wrote in the nineteenth century, based on his early tagging work on the rivers of the Kyle of Sutherland in Scotland, that '... the fish of all these rivers come up the estuary for twenty miles promiscuously together, each river has its own peculiar race of fish, and each race finds its own river with most perfect decision'. This work was in part to respond to the statement that '... those who deal in stake-nets and illegal fixtures endeavour to assert that salmon spawn in the sea, and that they only catch the fish that are bred on their own coast, and not the fish that are bred in rivers'.
>
> Young's views were echoed by many, including Calderwood (1934). Yet Huntsman (1938), a leading salmon biologist of the time, wrote '... Mr. Calderwood replies to my scepticism concerning homing instinct in salmon and the theory of separate races in the Atlantic salmon' and goes on to say 'These theories have come to be so firmly believed that they almost constitute a "religion" to many anglers ...'. However, 20 years later, Jones (1959) found the evidence on the subject inconclusive and the debate muddied by an inconsistent use of the term 'race'.
>
> In the 1990s, debate raged over genetic issues associated with the listing of Atlantic salmon in the rivers of Maine, USA, as endangered under the Endangered Species Act (Committee on Atlantic Salmon in Maine 2002). State officials contended that decades of stocking foreign fish meant that the state's Atlantic salmon were not a unique species deserving of endangered status, while the National Marine Fisheries Service and the US Fish and Wildlife Service argued they were.
>
> Currently stocking with non-native fish remains highly contentious in the angling press on both sides of the Atlantic Ocean, as the following quotes cited in a recent article in the UK press (Mole 2001) attest: 'There seems to be universal agreement that when stocked fish interbreed with wild fish there is a loss of diversity and the subsequent stock performs less well' and 'The notion of discrete populations so finely tuned to their environments that introduced individuals represent a genetic threat is to turn genetics on its head'.

context, refers to the internal biological processes and molecular elements responsible for, and controlling, inheritance, reproduction and development.

Studies of the genetics of various behaviours and traits such as timing of returns and homing to natal rivers to spawn have been undertaken since the nineteenth century (Box 1.1). However, their importance has been controversial, as has been the importance of genetics in many spheres of biology. It is only since the mid-1960s, with advances in the field of genetics arising from the discovery of the molecular basis of inheritance (i.e. DNA) by Watson and Crick in 1953, that the tools for advancing understanding have gradually become available. Not surprisingly most current knowledge of the genetics of the Atlantic salmon comes from work carried out in the last 20 years. A major driver of recent genetic work has been the recognition of the need to understand the threat posed to wild salmon by the documented genetic interactions resulting from their interbreeding with escaped farm salmon (Fig. 1.2) in those parts of the species range where the species is now farmed.

Most people today acknowledge the importance of genetics to the biology of species though, as the cartoon in Fig. 1.3 highlights, the fundamental connection of this aspect to the biological productivity of a species is not always made. Yet its importance has long been recognised in relation to the development of the cultured strains of species such as cattle, sheep, wheat and rice with superior productive performance on which world food production is so dependent. Also, more recently, its importance in human biology is increasingly accepted as a wide variety of medical conditions which affect human survival and fitness are being shown to be inheritable. Furthermore, in medicine, the application of DNA-based molecular genetics approaches has had a revolutionary impact on the treatment of inherited conditions.

Fig. 1.2 (a) An escaped farm salmon female spawning with a wild male on the River Polla, Scotland, in autumn 1989 following the inadvertent loss of fish from a sea cage in Loch Eriboll, into which the river flows. Note the milt cloud behind the salmon. (b) The fertilised eggs from the mating, left exposed due to disturbance of the fish by the observers. (Photos credit: J. Webb.)

Yet despite this, a general acceptance has been slow to emerge that genetic processes are of central importance to a species' biological character and recruitment success, and thus to conservation and restoration programmes.

Though acceptance of this view is increasingly widespread, management practice relating to Atlantic salmon still, for the most part, fails to take into account genetic principles and the specific genetic knowledge which is available. This is perhaps not surprising as the genetic nature of most species is not visible; the only obvious manifestation of genetics in most species is their mating. In aquatic species such as the Atlantic salmon even this is often difficult to

Oh Gladys! Just how do you do it?
You've got so many healthy little ones!

Fig. 1.3 The importance of genetics. (Image credit: E. M. Verspoor.)

observe and can provide only a very limited insight. The inclusion of genetics in Atlantic salmon conservation and restoration is further set back by the perception among those concerned that there is little agreement on the implications of the principles and available knowledge for species management. This situation is an unfortunate legacy of public debate among salmon geneticists over some issues, and a misunderstanding of key genetic issues. It is also fed by the reluctance of practitioners to make what they consider to be the often radical changes to management practices that genetic considerations indicate are needed, particularly where they do not understand the underlying science. However, probably the biggest obstacle to incorporating genetics into Atlantic salmon management policy and programmes is that resource managers and biologists do not know how to do so.

Few managers and policy makers have even a basic training in this field of science. Genetics is a complex subject that arguably makes rocket science simple by comparison; certainly we were able to put a rocket into space long before we were able to sequence the human genome! Yet few management programmes employ geneticists and much of what is known about the genetics of Atlantic salmon and its implications for species management is inaccessible to the non-specialist; it is both scattered throughout the scientific literature and couched in technical jargon.

However, even if this obstacle is addressed, existing knowledge is far from perfect and those involved in species conservation and management need to take Ernst Mayr's (1982) advice (p. 2) when dealing with genetic issues. Strangely, this advice seems not to be accepted when it comes to genetics though it is generally followed with regard to other aspects of the species' biology. This may be because of the discomfort of dealing with something which is both complex and largely a hidden dimension; it may seem that the safer option is to ignore it. However, as should become clear in the pages which follow, to ignore genetics is to court failure.

1.2 Genetics, management and conservation

Genetic processes, which encompass interactions with the environment, lie at the heart of a species' biology and the biological diversity seen within and among localities. This encompasses morphological as well as behavioural and life-history diversity, whose extent in most species remains to be fully explored. In turn, the foundation of a species' genetic character lies in its DNA, the molecular store of the biological information found in each cell and passed on from parents to offspring. This molecular store contains what are glibly referred to as genes, the largely mysterious, functional physical units of inheritance. In interaction with the environment, genes set the pattern and tempo of the developmental and the physiological processes; these determine an individual's character, its chances of survival, and the contribution it makes to the abundance of subsequent generations. A species' DNA, and the genes it defines, are integral to reproduction; without it the species would not exist. As such, the conservation of a species requires the conservation of its genes, and those aspects of gene character and organisation which are important to individual survival and reproductive fitness. Thus, understanding of a species' genetic character, and how it is affected by various activities and interventions, is essential to the development of effective conservation and restoration programmes. Central to any species' genetic character is how variation in the nature of genes is organised among, as well as within, individuals. Genes exist in individuals and, in sexually reproducing species, individuals are part of groups within which interbreeding occurs more or less at random, but among which interbreeding is more or less constrained. These groups are what geneticists refer to as demes or genetic populations, and what some in the fisheries field refer to as genetic stocks. The collection of genes in these genetic populations (referred to as gene pools) are replicated and reorganised in each generation by sexual reproduction into new individuals with new gene combinations. Whatever their name, such groups are the basic biological units underpinning reproduction and the recruitment of new individuals and must thus be the target for conservation and management.

A basic understanding of genetic processes operating at the population level is needed to understand the relevance of genetics to conservation and management. Each genetic population represents a more or less distinct set of genealogical lineages, and collectively its constituent populations define the species' biology and distribution. At the same time, the particular genetic character of the population(s) in a given locale affects the local abundance and biology of the species. These genetic populations will often not be coincident with what management calls stocks, a point which is crucial for managers to recognise. Stocks, in so far as they are generally defined for management purposes, are arbitrary groups defined on the basis of geography, or a particular trait such as time of migration or life-history pattern. They may or may not be of biological significance.

Though used differently here, the terms 'genetic population' and 'stock' are used interchangeably in the literature though attempts have been made to restrict the use of the term 'stock' to refer to a genetic population (e.g. 1981 STOCS Symposium: Ihssen *et al.* 1981).

However, while laudable in principle, this has led to problems, because the need for a term to refer to arbitrary management groups has resulted, at least among salmon biologists and managers, in the continued use of the term 'stock' for groups of fish other than genetic populations. While the choice of term is not important per se, it is important that arbitrary groups of fish defined by managers are not confused with genetic populations, or vice versa. Given its historical application to an arbitrarily defined collection of fish, it is arguably better to reserve the use of the term 'stock' to such groups and use the term 'genetic population', or simply 'population', for a genetic population. Regardless of the merits of the case, to avoid confusion these terms have been used in the present book in this manner. The structuring of a species into populations, and the existence of differences among them in their genetic nature, is easy to understand in concept. However, the resolution of structure and the characterisation of differences is in practice fraught with difficulties. The molecular basis of inheritance is hidden from view and the reproductive interactions among individuals are normally difficult to study. Thus we are required to gain most of our insights from inferences drawn from indirect studies.

What has been learned generally about the genetics of organisms confirms that the genetic nature of a species is highly complex, and arises from interactions of processes operating at both the individual and population levels. This nature has been moulded by processes such as natural selection (i.e. the differential survival or reproductive success of genetically different individuals) over the course of the species' history. To at least some degree, this is expected to result in a unique genetic character for each species and, to a lesser extent, for each population within a species. Understanding how differences in the genetic character of populations are important to their local survival and reproduction is crucial for identifying which management and conservation approaches are likely to be most effective. If they are important, alteration of a species' genetic state, or changes to its historical environment, risks changing its character, reducing local reproductive success, and threatening local survival.

Genetic diversity is at the core of biodiversity, both within as well as among species (Avise 1996). It is in part encapsulated by the systematic classification of organisms into species and higher taxa, but evolution presents us with a continuum of genetic divergence among organisms. Taxonomic classifications describe only the part of this continuum where genetic divergence coincides with visually obvious morphological differentiation and easily detected reproductive isolation. The latter are the basis of most species designations and remain the main focus of resource managers, as well as the general public. However, molecular studies leave no doubt that within most, if not all, designated species there exists a vast abundance of genetic diversity (Magurran and May 1999).

Genetic diversity among species accounts for the heritable differences seen in their development, morphology and behaviour. Breeding studies indicate that these can also differ due to genetic variation among individuals within and among locations. It is likely that such variation affects the local character and abundance of species, and the ability of populations, and species as a whole, to evolve and adapt to environmental change. As such the management of species, from the perspective of either conservation or the restoration of local populations, must take genetics into consideration. It requires the conservation of a species' genes, and those aspects of gene variation and organisation at the individual and population level, important in both the short and long term, to species survival and reproductive fitness. The nature of this genetic character in the Atlantic salmon is largely conditioned by the homing of the species to its natal rivers to spawn, something which in many rivers represents a formidable physical challenge (Fig. 1.4).

Fig. 1.4 An Atlantic salmon jumping at Buchanty Spout on the River Almond, Scotland. (Photo credit: E. Verspoor.)

1.3 Purpose of this book

Seminal books such as *Genetics and Conservation: A reference for managing wild animal and plant populations*, edited by Christine Schonewald-Cox et al. (1983), have done much to draw attention, over the last two decades, to the importance of genetics in species conservation and management. The book *Population Genetics and Fisheries Management* edited by Nils Ryman and Fred Utter (1987) has done so specifically in relation to fish species and contains one of the first overviews of early molecular genetic work on Atlantic salmon. However, despite such books, views on the genetic character of the Atlantic salmon and its specific relevance to Atlantic salmon management and conservation remain controversial (Box 1.1). This is particularly the case as regards the extent of population structuring and the importance of genetic population differentiation to local recruitment success. At one extreme, genetic variation is seen as vitally important while at the other it is seen as largely trivial.

Controversy over genetics and fisheries management generally simmers in the background. However, it has recently gained a higher profile, as illustrated by recent debates surrounding the distinctiveness of the Atlantic salmon in Maine in the north-eastern USA and the merits of stocking non-native fish in Britain (Box 1.1). Thus, in the minds of many of those charged with species management, the extent to which genetics should be taken into account in the design and execution of rebuilding and conservation programmes is unclear. This situation exists despite the Atlantic salmon being one of the best-studied fish species and is made worse by the fact that the genetic workings are largely hidden at the molecular level and outside of the world of everyday experience. Any visible effects of genetics seen in relation to the biology of individuals and stocks are, with few exceptions, subtle and complex.

Developments in molecular biology over the last four decades associated largely with work on human genetics, combined with advances in genetics theory and salmon culture, have dramatically increased our understanding of Atlantic salmon genetics. However, much of what has been learned is inaccessible to all but geneticists, found as it is in the primary scientific literature shrouded in technical jargon. Unfortunately, no non-specialist's treatise currently exists which outlines current understanding of the genetics of Atlantic salmon and its relevance to species conservation and management.

The purpose of this book is to fill this communication gap. What is sought is to provide a thorough, objective scientific assessment of the implications of available genetic information on the Atlantic salmon, identifying what we do know, what it is reasonable to assume, and where existing information is inadequate and further research is required. It brings together available information and places it in the context of current genetics theory, assessing its implications and identifying how current practice can be improved by taking genetics into consideration.

1.4 Organisation of this book

The chapters of this book are organised into four parts to help achieve this aim. Part I, encompassing the first three chapters, provides up-to-date background information on the biology of the Atlantic salmon, on the species' general genetic nature at the level of the individual, and on the basic concepts related to the understanding of the genetics of its populations. Part II, also composed of three chapters, outlines what is specifically known about the genetics of the species at the population level. Next, in Part III, there are seven chapters. These look at how genetics is important with respect to the major management issues confronting the species. The final chapter, after Part III, provides a brief round-up and look at what the future is likely to hold as regards new techniques and new understanding.

The first chapter in Part I, Chapter 2, reviews the biology of the Atlantic salmon and highlights those aspects most relevant to understanding the genetic nature of the species. It is directed primarily at those unfamiliar with the biology of the Atlantic salmon. However, it should provide a useful update for those who know the species and provides some insights which have been gleaned from recent research. Following this, Chapter 3 gives a basic introduction to molecular genetics and to what is specifically known about the genetic character of the Atlantic salmon at the individual level. Understanding the basic genetic nature of individual salmon will help the reader come to grips with the material covered in Chapter 4. Chapter 4 introduces the basic concepts related to the genetic processes operating at the level of the population. It sets out what is currently known about the basic processes governing the genetic interaction between individuals in populations. It also describes the constraints and limitations faced in practice by scientists in their efforts to study genetic processes at the population level. These often dictate why many of the inferences drawn from studies can be accepted with little question while others remain inconclusive. Each of the chapters in the book stands on its own, but the background material in Part I, which may be familiar to some readers, is essential to properly appreciate much of the information presented in Part II.

Part II encompasses three chapters which present the results of studies of the genetic nature of the Atlantic salmon at the population level. This work is partitioned into three areas of investigation, though they represent a continuum of interacting processes. The first, addressed in Chapter 5, concerns what has been learned about how the species is geographically and evolutionarily structured into populations, including the general extent to which populations are differentiated. Following this, Chapter 6 considers what has been learned about the nature of genetic variation as it relates to demographic processes within populations, and has as its primary focus reproduction. Finally, Chapter 7 examines what has been learned about the genetic adaptation of Atlantic salmon populations to their local environmental conditions. This is a particularly crucial aspect of the species' genetic character which needs to be understood for the development of effective species conservation and management programmes.

The chapters in Part III set out the relevance of genetics theory and species-specific knowledge to particular conservation and management issues challenging managers and policy makers working with Atlantic salmon. Chapter 8 addresses what is arguably the most fundamental concern – the genetic consequences and impacts of population reductions, something which is central to other management issues addressed in Chapters 10–14. Following on, Chapter 9 deals with the use of genetics to assign individual Atlantic salmon to their river or region of origin, and to estimate the contributions made by different populations, or groups of populations, of fish found or exploited by fisheries at different locations at different times of the year. As will become clear, genetics offers a powerful and cost-effective tool. In many contexts it can provide key management information which is otherwise impractical, if not impossible, to obtain. Such information is often needed to support research into salmon population distribution and behaviour, as well as for monitoring abundance and exploitation.

The remaining chapters in Part III focus on the implications of genetics for the management of the species with regard to particular issues. Commonalities in genetic concerns give rise to some overlap in these chapters in Part III. For example, the genetic concerns related to stocking and ranching (Chapter 11) are similar to related to farm escapes (Chapter 12) as well as to the subject matter of Chapter 14, gene banking and supportive breeding. However, each of these represents a distinct issue from a management perspective and to have treated them together in one chapter would have been difficult. For example, Chapter 11 discusses the impacts of stocking per se on wild populations while Chapter 14 focuses specifically on genetic management in cultured populations developed for gene banking and supportive breeding programmes. This in itself represents a distinct area of management focus. Such programmes are increasingly being developed as Atlantic salmon populations in more and more rivers are faced with extinction. Taking fish into culture poses major challenges as regards preserving genetic diversity. However, it represents the only practical management option for conserving the gene pools of threatened populations until conditions allow self-sustaining wild populations to be restored.

Science as yet provides far from a complete or even sufficient understanding of the genetic character of the Atlantic salmon and how this needs to be managed and impinges on management. However, new developments in molecular biology are occurring on a daily basis and will offer new tools and opportunities for expanding what is known about the genetics of species. We can expect many of these to be exploited by researchers seeking to increase understanding of the genetics of Atlantic salmon populations. What the future might hold in this regard is considered briefly in Chapter 15.

The book does not attempt to catalogue the vast literature that exists on the biology and genetics of the Atlantic salmon. Detailed referencing of statements made would have dis-

One of many moments of Dr Smith's show
'It's the DNA' with his sidekick Jim the Fish.

Fig. 1.5 The communication challenge. (Image credit: E. M. Verspoor.)

tracted the intended audience for the book who are unlikely to want to delve deeply into the literature. This would have detracted from the purpose of the book, to communicate what for most intended readers will often be unfamiliar concepts and facts. Thus in the book a deliberate attempt has been made to keep numbers of references to the minimum by focusing on the most recent, key papers and, wherever possible, review articles are cited. However, key references are provided to support facts given and statements made, and these will allow anyone who wishes to delve further into the primary literature underlying the science presented.

The philosophy underlying the preparation of the chapters has also been to minimise the use of technical words, so as to make the book accessible to the widest audience. This represents a major challenge for geneticists when speaking to non-specialist audiences, something which may often not be appreciated. Genetics, unfortunately, is rife with technical terms for which there is no real common equivalent and it is often not possible to communicate concepts or points effectively without their use (even worse when the term means something else in everyday speech! Fig. 1.5). Unfortunately, in many cases, technical words can only be replaced by long and imprecise phrases if a concept is to be put in plain language. Where such terms are used frequently, excluding such words would result in long stretches of repetitive text. Thus the use of some technical words is unavoidable; for these a glossary is provided.

Finally, by way of reinforcement and to ensure clarity in the 'take-home' messages from each chapter, each includes a 'summary and conclusions' section in which these are highlighted. It is hoped that these will help to address what, next to understanding the genetic concerns relating to management issues, is the biggest challenge facing managers. This is finding ways of effectively and practically integrating the genetics dimension into management and conservation programmes. For example, molecular markers may now be able to resolve population structuring and assign all individuals to population of origin, but this will be of limited use unless it can be achieved within the limits of available resources. At the same time, it may be possible to identify all salmon populations but it might not be cost effective to do so or, even if it is, to manage each population independently. If not it will be essential to develop general management methods which though not targeting individual populations are still sensitive to population structuring. How this might be done may not always be obvious or simple. This remains a future challenge but it is hoped that this book will help with an important prerequisite to achieving this objective, providing a way in which biologists, managers and policy makers dealing with Atlantic salmon can gain a sound understanding of the genetic nature of the Atlantic salmon and its relevance to species conservation and management.

1.5 Summary and conclusions

- The Atlantic salmon, like many aquatic species throughout the world, is threatened by extinction in many parts of its range though particularly at its southern limit.
- Restoration and conservation programmes for Atlantic salmon have to date met with mixed success despite it being one of the most studied and managed of fish species.
- Genetics refers to the internal biological processes and molecular elements responsible for, and controlling, inheritance, reproduction and development.
- The genetic character of an Atlantic salmon, defined by its DNA, is central to its survival and reproductive success, and remains to be fully taken into account in its management.
- Genetic diversity among individuals is at the core of biodiversity within, as well as among, species and exists at both the individual and population level.
- Genetic populations, the fundamental units underpinning species recruitment, are groups of individuals within which interbreeding is more or less random but among which interbreeding is more or less constrained.
- To meet the conservation requirements of a species, in both the immediate and long term, it is essential to maintain genetic diversity and processes important to survival and reproductive fitness, both at the individual and population level.
- Development of effective management and conservation programmes requires, among other things, an understanding of the contribution made by intraspecific genetic diversity to the local viability and productivity of a species.
- Understanding of the genetics of Atlantic salmon has developed largely over the last 40 years and particularly over the last decade.
- The purpose of this book is to provide an accessible, up-to-date review of Atlantic salmon genetics and its implications for the species' management and conservation, as an aid to policy makers, managers and biologists for developing more effective management and conservation programmes.

Further reading

Avise, J.C. and Hamrick, J.L. (Ed.) (1996) *Conservation Genetics: Case histories from nature*. Chapman & Hall, New York.

Frankham, R. (1995) Conservation genetics. *Annual Review of Genetics*, **29**: 305–327.

Hedrick, P.W. (2001) Conservation genetics: where are we now? *Trends in Ecology and Evolution*, **16**: 629–636.

NRC (National Research Council) (1996) *Upstream: Salmon and society in the Pacific Northwest*. National Academy Press, Washington, DC.

References

Avise, J.C. (1996) The scope of conservation genetics. In: J.C. Avise and J.L. Hamrick (Ed.) *Conservation Genetics: Case histories from nature*, pp.1–9. Chapman & Hall, New York.

Calderwood, W.L. (1934) Is the seasonal habit of the salmon hereditary? *Salmon and Trout Magazine*, **76**: 227–233.

Committee on Atlantic Salmon in Maine (2002) *Genetic Status of Atlantic Salmon in Maine, Interim report*. National Academic Press, Washington, DC.

Huntsman, A.G. (1938) Sea behaviour in salmon: the case against an hereditary homing instinct. *Salmon and Trout Magazine*, **90**: 24–28.

Ihssen, P.E., Booke, H.E., Casselman, J.M., McGlade, J.M., Payne, N.R. and Utter, F.M. (1981) Stock identification: materials and methods. *Canadian Journal of Fisheries and Aquatic Sciences*, **38**: 1838–1855.

Jones, J.W. (1959) *The Salmon*. Collins, London.

Magurran, A.E. and May, R.M. (Ed.) (1999) *Evolution of Biological Diversity*. Oxford Scientific Press, Oxford.

Mayr, E. (1982) *The Growth of Biological Thought: Diversity, evolution, and inheritance*. Belknap Press, Cambridge, MA.

Mole, G. (2001) Save our stock fish. *Trout and Salmon*, **December 2001**: 10–11.

Netboy, A. (1968) *The Atlantic Salmon: A vanishing species*. Faber & Faber, London.

Netboy, A. (1980) *Salmon: The world's most harassed fish*. André Deutsch Ltd, London.

Parrish, D.L., Behnke, R.J., Gephard, S.R., McCormick, S.D. and Geeves, G.H. (1998) Why aren't there more Atlantic salmon (*Salmo salar*)? *Canadian Journal of Fisheries and Aquatic Sciences*, **55** (Supplement 1): 281–287.

Ryman, N. and Utter, F. (Ed.) (1987) *Population Genetics and Fisheries Management*. University of Washington Press, Seattle, WA.

Saunders, R.L. (1988) Salmon Science. *Atlantic Salmon Journal*, **Winter 1988**.

Schonewald-Cox, C.M., Chambers, S.M., MacBryde, B. and Thomas, L. (Ed.) (1983) *Genetics and Conservation: A reference for managing wild animal and plant populations*. Benjamin Cummings Publishing, London.

Watson, J.D. and Crick, F.H.C. (1953) Molecular structure of nucleic acids. *Nature*, **171**: 737–738.

WWF (2001) *The Status of Wild Atlantic Salmon: A river by river assessment*. WWF at http://www.panda.org/news_facts/publications/general/index.cfm

Young, A. (1854) *The Natural History and Habits of the Salmon*. Longman, Brown, Green & Longmans, London.

Part I
Background

2 The Atlantic Salmon

J. Webb, E. Verspoor, N. Aubin-Horth, A. Romakkaniemi and P. Amiro

Upper: eyed embryos in salmon ova, hatching eggs and newly hatched alevins. Lower: anadromous salmon in a river in Canada. (Photos credit: G. van Ryckevorsel, D. Summers.)

The observable biology of the Atlantic salmon, *Salmo salar*, is the visible face of its genetic character. It is why the species has a special place in human society, and why so much effort has been directed at its conservation and management. Over the centuries, thousands of scientific papers, popular articles and books have been written about Atlantic salmon behaviour, life history and habitats, and new insights into its biology are published almost daily. Its biology will be familiar to many readers, though the full spectrum of biological diversity the Atlantic salmon shows is not always appreciated. To other readers it will be unfamiliar. There is no up-to-date overview of Atlantic salmon biology to which readers, unfamiliar with the species or recent research findings, can be referred. Yet a basic understanding of the species' biology is essential to fully appreciate the material on the genetics of the species on which the book focuses. The overview provided here reviews those aspects of the species' biology most relevant to the discussions of the genetic character of the Atlantic salmon found in the chapters which follow.

2.1 Introduction

The Atlantic salmon (Fig. 2.1) is a salmonid fish, a group of species typified by a laterally compressed body form and a dorsal adipose fin, posterior to the main dorsal fin. The early

Fig. 2.1 A male anadromous Atlantic salmon in full spawning colours on its upstream migration to its natal stream to spawn, jumping a waterfall. (Photo credit: D. Hay.)

Fig. 2.2 Generalised life cycle of the Atlantic salmon showing the different life-history stages, from eggs, through alevins, fry, parr, smolts, post-smolts, migratory adults and spawning adults on completion of their return migration. (Atlantic Salmon Trust, image by R. Ade.)

life-history stages of the species, encompassing eggs, alevins, fry and parr, live in fresh water (Fig. 2.2). The latter older juvenile phase is characterised by vertical 'parr' marks and small red spots on the sides of the body (Fig. 2.2), features which are lost in older fish. In contrast, smolts, a later stage associated with the migration of fish from their natal rearing areas, are typically silver, and more elongated than parr with darker coloured fins (Fig. 2.2). The silvering derives from guanine crystals laid down in the skin, which obscure the finger marks and most spots on the body. The coloration of adult salmon is typically silver while at sea (Fig. 2.3). However, this coloration begins to be lost soon after river entry and the fish become increasingly reddish brown as they approach spawning. In the later stages of sexual maturation, adult males develop a characteristic exaggerated hooked lower jaw (Figs 2.2 and 2.3) referred to as a kype.

The Atlantic salmon occurs in temperate waters and shows a complex, diverse biology. The species is quintessentially anadromous, spawning (i.e. reproducing) in fresh water and

Fig. 2.3 Different body forms and coloration shown by maturing adult anadromous Atlantic salmon. The fish at the top is a fresh sea-run salmon seen as the type morph for the species. Males are distinguished from females by their brighter reddish coloration, hooked lower jaw ('kype') and large adipose fin. Note: the fish second from the top is a fully spawned adult female. (Atlantic Salmon Trust, image by R. Ade.)

followed by a freshwater juvenile phase and subsequent oceanic feeding migrations. This is combined with a tendency to home to natal areas to spawn and a variable capacity for iteroparity (i.e. to spawn in a number of different years). The former is a particularly key feature from a genetic point of view, and shared by all salmonid species. Homing results in the reproductive segregation of individuals into distinct groups whose members share a common natal origin. Within such groups of individuals, mating each year is more or less random. However, among such groups by virtue of their physical isolation, interbreeding is more or less constrained both within and across generations. These groups, known as genetic populations or demes, constitute one of the basic units of a species' biology. How Atlantic salmon are divided into populations, both between and within rivers, as a result of homing behaviour is considered in Chapter 5.

2.2 Taxonomy and geographic range

The Atlantic salmon was classified as the species *Salmo salar* by Linnaeus. Its Latin name *salar* means 'the leaper', a behaviour typically witnessed at waterfalls on its return spawning migration (Fig. 2.1). The species is one of 20 in the Salmoninae, a subfamily of the family Salmonidae (Philips and Oakley 1997). DNA differences among these salmonids show the Atlantic salmon to be most closely related to the brown trout, *Salmo trutta*, and these two species are now viewed as the only species in the genus *Salmo*. Rainbow and cutthroat trout, natives of western North America, share many traits with Atlantic salmon and brown trout and were formerly placed in the same genus but DNA analyses show them to be in the genus *Oncoryhynchus* with Pacific salmonids (Stearley and Smith 1993).

S. salar has in the past been viewed as polytypic (i.e. composed of a number of distinct evolutionary lineages) and variously divided into a number of species or subspecies (Scott and Crossman 1973). Most taxonomic classification of the species divided it into distinct anadromous and non-anadromous subspecies, *S. s. salar* and *S. s. sebago* Girard. However, the species has been considered monotypic since Wilder (1947) showed no consistent differences in the morphology and meristic character of the two forms (see also Claytor *et al.* 1991), though some still refer to the non-anadromous form as *S. salar sebago* Girard (e.g. Pursiainen *et al.* 1998). The latest genetic evidence bearing on this question is considered in Chapter 5.

The historical distribution of *S. salar*, comprehensively reviewed by MacCrimmon and Gots (1979), is the North Atlantic and associated coastal drainages (Fig. 2.4). In North America, the species was distributed in river systems and marine waters along the Atlantic coast from the Hudson River in New York state (41°N) north. It occurred in the Bay of Fundy, throughout the Gulf of St Lawrence and along the whole coast of Newfoundland and Labrador to the Fraser River (56°40′N). Isolated populations existed in rivers flowing into Ungava Bay in northern Quebec (59°N) and in the Kogaluk River, on the eastern side of Hudson Bay (59°30′N). The species occurred, in its non-anadromous form, inland in Labrador and northern Quebec, and west as far as Lake Ontario (79°30′W) in the St Lawrence drainage. However, self-sustaining populations of salmon no longer exist in many historical rivers at its southern distributional limits in the eastern United States and adjacent Maritime Provinces of Canada; native populations have also become extinct in the upper St Lawrence River, including Lake Ontario. Where the species does occur in these regions, remaining stocks are generally depressed and frequently supported by supplemental stocking programmes.

In Europe, the historical range of the Atlantic salmon extends from Iceland in the northwest (66°N), to the Barents and Kara Seas in the north-east (70°N, 83°E), and southward along the Atlantic coast, with only minor gaps, to the Douro River in Portugal (41°N). Eastward, it occurred in most rivers draining into the Baltic and North Seas. However, native, wild stocks are no longer found in the Elbe and the Rhine, two of Europe's largest rivers, or in many rivers draining into the Baltic Sea, which previously had abundant salmon runs. The species is also extinct or severely depressed in the rivers of France, Spain and Portugal, at the species' southern limit. As a result the species range has generally contracted and fragmented over the last century and a half (Fig. 2.4). This is largely attributable to industrialisation and adverse water management, particularly in the southerly parts of the species range where human activity is more concentrated (Parrish *et al.* 1998).

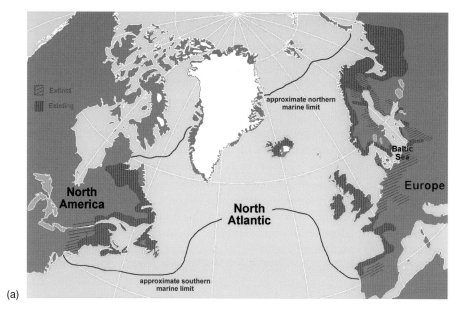

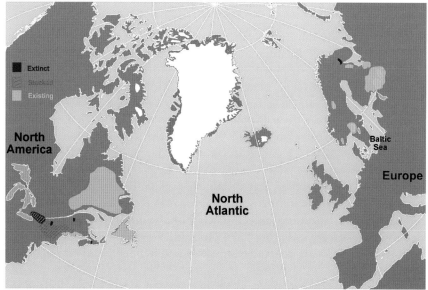

Fig. 2.4 The approximate distribution of wild anadromous (a) and non-anadromous (b) populations of Atlantic salmon in the North Atlantic region. In some southern parts of the species range where wild salmon are shown to still exist, some populations have gone extinct and others have severely depleted populations. In some areas where wild anadromous populations are shown to be extinct (e.g. the Baltic Sea), rivers have populations sustained by stocking (based on MacCrimmon and Gots 1979; IBSFC and HELCOM 1999).

2.3 Life-history variation

The Atlantic salmon, like most salmonids, shows life-history variations both within and between locations. The basic variations observed (Fig. 2.5) centre on whether or not salmon undergo a marine migration. In most rivers across the species range, individuals are anadrom-

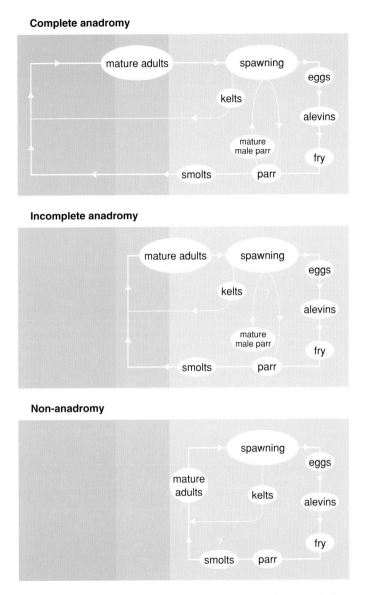

Fig. 2.5 Three basic types of life-history patterns seen in Atlantic salmon. Question marks indicate uncertainty or variation with regard to those aspects of the life cycle.

ous and migrate to sea. However, in some river systems, or parts thereof, return migrations are precluded by impassable waterfalls, and so-called 'landlocked' forms occur. In other cases, populations complete their entire life cycle in fresh water even though there are no barriers to their undertaking a marine migration. A further variation involves fish limiting their marine phase to estuaries. This behaviour occurs, for example, in some years in the rivers of Ungava Bay (northern Canada) where more extensive marine migrations are prevented by low sea temperatures (Power 1969).

In many locations, salmon appear to be either anadromous or non-anadromous, showing clearly that this life-history difference is associated with differences between rather than

Box 2.1 Inner Bay of Fundy Atlantic salmon.

The Bay of Fundy, in eastern Canada, is a 280 km (173 miles)-long arm of the Atlantic Ocean, situated between the provinces of New Brunswick and Nova Scotia. The salmon in the inner part of the Bay (iBoF), bounded by the Annapolis River in the south and the Mosher River in the north, show a biology which differs significantly from other salmon populations in the region and elsewhere (Fig. B2.1).

iBoF salmon are anadromous. Most become smolts after 2 years in fresh water and migrate to sea at a similar size to other salmon populations in the region (Jessop 1975; P. Amiro, unpubl.). However, seaward migration may begin in the autumn while actual movement to salt water usually occurs in late May or June (later than most other Atlantic coast salmon stocks). In the Little River, a tributary to the Stewiacke River, smolts have been observed migrating in July.

In contrast to anadromous salmon in Outer Bay rivers and elsewhere in North America, iBoF salmon have rarely been detected in the North Atlantic and tagging studies suggest they remain largely within the Bay itself during their marine phase. Post-smolts have been recovered throughout the Bay of Fundy as late as October of their smolt year (Amiro 1998) and, historically, recoveries of tagged post-smolts of iBoF origin were common in the traps set to catch juvenile herring in the western Bay, something unique to these salmon. Outside the iBoF, adult returns are dominated by a mixture of 2SW female salmon and male grilse. In contrast, iBoF rivers are dominated by grilse (1SW) and the proportion of repeat spawners is atypically high for the region. Reported survival rates among spawners range from 1% to 10%, and in some rivers, repeat spawners dominated and accounted for up to 70% of egg deposition.

Historical commercial catches of iBoF salmon were not correlated with the performance of Atlantic coast salmon stocks (Huntsman 1931; Amiro 1990). Annual recruitment to the angling fisheries was significantly correlated among almost all of the salmon populations of the iBoF but were not correlated with Atlantic coast rivers, providing further evidence of independence in their pattern of productivity and recruitment.

One notable exception to the above character of iBoF salmon relates to the Gaspereau River, in Kings County, Nova Scotia. In this river, the salmon have an age distribution that is more similar to Atlantic Coast stocks (Amiro and Jefferson 1996). At the same time, while the analysis of tag recovery information suggests that the river may also have a migration somewhat different from Atlantic coast rivers, they also appear to be different from the other inner Bay of Fundy populations.

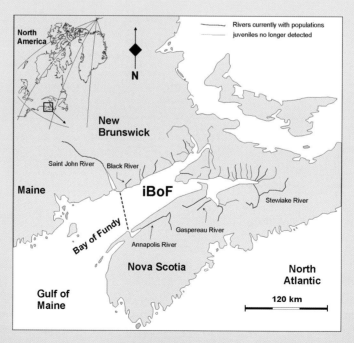

Fig. B2.1 Map of the Bay of Fundy region showing existing and historical Atlantic salmon rivers of the inner Bay area (iBoF).

within populations. Those populations with anadromous forms are the most widely familiar, and represent the archetypal Atlantic salmon. In general, they produce the largest and most economically valuable stocks of salmon and are thus the main focus of conservation and management concern. In contrast, populations with a non-anadromous life history have a more limited distribution and lower social and economic value; they are seen by many as rare and unusual. Yet in large parts of the species range non-anadromous salmon populations are an important part of the native fish community, with potentially important genetic implications, and equally deserving of conservation and management attention. The two population types have much in common. However, given their major difference in life history, it is perhaps easiest to discuss the biology of the two forms separately.

2.4 Biology of anadromous populations

Anadromous Atlantic salmon populations are distributed across most of the species range (Fig. 2.4) and most probably represent the ancestral form of the species (see, however, McDowall 2002). Non-anadromous populations most likely evolved from anadromous fish within the last 15 000 years, during the process of recolonisation of the salmon's modern distributional range after the last Pleistocene glaciation (Berg 1985; Chapter 5).

The life cycle of anadromous populations is characterised by variable, though normally large-scale, transitional migrations between fresh water and marine habitats (see Boxes 2.1 and 2.2 for examples of regional variation). The reproduction and nursery phases occur in fresh water and are followed by a period of feeding in the marine environment characterised by periods of rapid growth and the early stages of sexual maturation. Fish may grow to a very large size and the biggest, which have reached up to 70 lb (32 kg), are usually found in Russian and Norwegian rivers.

Anadromy involves major habitat changes by a sequence of directed movements or migrations. In order to reproduce, salmon must therefore perform adequately in each of a predictable sequence of disparate environments that fall into a series of main phases: egg development, hatch, dispersal and growth, smolting, migration to sea, movements at sea, around and back from marine feeding grounds, sexual maturation and a return migration as adults to their home rivers and spawning (Godin 1981; McCormick *et al.* 1998).

2.4.1 Distribution and life in fresh water

The freshwater phase of the life cycle encompasses adult spawning, egg deposition, embryonic development within ova and, following emergence from spawning areas, the dispersal and growth of juvenile fish. Conditions during all these stages determine individual and brood-group survivorship, and set limits on the production of migratory juveniles (Poff and Huryn 1998).

Across the species range, the rivers, streams and lake habitats occupied by salmon occur in a wide variety of landscapes and climates (Elliott *et al.* 1998). The species inhabits watercourses that are influenced by a broad range of geologies, ranging from granitic bedrock through volcanic and glacial substrates to limestone. On a local scale, this variation affects a wide range of variables important to salmon biology including water chemistry, hydrological regimes and the functional geomorphology of rivers (reviewed by Gibson 1993). Juvenile

Box 2.2 Baltic salmon.

Shallow and non-tidal, Europe's Baltic Sea is one of the world's largest brackish water basins. It is located on the transition between the temperate and boreal climatic zones (54°N–66°N, 10°E–30°E) with regular winter ice cover in the north and along the coast in the south. Historically, salmon spawned in 80–120 rivers with the largest populations in the larger rivers of the Gulf of Bothnia, Gulf of Riga and Gulf of Finland. The natural production of smolts from these rivers is estimated to have been around 8–10 million; today it is less than 2 million from 38 rivers (Fig. B2.2). A further 5.5 million smolts are released into the Baltic from hatcheries set up to mitigate the loss of production due to damming of rivers, the main factor behind the loss of production. The proportion of wild-born salmon is over 50% in many of the Baltic salmon fisheries, which indicates a superior performance (survival) of the wild smolts in comparison with reared smolts.

Spawning in Baltic rivers is confined to the main stems as smaller tributaries seem to be less suitable, perhaps due to their severe environmental conditions. In the River Tornionjoki the smallest tributaries holding regular spawning exceed an average annual flow of 30 $m^3 s^{-1}$ and the main spawning areas are located in sections of the river with average flows of 100–300 $m^3 s^{-1}$. The smallest contemporary spawning river flowing into the Gulf of Bothnia has an average annual flow of 12.4 $m^3 s^{-1}$. In contrast, south Baltic rivers with wild stocks are on average much smaller (IBSFC and HELCOM 1999). Baltic salmon are isolated from the North Sea and few migrate out through the narrow Denmark Straits (Christensen and Larsson 1979; Koljonen 2001).

Maturing salmon in the Baltic begin their 500–1500 km migration from the feeding areas to natal rivers in April. Migration routes northward follow mainly the eastern coast of the northern Baltic, with a typical migration speed of 20–40 km/day (Westerberg *et al*. 1999). The oldest and largest salmon migrate first, and the grilse follow 1–2 months later.

Wild fish tend to migrate earlier than reared fish (Ikonen and Kallio-Nyberg 1993; Karlsson *et al*. 1994; McKinnell *et al*. 1994) and start to ascend rivers in late May to early June, with the peak for MSW spawners in mid- or late June, after the peak of the spring floods. A few fish may ascend rivers in autumn after the main migration

Fig. B2.2 Map of the Baltic Sea area showing present Atlantic salmon rivers.

of the grilse in July–August (Nordqvist 1924). They spawn from late September in the northernmost rivers to early December in the southernmost rivers (Alm 1928; Christensen and Larsson 1979). Before World War II, the mean weights of grilse, 2SW, 3SW and 4SW spawning migrants were 1.5–1.8, 5–8, 11–13 and 15–18 kg, respectively. For several decades after the war, mean size-at-age of migrants was considerably smaller but it has increased again since the mid-1980s (Järvi 1938; Lindroth 1965; Karlsson and Karlström 1994). The trend has continued during the 1990s since the introduction of new fishery regulations (Romakkaniemi et al. 2003).

A typical female spawner produces 1200–1500 eggs/kg – about 10 000 eggs per fish (Romakkaniemi et al. 1995). Juvenile development is not exceptional but Baltic salmon appear to develop faster from eggs to the yolk sac stage than Atlantic salmon from Canada, at a given temperature (Brännäs 1988). Smolts enter the Baltic Sea in spring and early summer, when water temperature in the river normally ranges from 10° to 15°C. Their distribution within the sea is variable. Most smolts migrate to feeding areas in the central and southern Baltic Sea, but a varying proportion may remain in the southern Gulf of Bothnia or in the Gulf of Finland (Kallio-Nyberg and Ikonen 1992; Salminen et al. 1994; Kallio-Nyberg et al. 1999). Smolts from the River Neva (Russia, Gulf of Finland) feed closer to the home river than salmon stocks from the Gulf of Bothnia. Fish may remain in the sea between 1 and 4 years. About 90% of the grilse are males, and most of the females mature after 2 or 3 years in the sea. However, MSW fish of both sexes tend to dominate spawning runs. The proportion of repeat spawners varies from a few percent up to about 15% (Alm 1934; Järvi 1938; Romakkaniemi et al. 2003). After spawning, fish may either leave the rivers quite quickly or remain in fresh water until the following spring.

Baltic salmon are exploited by offshore, coastal, river mouth and river fisheries (Alm 1934; Christensen and Larsson 1979; Christensen et al. 1994) with salmon originating from the northernmost rivers most prone to exploitation, as they undergo the longest migrations and they are intercepted by every form of fishery on their migration route. Northern stocks started to decline after World War II due to overfishing, and were at their lowest levels in the 1980s. Since then many wild stocks have recovered with the river abundance of salmon commonly increasing by tenfold with a decline in fishing pressure and improvements in habitat conditions influencing survival and growth (Romakkaniemi et al. 2003). Annual catch limits have been in force in the early 1990s, and bans on early season fishing, started in the mid-1990s, have saved early-running migrants for spawning.

Short- and long-term synchronous trends in the abundance of the Baltic salmon stocks are well-known from the nineteenth and early twentieth century (Järvi 1938; Lindroth 1965) when sea fishing was less dominant than more recently. However, recently, covariation in abundance has also been observed (McKinnell and Karlström 1999; Romakkaniemi et al. 2003). This suggests a major component of the fluctuations to be of non-anthropogenic origin, probably due to natural climatic fluctuations.

Baltic salmon suffer from a syndrome termed M74, which may lead to the death of the majority of offspring at the late alevin stage (ICES 1994; Karlsson and Karlström 1994; Bengtsson et al. 1999). Although the physiological mechanisms and factors inducing M74 syndrome have not been resolved, its occurrence is directly linked to a low thiamine content of salmon eggs (Bylund and Lerche 1995; Amcoff et al. 1998) which appears to be linked to dietary factors in the sea. The syndrome has counteracted recovery of Baltic salmon stocks particularly during the early 1990s, when the incidence of M74 mortality was highest.

Atlantic salmon are viewed by some as having the most narrowly defined thermal requirements for survival, feeding and growth of all the species of salmonids (Elliott 1991). However, the range of water temperatures experienced by different populations is highly variable and some may experience the full range of thermal tolerance for the species (0–27.8°C) over the course of the year. Most streams with salmon are oligotrophic, i.e. poor in nutrients, and relatively unproductive but, in some regions, salmon occur naturally in highly productive, calcium-rich systems (e.g. southern United Kingdom and northern Spain).

Varying proportions of some anadromous populations do migrate between rivers and lakes. These migrations are probably adaptive and generate benefits in terms of more favourable environmental conditions, enhanced growth and increased reproductive success (Northcote 1978; Erkinaro and Niemelä 1995). However, in contrast to fluvial habitats, the key biophysical features of lacustrine habitats used by the species across its range are poorly understood.

The historical distribution and scale of adult production has been inferred from counts and fishery returns (Netboy 1968; Parrish *et al.* 1998). The main areas of juvenile production include Scotland, Norway, Russia (Kola/White Sea areas), Baltic Sea rivers, Iceland, Ireland and north-eastern North America. Although little is known of juvenile numbers within most rivers, numbers of adults returning annually to single watersheds vary from < 10 to > 100 000 individuals. Where conditions are suitable, larger rivers will have the most extensive and varied habitat suited to spawning and juvenile production. Larger rivers support larger numbers of juveniles and adults than smaller rivers. They are also likely to contain the largest number of distinct reproductive groups (i.e. genetic populations).

Some of the larger and better-known river systems with contemporary runs of anadromous Atlantic salmon in Europe are the Pechora (327 000 km^2) and Ponoi (15 467 km^2) in Russia; the Tornionjoki (40 010 km^2) and Tana (16 000 km^2) rivers which variously span Norway, Sweden and Finland; the Tay (6216 km^2) and Tweed (5000 km^2) in Scotland; the Shannon (11 700 km^2) and Moy (2086 km^2) in the Irish republic, and the Loire in France (155 000 km^2). In North America, there are the George (42 000 km^2) and Moisie (19 197 km^2) rivers in Quebec, the St John (54 500 km^2) and Miramichi (12 390 km^2) rivers in New Brunswick, and the Humber River (8124 km^2) in Newfoundland. Most of these large rivers produce large numbers of salmon though in some, such as the Loire and St John, numbers have been severely reduced by pollution and dam construction. However, the major part of wild global production comes from the much more numerous medium-sized or small drainages (Gibson and Cutting 1993).

2.4.2 Reproduction

Atlantic salmon have evolved significant behavioural diversity and morphological adaptations for reproduction to maximise reproductive success (reviewed by Fleming 1996, 1998). Spawning takes place between early autumn and early spring, and in most populations occurs over a period of weeks between October and December. Timing varies with latitude, and within and between rivers within a geographical region. In some high latitude populations, breeding may begin as early as September (Heggberget 1988), while it may extend into March in southern populations. Within a watershed, spawning may occur over a period of 10 weeks, or more, with little if any overlap in different locations. Indeed, in some rivers, spatial and temporal differences in the timing of spawning are sufficient to ensure reproductive segregation (Summers 1996a,b; Webb and McLay 1996).

Detailed studies have been made of spawning and nest-building behaviour in the wild and in quasi-natural conditions (White 1942; Jones 1959; Fleming 1996). Females deposit their eggs as discrete groups in one or more nests in composite structures called redds. Spawning sites are selected by the female based on locally variable criteria of water depth and velocity, gravel size, stability, compaction and porosity, and the availability of nearby cover. Redds are most often constructed in stable gravel at the tails of pools or on gravel bars where water flow is accelerating, and the gravel contains low concentrations of fine materials (i.e. silt and sand) and a good through-flow (by percolation and upwelling) of oxygenated water (Chapman 1988).

A female may construct nests in several redds in the same vicinity. Competition and aggression among nesting females is rare, but can occur where spawning habitat is limited. Interactions are most common during the early phase of spawning site selection and nest construction, but usually diminish as spawning nears. Nest construction consists of three main phases – site selection, excavation and burial – and can take between 1 and 48 hours. Females construct their nests by using their tail to excavate a pit in the riverbed, removing fine sediment by hydraulic displacement (by suction effects). The average depth of nests is positively correlated with female body length and larger females may deposit their eggs as much as 500 mm below the riverbed.

Adult males take no part in the acquisition of breeding sites or nest construction. Male behaviour in the run-up to spawning is directed towards seeking out and competing for females, and encompasses a range of aggressive behaviours associated with mate acquisition, courtship and guarding. Part of this process may involve extended periods of combat and display. The sex ratio among adult spawners can range from one to one, to a strong bias towards males. Where males are forced to compete for females, size-related dominance hierarchies develop and as a result, smaller, subordinate males may fail to gain access to breeding females. Subordinates often adopt 'satellite' positions downstream or to one side of dominant males, and attempt to gain access to the female (by 'sneaking') in an effort to fertilise some eggs as they are deposited. Sexually mature male parr (often referred to inappropriately as 'precocious' parr) are also often present, and they also use sneaking mating tactics to avoid confrontation with the much larger adult males (Jones and Hutchings 2001; Garant *et al.* 2002). Overall, the efficiency of ova fertilisation in the wild is typically high (Jones 1959). The genetic significance of mature parr spawning is discussed in Chapter 6.

Body size of adult fish is believed to be an adaptive trait in both sexes and an important determinant of reproductive success. Male secondary sexual characters such as body coloration and kype size are also considered to be adaptive (Fleming 1996; Garant *et al.* 2001). Females may exhibit mate choice, with aggression (pushing and biting) being most often directed towards smaller adult males and mature male parr. However, the main form of female choice is delay in spawning when courted by small or low-status males (de Gaudemar *et al.* 2000). Females are known to mate with more than one adult male (i.e. polyandry) and males may mate with more than one female (i.e. polygamy) (Chapter 6).

There is little information on the numbers of eggs deposited in single egg pockets or redds. What is known shows females may create several redds before depositing all their eggs. Excavation surveys suggest that some structures may be empty and others may contain the eggs of more than one female (Taggart *et al.* 2001; Chapter 6). Redd superimposition (i.e. the deposition of eggs by a female at a site previously used in that breeding season by another female) may therefore be quite common, particularly where spawning habitat is spatially or temporally limited.

Counts of redds have been shown in Atlantic salmon to be positively correlated with spawning escapement of adult females, and are therefore of some utility for management (i.e. stock assessment: Beland 1996). However, there does not appear to be a simple one-to-one relationship between number of redds and number of spawning females. Recent studies, based on DNA analyses of individual relatedness of eggs, indicate that natural spawning patterns are complex. Multiple spawning is very common and redds are often aggregates of a number of separate spawning events, often involving different males and females (Taggart *et al.* 2001; Chapter 6). Thus while perhaps broadly correct, field observations and counts of redds provide only a superficial view of spawning.

Breeding success may be affected by a number of factors. Adult females can maximise their reproductive success by choosing superior breeding sites among potential sites, which vary in quality. In many streams, spawning habitat is often discontinuous and patchy.

Females are probably not in a position to take account of the distribution of existing fry and parr. These, derived from spawning in previous years, are potential competitors for their progeny (Hendry *et al.* 2001). The distribution of ova is therefore determined solely by the distribution of suitable spawning habitat and the ability of spawners to gain access to it, rather than by the distribution and availability of potential rearing habitats (Webb *et al.* 2001). Interspecific competition between salmon and brown trout for spawning habitat may occur leading to the disturbance of existing redds or to the production of interspecific hybrids (Gephard *et al.* 2000). It has been known for more than a century that the two species can be crossed to produce hybrids but it is only from genetic studies that the occurrence of natural hybridisation and the factors affecting its extent have come to be appreciated (Chapter 6).

Following spawning, both male and female anadromous fish are known by local names, e.g. 'kelts', 'black salmon' or 'racers'. Rates of mortality among post-spawners vary depending on population but typically only a small proportion survives to return and spawn again (Belding 1934). Upon returning to the sea, the spawned fish resume feeding and recovers its silver colour. Post-spawning survival is often highest in northern latitude streams and is usually heavily biased towards females (Fleming 1996); the reasons for this are not clear. Survival may also be high among lake populations (Zelinsky 1985) and in short rivers where the duration of riverine migration and residence prior to spawning is limited (e.g. Belding 1934; Mills 1989). The proportion of repeat spawners in a stock generally varies. With regard to females, it ranges from 3% to 10%, but in some populations that inhabit short coastal streams with short river residence it may be as high as 70% (Mills 1989; Cunjak *et al.* 1998), with the incidence in males likely to be somewhat lower due to their higher post-spawning mortality rates. Following spawning, surviving fish may remain in the sea for as little as 3–5 months ('short absence'), or more than 1 year ('long absence'), before returning to spawn a second time. The proportions of each type tend to vary with the sea-age and run-timing characteristics of the maiden spawning stock in a river.

In contrast to many other species of bony fish, Atlantic salmon and other salmonids are characterised by the production of relatively small numbers of large eggs. Data from 28 locations across the species range indicate a mean body length at spawning of 671 mm and a mean fecundity of 5355 eggs. However, fecundity and egg size vary among populations, and larger and older females tend to produce more eggs (e.g. Mills 1989). Typically, eggs are 5–7 mm in diameter and total numbers range from 2000 to 15 000 per female (1500–1800 eggs per kg).

2.4.3 Egg size, development and survival

Egg diameter, weight and yolk content largely determine embryo and alevin size. Large eggs tend to produce larger fry than small eggs (Srivastava and Brown 1991), an effect which may become more pronounced as fish emerge and begin feeding on external food sources following yolk sac absorption (Kazakov 1981). However, the efficiency of yolk utilisation by alevins depends upon environmental conditions during incubation and the early phase of free-swimming life so that any initial size advantage gained through this 'maternal effect' may or may not be maintained (Meekan *et al.* 1998).

Post-spawning, levels of fry recruitment in streams depend on there being gravel substrate stability and hydrological and hydrochemical regimes conducive to egg survival. The processes of embryo development and hatching occur entirely within spawning gravels that afford some protection from displacement and predation. This is also true of the later stage of alevin development almost to the time of absorption of the yolk sac in the spring when the young fish emerge from the gravel. However, the long over-winter incubation (typically November to May) leads to the risk of nest deterioration through sediment infiltration (Montgomery and Buffington 1998), or release from gravels due to erosion. Eggs are most vulnerable to loss during the period from fertilisation to the eyed embryo stage. During this period they are particularly susceptible to mechanical shock caused by overcutting of existing redds by other spawners, washout during flood events (Lapointe *et al.* 2000) and scouring by ice (Cunjak *et al.* 1998). Eggs and alevins displaced from gravels may suffer higher levels of predation, due to increased exposure to salmon parr and other species such as brown trout, common eel (*Anguilla anguilla*) and grayling (*Thymallus thymallus*).

The rate at which eggs and alevins develop is primarily a function of water temperature. The thermal environment to which the eggs and alevins are exposed varies significantly within and between spawning locations, and within and between catchments. At 8°C, the incubation period for the embryo to develop eyes (i.e. eyed stage) is ~ 245 C degree days (above 0°C) and 510 to hatch. Depending on local conditions, the incubation period may range from 70 to 160 days.

The duration of incubation and the date at emergence are generally linked to spawning date, with higher temperatures causing a faster rate of development and consequently a shorter incubation period and earlier emergence date. However, fewer temperature 'units' are required at lower average temperatures than at higher ones, and particularly below 5–6°C the relationship becomes non-linear. This effect allows eggs to develop and hatch normally at high latitudes or in otherwise very cold environments. It may also serve to reduce the influence of year-to-year variation in water temperatures on inter-annual variation in subsequent hatch and emergence times (Brannon 1987).

Groundwater, when low in dissolved oxygen, may affect egg survival and hatching times particularly among egg pockets positioned near or within the hyporheic zone, where ground and surface water mix. In winter, groundwater tends to be warmer than surface water and this may lead to an acceleration of embryo development rates and the associated metabolic demands, resulting in smaller fry at emergence. Where present, complex groundwater and surface water interactions may therefore generate significant spatial and temporal variations in water quality and temperature and influence hatch and emergence times accordingly (Malcolm *et al.* 2003). Laboratory studies show that the key egg stages of fertilisation, water hardening and hatch can also be affected by low pH, particularly below pH 4.5. Chronic exposure to acidic conditions can lead to extended alevin development times and reduced growth (Sayer *et al.* 1993).

While in the gravel, density-dependent and density-independent mortality and mortality factors will therefore vary spatially and temporally on different groups of eggs and alevins within and among nests and redds. Density-dependent compensation for mortality during these stages is unlikely. There is likely to be variation among localities in the degree to which mortality at these life stages occurs, and in the impact it has on subsequent juvenile recruitment. In some locations, effects may be substantial and differentially affect the survival of individuals, families and populations (Chapter 7).

The eggs hatch in the spring and the resulting alevins (15–25 mm long) are nourished by their attached yolk sac for a further period of 3–8 weeks (290 degree days). For most of this time they remain buried in the gravel. However, as yolk utilisation nears completion they will emerge into the stream as free-swimming, exogenous feeding fry.

2.4.4 Emergence and dispersal of fry

Little is known about the movements and behaviour of alevins and fry within spawning nests as this is inherently difficult to study (Godin 1981). However, what is known shows that alevins respond negatively to light, positively to gravity and hard surfaces, until close to emergence when these tendencies weaken and even reverse. Shortly after hatch alevins may move downward into the spawning bed, and further lateral movements have been observed in response to the direction of water flow and dissolved oxygen levels (García de Leániz et al. 1993). By the end of the incubation period, the alevins have developed into fully finned and pigmented fry, but may still bear a small yolk sac. As the time of emergence approaches the young fry become increasingly active and move up to just below the gravel surface. Emergence from redds occurs mainly at night, a behaviour seen as adaptive and a means of avoiding predation (Gustafson-Marjanen and Dowse 1983). Patterns of emergence are influenced by temperature, stream discharge, moonlight and sediment quality, and there may be selective pressure for salmonid fry to emerge, disperse and begin feeding at a time when food availability is highest (e.g. Brannon 1987; Einum and Fleming 2000) and, in some areas, differences in predation on early and later components of brood groups of emerging fry may occur (Brännäs 1995).

Emergence from the gravel is the first of a number of directed movements leading to a series of habitat changes ending in reproduction (Groot and Margolis 1991). The duration of the emergence and subsequent fry dispersal tends to reflect the duration of the adult spawning period. In stable habitats with predictable seasonal quality, losses due to starvation and predation may select for increasing synchrony of spawning and emergence within and among redds in a population. In contrast, temporally and spatially unstable habitats may favour less predictable emergence behaviour. Sediment movement or dewatering may also result in the premature release of fry, causing mass strandings and associated mortality where spawning areas become disconnected from main channel areas.

The nature and scale of the processes involved in dispersal of fry from redds into suitable rearing habitats has potentially important implications for salmon population structure and productivity (Gustafson-Greenwood and Moring 1990). Fry exhibit territorial behaviour soon after emergence. This behaviour promotes dispersal (Beall et al. 1994), predominantly down- and, to a lesser extent, upstream (García de Leániz et al. 2000), depending on local channel gradients and associated riverbed features and factors such as river discharge (Crisp 1995; Jensen and Johnson 1999), ambient light levels (Gustafson-Marjanen and Dowse 1983) and redd position (Webb et al. 2001). Local variation in dispersal behaviour may exist in response to historical differences among locations in the distribution and quality of suitable rearing habitats. In many rivers, the distribution of spawning and rearing habitats is not continuous or overlapping. Variation in habitat suitability, patch size and distribution may act to limit dispersal and, thereby, recruitment due to inherited constraints on fry dispersal, while larger-scale distribution will be conditioned by adult migratory and spawning behaviour.

The majority of fry disperse within 2 weeks of emergence but how long they take to settle and establish a feeding territory is unknown. Dispersal up to 300 m downstream in a single night has been observed. Active feeding and growth may occur during dispersal and studies during the first summer of free-swimming life suggest settlement up to 1.5 km from the natal redd. The first individuals to emerge from redds may often obtain the best habitats. This may result in lower mortality and faster growth with later-emerging fry competitively displaced by fish which emerged earlier and have prior residence (O'Connor *et al.* 2000). These factors may act as primary drivers for fry dispersal because, both within and among redds, density-dependent interactions are likely to be most intense during the earliest period of free-swimming life. However, early-emerging fry may also face poor conditions in some years, and annually variable size-selective mortality associated with key hydroclimatic events (e.g. flooding) have been reported (Good *et al.* 2001). Competition among fry emerging from composite or closely spaced redds for space and resources is likely to have important implications for the genetic character of populations and for the spatial and temporal aspects of sampling of early life-history stages in genetic studies (Chapter 6).

2.4.5 Free-swimming juvenile life and production

Many environmental factors affect juvenile Atlantic salmon (e.g. Elliott 2001). As in most salmonids, natural mortality tends to be greatest during the early free-swimming juvenile stages (Mills 1989). Fry and parr feed on chironomids, stoneflies and caddis flies, aquatic annelids and molluscs as well as numerous terrestrial invertebrates that fall into the water (Scott and Crossman 1973). During the first summer, growth is generally rapid and fry may quadruple in length as they become parr. Depending on their growth rate, a function of food availability and temperature, they will spend 1–6 years in fresh water before descending to the sea as smolts. During this time, many factors may act to generate losses and limit the numbers of smolts produced (e.g. Letcher *et al.* 2002), including quantity and quality of habitat, food abundance and quality, the distribution and abundance of conspecifics, and interactions with other fish, birds and mammals. First winter survival can be size-dependent and the size class affected can vary among years (Aubin-Horth *et al.* 2005).

Habitat availability influences patterns of habitat preference and use. Habitats most commonly used by fry and parr in rivers and streams include riffle areas containing gravel or cobble substrate, but they may also use pools, ponds and slow-moving and weedy areas. Higher densities are usually associated with complex channels containing coarser grade substrates (Guay *et al.* 2000). Habitats preferred by parr tend to change with season. During the summer, parr usually prefer shallow riffles and runs, typically 5–90 cm deep and a mean water velocity of 10–80 cm sec^{-1}, and a gravel to boulder substratum (Heggenes 1990). In contrast, during the winter months when temperatures are below 9°C, many fish may move into deeper water or hide beneath larger stones (reviewed by Cunjak *et al.* 1998). Accordingly, winter survival may be poor where access to either suitably coarse substrate or pool habitat is limited (Hutchings 1986). Juvenile salmon may also move from natal areas to river main stems, other tributaries, lakes and even estuaries to exploit habitats more conducive to growth and survival (reviewed by McCormick *et al.* 1998). This strategy appears to underlie the use of lacustrine habitats in some areas of Newfoundland, Iceland, Norway, Finland and Ireland by reducing exposure to predators and competitors (Hutchings 1986), reducing generation times and increasing rates of parr maturity compared to nearby fluvial systems (Klemetsen *et al.* 2003). Overall, the growth, survival and density of juveniles are highly variable across the species range and can be as great within rivers as between rivers (e.g. see Mills 1989; Prévost *et al.* 1992), generating major differences in stock–recruitment functions (see Ricker 1954; Prévost and Chaput 2001).

The length of the freshwater phase, ending with smolting and migration to sea, varies with latitude. It also differs between North America and Europe (Metcalfe and Thorpe 1992). At low latitudes it is generally 1–2 years, at intermediate latitudes 2–4 years and further north 3–6 years. However, smolt age varies within and between river systems (Symons 1979). As a result, within a population, a single year class of eggs or a single mating may contribute to a number of different smolt cohorts. Therefore the different individuals of a family group or populations can be exposed to a very wide temporal range of environmental factors affecting freshwater growth and survival.

The social behaviour of parr is highly variable and site specific. In general, parr are strongly territorial and, in most rivers and streams, single parr may have a territory of just a few square metres within a more extensive 'home range'. The territories of different fish 'interconnect' to form a mosaic across the riverbed whose dimensions alter with pertaining conditions (Grant *et al.* 1998). Territory and home range size is to some degree proportional to body size and can vary to ensure a constant supply of food and, therefore, tends to increase as fish get older. Competition linked to territoriality can be intraspecific (Grant *et al.* 1998) or interspecific (Fausch 1998).

One of the main impacts of territoriality is to limit production by setting a maximum density that a habitat of a given productive capacity can support (Grant and Kramer 1990; Kocik and Ferreri 1998). Thus territoriality limits population size and density and subsequent smolt production by a complex interplay between the various density-dependent and density-independent factors affecting survival and growth.

2.4.6 *Sexual maturation of parr*

The life cycle of most anadromous fish encompasses a marine migration. However, in anadromous populations a proportion of males commonly become sexually mature at the parr stage, without leaving fresh water. These mature parr have the capacity to reproduce with anadromous females, even though they are an order of magnitude smaller and have never migrated to salt water (Myers and Hutchings 1987; Hutchings and Myers 1988). Mature female parr with small numbers of relatively large viable eggs have been detected in some anadromous populations, but are rare (Fleming 1996).

Mature male parr, distinguished from immature fish by their characteristic rotund shape and dark coloration, do not have a kype. They typically mature at between 1 and 5 years of age (Myers *et al.* 1986) though, in the south of the species range, some male parr have been found to attain sexual maturity only a few months after emergence as fry (Utrilla and Lobón-Cervia 1999). After maturing as parr, surviving males appear to become smolts, migrate to sea and return to spawn again as full-size salmon (C. Garcia de Leániz, pers. comm.). However, the extent to which they contribute to annual smolt and subsequent adult cohort groups is unknown.

Mature male parr have been detected within wild juvenile populations in Spain (Martinez *et al.* 2000), Iceland (Scarnecchia 1983), Norway (L'Abée-Lund 1989) and the US (Letcher *et al.* 2002). In Canada, their incidence among populations is highly variable, with the percentage of males for a given age class that are mature varying between 0 and 100%, and within a population older parr are the most likely to be sexually mature (Dalley *et al.* 1983; Myers *et al.* 1986). However, significant variations in the proportion of mature males have also been reported among samples gathered concurrently in different sites and tributaries of a common river basin (Aubin-Horth and Dodson 2004).

Maturity among resident parr is a very obvious manifestation of the phenotypic plasticity of the species (Thorpe *et al.* 1998). It is regarded as a 'conditional strategy', meaning that the expression of the phenotype is not predetermined genetically, but is probably most strongly linked to the individual growth rate and therefore size and condition at age (Hutchings and Myers 1994). Years of higher growth may therefore be associated with higher levels of juvenile maturation (Myers *et al.* 1986). Laboratory studies have shown a relationship between the tendency of individuals to become mature and body size and/or lipid reserves (e.g. Rowe *et al.* 1991; Berglund 1995).

In the wild, populations with larger individuals tend to have a higher proportion of mature male parr. Moreover, within a population, the initially fastest-growing individuals tend to mature in a higher proportion than slower-growing individuals, though the latter may overtake the former by the autumn (Murphy 1980). Studies following growth trajectories of individual parr suggest that size divergence of fish destined to mature begins soon after emergence from the spawning nests. Pre-maturing male parr may therefore exhibit a size advantage more than a year before maturation (Aubin-Horth and Dodson 2004). Furthermore, the size of maturing parr is often larger than the size of males that will not develop functional gonads in the spring preceding spawning (Whalen and Parrish 1999).

The detailed spawning behaviour of sexually mature male parr is described by Jones (1959). Aggressive displays and attacks are common between male parr attempting to exploit opportunities to spawn. However, they do not use aggressive behaviour to compete with the large anadromous males to gain access to spawning adult females, but swim quickly ('sneak') into the base of spawning nests during the period of construction prior to oviposition. During the last stages of nest construction, the parr move closer to the adult female's vent or hide among the larger stones among which eggs are destined to be laid. Milt release is synchronous with that of the attendant adult male. The genetic implications of early maturation among male parr and their spawning success are considered in Chapter 6.

2.4.7 Movements of parr

In some populations, some parr may emigrate downstream from natal nursery areas in the autumn (Cunjak *et al.* 1998). The significance of this behaviour is not fully known but see Huntingford *et al.* (1992). Autumn migrants are typically parr-like in appearance and often consist of a mixture of immature fish of both sexes and mature male parr. Field research suggests that both groups are linked closely with spring migrants that leave the same streams as smolts in the following spring (Youngson *et al.* 1983). However, in contrast to smolts, these fish do not enter the sea immediately and may remain in fresh water until at least the following spring. Downstream migration among immature fish may reflect a state-dependent requirement for more suitable overwintering habitat, which may be more readily available in the main stem areas of larger river channels characterised by extensive areas of coarse substrate and pool habitats.

Atlantic salmon parr derived from anadromous parents are known to occur in ponds and lakes. In Newfoundland, upstream migrations of parr from outlet streams into lakes where they remain and mature (males), or become smolts, have been recorded from May through to November (Hutchings 1986; Ryan 1986). These migrations differ from those movements of mature male parr in response to the presence and distribution of mature females, which increase their chance of reproductive success (Webb and Hawkins 1989).

2.4.8 Smolt migration

Surviving anadromous juveniles, after a variable number of years in fresh water, undergo physiological and morphological changes (i.e. smolting), and become smolts. The change comes in response to environmental cues following attainment of a minimum size-related developmental stage (McCormick *et al.* 1998). This size varies with latitude (Nicieza *et al.* 1994) though, in general in Europe, most parr which are 10 cm in length by the end of a growing season become smolts in the following spring. In Canada, the smolt sizes are generally larger.

The spring transformation involves a number of morphological, physiological and behavioural changes. Parr that become smolts lose their characteristic 'parr marks' and spots due to the deposition of guanine and hypoxanthine crystals, a process which gives smolts their characteristic silvery appearance. Furthermore, their bodies become more streamlined and buoyancy increases. Fish entering this life-history stage also undergo physiological changes, including the development of the ability to maintain a mineral balance in their body fluids in water of higher salinity. This involves cellular changes which reduce the time required to adjust to the ionic disturbance caused by movement into the more saline water of estuaries and the sea, thereby permitting uninterrupted feeding and improved swimming performance (McCormick *et al.* 1998).

The main primary 'priming' and 'releasing' factors linked to the initiation and maintenance of smolt migration are seasonal changes in photoperiod, water temperature, lunar phase and water flow (reviewed in McCormick *et al.* 1998). These factors often vary with latitude and altitude, in such a manner that patterns of smolt development and migration differ between and within rivers (Antonsson and Gudjonsson 2002). Fish may require quite different strategies in systems where smolts are required to migrate through large lakes that are thermally stable and with a largely imperceptible water flow.

The timing of smolt migration appears critical as survival may be dependent on smolts entering the sea over just a few key weeks in the spring (Hansen 1987). Success in the transition from fresh water to seawater depends on the interaction between a wide range of environmental and physiological processes. Genetic factors may influence the timing of migration and the associated physiological changes (Nicieza *et al.* 1994; Hansen *et al.* 2003) but the relative importance of genes and environmental factors probably varies among populations, locations and years (Klemetsen *et al.* 2003). Local responses to key physiological and behavioural stressors are referred to as 'smolt windows' which are required for survival and, thus, are adaptive (McCormick *et al.* 1998; Chapter 7).

2.4.9 Marine life and distribution

Atlantic salmon smolts normally enter the marine environment during the late spring and early summer where they remain for between 1 and 5 years (Klemetsen *et al.* 2003). The post-smolt stage is probably the least understood period during the life history of the species; once smolts leave fresh water very little is known about their behaviour and the factors that affect them, and details of subsequent movements away from coastal regions is very limited. Nevertheless, in many areas they make limited use of estuaries, and the timing and speed of movement into coastal waters can vary among populations and regions (Thorpe 1994). In the early stages of their marine migration post-smolts move in small schools and loose aggregations (Dutil and Coutu 1988) that remain close to the surface and are strongly influenced by water currents, wind direction and tides (Lacroix and McCurdy 1996). Among post-smolts in the Gulf of St Lawrence, movement away from the nearshore seems to be triggered by decreasing temperatures in the autumn (Dutil and Coutu 1988).

Our knowledge of how Atlantic salmon are distributed in the sea is based largely on studies of tag recaptures in marine fisheries, in West Greenland, the Faroe Islands and in the northern Norwegian Sea, associated with major marine feeding areas. Outside these areas both post-smolts and immature adults are thought to be widely but not evenly distributed (Hansen et al. 2003). Little detailed seasonal distribution and abundance data exist on a broader scale, though recently, post-smolts have been caught during pelagic trawl and drift-net surveys off the north of Scotland (Shelton et al. 1997), and in the Norwegian (Holm et al. 2000) and Labrador (Reddin and Short 1991) seas. Post-smolts have also been recorded in catches in herring and mackerel weirs in the Bay of Fundy and south of Nova Scotia (Meister 1984).

At sea, Atlantic salmon appear to be associated with prominent ocean currents, continental shelf features and feeding areas. Actual information on marine distribution is limited but a number of possible feeding and migration routes have been inferred for the main regional groupings of populations. Fish derived from rivers of eastern North America remain predominantly in the West Atlantic area in the waters off Western Greenland, the shelf off Newfoundland, and in the Labrador Sea. In contrast, European stocks move far into the North and north-east Atlantic to feed. Salmon from Canada and southern Europe are major contributors to the stocks off the west coast of Greenland (Hansen and Jacobsen 2000). Ocean feeding areas to the north of the Faroe Islands are thought to be populated by salmon from Norway, Scotland and Russia. Areas to the south of the Faroe Islands are thought to be utilised by fish from the same rivers as the north, but with fish from Ireland also present (Hansen and Jacobsen 2000). Within these key areas, the representation of the main regional source groups of fish is thought to be variable, and may vary seasonally (Jacobsen et al. 2001). However, very little is known of the post-smolt patterns of migration that lead to their formation (Holst et al. 2000) and the extent to which movements at sea are subject to genetic control (i.e. stock-specific behavioural traits) is not known. However, given the frequency of anadromous migrations directed by homing that are displayed by many different species of salmonids it would be surprising if it were not based on variable, but locally adaptive, fitness strategies (Dittman and Quinn 1996).

At sea, Atlantic salmon appear to be opportunistic pelagic feeders that exploit a wide range of crustacean and fish prey. Typically, their diet consists of euphausid shrimps and prawns, squid, and a range of fish species such as sandeels (*Ammodytes* spp.), herring (*Clupea harengus*), sprat (*Sprattus sprattus*), lantern fish (Myctophidae), capelin (*Mallotus villosus*) and pearlside (*Maurolicus mülleri*) (reviewed by Hislop and Shelton 1993). The distribution and abundance of these species may vary seasonally, spatially and from year to year but salmon growth is generally rapid and fish typically begin to mature after 1–3 years, with cases of 5 years only rarely seen. Salmon that return after 1 year at sea are called grilse or 1 sea-winter salmon (1SW), while those that remain at sea for longer periods are multi-sea-winter (MSW) salmon. The factors that may influence the sea age at return and their impact on survival and fecundity are reviewed by Gardner (1976) and Hutchings and Jones (1998). In most rivers today, the majority of wild fish return as grilse or 2SW salmon.

Marine survival varies between years and stocks (Shearer 1992). Losses are thought to be principally due to the temporally and spatially variable effects of predation, starvation, parasites and disease, though freshwater influences may impact marine survival (Klemetsen *et al.* 2003). The highest rates of mortality are believed to arise from events during the first few months after the fish leave fresh water and, in Scotland, growth in the first year at sea has been linked to subsequent adult returns (Shearer 1992). However, patterns of marine survival among smolts suggest that factors affecting the fish later in the marine phase may also be a significant source of mortality (Potter and Crozier 2000). Sea surface temperature may be one of the key factors affecting natural salmon mortality by influencing the distribution of plankton assemblages and associated dependent prey species (Mills 2000, 2003). A number of species of predatory fish, birds and marine mammals are thought to contribute to mortality (NASCO 1996), though variably in space and time. Predation pressure may be affected by wide-scale changes in overall marine productivity; for example, oceanic warming may force some predators to begin exploiting novel sources of prey (Pearcy 1992). In contrast to Pacific salmon, evidence of density-dependent marine mortality has not been reported, and the underlying mechanisms linking ocean climate to salmon survival remain unclear.

Very little is known about the marine life of Atlantic salmon (Friedland 1998; Hansen and Quinn 1998; Mills 2000, 2003). However, based on what is known, a number of features stand out. Most anadromous Atlantic salmon undergo long marine migrations to and from their natal rivers to ocean feeding grounds, and survival and production is heavily dependent on the ocean environment, where most growth and mortality occurs (Groot and Margolis 1991). Oceanic influences on salmon growth, behaviour and survival are complex and difficult to study because of the large temporal and spatial scales over which they operate (e.g. Friedland 1998) and the dynamic nature of the marine environment in both time and space. However, changes in marine environmental conditions affect the growth and survival of many fish species (Drinkwater 2000) and broad-scale fluctuations in marine survival have been correlated with large-scale changes in the North Atlantic Ocean climate (Reddin and Friedland 1993).

Reliable data on changes in numbers of returning salmon are available for only a small number of North Atlantic rivers. However, the few for which good data sets are available indicate that survival at sea has declined by ~ 50% over the past 10–30 years (Reddin *et al.* 2000). Typically, natural mortality in the sea of wild and reared smolts currently ranges from 70% in the River Bush, Northern Ireland, to 99% in the Penobscot River, Maine, USA (Potter and Crozier 2000). Therefore, though much remains to be learned about the marine life of the Atlantic salmon, it is clear that marine conditions are likely to exert important influences on the species. Against a background of increasing concern about the decline of the species, it is urgent that these effects are factored into the development of plans for the future conservation and management of stocks (Mills 2000, 2003).

2.4.10 Homing and return marine migration

Atlantic salmon return to spawn in the rivers in which they hatched and spent their juvenile lives. How this precise homing to natal rivers works remains largely a mystery but the homing of farm salmon to natal rearing facilities can be very precise (Fig. 2.6). In part, it appears to rely on imprinting on the natal stream during the juvenile phase via chemical olfactory cues and an acute ability to detect chemical odours of the natal stream (Hasler and Scholz 1983; Hansen *et al.* 1993). Homing has been reported over a range of geographical scales, and levels of precision may encompass distances of less than a few kilometres from natal rearing areas (Foster and Schom 1989; Youngson *et al.* 1994), enabling returning adult fish to locate not

Fig. 2.6 Mature farmed salmon homing in October 1989 to the hatchery outfall in the River Polla, Scotland, where they were reared as juveniles; they escaped from sea cages in Loch Eriboll, into which the river runs, the previous February. (Photo credit: J. Webb.)

only their home river system but also specific spawning areas or tributary streams within them.

The return migration of salmon to their natal rivers can be split into two main phases, the first phase involving orientation from the feeding areas to the home region and coastline, and a second, more directed, homing phase in coastal and estuarine areas (Hawkins *et al*. 1979; Hansen *et al*. 1993). It is unknown exactly what cues start maturing Atlantic salmon on their return to natal areas from their ocean feeding grounds after up to 5 years at sea. However, most likely one of the main drivers is the onset of sexual maturation, a complex physiological process influenced by factors such as growth and photoperiod.

Upon river entry, Atlantic salmon largely cease feeding and the important physiological transition between saline and fresh water is once again made. The timing of river entry varies greatly among and within rivers and is thought to be population-specific and heritable (Stewart *et al*. 2002; Chapter 7). However, it also appears to vary with a wide range of environmental influences (Gardner 1976). This variation, one of the most dramatic displayed by the species, influences exposure to fisheries, predators and other stressors and, thereby, can have an important impact on the abundance and character of spawners in a river (Smith *et al*. 1994). In so far as this influences exploitation by humans, in many areas, salmon rivers tend to be classified on the basis of the timing and character of their returning adults.

The phenomenon of adult run timing encompasses two important levels of variation: sea age at return and seasonal timing of river entry. The behaviour of adults in different rivers is highly variable and specific as regards the presence, proportion and timing of the different behavioural components, and shows some degree of correlation with river size and environmental variables such as river flow (Scarnecchia 1983; Jonsson *et al*. 1991). Older individuals tend, in general, to return earlier than younger ones (Shearer 1992), so that MSW salmon tend to begin to return before grilse. In some of the larger Scottish rivers (e.g. Tay, Tweed, and

Dee), MSW salmon may enter rivers every month of the year, while grilse enter rivers from about May to mid-January. Therefore in these rivers, although displaced by calendar date, the two main sea-age groups may also each contain early, middle and later-running components. Among previous spawners, patterns of return tend to follow a fixed pattern, with individuals tending to return to sites, from where they previously emigrated to sea as smolts, at approximately the same date on each spawning migration (Foster and Schom 1989; Shearer 1992).

Most insight into the timing of river entry derives from fisheries that operate within limited statutory seasons and from counting fences, and many larger rivers across the species range have salmon runs outside the dates of local fisheries. Nevertheless, observed patterns of return in most cases will reflect the time of return of the majority of adults that enter a system or, at least, the timing of return of the most numerically or economically important stock components in rivers. Based on this information, rivers are most often classified as being 'early' or so-called spring rivers, summer rivers, or late (autumn) rivers. Most rivers with anadromous salmon are characterised by either one or a combination of these broad patterns (Nordqvist 1924). Rivers in eastern Canada (Saunders 1967), Iceland (Gudjonsson and Mills 1982) and Norway (Jensen 1990) are characterised by returns during the summer and autumn. In contrast, many of the larger rivers in the UK (particularly Scotland), Ireland and northern Russia (Kola) have the greatest diversity of patterns of run timing, with MSW fish returning over 12 months of the year (Nordqvist 1924; Shearer 1992; Webb and Campbell 2000).

River migration, leading up to spawning, in many rivers appears to be composed of two or three phases. First there is a period of steady progress upstream, with periods of active movement alternating with stationary periods, followed by a long residence, perhaps in a single pool. A final phase of further upstream migration may also take place just prior to spawning (e.g. Hawkins and Smith 1986; Økland et al. 2001). Typically, fish from a common natal source tend to regroup at or near their 'home' areas prior to spawning (Youngson et al. 1994). The observed patterns are likely to be influenced by changes in environmental conditions such as water flow and temperature, natural and man-made obstacles, and tributary junctions (Stuart 1962; Banks 1969). River behaviour is likely to be influenced by spatial and temporal patterns of predation and disease, and the impact of fisheries (Randall et al. 1991).

The patterns of sea-age at return and associated run timing for rivers and tributaries are increasingly seen as important traits which characterise and differentiate anadromous Atlantic salmon populations. In many UK rivers, run timing and the timing and location of spawning often appear linked, providing fishery managers with a potentially obvious and practical indicator of how regional and individual river stocks are structured into genetic populations (Webb and Campbell 2000; Youngson et al. 2003; Chapter 5). Furthermore, in most situations, individuals migrating back to a given locale will be of varying sea and freshwater age such that breeding in populations in a given year will generally involve individuals born in different years. The result is that generations are overlapping, a life-history trait which has major implications for the genetic character of Atlantic salmon populations (see Chapter 6).

2.5 Biology of non-anadromous populations

Non-anadromous, freshwater resident populations of Atlantic salmon are more restricted in their distribution than anadromous forms (Fig. 2.4) and confined to parts of the species range covered by ice during the Pleistocene glaciation, prior to 15 000 yrs BP. As many areas they now inhabit could only be reached by marine migration, e.g. the island of Newfoundland, it is believed that non-anadromous populations have evolved postglacially from anadromous populations (Power 1969; Berg 1985). This view is in line with the available genetic evidence (Chapter 5).

The potential for both freshwater resident and estuarine population types to evolve from anadromous populations is evidenced by the observations of Power (1969) on salmon in the Ungava Bay region of northern Quebec. In these rivers, in years when marine migrations are blocked by lethally cold sea temperatures, salmon will migrate to the river mouths and estuaries where they then feed and mature before migrating back up rivers to spawn. Such conditions are likely to have existed for periods in some parts of the species range during the Younger Dryas period ~ 10 000 years BP (Berg 1985). Interestingly, freshwater maturation is still within the physiological capacity of most anadromous populations as it can usually be induced in anadromous fish cultured in fresh water for their whole life cycle.

The biological nature of non-anadromous populations has been investigated far less intensively than that of its anadromous cousins, though many aspects of the biology of populations in Maine have been studied (Warner and Havey 1985). Thus the extent to which the detailed aspects of the biology of non-anadromous populations varies and differs from anadromous populations remains incompletely understood. The sections that follow serve to highlight what is known about non-anadromous populations. What is particularly noteworthy, even given our more limited knowledge of this component of the species' diversity, is the high degree of ecological plasticity shown by freshwater populations.

2.5.1 Geographic distribution

Within postglacial areas of the species' distribution, non-anadromous populations of Atlantic salmon inhabit both lacustrine and fluvial habitats, often in parts of river systems above waterfalls, which are impassable to anadromous migrants. In Europe they occur in 14 river systems (Kazakov 1992). In 11 of these they are associated with lakes, including Europe's largest – Ladoga and Onega in Russia and in Lake Vänern in Sweden. A fishery that operated in Lake Ladoga in the 1950s and 1960s was based entirely on non-anadromous forms and exploited spawners from 40 different rivers.

Despite the absence of any physical barrier to migration, Ladoga Lake salmon do not migrate to the Baltic Sea as is the case for the anadromous populations of the lake's outlet river, the Neva. Salmon in Lake Onega, connected to Lake Ladoga by the Svir River, are landlocked by virtue of an impassable waterfall, as are the salmon of Lake Jänisjärvi and Lake Sandal, which drain into Lake Onega. Historically salmon spawned in 34 of Lake Onega's rivers and in four rivers draining into Lake Jänisjärvi. In the Lake Saimaa basin in southern Finland, which drains into Lake Ladoga, resident forms occur in at least three rivers and one lake.

Non-anadromous forms in Europe also occur in lakes of the River Kem and River Vig, north of Lake Onega, which drain into the White Sea, and were historically found in Lake Imandra on the Kola Peninsula. Outside this region, non-anadromous salmon are found in four other areas in Europe. They occur in Sweden in Lake Vänern and spawn in its tributary rivers as well as in three Norwegian rivers situated above impassable falls. They also occur in the upper reaches of the River Namsen in central Norway in the main river and about 20 tributaries, and in southern Norway in the River Otra and its tributary streams, primarily around Bygglansfjord Lake, and in the neighbouring River Nidelva and associated lakes.

The distribution of non-anadromous populations in eastern North America is more widespread but less well documented. Populations were historically present in many of the lakes of New York, New Hampshire, Maine, New Brunswick and Nova Scotia as well as being widely dispersed throughout south and eastern Quebec, Labrador and across the island of Newfoundland (Scott and Crossman 1964; MacCrimmon and Gots 1979). They were also present historically in the Great Lakes in Lake Ontario and have more recently been found in the Kogaluk River in the north-eastern Hudson Bay. In Newfoundland non-anadromous forms occur in most watersheds. In many cases, they exist in the absence of any physical barriers to marine migration. Recent work shows they can co-habit as distinct populations and remain distinct from anadromous populations (Verspoor and Cole 1989, 2005), despite being able to be interbred successfully in culture (Hutchings and Myers 1985).

The total number of non-anadromous populations in North America is unknown. Based on the reported occurrence in the literature and unpublished evidence (L. Cole, pers. comm.; Verspoor, unpubl.), it is likely that many hundreds of non-anadromous populations are present in North America. Multiple populations are likely to exist in many river systems, e.g. the Gander River in Newfoundland (Verspoor and Cole 1989, 2005) and Saguenay River in Quebec (Tessier and Bernatchez 1999), as is the case in parts of Europe where non-anadromous forms are found, e.g. Lake Ladoga (Kazakov 1992) and Lake Vänern (Ros 1981).

In general, populations of non-anadromous salmon in Europe have declined, and many appear to have gone extinct, mostly due to dams, river dredging, pollution and overfishing (Ros 1981; Kazakov 1992). A number of populations in the southern part of the species' North American range have also been lost through habitat destruction from dams and pollution (Warner and Havey 1985). In Lake Ontario, large populations once spawned in many of the lake's tributary rivers (Christie 1972). In other parts of the North American range they are generally undisturbed and see little or no exploitation. However, in some areas, local populations are potentially threatened by interbreeding with anadromous forms due to the stocking of anadromous fish into areas previously inaccessible to the latter or the building of fish passes which allow access to anadromous migrants. Non-anadromous salmon, known locally as Sebago salmon, are widely stocked into the lakes of Maine outside their previous range (e.g. Warner and Havey 1985).

2.5.2 Life history and behaviour

Non-anadromous salmon show one of three basic life histories. The most widespread appears to be a simple contraction of the anadromous life cycle where the marine migration from river rearing areas is replaced by a migration into large lakes followed by a return migration into natal rivers to spawn. This is the case for the populations in Lake Vänern, Lake Saimaa, Lake Ladoga and Lake Onega in Europe (Kazakov 1992) and in Maine (Warner and Havey 1985) and in Lake Ontario (Christie 1972) in North America. In Lake Vänern (Kazakov 1992), fish from the Klarälven River spend 2–4 years in the river before becoming smolts at around 18 cm and migrating into the lake where they stay for 2–5 years before returning to spawn at a mean length of 70 cm and mean weight of 3.1 kg. Historically, they underwent a long migration hundreds of kilometres upstream from June to early September and spawned from October to early November. In contrast, salmon stay in the Gullspångälven for only 1–2 years until about 14 or 19 cm, respectively, after which they move into the lake and feed for 4–5 years before returning to spawn, a few tens of kilometres at most, up the river at a mean size of 84 cm and 7.2 kg. A similar life-history pattern is seen in non-anadromous populations in Maine (Warner and Havey 1985), Lake Ontario (Christie 1972) and in some lakes in Quebec (Tessier and Bernatchez 1999). While most populations spawn in rivers, some also appear to utilise gravel shoals in lakes (Kendal 1935).

In systems such as lakes Ladoga, Onega and Vänern (Kazakov 1992; Ros 1981) and many lakes in Maine (Kendal 1935), populations will utilise both lakes and rivers, migrating into lakes to feed before returning, sometimes up to 350 km, to spawn. In some cases they also utilise gravel shoals in lakes (Kendal 1935) or gravel areas near the lake outflow (Couturier *et al.* 1986). In Little Gull Lake in Newfoundland, resident fish appear to be exclusively lake dwellers and spawners (Verspoor and Cole 2005), a habit which would account for their reproductive isolation from river-spawning anadromous forms and for differences in meristic traits (Claytor and Verspoor 1991). As with anadromous salmon, spawning occurs in the autumn.

A freshwater migratory habit appears to be common and widespread in non-anadromous populations but in some, the migratory habit has been lost. Salmon in the Mellingselva, a tributary of the River Namsen in Norway (Berg and Gausen 1988), and in Ouananiche Beck in Newfoundland (Gibson *et al.* 1996) are stream resident. In contrast, populations found in Little Gull Lake (Verspoor and Cole 2005) and Big Triangle Pond (Couturier *et al.* 1986) on the island of Newfoundland appear to be lake resident, with spawning occurring in the lake near inlet streams or on gravel shoals. In this part of the species range, some populations of fish also appear to migrate into river estuaries to feed during the summer before returning to fresh water (P. Downton, pers. comm.). This is also seen with regard to anadromous populations in the rivers of Ungava Bay in years when sea temperatures are too low and marine migrations are curtailed.

The early juvenile phase of the life cycle of non-anadromous fish does not appear to be markedly different from anadromous forms where it has been studied (Scott and Crossman 1964; Warner and Havey 1985; Gibson *et al.* 1996). However, there appears to be variation among non-anadromous populations with regard to the extent to which fish smolt, the physiological changes associated with the migration to salt water and the adjustment to a high salinity environment. Fish from the Byglandsfjorden population in Norway undergo a normal parr–smolt transformation as regards coloration and changes in the activity of enzymes such as Na^+,K^+-ATPase associated with ionic regulation, as would appear to be the case for salmon in Lake Vänern, Sweden (Staurnes *et al.* 1992). However, at least in some non-anadromous populations in Newfoundland, the transformation seems to be only partial (Birt and Green 1986).

Life-cycle variation among non-anadromous populations, and associated loss of smolting and ability to deal with changes in salinity, probably represent different degrees of evolutionary modification of the species' ancestral anadromous life-history pattern. The degree of modification of the ancestral anadromous pattern shown by any given population is most likely a reflection of the particular constraints of the habitat and the time since the populations were established, from anadromous colonisers, following local deglaciation.

2.5.3 *Maturation and reproduction*

In many non-anadromous populations in North America and Europe, sexually mature fish are dark in colour (Fig. 2.7a) though some are silver as is characteristic of anadromous fish (Scott

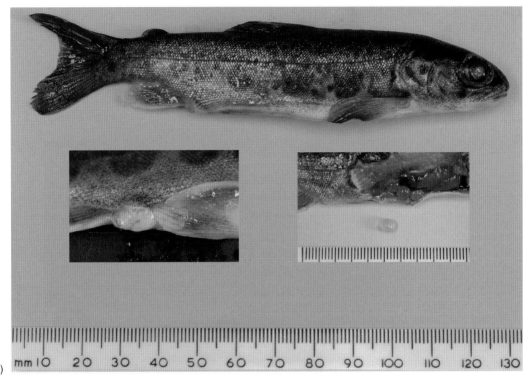

Fig. 2.7 Mature non-anadromous Atlantic salmon from Newfoundland showing coloration. (a) A male from Little Gull Lake. (b) A post-spawning female from Bristol Cove River. Inset: the vent, ovaries and unspawned eggs to scale; the lower specimen has been frozen. (Photos credit: E. Verspoor, D. Hay.)

and Crossman 1964). However, the small fish of the upper Namsen River in Norway (Berg and Gausen 1988), and from the small coastal Bristol Cove River in Newfoundland (Gibson *et al.* 1996), retain characteristic parr markings on their bodies when mature (Fig. 2.7b). This appears to some extent, at least, to be the result of their small size at maturity, something which varies considerably in non-anadromous populations. Males and females in Ouananiche Beck, a landlocked tributary of the Bristol Cove River, had a mean length of 9.2 cm at age 1+ and 10.2 cm at 2+/3+, respectively (Gibson *et al.* 1996) while in the main stem of the river the length averages 12–13 cm (Fig. 2.7b; Verspoor, unpubl.). Maturing females in the Mellingselva, Norway, average 17.6 cm in length with a growth rate after age 2 of about 2.0 cm per year and an average age of maturity of 3.9 years (Berg and Gausen 1988). In contrast, in Little Gull Lake, Newfoundland, mature fish range in size from 17 to 30 cm, with a maximum weight of 0.29 kg with males and females maturing at age 4–6 and 5–7, respectively (Verspoor and

Cole 1989). In other lakes in Newfoundland adult weights of up to 5 kg occur (Leggett and Power 1967) while mean weights of 7–8 kg have been recorded for non-anadromous fish in lakes elsewhere in North America such as in Maine and Quebec (Scott and Crossman 1964). The largest reported is 20.4 kg from Lake Ontario. In Lake Vänern, Sweden, and Lake Ladoga, Russia, individuals of 18 kg or more have been caught (Kazakov 1992) though significant variation occurs among populations within these river systems. Lake Vänern fish spawning in the Klarälven River average 1.5–2 kg while those spawning in the Gullspång River average 4–9 kg (Ros 1981).

Spawning of non-anadromous salmon, as with anadromous populations, occurs in the autumn and appears to be behaviourally similar to that of anadromous fish. As with anadromous stocks, there appears to be considerable variation among localities in spawning time, even among populations within river systems. In the Lake Vänern watershed (Ros 1981), spawning historically occurred in October to early November in the Klarälven River. In contrast, in the lower reaches of the Gullspangälven River spawning by larger fish takes place from November to early December. In Ounaniche Beck and the Bristol Cove River, in Newfoundland, the small non-anadromous salmon spawn in mid to late October (Gibson *et al.* 1996) as many other populations across the island appear to do (Hutchings and Myers 1985; E. Verspoor, pers. obs.). Male fish in some lakes in Maine mature at 1 year of age as parr, presumably participating in spawning, similar to many anadromous populations (Warner and Havey 1985).

Mean numbers of eggs range from just 33 eggs for a 10.6 cm female in the Ouananiche Beck and an average of about 95 eggs for females with a mean size of 17.9 cm in the Namsen (Berg and Gausen 1988) to 268 eggs for a 25 cm fish in Candlestick Pond, Newfoundland (Barbour *et al.* 1979). In Little Moose Lake, New York, a 1.27 kg fish produced 1633 eggs (Warner and Havey 1985). Fish from Lake Vänern spawning in the Klarälven River produce small eggs (8000 per litre) while those spawning in the Gullspång River produce eggs that are significantly larger (4800 per litre) (Ros 1981). Repeat spawning has been reported (e.g. Barbour *et al.* 1979; Leggett and Power 1967; Warner and Havey 1985) and is likely to be common given the age range of mature fish seen in most populations; however, mortality associated with spawning appears to be high though variable among locations (e.g. Leggett and Power 1967; Gibson *et al.* 1996). Spawning aggregations usually encompass a number of different age classes of fish and, in Maine, up to 10 different age classes have been recorded (Warner and Havey 1985).

2.6 Summary and conclusions

- The Atlantic salmon is a salmonid fish of the temperate and subarctic regions of the North Atlantic assigned to a single subspecies within the taxon *Salmo salar* L.
- Atlantic salmon are fundamentally anadromous, but can mature in fresh water. Non-anadromous populations have evolved which do not go to sea, giving rise to a major biological dichotomy within the species.
- Atlantic salmon show a strong tendency to home to natal areas to spawn, at both the river system level and the tributary level within river systems.
- In anadromous populations, distributed across the species range, spawning and juvenile development (1–6 years) occur in fresh water after which the physiological transformation

called 'smolting' leads to migration of juveniles to the sea, where they feed and mature for 1–4 years before returning to their natal rivers to spawn.
- Sexual maturation of a proportion of anadromous male parr in fresh water and their involvement in spawning is common. A proportion of anadromous spawners migrate back to the sea, and return to spawn again.
- Non-anadromous, freshwater populations show a restricted distribution. Individuals may live 10 or more years and spawn over a number of consecutive years.
- Non-anadromous populations may undergo freshwater migrations from rivers to lakes and back to rivers to spawn or may be river or lake resident. They also occur in situations where return marine migrations are prevented by physical barriers such as waterfalls or in situations where no such barriers exist.
- Anadromous and non-anadromous populations are similar in body morphology but biologically distinct, and both show variation within and among populations in growth, age of maturity, migration behaviour and habitat utilisation.
- Atlantic salmon are now extinct in many rivers which previously had healthy populations; in many more rivers stocks are in decline as a result of decreased marine survival, over-exploitation and human-induced habitat changes such as pollution, dam building, water extraction and unsympathetic methods of land use.
- Anadromous and non-anadromous Atlantic salmon show a high level of variation in life history, behaviour and habitat utilisation both within and among locations.
- The species' homing behaviour and habitation of a wide range of diverse and fragmented environments provide conditions favourable for the subdivision of the species into a large number of reproductively distinct and adaptively differentiated genetic populations.

Further reading

Baum, E. (1997) *Maine Atlantic salmon: a national treasure*. Atlantic Salmon Unlimited, Hermon, ME.
IBSFC and HELCOM (1999) *Baltic Salmon Rivers: Status in the late 1990s as reported by the countries in the Baltic Region*. The Swedish Environmental Protection Agency; the Swedish National Board of Fisheries.
Jones, J.W. (1959) *The Salmon*. Collins, London.
Mills, D. (1989) *Ecology and Management of Atlantic Salmon*. Chapman & Hall, London.
Mills, D. (2000) *The Ocean Life of Atlantic Salmon: Environmental and biological factors influencing survival*. Fishing News Books, Oxford.
National Research Council (1996) *Upstream: Salmon and society in the Pacific Northwest*. Committee on Protection and Management of Pacific Northwest Anadromous Salmonids. National Academy Press, Washington, DC.
Youngson, A.F. and Hay, D.W. (1996) *The Lives of Salmon: An illustrated account of the life-history of Atlantic salmon*. Swanhill Press, Shrewsbury, UK.

References

Alm, G. (1928) The salmon in the Baltic area of Sweden. *Rapports et Procès-verbaux des Réunions du Conseil International pour l'Exploration de la Mer*, 48: 81–99.

Alm, G. (1934) Salmon in the Baltic precincts. *Rapports et Procès-verbaux des Réunions du Conseil International pour l'Exploration de la Mer*, **92**: 1–63.

Amcoff, P., Börjeson, H., Lindeberg, J. and Norrgren, L. (1998) Thiamine concentrations in feral Baltic salmon exhibiting the M74 syndrome. In: G. McDonald, J.D. Fitzsimons and D.C. Honeyfield (Ed.) *Early Life Stage Mortality Syndrome in Fishes of the Great Lakes and Baltic Sea*, pp. 82–89. American Fisheries Society, Symposium 21, Bethesda, MD.

Amiro, P.G. (1990) Recruitment variation in Atlantic salmon stocks of the inner Bay of Fundy. *CAFSAC Research Document*, **90/41**: 1–26.

Amiro, P.G. (1998) An assessment of the possible impact of salmon aquaculture on inner Bay of Fundy Atlantic salmon stocks. *DFO Atlantic Fisheries Research Document*, **98/163**: 1–17.

Amiro, P.G. and Jefferson, E.M. (1996) Status of Atlantic salmon in Salmon Fishing Area 22 and 23 for 1995, with emphasis on inner Bay of Fundy stocks. *DFO Atlantic Fisheries Research Document*, **96/134**: 1–16.

Antonsson, T. and Gudjonsson, S. (2002) Variability in timing and characteristics of Atlantic salmon smolts in Icelandic rivers. *Transactions of the American Fisheries Society*, **131**: 643–655.

Aubin-Horth, N. and Dodson, J.J. (2004) Influence of individual body size and variable thresholds on the incidence of a sneaker male reproductive tactic in Atlantic salmon. *Evolution*, **58**: 136–144.

Aubin-Horth, N., Ryan, D.A.J., Good, S.P. and Dodson, J.J. (2005) Importance of spatial scale on the evaluation of a life-history trait in Atlantic salmon and its consequences for management. *Evolutionary Ecology Research*, **7**: 1171–1182.

Banks, J.W. (1969) A review of the literature on the upstream migration of adult salmonids. *Journal of Fish Biology*, **1**: 85–136.

Barbour, S.E., Rombough, P.J. and Kerekes, J.J. (1979) A life history and ecologic study of an isolated population of 'Dwarf' Ouananiche, *Salmo salar*, from Gros Moren National Park, Newfoundland. *Naturaliste Canadiene*, **106**: 305–311.

Beall, E., Dumas, J., Claireaux, D., Barrierie, L. and Marty, C. (1994) Dispersal patterns and survival of Atlantic salmon (*Salmo salar* L.) juveniles in a nursery stream. *ICES Journal of Marine Science*, **51**: 1–9.

Beland, K.F. (1996) The relation between redd counts and Atlantic salmon (*Salmo salar*) parr populations in the Dennys River, Maine. *Canadian Journal of Fisheries and Aquatic Sciences*, **53**: 513–519.

Belding, D.L. (1934) The cause of the high mortality in the Atlantic salmon after spawning. *Transactions of the American Fisheries Society*, **64**: 219–224.

Bengtsson, B-E., Hill, C., Bergman, Å., Brandt, I., Johansson, N., Magnhagen, C., Södergren, S. and Thulin, J. (1999) Reproductive disturbances in Baltic fish: a synopsis of the FiRe project. *Ambio*, **28**: 2–8.

Berg, O.K. (1985) The formation of non-anadromous populations of Atlantic salmon, *Salmo salar* L. in Europe. *Journal of Fish Biology*, **27**: 805–815.

Berg, O.K. and Gausen, D.A.F. (1988) Life history of a riverine, resident Atlantic salmon *Salmo salar* L. *Fauna Norvegica Series A*, **9**: 63–68.

Berglund, I. (1995) Effects of size and spring growth on sexual maturation in 1+ Atlantic salmon (*Salmo salar*) male parr: interactions with smoltification. *Canadian Journal of Fisheries and Aquatic Sciences*, **25**: 2682–2694.

Birt, T.P. and Green, J.M. (1986) Parr-smolt transformation in sexually mature male anadromous and non-anadromous Atlantic salmon, *Salmo salar*. *Canadian Journal of Fisheries and Aquatic Sciences*, **43**: 680–686.

Brannon, E.L. (1987) Mechanisms stabilising salmonid fry emergence timing. *Canadian Special Publication Fish and Aquatic Sciences*, **96**: 120–124.

Brännäs, E. (1988) Effects of abiotic and biotic factors on hatching, emergence and survival in Baltic salmon (*Salmo salar* L.). University of Umeå, Department of Animal Ecology. Doctoral dissertation.

Brännäs, E. (1995) First access to territorial space and exposure to strong predation pressure: a conflict in early emerging Atlantic salmon (*Salmo salar* L.) fry. *Evolutionary Ecology*, 9: 411–420.

Bylund, G. and Lerche, O. (1995) Thiamine therapy of M74 affected fry of Atlantic salmon *Salmo salar*. *Bulletin of the European Association of Fish Pathologists*, 15: 93–97.

Chapman, D.W. (1988) Critical review of the variables used to define effects of fines in redds of large salmonids. *Transactions of the American Fisheries Society*, 117: 1–21.

Christensen, O. and Larsson, P.O. (1979) Review of Baltic salmon research. *ICES Cooperative Research Report*, 89: 1–124.

Christensen, O., Eriksson, C. and Ikonen, E. (1994) History of the Baltic salmon, fisheries and management. *ICES Cooperative Research Report*, 197: 23–39.

Christie, W.J. (1972) Lake Ontario: effects of exploitation, introductions, and eutrophication on the salmonid community. *Journal of the Fisheries Research Board of Canada*, 29: 913–929.

Claytor, R.R. and Verspoor, E. (1991) Discordant phenotypic variation in sympatric resident and anadromous Atlantic salmon (*Salmo salar*) populations. *Canadian Journal of Zoology*, 69: 2846–2852.

Claytor, R.R., MacCrimmon, H.R. and Gots, B.L. (1991) Continental and ecological variance components of European and North American Atlantic salmon (*Salmo salar*) phenotypes. *Biological Journal of the Linnean Society*, 44: 203–229.

Couturier, C.Y., Clarke, L. and Sutterlin, A.M. (1986) Identification of spawning areas of two forms of Atlantic salmon (*Salmo salar* L.) inhabiting the same watershed. *Fisheries Research*, 4: 131–144.

Crisp, D.T. (1995) Dispersal and growth rate of 0-group salmon (*Salmo salar* L.) from point stocking together with some information from scatter stocking. *Ecology of Freshwater Fish*, 4: 1–8.

Cunjak, R.A., Prowse, T.D. and Parrish, D.L. (1998) Atlantic salmon (*Salmo salar*) in winter: 'the season of parr discontent'? *Canadian Journal of Fisheries and Aquatic Sciences*, 55 (Supplement 1): 161–180.

Dalley, E.L., Andrews, C.W. and Green, J.M. (1983) Precocious male Atlantic salmon parr (*Salmo salar*) in insular Newfoundland. *Canadian Journal of Fisheries and Aquatic Sciences*, 40: 647–652.

Dittman, A.H. and Quinn, T.P. (1996) Homing in Pacific salmon: mechanisms and ecological basis. *Journal of Experimental Biology*, 199: 83–91.

Drinkwater, K.F. (2000) Changes in ocean climate and its general effect on fisheries: examples from the North-west Atlantic. In: D. Mills (Ed.) *The Ocean Life of Atlantic Salmon: Environmental and biological factors influencing survival*, pp. 116–135. Fishing News Books, Oxford.

Dutil, J.D. and Coutu, J.M. (1988) Early marine life of Atlantic salmon, *Salmo salar*, post-smolts in the northern Gulf of St Lawrence. *Fishery Bulletin*, 86: 197–212.

Einum, S. and Fleming, I.A. (2000) Selection against late emergence and small offspring in Atlantic salmon (*Salmo salar*). *Evolution*, 54 (2): 628–639.

Elliott, J.M. (1991) Tolerance and resistance to thermal stress in juvenile Atlantic salmon, *Salmo salar*. *Freshwater Biology*, 25: 61–70.

Elliott, J.M. (2001) The relative role of density in the stock-recruitment relationship of salmonids. In: E. Prévost and G. Chaput (Ed.) *Stock, Recruitment and Reference Points: Assessment and management of Atlantic salmon*, pp. 25–66. INRA Éditions, Paris.

Elliott, S.R., Coe, T.A., Helfield, J.M. and Naiman, R.J. (1998) Spatial variation in environmental characteristics of Atlantic salmon (*Salmo salar*) rivers. *Canadian Journal of Fisheries and Aquatic Sciences*, 55 (Supplement 1): 267–280.

Erkinaro, J. and Niemelä, E. (1995) Growth differences between the Atlantic salmon parr, *Salmo salar*, of nursery brooks and natural rivers in the River Teno watercourse in northern Finland. *Environmental Biology of Fish*, 42: 277–287.

Fausch, K.D. (1998) Interspecific competition and juvenile Atlantic salmon (*Salmo salar*): on testing effects and evaluating the evidence across scales. *Canadian Journal of Fisheries and Aquatic Sciences*, 55 (Supplement 1): 218–231.

Fleming, I.A. (1996) Reproductive strategies of Atlantic salmon: ecology and evolution. *Reviews in Fish Biology and Fisheries*, **6**: 379–416.

Fleming, I.A. (1998) Pattern and variability in the breeding system of Atlantic salmon (*Salmo salar*), with comparisons to other salmonids. *Canadian Journal of Fisheries and Aquatic Sciences*, **55** (Supplement 1): 59–76.

Foster, J.R., and Schom, C.B. (1989) Imprinting and homing of Atlantic salmon (*Salmo salar*) kelts. *Canadian Journal of Fisheries and Aquatic Sciences*, **46**: 714–719.

Friedland, K.D. (1998) Ocean climate influences on critical Atlantic salmon (*Salmo salar*) life history events. *Canadian Journal of Fisheries and Aquatic Sciences*, **55** (Supplement 1): 119–130.

Garant, D., Dodson, J.J. and Bernatchez, L. (2001) A genetic evaluation of mating system and determinants of individual reproductive success in Atlantic salmon (*Salmo salar*). *Journal of Heredity*, **92**: 137–145.

Garant, D., Fontaine, P-M., Good, S.P., Dodson, J.J. and Bernatchez, L. (2002) Influence of male parental identity on growth and survival of offspring in Atlantic salmon (*Salmo salar*). *Evolutionary Ecology Research*, **4**: 537–549.

García de Leániz, C., Fraser, N. and Huntingford, F. (1993) Dispersal of Atlantic salmon fry from a natural redd: evidence for undergravel movements? *Canadian Journal of Zoology*, **71**: 1454–1457.

García de Leániz, C., Fraser, N. and Huntingford, F. (2000) Variability in performance in wild Atlantic salmon, *Salmo salar* L., fry from a single redd. *Fisheries Management and Ecology*, **7**: 489–502.

Gardner, M.L.G. (1976) A review of factors which may influence the sea-age of maturation of Atlantic salmon *Salmo salar* L. *Journal of Fish Biology*, **9**: 289–327.

de Gaudemar, B., Bonzom, J.M. and Beall, E. (2000) Effects of courtship and relative mate size on sexual motivation in Atlantic salmon. *Journal of Fish Biology*, **57**: 502–515.

Gephard, S., Moran, P. and Garcia-Vazquez, E. (2000) Evidence of successful natural reproduction between brown trout and mature Atlantic salmon parr. *Transactions of the American Fisheries Society*, **129**: 301–306.

Gibson, R.J. (1993) The Atlantic salmon in freshwater: spawning, rearing and production. *Reviews in Fish Biology and Fisheries*, **3**: 39–73.

Gibson, R.J. and Cutting, R.E. (Ed.) (1993) Production of juvenile Atlantic salmon, *Salmo salar*, in natural waters. *Canadian Special Publication of Fisheries and Aquatic Sciences*, **118**: 1–262.

Gibson, R.J., Williams, D.D., McGowan, C. and Davidson, W.S. (1996) The ecology of dwarf fluvial Atlantic salmon, *Salmo salar* L., cohabiting with brook trout, *Salvelinus fontinalis* (Mitchill), in southeastern Newfoundland, Canada. *Polish Archives for Hydrobiology*, **43**: 145–166.

Godin, J.J. (1981). Migrations of salmonid fishes during life history phases: daily and annual timing. In: E.L. Brannon and E.O. Salo (Ed.) *Proceedings of the Salmon and Trout Migratory Behaviour Symposium*, pp. 22–50. School of Fisheries, University of Washington, Seattle, WA.

Good, S.P., Dodson, J.J., Meekan, M.G. and Ryan, D.A.J. (2001) Annual variation in size-selective mortality of Atlantic salmon (*Salmo salar*) fry. *Canadian Journal of Fisheries and Aquatic Sciences*, **58**: 1187–1195.

Grant, J.W.A. and Kramer, D.L. (1990) Territory size as a predictor of the upper limit to population density of juvenile salmonids in streams. *Canadian Journal of Fisheries and Aquatic Sciences*, **47**: 1724–1737.

Grant, J.W.A., Steingrimsson, S.Ó., Keeley, E.R. and Cunjak, R.A. (1998) Implications of territory size for the measurement and prediction of salmonid abundance in streams. *Canadian Journal of Fisheries and Aquatic Sciences*, **55** (Supplement 1): 181–190.

Groot, C. and Margolis, L. (1991) *Pacific Salmon Life Histories*. University of British Columbia Press, Vancouver.

Guay, J.C.D., Boisclair, D., Rioux, M., Leclerc, M., Lapointe, M. and Legendre, P. (2000) Development and validation of numerical habitat models for juveniles of Atlantic salmon (*Salmo salar*). *Canadian Journal of Fisheries and Aquatic Sciences*, **57**: 2065–2075.

Gudjonsson, T. and Mills, D. (1982) *Salmon in Iceland*. Atlantic Salmon Trust, Pitlochry, UK.

Gustafson-Marjanen, K.I. and Dowse, H.B. (1983) Seasonal and diel patterns of emergence from the redd of Atlantic salmon (*Salmo salar*) fry. *Canadian Journal of Fisheries and Aquatic Sciences*, **40**: 813–817.

Gustafson-Greenwood, K.I. and Moring, J.R. (1990) Territory size and distribution of newly-emerged Atlantic salmon (*Salmo salar*). *Hydrobiologia*, **206**: 125–131.

Hansen, L.P. (1987) Growth, migration and survival of lake reared juvenile anadromous Atlantic salmon *Salmo salar*. *Fauna Norvegica Series A*, **8**: 29–34.

Hansen, L.P. and Jacobsen, J.A. (2000) Distribution and migration of Atlantic salmon *Salmo salar* L., in the sea. In: D. Mills (Ed.) *The Ocean Life of Atlantic Salmon: Environmental and biological factors influencing survival*, pp. 75–87. Fishing News Books, Oxford.

Hansen, L.P. and Quinn, T.P. (1998) The marine phase of Atlantic salmon (*Salmo salar*) life cycle, with comparisons to Pacific salmon. *Canadian Journal of Fisheries and Aquatic Sciences*, **55** (Supplement 1): 104–118.

Hansen, L.P., Jonsson, N. and Jonsson, B. (1993) Oceanic migration of homing salmon, *Salmo salar*. *Animal Behaviour*, **45**: 927–941.

Hansen, L.P., Holm, M., Holst, J.C. and Jacobsen, J.A. (2003) The ecology of post-smolts of Atlantic salmon. In: D. Mills (Ed.) *Salmon at the Edge*, pp. 25–39. Blackwell Science, Oxford.

Hasler, A.D. and Scholz, A.T. (1983) *Olfactory Imprinting and Homing in Salmon*. Springer-Verlag, Berlin.

Hawkins, A.D. and Smith, G.W. (1986) Radio-tracking observations of Atlantic salmon ascending the Aberdeenshire Dee. *Scottish Fisheries Research Report*, **36**: 1–24.

Hawkins, A.D., Urquhart, G.G. and Shearer, W.M. (1979) The coastal movements of returning Atlantic salmon, *Salmo salar* L. *Scottish Fisheries Research Report*, **15**: 1–15.

Heggberget, T.G. (1988) Timing of spawning in Norwegian Atlantic salmon (*Salmo salar*). *Canadian Journal of Fisheries and Aquatic Sciences*, **45**: 845–849.

Heggenes, J. (1990) Habitat utilisation and preferences in juvenile Atlantic salmon (*Salmo salar*) in streams. *Regulated Rivers*, **5**: 341–354.

Hendry, A.P., Berg, O.K. and Quinn, T.P. (2001) Breeding location choice in salmon: causes (habitat, competition, body size, energy stores) and consequences (life span, energy stores). *Oikos*, **93**: 407–418.

Hislop, J.R.G. and Shelton, R.G.J. (1993) Marine predators and prey of Atlantic salmon (*Salmo salar* L.). In: D. Mills (Ed.) *Salmon in the Sea and New Enhancement Strategies*, pp. 104–118. Fishing News Books, Oxford.

Holst, J.C., Couperus, B., Hammer, C., Jacobsen, J.A., Jákupsstovu, S.H., Krysov, A., Melle, W., Mork, K.A., Tangen, Ø., Vilhjálmsson, H. and Smith, L. (2000) Report on surveys of the distribution, abundance and migrations of the Norwegian spring-spawning herring, other pelagic fish and the environment of the Norwegian Sea and adjacent waters in late winter, spring and summer of 2000. *ICES CM*, **2000/D:03**.

Holm, M., Holst, J.C. and Hansen, L.P. (2000) Spatial and temporal distribution of post-smolts of Atlantic salmon (*Salmo salar* L.) in the Norwegian Sea and adjacent areas. *ICES Journal of Marine Science*, **57**: 955–964.

Huntingford, F.A., Thorpe, J.E., Garciá de Leániz, C. and Hay, D.W. (1992) Patterns of growth and smolting in autumn migrants from a Scottish population of Atlantic salmon, *Salmo salar* L. *Journal of Fish Biology*, **41** (Supplement A): 43–51.

Huntsman, A.G. (1931) Periodical scarcity of salmon. *Fisheries Research Board of Canada, Progress Report*, **2**: 16–17.

Hutchings, J.A. (1986) Lakeward migrations of juvenile Atlantic salmon (*Salmo salar*). *Canadian Journal of Fisheries and Aquatic Sciences*, **43**: 732–741.

Hutchings, J.A. and Jones, M.E.B. (1998) Life history variation and growth rate thresholds for maturity in Atlantic salmon, *Salmo salar*. *Canadian Journal of Fisheries and Aquatic Sciences*, **55**: 22–47.

Hutchings, J.A. and Myers, R.A. (1985) Mating between anadromous and nonanadromous Atlantic salmon, *Salmo salar*. *Canadian Journal of Zoology*, **63**: 2219–2221.

Hutchings, J.A. and Myers, R.A. (1988) Mating success of alternative maturation phenotypes in male Atlantic salmon *Salmo salar*. *Oecologia*, **75**: 169–174.

Hutchings, J.A. and Myers, R.A. (1994) The evolution of alternative mating strategies in variable environments. *Evolutionary Ecology*, **8**: 256–268.

IBSFC and HELCOM (1999) *Baltic Salmon Rivers: Status in the late 1990s as reported by the countries in the Baltic Region*. The Swedish Environmental Protection Agency; the Swedish National Board of Fisheries, 69 pp.

ICES (1994) *Report of the study group on occurrence of M-74 in fish stocks. ICES CM*, **1994/ENV: 9**: 24.

Ikonen, E. and Kallio-Nyberg, I. (1993) The origin and timing of the coastal return migration of salmon (*Salmo salar*) in the Gulf of Bothnia. *ICES CM*, **1993/M: 34**: 9.

Jacobsen, J.A., Lund, R.A., Hansen, L.P. and O'Maoileidigh, N. (2001) Seasonal differences in the origin of Atlantic salmon (*Salmo salar* L.) in the Norwegian sea based on estimates from age structures and tag recaptures. *Fisheries Research*, **52**: 169–177.

Järvi, T.H. (1938) Fluctuations in the Baltic stock of salmon. *Rapports et Procès-verbaux des Réunions du Conseil International pour l'Exploration de la Mer*, **106**: 1–114.

Jensen, A.J. (1990) The effects of water temperature on early life history, juvenile growth and prespawning migrations of Atlantic salmon (*Salmo salar*) and brown trout (*Salmo trutta*). PhD Thesis. University of Trondheim.

Jensen, A.J. and Johnson, B.O. (1999) The functional relationship between peak spring floods and survival and growth of juvenile Atlantic salmon (*Salmo salar*) and brown trout (*Salmo trutta*). *Functional Ecology*, **13**: 778–785.

Jessop, B.M. (1975) *Investigation of the salmon (*Salmo salar*) smolt migration of the Big Salmon River, New Brunswick, 1966–72*. Fisheries and Marine Service, Department of the Environment, Technical Report. Series No. MAR/T-75-1.

Jones, J.W. (1959) *The Salmon*. Collins, London.

Jones, M.W. and Hutchings, J.A. (2001) The influence of male parr body size and mate competition on fertilisation success and effective population size in Atlantic salmon. *Heredity*, **86**, 675–684.

Jonsson, N., Hansen, L.P. and Jonsson, B. (1991) Variation in age, size and repeat spawning of adult Atlantic salmon in relation to river discharge. *Journal of Animal Ecology*, **60**: 937–947.

Kallio-Nyberg, I. and Ikonen, E. (1992) Migration pattern of two salmon stocks in the Baltic Sea. *ICES Journal of Marine Science*, **49**: 191–198.

Kallio-Nyberg, I., Peltonen, H. and Rita, H. (1999) Effects of stock-specific and environmental factors on the feeding migration of Atlantic salmon (*Salmo salar*) in the Baltic Sea. *Canadian Journal of Fisheries and Aquatic Sciences*, **56**: 853–861.

Karlsson, L. and Karlström, Ö. (1994) The Baltic salmon (*Salmo salar* L.): its history, present situation and future. *Dana*, **10**: 61–85.

Karlsson, L., Karlström, Ö. and Hasselborg, T. (1994) Timing of the Baltic salmon run in the Gulf of Bothnia: influence of environmental factors on annual variation. *ICES CM*, **1994/M: 17**: 15.

Kazakov, R.V. (1981) The effect of the size of Atlantic salmon, *Salmo salar*, eggs on embryos and alevins. *Journal of Fish Biology*, **19**: 353–360.

Kazakov, R.V. (1992) Distribution of Atlantic salmon, *Salmo salar* L., in freshwater bodies of Europe. *Aquaculture and Fisheries Management*, **23**: 461–475.

Kendall, W.C. (1935) The Fishes of New England. The salmon family, Part 2, The salmons. *Memorials of the Boston Society of Natural History*, **9**: 1–166.

Klemetsen, A., Amundsen, P.A., Dempson, J.B., Jonsson, B., Jonsson, N., O'Connell, M.F. and Mortensen, E. (2003) Atlantic salmon *Salmo salar* L., brown trout *Salmo trutta* L. and Arctic charr *Salvelinus alpinus* L.: a review of aspects of their life histories. *Ecology of Freshwater Fish*, **12**: 1–59.

Kocik, J.F. and Ferreri, C.P. (1998) Juvenile production variation in salmonids: population dynamics, habitat, and the role of spatial relationships. *Canadian Journal of Fisheries and Aquatic Sciences*, **55** (Supplement 1): 191–200.

Koljonen, M.L. (2001) Conservation goals and fisheries management units for Atlantic salmon in the Baltic Sea area. *Journal of Fish Biology*, **59** (Supplement A): 269–288.

L'Abée-Lund, J.H. (1989) Significance of mature male parr in a small population of Atlantic salmon (*Salmo salar*). *Canadian Journal of Fish and Aquatic Sciences*, **46**: 928–931.

Lacroix, G.L. and McCurdy, P. (1996) Migratory behaviour of post-smolt Atlantic salmon during initial stages of seaward migration. *Journal of Fish Biology*, **49**: 1086–1101.

Lapointe, M., Eaton, B., Driscoll, S. and Latulippe, C. (2000) Modelling the probability of salmonid egg packet scour due to floods. *Canadian Journal of Fisheries and Aquatic Sciences*, **57**: 1120–1130.

Leggett, W.C. and Power, G. (1967) Differences between two populations of landlocked Atlantic salmon (*Salmo salar*) in Newfoundland. *Journal of the Fisheries Research Board of Canada*, **26**: 1585–1596.

Letcher, B.H., Gries, G., and Juanes, F. (2002) Survival of stream-dwelling Atlantic salmon: effects of life history variation, season and age. *Transactions of the American Fisheries Society*, **131**: 838–854.

Lindroth, A. (1965) The Baltic salmon stock. Its natural and artificial regulation. *Mitteilungen, Internationale Vereinigung für Theoretische und Angewandte Limnologie*, **13**: 163–192.

Malcolm, I.A., Youngson, A.F. and Soulsby, C. (2003) Survival of salmonid eggs in gravel bed streams: effects of groundwater-surface water interactions. *River Research and Applications*, **19**: 303–316.

Martínez, J.L., Moran, P., Perez, J., de Gaudemar, B., Beall, E. and Garcia-Vazquez, E. (2000) Multiple paternity increases effective size of southern Atlantic salmon populations. *Molecular Ecology*, **9**: 293–298.

McCormick, S.D., Hansen, L.P., Quinn, T.P. and Saunders, R.L. (1998) Movement, migration, and smolting of Atlantic salmon (*Salmo salar*). *Canadian Journal of Fisheries and Aquatic Sciences*, **55** (Supplement 1): 77–92.

MacCrimmon, H.R. and Gots, B.L. (1979) World distribution of Atlantic salmon, *Salmo salar*. *Journal of the Fisheries Research Board of Canada*, **36**: 422–457.

McDowall, R.M. (2002) The origin of the salmonid fishes: marine, freshwater . . . or neither? *Reviews in Fish Biology and Fisheries*, **11**: 171–179.

McKinnell, S., and Karlström, Ö. (1999) Spatial and temporal covariation in the recruitment and abundance of Atlantic salmon populations in the Baltic Sea. *ICES Journal of Marine Science*, **56**: 433–443.

McKinnell, S., Lundqvist, H. and Johansson, H. (1994) Biological characteristics of the upstream migration of naturally and hatchery-reared Baltic salmon, *Salmo salar* L. *Aquaculture and Fisheries Management*, **25** (Supplement 2): 45–63.

Meekan, M.G., Dodson, J.J., Good, S.P. and Ryan, D.A.J. (1998) Otolith- and fish-size relationships, measurement error and size-selective mortality during the early life of Atlantic salmon, *Salmo salar*. *Canadian Journal of Fisheries and Aquatic Sciences*, **55**: 1663–1673.

Meister, A.L. (1984) The marine migrations of tagged Atlantic salmon (*Salmo salar* L.) of USA origin. *ICES CM*, **1984/M:27**.

Metcalfe, N.B. and Thorpe, J.E. (1992) Early predictors of life-history events: the link between first feeding date, dominance and seaward migration in Atlantic salmon, *Salmo salar* L. *Journal of Fish Biology*, **41** (Supplement B): 93–99.

Mills, D. (1989) *Ecology and Management of Atlantic Salmon*. Chapman & Hall, London.

Mills, D. (Ed.) (2000) *The Ocean Life of Atlantic Salmon: Environmental and biological factors influencing survival*. Fishing News Books, Oxford.

Mills, D. (Ed.) (2003) *Salmon at the Edge*. Blackwell Science, Oxford.

Montgomery, D.R. and Buffington, J.M. (1998) Channel processes, classification, and response potential. In: R.J. Naiman and R.E. Bilby (Ed.) *River Ecology and Management*, pp. 13–42. Springer-Verlag, New York.

Murphy, T.M. (1980) Studies on precocious maturity in artificially reared 1+ Atlantic salmon parr (*Salmo salar* L.). PhD Thesis, University of Stirling.

Myers, R.A. and Hutchings, J.A. (1987) Mating of anadromous Atlantic salmon, *Salmo salar* L. with mature male parr. *Journal of Fish Biology*, **31**: 143–146.

Myers, R.A., Hutchings, J.A. and Gibson, R.J. (1986) Variation in male parr maturation within and among populations of Atlantic salmon, *Salmo salar*. *Canadian Journal of Fisheries and Aquatic Sciences*, **43**: 1242–1248.

NASCO (1996) *The Atlantic salmon as the predator and the prey. Report of the special session of the Council*. Publication CNL (96) **59**. North Atlantic Salmon Conservation Organisation.

Netboy, A. (1968) *The Atlantic Salmon: A vanishing species?* Faber & Faber, London.

Nicieza, A.G., Reyes-Gavilán, F.G. and Braña, F. (1994) Differentiation in juvenile growth and bimodality patterns between northern and southern populations of Atlantic salmon (*Salmo salar* L.). *Canadian Journal of Zoology*, **72**: 1603–1610.

Nordqvist, O. (1924) Times of entering of the Atlantic salmon (*Salmo salar* L.) in the rivers. *Conseil Permanent International pour L'exploration de la Mer. Rapports et Procès Verbaux*, Vol. XXXIII, 57 pp.

Northcote, T.G. (1978) Migratory strategies and production in freshwater fishes. In: S.D. Gerking (Ed.) *Ecology of Freshwater Fish Production*, pp. 326–359. Blackwell Science, Oxford.

O'Connor, K.L., Metcalfe, N.B. and Taylor, A.C. (2000) The effects of prior residence on behaviour and growth rates in juvenile Atlantic salmon (*Salmo salar*). *Behavioral Ecology*, **11**: 13–18.

Økland, F., Erkinaro, J., Moen, K., Niemelä, E., Fiske, P., McKinley, R.S. and Thorstad, E.B. (2001) Return migration of Atlantic salmon in the River Tana: phases of migratory behaviour. *Journal of Fish Biology*, **59**: 862–874.

Parrish, D.L., Behnke, R.J., Gephard, S.R., McCormick, S.D. and Reeves, G.H. (1998) Why aren't there more Atlantic salmon (*Salmo salar*)? *Canadian Journal of Fisheries and Aquatic Sciences*, **55** (Supplement 1): 281–287.

Pearcy, W.G. (1992) *Ocean Ecology of North Pacific Salmonids*. Washington Sea Grant, University of Washington Press, Seattle, WA.

Philips, R.B. and Oakley, T.H. (1997) Phylogenetic relationships among the Salmoninae based on nuclear and mitochondrial DNA sequences. In: T.D. Kocher and C.A. Stepien (Ed.) *Molecular Systematics of Fishes*. Academic Press, London.

Poff, N.K. and Huryn, A. (1998) Multi-scale determinants of secondary production in Atlantic salmon (*Salmo salar*) streams. *Canadian Journal of Fisheries and Aquatic Sciences*, **55** (Supplement 1): 201–217.

Potter, E.C.E. and Crozier, W.W. (2000) A perspective on the marine survival of Atlantic salmon. In: D. Mills (Ed.) *The Ocean Life of the Atlantic Salmon: Environmental and biological factors influencing survival*, pp. 19–36. Fishing News Books, Oxford.

Power, G. (1969) The salmon of Ungava Bay. *Arctic Institute of North America Technical Paper*, **22**: 1–73.

Prévost, E. and Chaput, G. (Ed.) (2001) *Stock, Recruitment and Reference Points: Assessment and management of Atlantic salmon. Hydrobiologie et Aquaculture*. INRA Éditions, Paris.

Prévost, E., Chadwick, E.M.P. and Claytor, R.R. (1992) Influence of size, winter duration, and density on sexual maturation of Atlantic salmon (*Salmo salar*) juveniles in Little Codroy River (south west Newfoundland). *Journal of Fish Biology*, **41**: 1013–1019.

Pursiainen, M., Makkonen, J. and Piironen, J. (1998) Maintenance and exploitation of landlocked salmon, *Salmo salar* m. *sebago*, in the Vuoksi watercourse. In: I.G. Cowx (Ed.) *Stocking and Introduction of Fish*. Fishing News Books, Oxford.

Randall, R.G., Wright, J.A. and Pickard, P.R. (1991) Effect of run timing on the exploitation by anglers of Atlantic salmon in the Miramichi river. *Canadian Technical Report of Fisheries and Aquatic Sciences*, **1790**.

Reddin, D.G. and Friedland, K.D. (1993) Marine environmental factors influencing the movement and survival of Atlantic salmon. In: D. Mills (Ed.) *Salmon in the Sea and New Enhancement Strategies*, pp. 79–103. Fishing News Books, Oxford.

Reddin, D.G. and Short, P.B. (1991) Post-smolt Atlantic salmon (*Salmo salar*) in the Labrador Sea. *Canadian Journal of Fisheries and Aquatic Sciences*, **48**: 2–6.

Reddin, D.G., Heilbig, J., Thomas, A., Whitehouse, B.G. and Friedland, K.D. (2000) Chapter 8. In: D. Mills (Ed.) *Ocean Life of Salmon: Environmental and biological factors influencing survival*, pp. 88–91. Fishing News Books, Oxford.

Ricker, W.E. (1954) Stock and recruitment. *Journal of Fisheries Research Board of Canada*, **11**: 559–623.

Romakkaniemi, A., Karlsson, L. and Karlström, Ö. (1995) Wild Baltic salmon stocks: fecundity and biological reference points concerning their status. *ICES CM*, **1995/M: 28**: 11.

Romakkaniemi, A., Perä, I., Karlsson, L., Jutila, E., Carlsson, U. and Pakarinen, T. (2003) Development of wild Atlantic salmon stocks in the rivers of the northern Baltic Sea in response to management measures. *ICES Journal of Marine Science*, **60**: 329–342.

Ros, T. (1981) Salmonids in the Lake Vänern area. *Ecological Bulletin (Stockholm)*, **34**: 21–31.

Rowe, D.K., Thorpe, J.E. and Shanks, A.M. (1991) Role of fat stores in the maturation of male Atlantic salmon (*Salmo salar*) parr. *Canadian Journal of Fisheries and Aquatic Sciences*, **48**: 405–413.

Ryan, P.M. (1986) Lake use by wild anadromous Atlantic salmon, *Salmo salar*, as an index of subsequent adult abundance. *Canadian Journal of Fisheries and Aquatic Sciences*, **43**: 2–11.

Salminen, M., Kuikka, S. and Erkamo, E. (1994) Divergence in the feeding migration of Baltic salmon (*Salmo salar* L.): the significance of smolt size. *Nordic Journal of Freshwater Research*, **63**: 32–42.

Saunders, R.L. (1967) Seasonal patterns of return of Atlantic salmon in the Northwest Miramichi river, New Brunswick. *Journal of Fisheries Research Board of Canada*, **24**: 21–32.

Sayer, M.D.J., Reader, J.P. and Dalziel, T.R.K (1993) Freshwater acidification: effects on the early life stages of fish. *Reviews in Fish Biology and Fisheries*, **3**: 95–132.

Scarnecchia, D.L. (1983) Age at sexual maturity in Icelandic stocks of Atlantic salmon (*Salmo salar*). *Canadian Journal of Fisheries and Aquatic Sciences*, **40**: 1456–1468.

Scott, W.B. and Crossman, E.J. (1964) *Fishes Occurring in the Freshwaters of Insular Newfoundland*. Department of Fisheries, Canada.

Scott, W.B. and Crossman, E.J. (1973) *Freshwater Fishes of Canada*. Fisheries Research Board of Canada, Bulletin **184**, Department of Fisheries and Oceans, Scientific Information and Publications Branch, Ottawa.

Shearer, W.M. (1992) *The Atlantic Salmon: Natural history, exploitation and future management*. Fishing News Books, Oxford.

Shelton, R.G.J., Turrell, W.R., MacDonald, A., McLaren, I.S. and Nicoll, N.T. (1997) Records of post-smolt Atlantic salmon, *Salmo salar* L., in the Faroe-Shetland Channel in June 1996. *Fisheries Research*, **31**: 159–162.

Smith, G.W., Smith, I.P. and Armstrong, S.M. (1994) The relationship between flow and entry to the Aberdeenshire Dee by returning adult Atlantic salmon. *Journal of Fish Biology*, **45**: 953–960.

Srivastava, R.K. and Brown, J.A. (1991) The biochemical characteristics and hatching performance of cultured and wild Atlantic salmon (*Salmo salar*) eggs. *Canadian Journal of Zoology*, **69**: 2436–2441.

Staurnes, M., Lysfjord, G. and Berg, O.K. (1992) Parr-smolt transformation of a nonanadromous population of Atlantic salmon (*Salmo salar*) in Norway. *Canadian Journal of Zoology*, **70**: 197–199.

Stearley, R.F. and Smith, G.R. (1993) Phylogeny of the Pacific trouts and salmon (*Oncorhynchus*) and genera of the family Salmonidae. *Transactions of the American Fisheries Society*, **122**: 1–33.

Stewart, D.C., Smith, G.W. and Youngson, A.F. (2002) Tributary-specific variation in run timing of adult Atlantic salmon (*Salmo salar*) has a genetic component. *Canadian Journal of Fisheries and Aquatic Sciences*, **59**: 276–281.

Stuart, T.A. (1962) The leaping behaviour of salmon and trout at falls and obstructions. *Freshwater and Salmon Fisheries Research*, 28: 1–46.

Summers, D.W. (1996a) Environmental influences on the timing of spawning of Atlantic salmon, *Salmo salar* L., in the River North Esk. *Fisheries Management and Ecology*, 3: 281–283.

Summers, D.W. (1996b) Differences in the time of river entry of Atlantic salmon, *Salmo salar* L., spawning in different parts of the River North Esk. *Fisheries Management and Ecology*, 3: 209–218.

Symons, P.E.K. (1979) Estimated escapement of Atlantic salmon (*Salmo salar*) for maximum smolt production in rivers of different productivity. *Journal of Fisheries Research Board of Canada*, 36: 132–140.

Taggart, J.B., McLaren, I.S., Hay, D.W., Webb, J.H. and Youngson, A.F. (2001) Spawning success in Atlantic salmon (*Salmo salar* L.): a long-term DNA profiling based study conducted in a natural stream. *Molecular Ecology*, 10: 1047–1060.

Tessier, N. and Bernatchez, L. (1999) Stability of population structure and genetic diversity across generations assessed by microsatellites among sympatric populations of landlocked Atlantic salmon (*Salmo salar* L.). *Molecular Ecology*, 8: 169–179.

Thorpe, J.E. (1994) Salmonid fishes and the estuarine environment. *Estuaries*, 17: 73–93.

Thorpe, J.E., Mangel, M., Metcalfe, N.B., Huntingford, F.A. (1998) Modelling the proximate basis of salmonid life-history variation, with application to Atlantic salmon, *Salmo salar* L. *Evolutionary Ecology*, 12: 581–599.

Utrilla, C.G. and Lobón-Cervia, J. (1999) Life-history patterns in a southern population of Atlantic salmon. *Journal of Fish Biology*, 55: 68–83.

Verspoor, E. and Cole, L.C. (1989) Genetically distinct sympatric populations of resident and anadromous Atlantic salmon *Salmo salar*. *Canadian Journal of Zoology*, 67: 1453–1461.

Verspoor, E. and Cole, L.C. (2005) Genetic evidence for lacustrine spawning of the non-anadromous Atlantic salmon *Salmo salar* L. population of Little Gull Lake, Newfoundland. *Journal of Fish Biology*, 67 (Supplement A): 200–205.

Warner, K. and Havey, K.A. (1985) *Life history, ecology and management of Maine landlocked salmon (Salmo salar)*. Maine Department of Inland Fisheries and Wildlife, ME.

Webb, J.H. and Campbell, R.N.B. (2000) Patterns of run timing in adult Atlantic salmon returning to Scottish rivers: some new perspectives and management implications. In: F.G. Whoriskey, Jr. and K.F. Whelan (Ed.) *Managing Wild Atlantic Salmon: New challenges, new techniques*, pp. 100–138. Atlantic Salmon Federation, St Andrews, New Brunswick.

Webb, J.H. and Hawkins, A.D. (1989) The movements and spawning behaviour of adult salmon in the Girnock Burn, a tributary of the Aberdeenshire Dee: 1986. *Scottish Fisheries Research Report* 40.

Webb, J.H. and McLay, H.A. (1996) Variation in the time of spawning of Atlantic salmon (*Salmo salar*) and its relationship to temperature in the Aberdeenshire Dee, Scotland. *Canadian Journal of Fisheries and Aquatic Sciences*, 53: 2739–2744.

Webb, J.H., Fryer, R.J., Taggart, J.B., Thompson, C.E. and Youngson, A.F. (2001) Dispersion of Atlantic salmon (*Salmo salar*) fry from competing families as revealed by DNA profiling. *Canadian Journal of Fisheries and Aquatic Sciences*, 58: 2386–2395.

Westerberg, H., Sturlaugson, J., Ikonen, E. and Karlsson, L. (1999) Data storage tag study of salmon (*Salmo salar*) migration in the Baltic: behaviour and the migration route as reconstructed from SST data. *ICES CM*, **1999/AA: 06**: 18.

Whalen, K.G. and Parrish, D.L. (1999) Effect of maturation on parr growth and smolt recruitment of Atlantic salmon. *Canadian Journal of Fisheries and Aquatic Sciences*, 56: 79–86.

White, H.C. (1942) Atlantic salmon redds and artificial spawning beds. *Journal of Fisheries Research Board of Canada*, 6: 37–44.

Wilder, D.G. (1947) A comparative study of the Atlantic salmon, *Salmo salar* Linnaeus, and the lake salmon, *Salmo salar sebago* (Girard). *Canadian Journal of Research*, D25: 175–189.

Youngson, A.F., Buck, R.J.G., Simpson, T.H. and Hay, D.W. (1983) The autumn and spring emigrations of juvenile Atlantic salmon, *Salmo salar* L., from the Girnock burn, Aberdeenshire, Scotland: environmental release of migration. *Journal of Fish Biology*, **23**: 625–639.

Youngson, A.F., Jordan, W.C. and Hay, D.W. (1994) Homing of Atlantic salmon (*Salmo salar* L.) to a tributary stream in a major river catchment. *Aquaculture*, **121**: 259–267.

Youngson, A.F., Jordan, W.C. Verspoor, E., McGinnity, P., Cross, T. and Ferguson, A. (2003) Management of salmonid fisheries in the British Isles: towards a practical approach based on population genetics. *Fisheries Research*, **62**: 193–210.

Zelinsky, Y.P. (1985) *Structure and Differentiation of Populations and Forms in Atlantic Salmon.* Nauka, Leningrad.

3 The Atlantic Salmon Genome

P. Morán, E. Verspoor and W. S. Davidson

A contemporary laboratory facility used for the study of DNA variation in salmonid fishes, and researchers carrying out molecular genetic analyses. Inset: a set of chromosomes from the cell of an Atlantic salmon as seen under a light microscope. (Photos credit: J. Hammar, E. Verspoor and R. Johnstone.)

A basic understanding of the molecular basis of inheritance is needed to appreciate properly the genetic nature of the Atlantic salmon, *Salmo salar*. This determines how genetics can affect the diversity in development, appearance and life history we see in the species. It is through such effects that genetics becomes an important variable in the management of the species and in the development of effective conservation programmes. A summary of current understanding of the molecular machinery responsible for inheritance in the Atlantic salmon is provided here.

3.1 DNA

Inheritance in Atlantic salmon, as in almost all organisms, is controlled by DNA (deoxyribonucleic acid), the molecule identified over 50 years ago by Watson and Crick (1953) to underpin inheritance; in a few viruses, it is controlled by RNA (ribonucleic acid). DNA resides in the eggs, sperm and tissue cells in two locations – the nucleus and mitochondria (Fig. 3.1). The ability of this molecule to carry the genetic information responsible for an organism's development, survival and reproduction derives from the nature of its structure (Box 3.1). The total DNA complement of a cell is known as the genome.

The nuclear DNA (nDNA) in sexually reproducing organisms such as the Atlantic salmon derives from the fusion of an egg and a sperm such that it is composed of two copies, one maternal and the other paternal. Cells, as well as organisms, which contain two copies of nDNA are termed 'diploid', a state which exists for all body cells except eggs or sperm. The latter contain a single copy and are in a 'haploid' state.

The second location for DNA in the cell is the mitochondria (Fig. 3.1), small rod-like organelles found within the cell, ~ 1 μm (micrometre, i.e. 10^{-6} m) in diameter and 7 μm in length, where important molecular processes related to energy production take place. In most species this DNA is inherited only from the maternal parent. This represents a small fraction of 1% of the total cellular DNA. In mammals, 99.99% of mitochondrial DNA

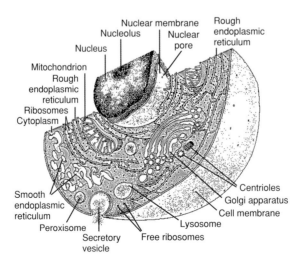

Fig. 3.1 A generalised model of the structure of an animal cell. Most DNA resides in the nucleus in the form of chromatin or chromosomes, complexes of DNA and proteins, but it is also found in mitochondria, energy-producing organelles which occur in large numbers in all cells. (Image credit: National Human Genome Research Institute.)

Box 3.1 Basic molecular structure of DNA.

Each chromosome, or piece of chromatin, contains a single long molecule of DNA, formed by two antiparallel chains of monomers called nucleotides (i.e. chains running in opposite directions) which form a double helix. The DNA is associated with histones, a type of round protein (Fig. B3.1), wound around the DNA like sewing thread around a spool, which condenses the DNA. There are five types of histone – H1, H2a, H2b, H3 and H4 – and the spool-like structure is known as a nucleosome. Each nucleotide has three parts: a sugar (2′-deoxyribose), a phosphate group and a nitrogen-containing ring structure called a base, of which there are different types – adenine (A), cytosine (C), guanine (G) and thymine (T). Nucleotides are linked to one another by phosphodiester bonds to form a polynucleotide which might be several million nucleotide bases. The bases of the two-polynucleotide chains interact with each other forming pairs; thymine always interacts with adenine and guanine with cytosine. This is known as complementary base pairing and, because the same bases always pair, if the base composition of one chain (i.e. its sequence) is known, so is that of the other. The amount of DNA is measured in terms of the number of base pairs (bp): a 1000 bp is a kilobase (kb) and 1000 kb is a megabase (Mb).

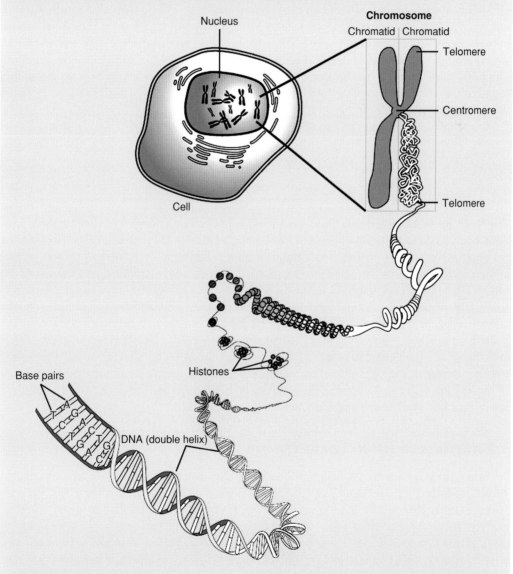

Fig. B3.1 Basic organisation and structure of nuclear DNA in the cell. (Image credit: National Human Genome Research Institute.)

(mtDNA) is inherited from the mother as sperm carry mitochondria in their tail and have only ~ 100 mitochondria compared to 100 000 in an egg; liver cells have ~ 1000 mitochondria. Furthermore, as the cells develop and divide, more and more of the male mtDNA is diluted out. Male mitochondria may also be preferentially destroyed by the egg after fertilisation, possibly due to incompatibilities with the maternal genome which controls the cellular environment of the egg. Thus the mtDNA haplotypes in most, if not all, mitochondria in an individual's cells are identical.

The total amount of DNA per haploid cell (i.e. the DNA in one gamete) is called the C value of an organism and varies among species. Among fish, it ranges from ~ 400 million base pairs (Mb) (4.3 pg – picograms, i.e. 10^{-12} grams) in puffer fish (Tetraodonitae) up to over 7000 Mb (7.0 pg) in some salmonid fish species (Lockwood and Derr 1992). The value for the Atlantic salmon is 5.7 pg, equivalent to around 6000 Mb of DNA and nearly twice the content of the human genome. Humans have ~ 3300 Mb per haploid genome in each cell, an amount which would stretch over 2 m in length if put end to end. The larger size of the salmonid genome is curious as salmonids are considered to be more ancient and less 'evolved' than puffer fish, suggesting that complexity is not a simple function of genome size. This finding is known as the C value paradox.

3.2 Chromatin and chromosomes

3.2.1 Nature and structure

The DNA in the cell nucleus in most organisms is present as a number of distinct molecules. Also, the DNA is structurally organised (Box 3.1); to function, it cannot simply be stuffed into the nucleus like a ball of string. This organisation is achieved with the help of proteins which help to form each DNA molecule into a precise, compact, linear structure, a dense string-like fibre referred to as chromatin. In the chromatin, the DNA molecule wraps around groups of small protein molecules, called histones, to form a series of structures called nucleosomes, connected by the DNA strand. Under the microscope, uncondensed chromatin has a 'beads on a string' appearance. The string of nucleosomes, already compacted by a factor of six, is then coiled into an even denser structure which compacts the DNA by a factor of 40. The overall negative charge of the DNA due to the phosphate groups is neutralised by the positive charge of the histone molecules, allowing the DNA to take up much less space and to be folded and densely packed until genes need to be activated.

3.2.2 Replication, cell division and growth

An organism is formed by the fusion of two haploid gametes to give a single diploid cell. The whole organism then develops by cell division and growth. The cell cycle is the period between two divisions. Throughout the cell cycle, chromatin fibres take on different forms inside the nucleus. Non-dividing cells are said to be in interphase. During this period, the cell is carrying out its normal metabolic functions and the chromatin is dispersed throughout the nucleus in what appears to be a tangle of fibres. When cells get ready to divide, their chromatin changes dramatically. First, chromatin strands make copies of themselves through the process of DNA replication (Box 3.2). Once this is done, a cell is ready to divide into two cells.

Box 3.2 DNA replication.

During the process of DNA replication, the DNA molecule unwinds, with each single strand becoming a template for the enzyme polymerase to synthesise a complementary strand (Fig. B3.2). Each of the two resulting molecules will have one old and one new DNA strand but, barring replication errors (i.e. mutations), will be an exact copy of the parent molecule.

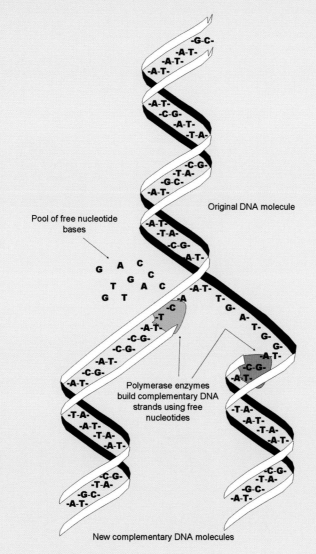

Fig. B3.2 The basic process of DNA replication.

The process of nuclear division is termed mitosis and has four steps or phases (Box 3.3). During the first prophase, the nuclear membrane disappears and the chromatin begins to be compacted. During metaphase the chromatin is compressed to an even greater degree, a 10 000-fold compaction, into specialised structures for reproduction, termed chromosomes,

Box 3.3   A comparison of the processes of mitosis and meiosis.

Mitosis, the duplication of nuclear DNA, is an essential feature of normal cell division in diploid cells (Fig. B3.3). It is divided into four phases. In the first, prophase, the chromosomes condense. Then in metaphase they are aligned on the equator of the cell. This is followed by anaphase, when the centromeres split, the two sister chromatids separate towards opposite poles. When they near the poles, telophase begins and the nuclear membranes form around the two nuclei and the chromosomes begin to decondense. At the same time cytokinesis i.e. division of the cytoplasm, is initiated at the equator of the cell.

Meiosis, the process which leads to the production of haploid gametes from diploid cells, is a two-stage division process that yields four daughter cells (Fig. B3.3), each of which is haploid. In the first division, meiosis I, homologous chromosomes pair to form bivalents. As the cells progress through prophase I, the chromosome pairs shorten and individual chromatids become visible, along with the chiasmata. The latter are chromosomal locations where crossing-over between non-sister chromatids (i.e. recombination) occurs. At metaphase I the paired chromosomes separate. One member of each pair moves to the opposite poles of the cell. Anaphase I and telophase I follow and result in the formation of two haploid cells. The second division is also composed of four steps. As a result, chromatids separated at metaphase II result in the formation of four haploid cells.

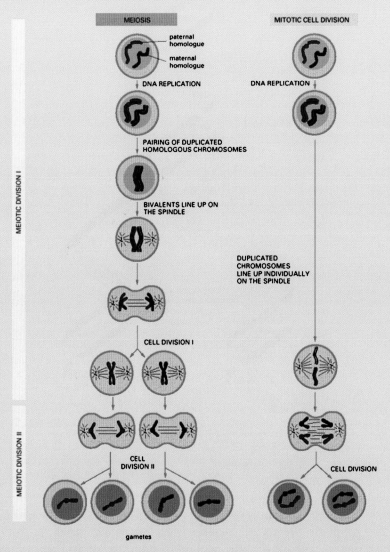

Fig. B3.3   Schematic for one pair of homologous chromosomes comparing the basic genetic processes of mitosis and meiosis. (Image credit: after Alberts *et al.* 2002.)

which can easily be visualised by microscopy. Each chromosome consists of two identical daughter chromatids attached at a single point called the centromere. The positions of the centromeres are used to assign chromosomes into morphological groups (i.e. metacentric, submetacentric, acrocentric and telocentric) depending on the relative size of the chromosome arms. During metaphase chromosomes are aligned on the equator of the cell. This is achieved by the action of tubulin-containing spindle fibres that run from both poles of the cell and attach to the centromeres of the chromosomes. During the next step, anaphase, the daughter chromatids are pulled to opposite poles of the cell by the spindle. Once chromatids are in each pole, the chromatin partially decondenses, nuclear membranes form around the DNA, and a cleavage furrow forms across the equator of the cell, a step referred to as telophase. This is then followed by division of the cell to produce two identical cells.

3.2.3 Number and ploidy level

The DNA in the mitochondria of the Atlantic salmon, as is the case for all vertebrates, is a single molecule formed into a circular loop. It has been fully analysed in many species. In Atlantic salmon it is composed of 16 665 base pairs (bp) (Hurst *et al.* 1999) very similar in size and organisation to that of other fish species.

The number of nuclear chromosomes is the same in all cells of a given individual and is more or less constant in all the individuals of the same species, with most species having a distinct number, size and morphology of its chromosomes (i.e. a distinct karyotype). As diploid individuals are the product of the fusion of two haploid gametes, each with n chromosomes, diploid individuals are said to be $2n$. In principle at least, these $2n$ chromosomes can be arranged in pairs. The two members of a pair are essentially identical with the exception of chromosomes involved in sex determination, though not all species have differentiated sex chromosomes. In mammals, females have two X chromosomes and are referred to as the homogametic sex. Male mammals have one X chromosome and one Y chromosome and are the heterogametic sex. In birds it is the female which is the heterogametic sex (with chromosomes called ZW) and males are homogametic (ZZ) (Box 3.4). The total number of chromosomes ($2n$) varies among species; for example, humans have 46 chromosomes (23 pairs) while in a subspecies of the ant, *Myrmecia pilosula*, females have a single pair of chromosomes and males have only one! The size of chromosomes is also highly variable. In general, bigger chromosomes contain more DNA and smaller chromosomes contain less. For example, in humans the biggest chromosome contains 200 Mb and the smallest just 35 Mb.

nDNA in the Atlantic salmon is typically organised into 58 chromosomes (29 pairs) (Hartley 1987) consisting of 16 metacentric, 24 large acrocentric and 18 small acrocentric pairs (Fig. 3.2). The Atlantic salmon is genetically most closely related to the brown trout, *S. trutta*, the only other species within the genus *Salmo*. A typical brown trout karyotype has 80 chromosomes consisting of 10 metacentric and 30 acrocentric pairs.

No morphologically differentiated sex chromosomes have been identified in Atlantic salmon. However, sex determination is the same as in humans, with females the homogametic sex and males the heterogametic sex. This is supported by breeding studies of normal and sex-reversed individuals, and is true for all salmonids so far studied. Heteromorphic sex chromosomes have been identified in some salmonids (Box 3.4; see also Devlin and Nagahama 2002). Further support for this model for sex determination comes from the discovery of male specific sex-linked microsatellite markers (Woram *et al.* 2003; Gilbey *et al.* 2004).

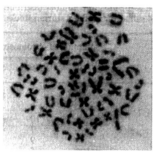
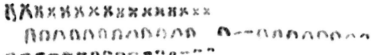

Fig. 3.2 Metaphase karyotype and reconstructed chromosome pairing for a cell from gills, with $2n = 58$. (Image credit: García-Vazquez *et al.* 1988.)

Box 3.4 X/Y chromosome sex determination.

The pattern of sex determination in animals such as humans was found to be associated with the inheritance of unusual chromosomes. These were called sex chromosomes and are the exception to the rule that chromosomes occur in pairs in diploid cells. However, females of most species, including Atlantic salmon, carry two sex chromosomes that are the same. These are called X chromosomes. In these species, the male has one X chromosome as well as a different chromosome known as the Y chromosome. In humans the Y chromosome is visually distinct from the X while in Atlantic salmon the two cannot be differentiated on appearance.

The correlation of sex with either the XX or XY diploid state was strong evidence that the genes that controlled sex determination were carried on these chromosomes. During meiosis (Fig. B3.4), the two X chromosomes of the female pair and segregate into gametes just like any other chromosome pair (non-sex chromosomes are called autosomes). In males, the X chromosome and the Y chromosome are able to pair because the Y chromosome carries a small region that is similar to the X chromosome. In this system, males produce two types of gamete, those with an X chromosome and those with a Y, in equal frequency. If the egg is fertilised by an X-carrying sperm, the resulting zygote has two X chromosomes and will produce a female; if the egg is fertilised by a Y-carrying sperm, the zygote will have one X and one Y chromosome and will produce a male.

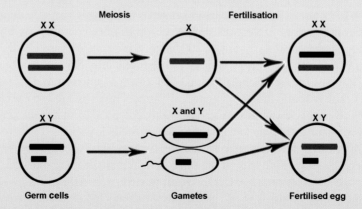

Fig. B3.4 Schematic of sex chromosome transmission starting with germ cells and ending with the production of newly fertilised male and female eggs.

Box 3.5 Triploidy.

The genetic state where the cells of an animal have three copies of the nuclear genes (i.e. three sets of chromosomes) is rare in nature. However, this genetic state is easily induced (Tave 1993) and has been studied in fish in culture. One of the main interests in this state of triploidy is because it interferes with maturation, and almost all triploids appear to be infertile. Thus they have the potential to be able to redirect energy, which would otherwise have been used to produce gonads, into growth. Triploids are produced by shocking newly fertilised eggs using temperature, pressure or chemicals (see Fig. B3.5). The shock causes the second polar body containing a duplicate set of maternal chromosomes, which is normally expelled at the time of fertilisation, to be retained in the egg. The three haploid nuclei, two maternal and one paternal, then fuse to form a single triploid nucleus. Variations on these manipulations can also be used to artificially produce tetraploids as well as gynogenetic (diploid but with two copies of maternal genes) or androgenic (diploid but with two copies of paternal genes) fish.

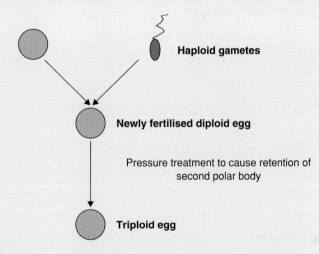

Fig. B3.5 Schematic of one common method used to produce triploid Atlantic salmon.

The nuclear genomes of many vertebrate species appear to have evolved by a series of genome duplications involving the doubling of the number of chromosomes possessed by ancestral species. This doubling of the number of chromosomes is known as tetraploidisation, and can also be artificially induced. Such a chromosome manipulation is a common practice used in the aquaculture industry to produce all female lines and/or sterile fish (Box 3.5).

Tetraploidisation has occurred naturally in several fish genera (Ohno 1970). Why this has occurred at various points during the course of evolution is not certain but it is generally associated with increased organismal complexity (Volff 2005). Evidence for tetraploidisation in the different species of salmonids comes from a higher DNA content per cell (i.e. twice the DNA content of the most closely related non-salmonid species such as the smelt), the existence of four rather than two associated chromosomes (i.e. multi-valents) in male meiosis, and the presence of duplicated gene loci in many species in the family. This has led to a phenomenon known as pseudo-linkage where there is a non-random association of duplicated groups of genes found on different, physically unlinked chromosomes (Wright *et al.* 1983). However, many duplicated gene loci have also diverged and the genome generally is in the process of evolving back into a diploid state (Allendorf and Thorgaard 1984). This re-diploidisation has involved chromosomal rearrangements (e.g. the apparent loss of a chromosome due to two chromosomes fusing) and the differential mutation of base sequences in homologous genes. Genome duplication is believed to have been particularly important in the evolution of three orders of teleost fishes: the Cypriniformes, especially the families Catastomidae and Cyprinidae (e.g. Ferris 1984; Collares-Pereira and Moreira da Costa 1999), Salmoniformes, particularly the Salmonidae (e.g. Allendorf and Thorgaard 1984) and Siluriformes, most notably the Callichthyidae (e.g. Oliveira *et al.* 1992).

3.3 Genes and genome organisation

3.3.1 Molecular nature and structure

A *gene* represents a unit of specific molecular information and corresponds to a discrete segment of DNA. The information is given by the sequence of the different bases (A, C, G, T) found in that segment. In the case of genes coding for proteins, this sequence is transcribed (i.e. copied) by an enzyme called RNA polymerase which uses one of the DNA strands as a template to generate a complementary copy made of RNA (ribonucleic acid) called messenger RNA, or mRNA (Box 3.6). The mRNA is in turn translated by ribosomes, complexes consisting of ribosomal RNA and proteins, into a protein with a specific sequence of amino acids. The translation of mRNA uses the genetic code whereby a sequence of three bases (a codon) specifies a particular amino acid. This one-way flow of genetic information is called the 'central dogma of molecular biology' (i.e. DNA $\rightarrow$ RNA $\rightarrow$ protein) which appears to be mostly but not always true.

Genes vary greatly in size from less than 100 base pairs to several millions of base pairs, but not all the information in the gene is used to produce the final product. Nuclear genes are often split into a series of coding segments called exons which are separated by non-coding sequences called introns. Note that this is not the case for mitochondrial genes, where introns are not present.

Gene expression is the term used to describe the information processing which occurs when DNA sequences are translated into protein. Gene expression is highly regulated. Not all of the genes present in a cell are active at any given time and the various types of cells express different genes. A cell type is defined by the proteins that it produces (e.g. muscle cells have a different suite of proteins from a red blood cell). The expression of a gene is regulated by a segment of DNA upstream of the coding sequence, called a promoter. The regulation of gene expression is achieved by the interaction of gene promoters and DNA-binding proteins called transcription factors. There are several types of these factors. Basal transcription factors are always needed for a gene to be transcribed. In contrast, transcription factors known as enhancers stimulate gene expression and those referred to as silencers or repressors restrict gene expression.

3.3.2 Number and molecular distribution

In mtDNA, only a small portion of the molecule is composed of non-coding DNA. Such DNA is largely associated with the D-loop region, the site where DNA replication is initiated. The remainder of the mtDNA molecule codes for the 16S and 12S ribosomal RNAs and

Box 3.6 Cellular processes associated with expression of genes and the production of proteins.

When genes are expressed, the genetic information coded by the DNA base sequence is first transcribed (copied) to a molecule of messenger RNA (mRNA) in a process similar to DNA replication (Fig. B3.6). The mRNA molecules go from the cell nucleus to the cytoplasm, where the DNA is read three bases at a time (a codon) according to what is known as the genetic code, with particular base triplets specifying particular amino acids in a process known as translation. This is carried out by ribosomes (structures made of proteins and another class of RNA termed rRNA). Transfer RNAs (tRNAs) bring amino acids to the ribosomes and they are enzymatically combined to form a growing protein whose length is defined by the number of triplet codons. Subsequently, proteins will be variously folded and combined with other proteins or molecules in a way which allows them to carry out their specific cellular functions.

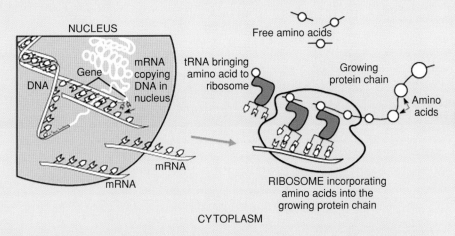

Fig. B3.6 The basic process of translating genetic information into proteins which then control cellular and developmental processes. (Image credit: US Department of Energy Human Genome Program, http://www.ornl.gov/hgmis)

22 transfer RNAs, involved in protein synthesis (Fig. 3.3), and cytochrome *b*, two subunits of the ATPase enzyme, three subunits of the cytochrome *c* oxidase enzyme, and seven subunits of the NADH dehydrogenase enzyme, all involved in energy production in the cell. The same coding and non-coding regions are found in Atlantic salmon and other vertebrates, and the structural arrangement of the molecule has been shown by sequencing to be conserved in more than 100 fish mitochondrial genomes (Miya *et al.* 2003).

The complete nuclear DNA sequence for the human genome has now been determined along with that of several other organisms such as the fruit fly, *Drosophila melanogaster*, the nematode worm, *Caenorhabditis elegans*, the mouse, *Mus musculus*, and the mustard plant, *Arabidopsis thaliana*. However, less is known about the number, nature and distribution of nuclear genes on chromosomes. In human chromosomes, genes tend to be spaced about 1000 base pairs apart, though their distribution is not random and clusters containing genes that are expressed in a tightly coordinated manner (gene families) appear to be common. However, the number of genes is not yet known for most genomes. For example, for humans, which are one of the best-studied animals, it was first believed that there were more than 100 000 genes. However, new work suggests that the number is more likely to be in the range

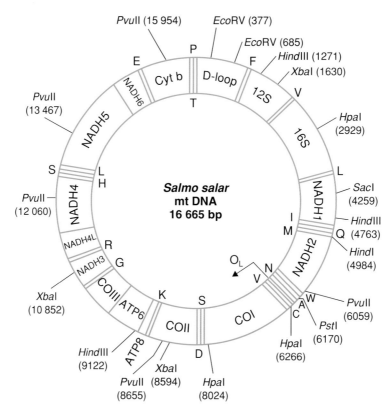

Fig. 3.3 A gene map of the Atlantic salmon mitochondrial DNA showing sites where restriction enzymes, e.g. PvuI, EcoRV, etc., were able to cut the molecule. Letters refer to tRNA genes. (Image credit: Hurst *et al.* 1999.)

30 000–50 000. The number of genes in the Atlantic salmon is not yet known but a recent estimate based on sequencing mRNAs is ~ 40 000 (Rise *et al.* 2004).

3.3.3 Extragenic DNA

The number of genes appears to be unrelated to the amount of DNA possessed by an organism, or to its phenotypic complexity. Genes make up only a small fraction of most eukaryotic genomes, the genomes found in higher organisms with complex cellular structures. In these genomes, estimates put the proportion of DNA that encodes genes to be between 10% and 50%. The function of most of the large amount of extragenic DNA is unclear but that associated with chromosome structure important in replication processes associated with cell division, such as the centromeres and telomeres, has a well-defined function. The rest is often referred to as 'junk DNA' because no function could be assigned to it. However, there is increasing evidence that this DNA may be vitally involved in the regulation of gene expression during development in some as yet unknown way. One possibility is that it is involved in determining the structure of the chromatin and controlling the access of enzymes such as RNA polymerase in relation to transcription.

The non-coding DNA is composed of three main types of DNA – single copy, moderately repeated and highly repeated – which, depending on the species, may constitute from 3% to 80% of the total DNA in the nucleus. In humans, 70–80% of the genome is made up of unique, single-copy sequences of DNA of which only a small fraction codes for proteins. The remaining 20–30% comprises moderately or highly repeated sequences that can be tandemly repeated or dispersed throughout the genome. The tandem arrays are called satellite DNA. Satellite DNA is classified depending on the length of the repetitive sequence. Highly repetitive satellite DNA consists of clusters of repetitive sequences of between 100 and 5000 base pairs. Minisatellite DNA contains tandem repeats of 10–100 bp, while the repeats in microsatellite DNA are 2–8 bp. Some of the repetitive elements in Atlantic salmon have been characterised (Goodier and Davidson 1994, 1998).

3.4 Genes and development

3.4.1 Genotypes, alleles and loci

An organism's phenotype (i.e. its observable character with regard to colour, shape, size, behaviour, etc.) is the product of the interaction of its overall genotype with the internal and external environment during the course of its development. Its genotype can be considered its genetic state with regard to a single gene, a number of genes, or all of them.

The physical position of a specific DNA fragment is called a locus (plural: loci). A locus may encompass a specific gene or simply refer to a particular stretch of DNA. If loci occur on different chromosomes, they are said to be independent, and combinations of these loci in the individuals will be a matter of chance. The same is true if loci are located on the same chromosome and the incidence of recombination between them is high, as often occurs when the physical separation between them is large. In contrast, if genes are closely situated on the same chromosome and recombination between them is infrequent or absent, they are referred to as being linked. The allele combinations for very tightly linked genes in the gametes produced by an individual, and thus in its relatives and offspring, will nearly always be the same. In such a case, they are effectively transmitted from one generation to the next as a single unit.

For a given nuclear gene or sequence (i.e. locus), an individual will have one copy received from its mother and one received from its father. These different copies are called alleles. As a result of mutation (a change in the DNA) and recombination (the crossing-over and recombining of paired DNA strands) events in either the maternal or paternal line, these alleles may be genetically different. The genotype at a particular locus can be either homozygous or heterozygous. Homozygotes are individuals where the maternal and paternal DNA is identical; heterozygotes are those individuals where the maternal and paternal alleles are different. Individuals differ genetically by having different homozygous or heterozygous states across their entire genome. The phenotypic effect, if any, that results from an individual having different maternal and paternal alleles depends on the functional relationship between the two alleles. An allele is said to be dominant when the phenotype is directly specified by one allele, independent of the other allele. It is said to be recessive if its effect is only manifest if an individual has two copies of this allele. Sometimes alleles are codominant and the phenotype is an average of the two alleles.

3.4.2 Genes and traits

Mendelian genes (named after the Czech monk, Gregor Mendel, who first uncovered the laws of heredity in 1865) control what are known as qualitative traits. For these traits, the phenotype has one or more discrete states which are determined by the relationship between the alleles of a gene (e.g. eye colour). Other aspects of the phenotype, however, vary in a continuous fashion (e.g. body weight, skin colour) and are referred to as quantitative traits. The nature of such traits is determined by the combined action of many genes or loci. Each gene is inherited as set out by Mendel but the contribution of the genes to a particular character will vary, with some being more influential than others. In addition, the environment experienced by the individuals will play a role in determining the phenotype. Quantitative traits include most of those important to survival and reproductive success (i.e. to Darwinian fitness). Physical locations in the genome containing genes which affect quantitative traits are known as quantitative trait loci.

3.4.3 Gene expression

Development is the process by which cells specialise, divide and organise into a complex organism. In the early stages of development an individual is called an embryo and the process of its development is called embryogenesis. Development can be viewed as being composed of four major components. The first is growth (i.e. the increase in size) and the second is cell division, which may occur at the same time or independently. The third component is cellular differentiation (i.e. the generation of specialised cells with a specific structure and function). This happens after the genetic commitment of a cell to a particular fate by its cellular milieu. The final component is morphogenesis, which is the differentiation of the collection of cells into a body and its organs.

The development of an organism from a fertilised egg involves a complex chronological sequence of gene activation and gene regulation (i.e. some genes are 'switched on' and some are 'switched off'). All cells within an individual have the same genetic constitution but each cell differs with regard to which genes are active and which are turned off, and the timing and extent of their expression. Molecules such as hormones and cytokines regulate differences in gene expression. These are in turn the products of genes whose production is also regulated. Environmental factors such as temperature or disease vectors are also often the stimuli for gene activation.

Some genes are permanently activated because they code for vital cell components (e.g. enzymes involved in energy production). However, in multicellular organisms, where cells become specialised in different tissues and organs, some genes are permanently inactivated because their function is irrelevant to that particular cell or tissue. An example of this in humans is the insulin gene which is switched on in the pancreas and switched off in all other cells of the body. Some other genes will be functional only at particular times such as in early development when they may determine the physical characteristics of cell lineages, for example whether they give rise to skin or muscle cells.

During development the processes controlled by the genes must achieve the coordination of the creation of different types of tissues and organs. To achieve this there must be communication between the different regions of the embryo, different tissue types and different organs.

Cells receive information in a variety of different ways. There may be unequal inheritance of molecules in the cell during cell division or exposure to different extracellular signals due to spatial positioning in the developing embryo. Such signals may change cell functions or stimulate cells to migrate to different positions in the embryo.

Gene function is regulated at many levels and operates differently in different cells. The expression of a gene producing a functional protein can potentially be controlled at any one of a number of biochemical levels. The most important, but by no means the only, level is at the initiation of transcription of the gene. The control of this process involves a complicated set of transcription factors interacting with enhancers and suppressers. This control may be achieved by determining the physical access of transcription factors to the promoter of a specific gene by changes in the chromatin structure. The full nature of gene regulation and the coordination of gene expression throughout development is only partially understood in multicellular organisms, even for the most extensively studied of species such as the fruit fly *Drosophila melanogaster* and the nematode *Caenorhabditis elegans*.

3.5 Variation among individuals

3.5.1 Origin

The quantitative DNA content of different members of the same species is usually the same, with the exception of differences due to sex chromosomes. However, major variation within a species with regard to the amount of DNA has been observed and may arise due to individuals varying in their numbers of chromosomes or chromosome structure. For example, two subspecies of the cutthroat trout, *Oncorhynchus clarki*, differ in the sizes of their genomes by 4% and two different life-history types of the whitefish, *Coregonus autumnalis*, differ by 7.5% (Lockwood and Derr 1992). The extent to which variation in the total DNA content of individuals of a species affects their individual characters is unclear. In humans the possession of an extra chromosome (known as trisomy) can have major phenotypic effects (e.g. Down's syndrome). The greatest amount of genetic variation detected among individuals comes from alterations in the sequence of the DNA present and its organisation. Included are base changes and minor variation among individuals in the amount of DNA present caused by a deletion or an insertion of a small number of bases. This variation arises over the course of evolution of a species through two main processes known as mutation and recombination.

Mutation
The DNA content of two individuals of the same species may be the same but it does not mean that their DNA sequences are identical. This is because of the process of mutation. Mutations are alterations to the usual (i.e. most common or wild-type) DNA sequence as a result of the action of chemical and physical agents or errors in DNA replication. They can occur in both nuclear and mitochondrial components of the genome. Mutations are perpetuated by cell division but only those that occur in the gametes will be passed on from one generation to the next. An organism whose wild-type genome has changed as a result of a mutation is called a mutant. Mutations occur in two forms. Point mutations involve a change in the base present

at a locus; gross mutations involve alterations of longer stretches of DNA sequences, usually insertions and deletions. Mutations that occur within the coding region of a gene may affect the amino acid sequence of the protein that it encodes. The effects of mutations in non-coding or intergenic regions are unclear; possibly the functioning of genes is affected. From a population point of view, only mutations that are inherited are important.

In humans, mutations are the underlying cause of genetic diseases and cancer, but this does not mean that mutation is synonymous with illness. Mutation is the basis of variation and variation is required for evolution. Differences between individuals of the same species arise from mutations but these differences may be good, bad or neutral. Whether an allele is called the wild type or a mutant in common usage depends on which population is being considered. For example, blue eyes and brown eyes in humans are due to a number of variant genes. In Mediterranean populations the common gene is the one responsible for brown eyes whereas in Baltic populations the normal gene is that responsible for blue eyes.

Recombination
Genetic variation can also be produced by recombination. The information carried by an individual's gametes is a combination of the information that an individual has inherited from its parents when their gametes fused to form a fertilised egg. The process of gametogenesis (Box 3.7) whereby gametes are produced in an individual, involves meiosis, a cell division process similar to mitosis. The process includes the rearrangement of DNA information by two processes: independent assortment of chromosomes and crossing over (i.e. DNA interchange). Crossing over within genes can produce new genetic variants at a particular locus. When it occurs between genes or when chromosomes are reassorted, it can change the combinations of genes that are passed on to the next generation. This means that each of the gametes produced by that individual will have a different mix of maternal and paternal DNA.

Barring mutation, the mtDNA of an individual is identical to that of its mother. However, this is often not the case as the rate of mutation in mtDNA is high compared with that for nDNA and gives rise to a much higher level of variation between individuals than for the same amount of nDNA. Furthermore, evidence suggests that recombination between different mtDNA molecules is either absent or extremely rare such that the molecule behaves as a single unit with regard to its inheritance. These characteristics, and the ease with which mtDNA could be analysed separately from nDNA, have made this genome extremely popular for population and evolutionary studies.

3.5.2 Scope

The scope for genetic variation among individuals as a result of mutation, recombination and chromosomal rearrangements is huge, even taking into account that a very large number of potential genome states are likely to be non-viable. This is easy to see by considering the number of genetic states that can be generated by variation at just five nuclear DNA base positions. At each position, four different allelic states will be possible – C, G, A or T. At the same time, there will be ten different genotypes that a given individual could have at each of these locations. Assuming that these loci are independent, then there are 5^{10} (~ 10 000 000) possible genotypes (or genotype combinations). Thus the scope for individual variation is effectively limitless.

Box 3.7 Gametogenesis.

Prior to reproduction and the development of a new individual, the gametes have to be formed. This process starts in the parent at the embryo stage when precursors to the gamete cells develop and migrate to the gonads. At this point the cells divide and form oogonia and spermatogonia (Fig. B3.7). All spermatogonia that will ever exist are formed in a male embryo. The first step of spermatogenesis, the process of sperm production, is a mitosis to duplicate the spermatogonia. One of the daughter cells replaces the original spermatogonium and the other becomes the primary spermatocyte. Although the names are different, these two cells are identical and the only difference is their fate. The diploid primary spermatocyte undergoes meiosis I and forms two haploid secondary spermatocytes. These in turn undergo meiosis II and form four spermatids. The oocyte is also produced via meiosis, but in a different way. Cells called oogonia, produced before in the embryo, develop into primary oocytes. These diploid cells will remain in meiosis I until they mature in the female ovary, beginning at maturation. Oocytes that mature enter meiosis II, but their development is suspended until fertilised by a sperm. The unequal meiotic division of cellular material causes three polar bodies and one egg to be produced from the original oocyte, of which only the egg can go on to be fertilised while the polar bodies are normally degraded.

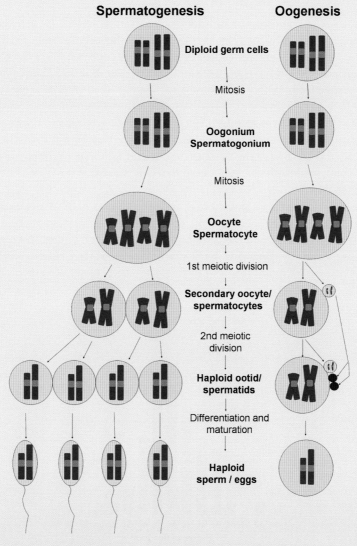

Fig. B3.7 Genetic processes leading to the production of eggs and sperm.

3.5.3 Detection

Characterisation of the actual variation which exists among individuals or between species presents a major challenge. With the exception of such traits as chromosome number and gross chromosome structure, the character of a species' genome, by virtue of its molecular nature and extreme complexity, is not amenable to direct analysis. Thus, to understand the nature of the Atlantic salmon genome and how it varies among individuals and differs from other species, researchers have had to employ various indirect methods. A basic understanding of these methods is essential to appreciate what they are able to tell us and, just as importantly, what they cannot tell us. This is essential to evaluate many of the inferences drawn from their use.

Phenotypic variation

Studies of genetic variation in Atlantic salmon and most other species were first carried out by analysing the inheritance by offspring of observed or phenotypic traits such as body size, colour, morphology, or the number of meristic characters such as fin rays, from parents. In general, two types of phenotypic traits are studied (Kirpichnikov 1981; Tave 1993). The first comprises qualitative traits where individuals can be classified into discrete types. A classic example of this in fish is the discontinuous colour variation displayed by the common goldfish *Carassius auratus* (e.g. see Kirpichnikov 1981). The second group contains quantitative traits which are measured rather than described, most often in a continuous way (e.g. body size – Figs 3.4 and 3.5) and are by far the most common and important type of trait biologically. The discipline concerned with the genetic analysis of the latter types of traits is known as quantitative genetics.

Quantitative phenotypes vary continuously and most are normally distributed. An example for salmon is shown in Fig. 3.4. This means that quantitative traits must be assessed by the analysis of their average value and the distribution of individual values. Where a normal distribution exists, or can be generated by a mathematical transformation of the data, the analysis can make use of standard statistical parameters (e.g. mean, variance, etc.).

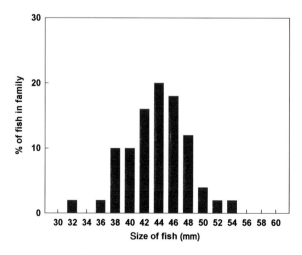

Fig. 3.4 Size differences are an example of quantitative variation in Atlantic salmon.

Fig. 3.5 Atlantic salmon siblings in a tank showing size variation. (Photo credit: E. Verspoor.)

The continuous variation of quantitative traits is a reflection of two things that complicate analysis of their genetic basis. The first is that most are likely to be controlled by the combined effects of many genes. For example, the more or less normal distribution for traits such as body size at age, requires the additive effects of variation at at least 6–12 loci. The second factor contributing to the continuous nature of the variation is that most quantitative traits are also influenced by the nature of the environment an individual or population experiences. When these two factors act in concert, and the phenotypic variation observed becomes continuous.

It is almost impossible to determine the number of genes and their mode of expression (i.e. dominant, recessive or codominant) which form the genetic basis of a quantitative trait. Therefore, the genetic basis of the traits is investigated by using controlled breeding experiments to partition the observed variation within and among families, or populations, into genetic and environmental components. The genetic component is further subdivided into its component parts which are the dominance, epistatic and additive variances. Dominance genetic variance is the component of the observed genetic variation that is due to the interaction between alleles as determined by their mode of expression. Epistatic genetic variance is the variation caused by epistasis (i.e. the interaction between different loci). While both of these components are genetic, their effect on the phenotype is not passed on from parents to offspring as it depends on which combination of genes is inherited (i.e. on the genotype) and this is often determined by chance, given that most genes are not tightly linked and most allelic variation assorts independently in offspring. Parents transmit their alleles to their offspring, not their genotype. In contrast, the third component, additive genetic variation, is inherited. This is because it is determined by the alleles themselves rather than the genotype as a whole.

Quantitative genetic analyses focus on measuring the amount of additive genetic variation present for a given trait. The proportion of phenotypic variation which is attributable to additive genetic variation is called the heritability for that trait, usually designated h^2, which ranges from 0 (no genetic basis at all to the variation) to 1 (all variation has a genetic basis). Heritability can be calculated in a variety of ways but it is normally measured by the relative amount of variation observed within and between families or populations in a given environment (see

Falconer and Mackay 1996, and references therein for a detailed description of how this is done). The latter is an important qualification. As the effect of the environment can vary tremendously (e.g. whether the study is carried out under warm or cold conditions) estimates of heritability are not easily generalised; for example, 80% of the variation observed in one environment might be genetic in origin, while it might only be 10% in another. The observed estimates show the proportion of variation attributable to additive genetic variation to range from 0.01 for traits such as meat colour score to 0.89 for weight at 12 weeks of age (Tave 1993).

Many studies of heritability across a range of environments are required before firm conclusions regarding the importance of genetics to observed variation in a trait can be drawn from quantitative studies. Furthermore, most quantitative genetic studies to date have been carried out in controlled conditions in culture and are therefore of relatively limited value for understanding the genetics of quantitative traits in wild populations. This is because quantitative studies require being able to track families or populations and are most easily carried out for animals born at the same time and reared under the same conditions. However, with the advent of molecular markers which allow families and stocks to be discriminated without the need for physical tagging, quantitative genetic analysis of animals in the wild is now possible and becoming increasingly common.

Chromosomes

The nDNA of Atlantic salmon can be visualised directly by microscopy at the gross level of whole chromosomes or, using electron microscopy, at the level of chromatin strands. With a normal light microscope, chromosomes can be clearly visualised when the cell is in the process of dividing and, particularly, during the metaphase stage of the cycle. The chemical colchicine, or a similar chemical that arrests the migration of the chromatids, can be used to interrupt the replicating cell and then visualise the chromosomes by staining so that they can more easily be seen and photographed. Pictures of the chromosomes are then cut out of photographs, sorted into homologous pairs, and set out in order of size and type (e.g. telocentric, acrocentric, etc.). The end result is known as an ideogram (Fig. 3.2) and is the conventional way to display karyotypes.

The process of generating an ideogram is tedious, particularly for a species such as the Atlantic salmon which has a large number of chromosomes. A procedure referred to as chromosome banding can aid chromosome identification and gives additional information about the underlying organisation of the chromosome. Chromosome banding involves staining of chromosome preparations with chemicals that differentially bind to regions of the DNA and their associated proteins, depending on their particular characteristics. G-banding and R-banding give a series of light and dark bands along the length of the chromosome while C-banding produces dark bands in the region where the chromatin is more compacted (called heterochromatin).

Atlantic salmon display polymorphism in chromosomal number even between individuals of the same population. The chromosomal number may vary from 56 up to 59. In European populations the standard number is usually 58 whereas in North American populations the standard number is 56. Variation in chromosome morphology in Atlantic salmon has also been reported (Hartley and Horne 1984; Pendás *et al.* 1993). This type of polymorphism has been found for a number of salmonids (e.g. Loudenslager and Thorgaard 1979; Thorgaard 1983).

The organisational structure of chromosomes themselves is as yet largely uncharacterised. Fluorescent in situ hybridisation (FISH) has been used to physically localise some repeated non-coding DNA sequences like telomeric sequences, microsatellites and minisatellites on Atlantic

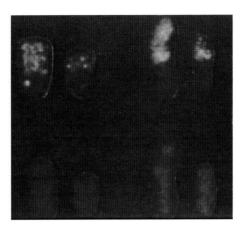

Fig. 3.6 Chromosomal variation in the Atlantic salmon detected by C-banding involving variation in the length of the heterochromatic arm of chromosome pair number 8 (Pendás *et al.* 1993), the chromosome which carries the nucleolar organiser region. Three different arm sizes, small (S), medium (M) and large (L), are distinguished by FISH hybridisation with a major ribosomal probe giving rise to six different genotypes. (Photo credit: E. García-Vázquez.)

salmon chromosomes (Fig. 3.6). Some coding sequences like those for ribosomal DNA genes, histone genes and transfer RNA genes have also been physically assigned to chromosomes pairs (Perez 1999). As a result researchers can begin to discriminate among the large number of chromosomes, which are otherwise largely indistinguishable when viewed under a microscope. This work is helping to understand the extent to which the species has so far reverted to a diploid chromosome state, by comparing the extent to which sequences for known genes are localised to one or more chromosome pairs. At the same time the results are beginning to allow the development of a physical map of the positions of particular genes on the chromosomes and an understanding of their potential organisation into linkage groups on each chromosome pair.

Proteins

Protein electrophoresis, analysis of the differential rate of migration of proteins under the influence of an electrophoretic field, was the first method available to study genetic variation at the molecular level in many species including Atlantic salmon (Verspoor *et al.* 2005). Proteins are made of amino acids joined by covalent peptide bonds to form polypeptides. Depending on the amino acid side-chains and the pH of the environment, proteins carry a positive or negative charge and therefore they migrate towards the positive or negative pole in an electrical field. Experimentally this is carried out in a semi-solid support (e.g. starch gel, acrylamide gel). Proteins can be visualised in the gel after electrophoresis using a general protein stain or by employing specific activity stains for enzymes. Information on two general forms of proteins can be gathered simultaneously using electrophoretic methods. One is isozymes, which are functionally similar forms of enzymes and are the products of gene duplications. The other is allozymes (i.e. allelic enzymes) which correspond to products from the same gene or locus. Mutations in the DNA sequence that code for a polypeptide may change the amino acid sequence and subsequently the net charge of the protein. Allozymes are codominant and, for a given protein locus that is inherited in a Mendelian fashion, an individual can be classified as homozygous or heterozygous. The genome of an individual can be defined on the basis of the variation detected at several protein loci.

Box 3.8 Sequencing.

The biochemical methods for DNA sequencing, developed in the 1970s by Sanger, are now the standard procedure used by researchers and have been highly automated. Single-stranded DNA to be sequenced is first 'primed' for replication with a short complementary strand at one end and then divided into four aliquots. To each of these is added a replication mixture, each of which has a different nucleotide, some of which are in a form which halts replication (depicted in Fig. B3.8a by a diamond shape). Each strand is replicated by the enzyme polymerase and terminated at random when a sequence-halting nucleotide is incorporated, e.g. the 'C' reaction produces strands terminated at positions corresponding to the Gs in the template strand being sequenced. This produces strands of different lengths. The reaction mixture is then subject to size characterisation by electrophoresis – classically one lane per reaction mixture as shown when the termination products cannot be distinguished – from which the sequence of the original single strand can be inferred. Recent developments now allow all four reaction mixtures to be run in the same lane by labelling each of the terminating nucleotides with a differently coloured fluorescent dye and the coloured bands screened using an automated scanner. A chromatographic trace from an automated DNA sequencer is shown in Fig. B3.8b. Using these methods DNA fragments of up to 1000 bp or more can be sequenced.

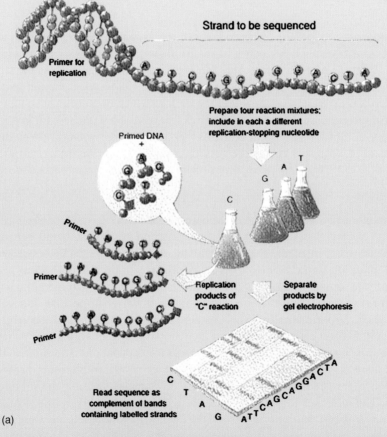

Fig. B3.8 DNA sequencing: (a) basic process and (b) a chromatographic trace from a modern automated sequencing machine based on differential fluorescent labelling of the terminating nucleotides. (Image credit: US Department of Energy Human Genome Program, http://www.ornl.gov/hgmis)

Box 3.8 (cont'd)

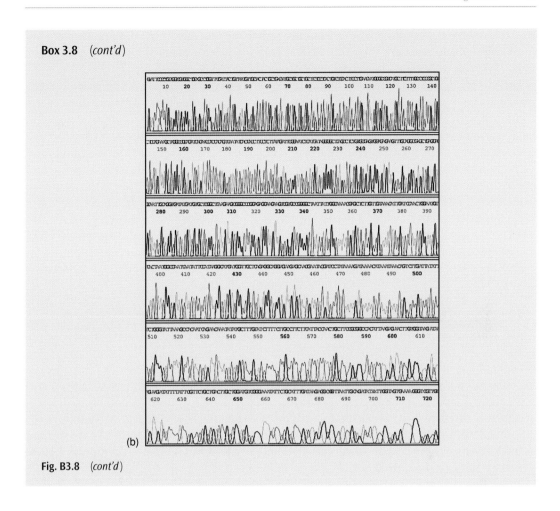

(b)

Fig. B3.8 (cont'd)

Before the advent of methods for visualising mutations directly in the DNA, protein electrophoresis was widely used to describe Atlantic salmon populations. Although this species is not very polymorphic compared to other salmonids, a considerable amount of information on this form of variation has accumulated (see Chapter 5).

DNA base sequences
Most genetic analysis is now centred on the direct characterisation of DNA sequence variation. Technology has advanced sufficiently over the past decade, largely as a result of research into the human genome, to make such analyses relatively easy and cost effective. DNA extraction is now straightforward and efficient methods for the determination of the base sequence of DNA have been developed (Box 3.8). Furthermore, to work on and investigate the genome of a given species, it is not necessary to deal with its entire genome of billions of base pairs all at the same time. The technologies developed allow the extent of similarities and differences between individuals to be investigated through the analysis of selected parts of the genome. Once a sequence is known, the corresponding sequence in individuals or related

species can be amplified for analysis. This has been made possible by the development of PCR (polymerase chain reaction; Box 3.9). PCR allows specific segments of DNA, ranging in size from 100 bp up to several kb, to be copied over a billion-fold.

Differences in a given amplified fragment can be detected using restriction enzyme digestion. This involves the use of a suite of enzymes called endonucleases each of which, when combined with DNA in a tube, will cut the double-stranded DNA where there is a specific base sequence (e.g. CCGG or GATC). Variation detected by this method is referred to as a restriction fragment length polymorphism or RFLP (Box 3.10). It results from the gain or loss of restriction enzyme sites caused by a change in DNA sequence. Other, more sophisticated methodologies such as single-strand conformation polymorphism (SSCP) allow the identification of point mutations and small insertions or deletions in the DNA. SNP (single nucleotide polymorphism) analysis is also becoming increasingly common, which involves the rapid detection of single base changes at known polymorphic loci. Alternatively, the cost of DNA sequencing and the time required has been reduced such that it is now practical to genetically type individuals based on the direct sequence analysis of complete fragments of hundreds of base pairs.

Mitochondrial DNA
Initial work using restriction enzyme analysis (see below) of the total mtDNA molecule to detect base sequence variation suggested that the level of variation in Atlantic salmon was of the order of one-third of that present in brown trout (Gyllensten and Wilson 1987). However, even though it appears to be lower than in other salmonids, recent work based on PCR analysis of some specific gene regions has revealed considerable variation among individuals (Chapter 5). To date direct work has focused on three regions, the ND1 gene and adjacent 16S ribosomal RNA gene, and to a lesser extent on the cytochrome *b* and ND4/ND5 gene regions. A full assessment of the extent of variation across the whole mitochondrial component of the genome remains to be carried out.

Nuclear DNA
Study of variation in the non-coding repetitive regions of the nuclear DNA is the most powerful element for revealing individual genetic differences. The variable sequences used are micro- and minisatellites. These are tandem repeats, with each unit being 2–30 bp long. The arrays frequently change their length (i.e. repeat number) by mutation, hence their name: variable number of tandem repeats (VNTRs). Differences between alleles in a given fragment (locus) are due to different numbers of the basic unit. By using sufficient loci with many alleles, one individual can be distinguished from another. This is the genetic basis of identification used in forensic science and a common method used in population genetics.

What is currently known about the nature of the DNA of nuclear genes of the Atlantic salmon is fragmented. Research groups working on the genetics of the species have done so from a variety of points of view. Some of our knowledge comes indirectly from the study of genetic variation in proteins, from which inferences about the genes coding for them can be made. With the advances in molecular DNA analysis technology over the last decade, this work has now been more or less superseded by direct studies of the DNA. These studies, including work carried out for individual identification and pedigree structure, population genetics, immunogenetics and phylogenetics now provide considerable information on specific nuclear DNA sequences in the species.

Box 3.9 Polymerase chain reaction.

Polymerase chain reaction or PCR is a quick, easy method for generating unlimited copies of any fragment of DNA from just a few copies of template (Fig. B3.9). It was developed in 1983 by Kary Mullis and represents one of the most important new technologies of the last century. PCR takes analysis of tiny amounts of DNA, even degraded DNA, to a new level of precision and reliability. PCR uses purified or unpurified template DNA and two primer molecules – short chains of 10–20 DNA bases which are complementary to the nucleotide sequences on either side of the piece of DNA of interest. Primers are normally specified from DNA sequencing information and constructed in the laboratory or by commercial companies. In the first step of PCR, the target DNA is denatured, i.e. the strands of its helix must be unwound and separated by heating to 90–96°C. In the second step, the primers are allowed to bind to their complementary bases on the now single-stranded DNA, while in the third the primers are extended by DNA synthesis by a polymerase enzyme starting from the primer. The result is two new DNA strands, each composed of one of the original strands and a new complementary strand. This process is then repeated until a sufficient number of copies of the targeted DNA have been produced, with the number of copies doubled with each cycle. PCR is carried out in a small test tube and involves a process called thermal cycling with a repeated controlled cycle of rapid heating and cooling. Each cycle takes a few minutes, and repeating the process for 30–40 cycles will generate millions of copies of a specific DNA strand.

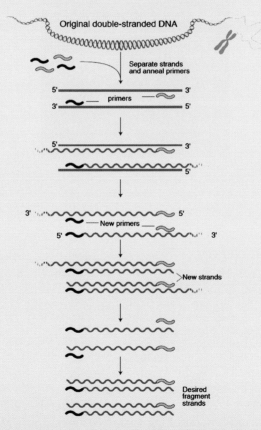

Fig. B3.9 Basic steps in the polymerase chain reaction. (Image credit: National Human Genome Research Institute.)

Box 3.10   DNA restriction enzyme analysis and electrophoretic separation.

DNA variation can be detected by sequencing (see Box 3.8) or by the digestion of DNA fragments, usually now generated by PCR, with any of a large number of restriction endonucleases, each of which cuts DNA at a point where a particular DNA sequence is found. Thus fragments with different DNA sequences will be cut into different numbers and sizes of fragments. These fragments are then usually identified by electrophoresis, which separates the fragments by their size and base composition in a gel matrix of agarose or acrylamide (Fig. B3.10). DNA has a negative charge, therefore in an electrical field will move to the positive pole. The speed will be proportional to the size of the DNA. Small fragments will migrate in the electric field faster than big fragments. The chemical, ethidium bromide, which binds preferentially with DNA and fluoresces, allows the visualisation of DNA under UV light.

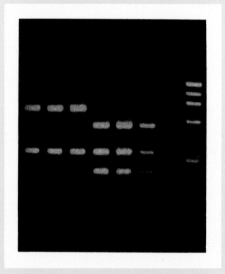

Fig. B3.10   Gel stained with ethidium bromide, showing DNA fragments.

To date, more than 40 000 DNA sequences (including both nuclear and mitochondrial) have been deposited in genetic databases. Many of the sequences correspond to the same DNA fragment of different individuals of the same species, but because of mutations, often just single base changes, the sequences are slightly different. Many more are known but have yet to be published.

Nearly half of the sequences listed correspond to non-coding regions. The most widely represented of the sequences are those relating to microsatellite loci and appear to have a moderate to high representation in the Atlantic salmon genome. Such loci, in common with most species, show a high incidence of allelic variation (i.e. polymorphism) with regard to the number of tandem repeats, which has made them very useful in studies of the genetics of populations and in the identification of individuals in studies of relatedness. Also, the high level of polymorphism they show and their apparently wide distribution throughout the nuclear genome make them useful in studies to identify the chromosomal location of loci controlling variation in quantitative traits, also known as quantitative trait loci or QTL. These studies look for microsatellites or other variable loci which are linked to character traits such as body weight or disease resistance. A number of microsatellite markers of QTL have been identified in salmonids linked to temperature tolerance, disease resistance and growth (e.g. Jackson *et al.* 1998; Palti *et al.* 1999; Robinson *et al.* 2001).

The complete amino acid sequence, a molecular trait which is genetically determined, has been elucidated and published in public databases for 607 proteins, as well as many partial sequences for these and other proteins. While this is a start, it represents only a small proportion of the over 10 000 proteins produced by a given cell, and an even smaller fraction of the total number of proteins which can be expected to be produced by the Atlantic salmon overall.

3.6 Summary and conclusions

- Variation in the sequence of the nucleotides adenosine, cytosine, guanine and thymine (A, C, G and T) in cellular DNA, as for most species, is the basis for inherited variation in the Atlantic salmon.
- The Atlantic salmon's genome, the collective name for the DNA found in a cell, consists of ~ 6 billion nucleotide bases of which > 99% are found in the cell nucleus (nDNA) and < 1% in cellular mitochondria (mtDNA).
- The Atlantic salmon's nDNA (i.e. the nuclear genome) is organised into 56 or 58 pairs of molecules, depending on the population; one of each pair is from the egg (maternal) and the other from the sperm (paternal).
- Each DNA molecule is associated with histone proteins into chromatin, which during the cell division cycle forms distinct structures referred to as chromosomes. The $2n$ value for Atlantic salmon (the number of chromosomes it possesses) is 56 or 58, i.e. 28 or 29 pairs of chromosomes.
- mtDNA is found in each cell in hundreds to thousands of identical copies of maternal origin.
- DNA is functionally divided into regions known as genes whose sequences, through their transcription and translation, control cellular and developmental processes.
- The exact number of genes in the Atlantic salmon genome is unknown but is estimated to be ~ 40 000.
- Genes compose only a small part of the DNA present. The function of the remainder of the DNA is poorly understood.
- Mutation of DNA bases and molecular recombination between DNA strands gives rise to genetically different forms at many locations (loci) in the genome. These variants are called alleles.
- At a given location (locus) an individual may be homozygous (have the same genetic forms) or heterozygous (have different forms), and may differ from other individuals in the allelic variants it possesses.
- Allelic variation has been found throughout both coding and non-coding regions of the DNA and, overall, each individual will be unique.
- Genetic variation at the molecular level underlies all heritable variation in the performance and character of individuals (i.e. in the phenotype).
- The analysis of molecular variation by its very nature is difficult, and is only directly visible with regard to variation in chromosome number and structure. Indirect methods such as the analysis of heritable electrophoretic variation in proteins have been used in the past; however, in the last decade work has been able to focus on the direct analysis of DNA

sequence variation using a range of newly developed methods for DNA isolation, manipulation and visualisation.
- Much less than 1% of the Atlantic salmon genome has so far been studied and characterised in detail and the molecular genetic basis of phenotypic variation in relation to development is not yet understood.
- The full sequence and gene structure of Atlantic salmon mtDNA is expected to be resolved in the next few decades, though it is likely to take much longer to gain a complete understanding of genome functioning in cellular and developmental processes.
- Despite the lack of specific understanding of the Atlantic salmon, it is clear that the processes leading from genotype to phenotype in the Atlantic salmon can be expected to be just as complex as in humans.

Further reading

Alberts, B., Johnson, A., Lewis, J., Raff, M., Roberts, K. and Walter, P. (2002) *Molecular Biology of the Cell* (4th edn). Garland Publishing, New York.
Brown, T.A. (1998) *Genetics: A molecular approach*. Chapman & Hall, London.
Goodsell, D.S. (1998) *The Machinery of Life*. Springer-Verlag, New York.
Tave, D. (1993) *Genetics for Fish Hatchery Managers* (2nd edn). Van Nostrand Reinhold, New York.

References

Allendorf, F.W. and Thorgaard, G.H. (1984) Tetraploidy and the evolution of salmonid fishes. In: B.J. Turner (Ed.) *Evolutionary Genetics of Fishes*, pp. 1–53. Plenum Press, New York.
Collares-Pereira, M.J. and Moreira da Costa, L. (1999) Intraspecific and interspecific genome size variation in Iberian Cyprinidae and the problem of diploidy and polyploidy with review of genome sizes within the family. *Folia Zoologica*, **48**: 61–76.
Devlin, R.H. and Nagahama, Y. (2002) Sex determination and sex differentiation in fish: an overview of genetic, physiological, and environmental influences. *Aquaculture*, **208**: 191–264.
Falconer, D.S. and Mackay, T.F. (1996) *Introduction to Quantitative Genetics* (4th edn). Longman Group, Harlow, UK.
Ferris, S.D. (1984) Tetraploidy and the evolution of catostomid fishes. In: B.J. Turner (Ed.). *Evolutionary Genetics of Fishes*, pp. 55–93. Plenum Press, New York.
García-Vázquez, E., Linde, A.R., Blanco, G., Sánchez, J.A., Vázquez, E. and Rubio, J. (1988) Chromosome polymorphism in farm fry stocks of Atlantic salmon from Asturias. *Journal of Fish Biology*, **33**: 581–587.
Gilbey, J., Verspoor, E., McLay, A. and Houlihan, D. (2004) A microsatellite linkage map for Atlantic salmon (*Salmo salar*). *Animal Genetics*, **35**: 98–105.
Goodier, J.L. and Davidson, W.S. (1994) Characterization of a repetitive element detected by NheI in the genomes of Salmo species. *Genome*, **37**: 639–645.
Goodier, J.L. and Davidson, W.S. (1998) Characterization of novel minisatellite repeat loci in Atlantic salmon (*Salmo salar*) and their phylogenetic distribution. *Journal of Molecular Evolution*, **46**: 245–255.
Gyllensten, U. and Wilson, A.C. (1987) Mitochondrial DNA of salmonids: inter- and intraspecific variability dectected with restriction enzymes. In: N. Ryman and F. Utter (Ed.) *Population Genetics and Fishery Management*, pp. 301–318. University of Washington Press, Seattle, WA.
Hartley, S.E. (1987) The chromosomes of salmonid fishes. *Biological Reviews*, **62**: 197–214.

Hartley, S.E. and Horne, M.T. (1984) Chromosome polymorphism and constitutive heterochromatin in Atlantic salmon, *Salmo salar*. *Chromosoma*, **89**: 377–380.

Hurst, C.D., Bartlett, S.E., Davidson, W.S. and Bruce, I.J. (1999) The complete mitochondrial DNA sequence of the Atlantic salmon, *Salmo salar*. *Gene*, **239**: 237–242.

Jackson, T.R., Ferguson, M.M., Danzmann, R.G., Fishback, A.G., Ihssen, P.E. and Crease, T.J. (1998) Identification of two QTL influencing upper temperature tolerance in rainbow trout (*Oncorhynchus mykiss*). *Heredity*, **80**: 143–151.

Kirpichnikov, V.S. (1981) *Genetic Bases of Fish Selection*. G.G. Gause (trans.). Springer-Verlag, Berlin.

Loudenslager, E.J. and Thorgaard, G.H. (1979) Karyotypic and evolutionary relationships of the Yellowstone (*Salmo clarki bouvieri*) and West-Slope (*S. c. lewisi*) cutthroat trout. *Journal of the Fisheries Research Board of Canada*, **36**: 630–635.

Lockwood, S.F. and Derr, J.N. (1992) Intra- and interspecific genome-size variation in the Salmonidae. *Cytogenetic and Cell Genetics*, **59**: 303–306.

Miya, M., Takeshima, H., Endo, H., Ishiguro, N.B., Inoue, J.G., Mukai, T., Satoh, T.P., Yamaguchi, M., Kawaguchi, A., Mabuchi, K., Shirai, S. and Nishida, M. (2003) Major patterns of higher teleostean phylogenies: a new perspective based on 100 complete mitochondrial DNA sequences. *Molecular Phylogenetics and Evolution*, **26**: 121–138.

Ohno, S. (1970) The enormous diversity in genome sizes of fish as a reflection of Nature's extensive experiments by gene duplication. *Transactions of the American Fisheries Society*, **99**: 120–130.

Oliveira, C., Almida-Toledo, L.F., Mori, L. and Toledo-Filho, S.A (1992) Extensive chromosomal rearrangements and nuclear DNA content changes in the evolution of the armoured catfishes genus *Corydoras* (Pisces, Siluriformes, Callichthyidae). *Journal of Fish Biology*, **40**: 419–431.

Palti, Y., Parsons, J.E. and Thorgaard, G.H. (1999) Identification of candidate DNA markers associated with IHN virus resistance in backcrosses of rainbow (*Oncorhynchus mykiss*) and cutthroat trout (*O. clarki*). *Aquaculture*, **173**: 81–94.

Pendás, A.M., Morán, P. and García-Vázquez, E. (1993) Ribosomal RNA genes are interspersed throughout a heterochromatic chromosome arm in Atlantic salmon. *Cytogenetics and Cell Genetics*, **63**: 128–130.

Perez, J. (1999) Aislamiento, caracterización y localización cromosómica de secuencias y familias génicas repetidos en salmónidos. PhD thesis. Universidad de Oviedo, Spain.

Rise, M.L. and 23 authors (2004) Development and application of a salmonid EST database and cDNA microarray: data mining and interspecific hybridization characteristics. *Genome Research*, **14**: 478–490.

Robinson, B.D., Wheeler, P.A., Sundin, K. and Thorgaard, G.H. (2001) Composite interval mapping reveals a major locus influencing embryonic development rate in rainbow trout (*Oncorhynchus mykiss*). *Journal of Heredity*, **92**: 16–22.

Tave, D. (1993) *Genetics for Fish Hatchery Managers* (2nd edn). Van Nostrand Reinhold, New York.

Thorgaard, G.H. (1983) Chromosomal differences among rainbow trout populations. *Copeia*, **3**: 650–662.

Verspoor, E., Beardmore, J.A., Consuegra, S., Gariá de Leaníz, C., Hindar, K., Jordan, W.C., Koljonen, M-L., Makhrov, A.A., Paaver, T., Sánchez, J.A., Skaala, O., Titov, S. and Cross, T.F. (2005) Population structure in the Atlantic salmon: insights from 40 years of research into genetic protein variation. *Journal of Fish Biology*, **67** (Supplement A): 3–54.

Volff, J.N. (2005) Genome evolution and biodiversity in teleost fish. *Heredity*, **94**: 280–294.

Watson, J.D. and Crick, F.H.C. (1953) Molecular structure of nucleic acids. *Nature*, **171**: 737–738.

Woram, R., Gharbi, K., Sakamoto, T., Hoyheim, B., Holm, L.E., Naish, K., McGowan, C., Ferguson, M.M., Phillips, R.B., Stein, J., Guyomard, R., Cairney, M., Taggart, J.B., Powell, R., Davidson, W. and Danzmann, R.G. (2003) Comparative genome analysis of the primary sex-determining locus in salmonid fishes. *Genome Research*, **13**: 272–280.

Wright, J.E., Johnson, K., Hollister, A. and May, B. (1983) Meiotic models to explain classical linkage, pseudolinkage, and chromosome pairing in tetraploid derivative salmonid genomes. *Isozymes: Current Topics in Biological and Medical Research*, **10**: 239–260.

4 Investigating the Genetics of Populations

M. M. Hansen, B. Villanueva, E. E. Nielsen and D. Bekkevold

Upper: returning anadromous Atlantic salmon in a holding pool in a river, waiting until the time is right to move into spawning areas to mate and lay their eggs. (Photo credit: G. van Ryckevorsel.) Lower: electrofishing for Atlantic salmon parr in the Flowers River, Labrador, to collect fish for population genetic studies. (Photo credit: E. Verspoor.)

Species, such as the Atlantic salmon, are collections of individuals which are born, develop and die. To survive over time they must reproduce and recruit new individuals. In a few species, including some fish (e.g. guppies: *Poecilia* spp.), recruitment results from asexual reproduction, i.e. clonal reproduction or parthenogenesis (a type of reproduction where the egg develops into a new individual without fertilisation). However, in most species, recruitment occurs from sexual reproduction. As discussed in Chapter 3, this is a genetic process with important implications for the genetic character of the individual. However, it also has important implications for the genetic dynamics of the species as a whole and, thereby, for the overall survival and character of a species.

The distribution of sexual reproduction in time and space determines which genetic types are produced and where. In most species, simply though not exclusively by virtue of geographic isolation, there is a subdivision into groups of individuals within which reproduction is more or less random but among which interbreeding is more or less constrained. As they are defined by their interbreeding, and interbreeding is necessary for reproduction, these groups represent the basic biological units underlying species survival. Such a biological unit is referred to by geneticists as a deme, a genetic population, or simply a population.

The term 'population' is used throughout the biological sciences, including fisheries biology, but there is no consensus about its use. It will be used by some to denote individuals of a species within national borders, e.g. 'Norwegian salmon'. By others it will be used to denote individuals of a species within a well-defined area such as a bay, a lake or a river. Here we use it in the genetic sense to denote a group of randomly mating individuals that is partly or totally reproductively isolated from other such groups within the same species. A population geneticist would strongly argue in favour of the latter definition as being the best use for the term, as its definition is not arbitrary but defined by the fundamental biological processes of reproduction and recruitment.

4.1 Overview

It is widely recognised that, apart from identical twins and organisms with clonal reproduction, all individuals are genetically different from each other. However, it is perhaps less generally accepted that nearly all species are to a smaller or larger extent subdivided into genetically different populations. There are very few examples available of truly panmictic species, i.e. where all individuals have an equal probability of mating and reproducing with any other individual within the species. Even in the case of the European eel, *Anguilla anguilla*, which was previously considered to be an example of a panmictic species, detailed studies using molecular genetic markers have now revealed that there is a significant, albeit small, degree of subdivision, possibly reflecting multiple populations (e.g. Wirth and Bernatchez 2001, but see also Dannewitz *et al.* 2005).

In nearly all species it turns out on closer inspection that there are some barriers that cause reproductive isolation partly or fully between some individuals. The factor that most commonly causes population subdivision is geographical distance. Some individuals may simply be so geographically remote from each other that interbreeding becomes unlikely. However, reproductive barriers can also be of other types, such as differences in time of reproduction, which precludes interbreeding between individuals from opposite ends of the period of reproduction.

The type of genetic population structure exhibited by an organism has very important biological consequences. *Within populations*, the number of reproducing individuals, combined with the number of migrants it receives from other populations, determines the amount of genetic variation found in the population. Thus, a small population will continuously lose genetic variation and, if this is not counterbalanced by immigrants reintroducing lost genetic diversity, the end result will be a genetically impoverished population. This has important consequences for the future prospects for survival of the population, as low levels of genetic variation may lead to loss of ability to adapt to future environmental changes. Also, consanguineous matings (i.e. inbreeding) are more likely to occur in small populations, leading to loss of fertility and viability.

Among populations, the degree of genetic uniqueness and local adaptations are determined by the combined effect of the number of reproducing individuals within populations, the number of migrants exchanged among populations, and the differences in environmental conditions experienced by different populations. Thus, in a species exhibiting large population sizes and with important differences in environmental conditions experienced by different populations, natural selection favouring specific alleles in the local environment is likely to be highly efficient. If, at the same time, the exchange of migrants among populations is low, then local natural selection is not counterbalanced by influx of different non-adapted alleles from other populations, and there is a strong possibility of local adaptation.

What is it then that we want to know about the genetic structure of Atlantic salmon? First of all, we need to know how the species in a given area, and across its range, is broken up into populations. Second, it is highly relevant to know to what extent populations are genetically distinct from each other. What is the magnitude of genetic differences? How much gene flow occurs among populations? This leads to the next key issue: what is the biological significance of genetic differentiation among populations? Does it imply that populations are adapted to local environmental conditions? The answers to all these questions are intimately linked to each other.

From a conservation perspective there are a number of questions that can be addressed using analysis of molecular genetic markers. For instance, what is the number of individuals within a population that successfully pass on their genes to the next generation (which represents one of the determinants of effective population size)? Have effective population sizes declined due to overexploitation or environmental degradation? To what extent have escaped farmed salmon interbred with indigenous salmon and contributed to indigenous gene pools? Molecular genetic markers can also be used to identify the parents of single offspring in experimental set-ups, which, for instance, can be used to analyse the reproductive output of a population or different components of the population.

In this chapter, we will first describe the basic concepts, some of which are mentioned above, and models of genetic population structure as developed by population geneticists. We then explain the differences between two important approaches to genetics, i.e. population and quantitative genetics, describe the conceptual differences between empirical studies of populations and experimental studies and provide a short description of the most commonly applied molecular genetic markers. Next, we list some of the most important issues and problems that can be addressed using molecular markers, and we end the chapter by considering the future perspectives of population genetic studies, where in particular a combination of molecular population genetics and quantitative genetics holds great promise.

4.2 Population genetics

4.2.1 Basic concepts

Population genetics can be defined as the science of studying how genetic variation is distributed among species, populations and individuals, and fundamentally it is concerned with how the evolutionary forces of mutation, natural selection, random genetic drift and migration, i.e. gene flow, affect the distribution of genetic variation.

The different evolutionary forces play different roles in the structuring and genetic composition of populations. Genetic variation in the form of different alleles at DNA loci is created by mutation, or recombination as discussed in Chapter 3. As discussed, this variation encompasses *point mutations*, where a single base in the DNA sequence is substituted by a different base, as *insertions/deletions*, where one or several bases are inserted or deleted from the DNA sequence, or as major chromosomal rearrangements, e.g. where a part of one chromosome breaks off and fuses with another chromosome. Once genetic variation is present, the frequencies of alleles can change due to selection, genetic drift and migration (gene flow).

Natural selection, or more commonly just selection, denotes a situation where an allele confers a survival and/or reproductive advantage/disadvantage on an individual compared to other alleles. There are many different types of selection. These include directional selection, where one allele is always favoured over others; overdominance, where heterozygotes, i.e. individuals possessing two different alleles, have a selective advantage compared to homozygotes (individuals possessing two identical alleles); and underdominance, where homozygotes have a selective advantage compared to heterozygotes.

In natural selection environmental circumstances determine which alleles are favoured over others. In contrast, sexual selection has also been shown to be a highly potent evolutionary force. Sexual selection occurs when individuals with specific heritable traits may be preferred by the opposite sex for mating. Examples of sexually selected traits in salmon may be male coloration or kype size. Selection can work in different ways on the genetic differentiation among populations. If selection pressure is the same in different populations, then it will tend to homogenise allele frequencies. Conversely, if selection differs among populations, e.g. in response to different environmental conditions, then selection increases genetic differentiation.

Genetic drift denotes the random changes of allele frequencies that occur when populations are of finite size. The best way to illustrate this is to think in terms of flipping a coin. If you do that an infinite number of times, then you should get heads or tails in exactly 50% of the cases. However, if you do it a finite number of times, say 50, then you may get, say, 45% heads even though the probability of heads is still 50%. In other words, the mean expected frequency of heads is always 50%, but the variance increases when the number of trials decreases. The same applies to individuals passing their genes on to the next generation. The fewer individuals that successfully reproduce, the more allele frequencies are expected to fluctuate in response to random genetic drift.

Migration, or gene flow, denotes when an individual disperses from one population to another and reproduces successfully there. Hence, gene flow acts to homogenise allele frequencies among populations.

Effective population size is one further crucial concept in population genetics which needs to be defined. It is usually denoted N_e (methods for estimating N_e are described in section 4.5.3).

Effective population size is a key parameter in conservation and population genetics, which determines the extent of genetic drift and inbreeding and measures the rate of loss in genetic variation in terms of number of individuals. It can be defined as the size of an ideal population, i.e. with equal sex ratio and variance of reproductive success among individuals approximated by a Poisson distribution, that would give rise to the same amount of inbreeding or genetic drift as occurs in the actual population of interest (see Frankham *et al.* 2002 for a thorough description of the concept). Keeping a sufficiently high N_e is a main priority in conservation biology, in the long term in order to maintain levels of genetic variation that ensure the evolutionary potential of populations and species, and in the short term to avoid a fitness decrease due to inbreeding depression (Franklin 1980). In particular, it is important to avoid sudden drastic decreases of N_e, so-called bottlenecks, where deleterious recessive alleles may become fixed at a rate at which their fixation cannot be counteracted by selection (Frankham *et al.* 2002).

4.2.2 Models of population structure

Relationships among populations can be of several different types, and many models of genetic population structure have been suggested over the years. Perhaps the most well-known is Wright's (1931) island model. In this model, populations are linked by gene flow, but gene flow is independent of the geographical distance between populations (Fig. 4.1). In other words, a migrant is just as likely to reproduce in a neighbouring population as in a geographically distant population. Another very important model is the stepping-stone model (Kimura and Weiss 1964). Here, it is assumed that gene flow occurs primarily or exclusively (a strict stepping-stone model) between neighbouring populations (Fig. 4.2). The model can be one-dimensional, e.g. salmon spawning in rivers along a coastline, or two-dimensional, a practical example of which could be salmon spawning in rivers situated on both sides of a narrow fjord. Overall, this must be considered a very realistic type of model to account for gene flow and genetic structuring in real populations, including salmonid fish. A consequence of the stepping-stone model is isolation by distance, i.e. there is a correlation between genetic

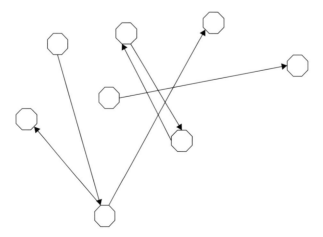

Fig. 4.1 Schematic representation of the island model.

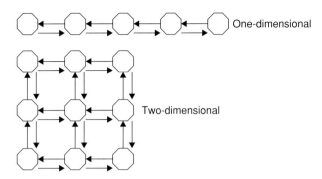

Fig. 4.2 Schematic representation of one- and two-dimensional stepping-stone models.

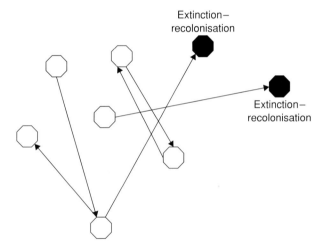

Fig. 4.3 Schematic representation of gene flow and extinction–recolonisation dynamics in metapopulations.

and geographical distance between populations, which can be tested using molecular markers (e.g. Rousset 1997).

Another concept in population genetics is that of the metapopulation. Metapopulations were originally defined as systems of populations linked by gene flow and experiencing extinction–recolonisation events, i.e. occasionally populations go extinct and are recolonised by migrants from other populations (Fig. 4.3). However, there has been a tendency to use a broader definition, which does not necessarily include extinction and recolonisation (e.g. see Hanski and Gilpin 1991 for an overview). The metapopulation concept has been applied to many organisms. However, whether or not extinctions and recolonisation of salmonid fish populations occur frequently under natural circumstances, thereby qualifying for use of the metapopulation term in a strict sense, remains to be seen. At least, the few long-term studies of genetic structure available, based on analysis of DNA from contemporary samples and from historical collections of fish scales, do not suggest frequent turnover of populations (Nielsen *et al.* 1999; Tessier and Bernatchez 1999; Hansen *et al.* 2002).

> **Box 4.1** Factors causing deviation from Hardy–Weinberg equilibrium.
>
> A number of factors can cause deviations from Hardy–Weinberg equilibrium in real populations. The most important can be summarised as follows:
>
> - **Non-random mating**
>   *Inbreeding*. Mating among close relatives will lead to a surplus of homozygotes.
>   *Assortative mating*. If individuals prefer to mate with other individuals with similar genotypes this may lead to a surplus of homozygotes.
>   *Disassortative mating*. If individuals prefer to mate with other individuals with different genotypes, then this may lead to a surplus of heterozygotes.
>   *A special case of non-random mating is termed the Wahlund effect*. This denotes a surplus of homozygotes due to the presence of individuals that represent different populations and do not interbreed. It can occur if, for instance, salmon from different populations are sampled in the sea. Since they are not part of a non-random unit, but will mate assortatively, i.e. with individuals from their own populations, the outcome will be a surplus of homozygotes.
> - **Migration**. Migration among genetically differentiated populations may lead to deviations from Hardy–Weinberg equilibrium. It could be argued that the 'Wahlund effect' described above is an outcome of migration.
> - **Selection**. Selection may lead to deviations from Hardy–Weinberg equilibrium. For instance, overdominance may lead to a surplus of heterozygotes.
> - **Small effective population size**. Small effective population sizes lead to random genetic drift, which again may result in genotype frequencies that do not conform to the expected Hardy–Weinberg proportions.
> - **Deviation from normal Mendelian segregation**. This could be a matter of concern, particularly in salmonid fishes, which have a tetraploid (i.e. four instead of two copies of each chromosome) ancestry and may still exhibit some residual tetraploid inheritance (e.g. Sakamoto *et al.* 2000).
> - **Technical artefacts**. Technical artefacts are well known, in particular in studies employing microsatellite markers. The occurrence of non-amplifying alleles, so-called null alleles, is a well-known problem (Pemberton *et al.* 1995), but other problems may also be important, such as the preferential PCR amplification of shorter alleles at the expense of larger alleles ('large allele drop-out'). Hence, in studies based on microsatellite analysis it is important to rule out technical artefacts before 'biological' interpretations (e.g. inbreeding or Wahlund effects) of deviations from Hardy–Weinberg equilibrium are invoked.

4.2.3 Population differentiation

In addition to specific models of genetic population structure it is also obvious that the genetic *differences* among populations may differ among species or even among different subsets of populations within the same species. It is therefore desirable to be able to quantify the genetic differences among populations. The most common measure of genetic differentiation is F_{ST}, originally defined by Wright (1931). However, perhaps an easier way to illustrate this is to focus on the measure G_{ST} (Nei 1973) which under most circumstances is the same as F_{ST}.

If we first consider a species composed of one population and focus on a locus with two alleles, A and a, occurring, respectively, at the frequencies p and q, then the expected frequencies or proportions of the three genotypes which will be observed are as follows: AA homozygotes $= p^2$, Aa heterozygotes $= 2pq$ and aa homozygotes $= q^2$. The expected frequency of the different genotypes is that which would occur if the association of alleles in gametes is random. When the expected and observed genotype frequencies in a sample from a population cannot be shown to be different, then they are said to be in Hardy–Weinberg (H-W) equilibrium, after the individuals who first described this genetic condition. Departures from H-W equilibrium can arise for a number of reasons (see Box 4.1).

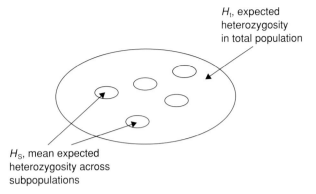

Fig. 4.4 Schematic representation of the distribution of expected heterozygosity in a subdivided population.

If we next consider a subdivided population (a population of populations; see Fig. 4.4), then the situation gets more complicated. The *total* expected heterozygosity, H_T, is $2p_T q_T$, where p_T and q_T denote the mean frequency of alleles A and a across all subpopulations. The expected heterozygosity *within* subpopulations is always lower than H_T, simply because individuals primarily mate with other individuals from the same subpopulation. In other words, there is a loss of expected heterozygosity due to the population subdivision. We can then define H_S as the mean of expected heterozygosity across all subpopulations. G_{ST} can then be calculated as:

$$G_{ST} = (H_T - H_S)/H_T.$$

G_{ST} thus measures the proportion of expected heterozygosity that is distributed between subpopulations. With two alleles G_{ST} can take any value between 0 and 1.

4.3 Quantitative genetics

4.3.1 How it differs from population genetics

Population genetics focuses on variation due to segregation of alleles at single loci at the population level. However, most traits of interest in natural populations do not show simple inheritance. Instead, most traits are complex quantitative traits that are influenced by many genes and by environmental factors and vary on a continuous scale rather than in discrete classes. A few examples of quantitative traits in salmonid fishes include growth rate, temperature tolerance, body weight, body length and yolk sac volume. Quantitative genetics focuses on this kind of variation (i.e. differences between individuals that are of a quantitative rather than a qualitative nature). Given that the trait on which natural selection acts directly (i.e. fitness, the combined ability to survive and reproduce) belongs to this category of traits, quantitative genetics theory is essential in the study of evolution and adaptation. Important processes involved in adaptation, such as reduction in fitness due to inbreeding, loss of variation due to selection and genetic drift, effects of crossing different populations, and genotype by environment interaction all have a substantial body of theory in quantitative genetics.

Quantitative genetics describes traits in statistical terms such as means, variances and covariances in groups of animals. Consequently, it has a strong mathematical emphasis and is often also referred to as 'biometrical genetics'. The value observed when the trait is measured (i.e. the phenotype) is the result of the contributions of many genes affecting the trait (i.e. the genotype) and environmental effects on these genes (i.e. the environment). For instance, the weight at a given age of a particular fish (e.g. 60 g; the phenotype) is the result of the effects of the particular assemblage of genes possessed by that particular fish that control weight at that age (for which the number and effects are unknown) and many non-genetic factors (e.g. water temperature and food available; the environment).

Given the nature of quantitative traits, genotypes cannot be directly inferred from phenotypic performance and phenotypic rather than genotypic variation can be observed. Statistical techniques, based on resemblance between relatives, are used to disentangle genetic and environmental variation. These techniques are based on comparisons of phenotypes of individuals with known degrees of relatedness. The more closely related two individuals are, the more similar will be their genetic makeup. Then a higher correlation in phenotypic performance for closer relatives than for distant relatives would indicate that there is heritable variation. Numerous techniques are available for estimating variance components from samples of pedigrees (e.g. Falconer and Mackay 1996). When data are available on a variety of relationships of different degrees (e.g. offspring, parents, grandparents, full-sibs, half-sibs, cousins) and environmental effects are unknown then restricted maximum likelihood (REML) is the most appropriate method for estimating variance components. This method is widely applied in animal breeding and is being used more recently in natural populations.

4.3.2 Quantitative genetic variation

A parameter of fundamental importance in quantitative genetics is the coefficient of heritability (h^2); the proportion of the total observed variation for a particular trait, in a given environment, which is attributable to the additive effects of genes. An h^2 of 0.4 would mean that 40% of the measured variation in performance is estimated to be inherited. It is a measure of paramount importance in evolution, as it determines how much of the variability is responsive to selection, thereby determining the evolutionary potential of a particular population. Heritability is generally found to be larger for morphological traits than for fitness-related traits, which is not surprising given that natural selection acts on differences in fitness (and thus the reduction in variability is stronger for fitness-related traits).

Estimates of heritabilities and other genetic parameters, such as genetic correlations between different traits, are difficult to obtain in natural populations because data are limited. Statistical methods for estimating the different components of the variance observed (e.g. REML) use phenotypic data and pedigree information, which is difficult to obtain in natural populations. Estimation of genetic correlations is particularly problematic given that large sample sizes are required to obtain accurate estimates. Molecular markers are, however, being used to overcome this problem (e.g. Kruuk *et al.* 2000).

The heritability of a particular trait depends on genes segregating in a particular population living under specific environmental conditions at a specific time (i.e. h^2 is specific for the population and environmental conditions in which it has been estimated). However, surprisingly, magnitudes for h^2 show a consistent pattern for similar traits among different species (Hill and Zhang 2004). For instance, the heritability of size is typically around 50% across species and that of fitness is around 10% (Frankham *et al.* 2002). Similarly, estimates from farmed populations of Atlantic salmon are of the same order of magnitude as those found in other terrestrial farmed species (e.g. Tave 1993).

4.3.3 Genotype by environment interaction

Genotype by environment interaction (G × E) occurs if genetically different fish differ in their response to environmental change (i.e. if genotypes show different performances in different environments). This is more likely to happen when the magnitude of both genotypic and environment differences are large. Thus, it is most likely to be detected in species inhabiting a variety of environmental ranges. Also, fitness-related traits seem to show more G × E interaction than traits more peripheral to fitness (Frankham *et al.* 2002). G × E interaction is of major concern in attempts to rehabilitate populations by stocking with fish from different locations, as fish raised in their native environment may perform better there than in other environments. The fitness of reintroduced fish cannot be predicted if there is significant G × E interaction. It is also of concern when results obtained with fish raised in simple experimental conditions (or in a laboratory environment) are extrapolated to natural populations. The success of reintroducing populations may be compromised by genetic adaptation, as superior genotypes under captive conditions may perform relatively poorly when released to the wild. Nevertheless, the action of natural selection may lead to adaptation in the longer term.

4.3.4 Integration of molecular and quantitative genetics

The traditional quantitative genetic approach has treated the genetic architecture of the trait itself as a black box, without knowledge on the number of genes that affect the trait, or the effects of each gene or their locations in the genome. This has primarily been due to the fact that the individual properties of the genes affecting quantitative traits have been largely unknown. With the tremendous recent advances in molecular genetics, a large amount of information is now becoming available on genes that affect quantitative traits. This molecular information is being integrated with existing quantitative genetic techniques in order to better understand the mechanisms of evolution. Statistical methods have been developed to identify genes that affect quantitative traits (QTL or quantitative trait loci) and to estimate their effects. Thus, the traditional distinction between molecular and quantitative genetics is likely to disappear, and the integration of the two fields will undoubtedly lead to much deeper insights into issues such as local adaptation in salmonid fishes.

4.4 The genetic characterisation of wild populations

The characterisation of wild populations of Atlantic salmon and other salmonids has relied largely on the analysis of molecular markers in samples of fish taken from the wild. The different sources of genetic information on population structure and genetic differentiation, and their limitations, are discussed below.

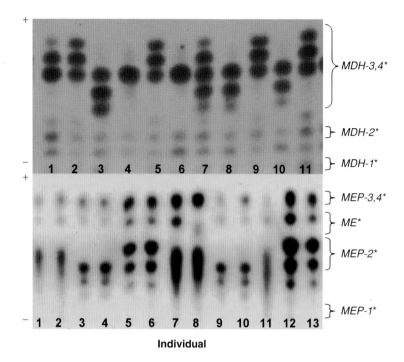

Fig. 4.5 Allozyme variation among individuals detected by starch gel electrophoresis and enzyme-specific histochemical staining. Top: variation observed with respect to malate dehydrogenase (MDH) in muscle tissue; banding is attributable to the gene products of four loci of which one of a pair of co-migrating loci (*MDH-3,4**) is variable with the seven variant genotypes attributable to the three alleles. Bottom: variation observed with respect to malic enzyme (MEP and ME) in muscle tissue with banding attributable to the interaction of the products of five loci, one of which is only very weakly expressed in muscle (*MEP-1**); different genotypes seen at *MEP-2** and *ME** are attributable to the presence of two alleles at each locus. (Image credit: E. Verspoor.)

4.4.1 Allozyme electrophoresis

It is difficult to study single loci based on morphological phenotypes, as only few morphological traits are encoded by single loci. However, biochemical and molecular markers have yielded vastly improved opportunities for empirical population genetics studies. The classical technique for studying genetic variation at single loci is allozyme electrophoresis (e.g. Utter *et al.* 1987), the principle of which is explained in Box 4.2. Typical examples of allozyme variation are shown in Fig. 4.5.

The selective neutrality of allozyme alleles has been debated for more than 20 years (Kimura 1983). Though a universal conclusion has never been reached, the present consensus is that most variation at allozymes is selectively neutral, but with some exceptions where selection has been suggested or demonstrated. One example is the malic enzyme (*MEP-2**) locus in Atlantic salmon where selection has been suggested to operate (Verspoor and Jordan 1989).

The major advantage of allozyme electrophoresis is that technically it is reasonably uncomplicated and may be applied to any organism with only relatively minor adjustments to experimental protocols. It allows for screening of a large number of loci, often more than 30–50. However, usually many of the loci studied turn out to be monomorphic. In a species

Box 4.2 Allozyme electrophoresis.

Allozyme electrophoresis denotes the technique for identifying genetic variation at the level of enzymes, which are directly encoded by DNA. The principle of the methodology is that mutations may lead to substitutions of amino acids in the enzyme, which may again result in a shift of conformation (i.e. shape) and net charge of the whole enzyme. Since allelic variation reflected in an enzyme may result in different properties, it is possible to identify different alleles by electrophoresis; tissue extracts are applied to a gel (usually starch) and an electric current is applied. Different allelic variants of an enzyme then migrate through the gel at a rate determined by the net charge and conformation of the enzyme. Finally, enzyme-specific histochemical staining is used to visualise specific enzymes, and different alleles are identified from different banding patterns (Fig. B4.2).

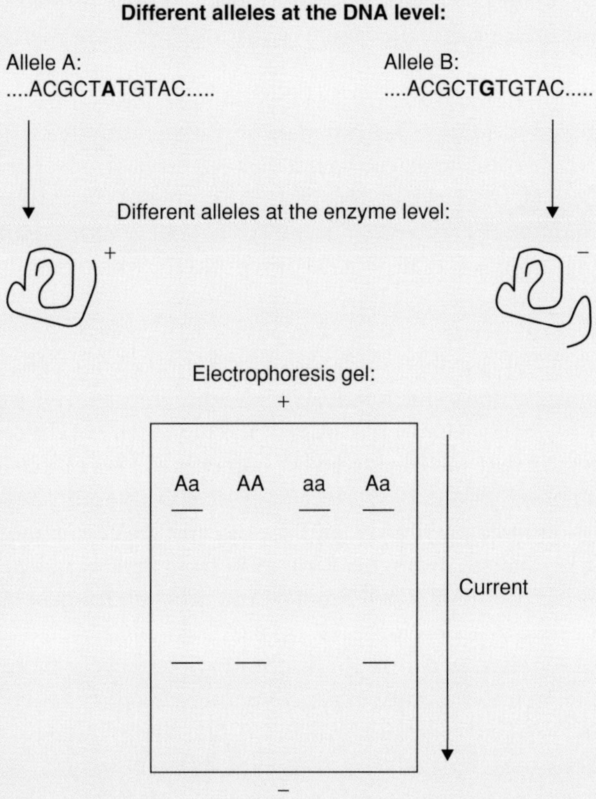

Fig. B4.2 Schematic representation of how genetic variation at the DNA level may translate into differences at the enzyme level, and how different enzyme alleles (allozymes) may be visualised using gel electrophoresis.

such as Atlantic salmon this is a particularly important problem with only a handful of loci exhibiting polymorphism (e.g. Ståhl 1987). Also, the level of variation at polymorphic loci is in most cases low, with individual loci usually exhibiting no more than two or three alleles. There are important demands concerning the freshness of tissue samples and many loci exhibit tissue-specific expression (e.g. some loci are only expressed in heart tissue) which renders it difficult to conduct non-destructive sampling.

Allozyme electrophoresis was previously the dominating technique for studies of the genetic structure of populations, but today it has to a large extent been replaced by DNA techniques, in particular, microsatellite DNA analysis.

4.4.2 Mitochondrial DNA

Even though most DNA in eukaryote organisms (i.e. organisms that have a nucleus: plants, fungi, animals) is found in the nucleus, organelles (mitochondria and, in plants, the chloroplasts) contain their own DNA. Mitochondrial DNA (hereafter abbreviated mtDNA) has the special feature that it is haploid (i.e. there is only one variant within an individual), maternally inherited, (mainly) selectively neutral and (mainly) non-recombining. Also, the mutation rate is higher (at least in some regions of the mtDNA molecule) than in most nuclear DNA regions.

The fact that mtDNA is haploid and maternally inherited has some very important implications for its use as a genetic marker for population studies. Among other things, mtDNA is subject to more genetic drift than nuclear loci, which results in stronger genetic differentiation as compared to nuclear DNA (Birky et al. 1989). Also, due to the maternal inheritance, one must be careful about drawing conclusions about the overall genetic population structure. For instance, if males migrate more than females, then mtDNA analysis alone does not give a correct picture of the true genetic structure of populations.

MtDNA is the classical genetic marker for phylogeographical studies (Avise 1994), i.e. studies on the geographical distribution of major evolutionary lineages. MtDNA is particularly useful for this purpose due to the relatively high mutation rate and the lack of recombination, which makes it relatively easy to reconstruct the phylogeny of different haplotypes. The assumption is then made that the phylogeny of haplotypes also reflects population history. Of course, this assumption should be subject to critical evaluation, both due to the special properties of mtDNA outlined previously, and because mtDNA essentially represents just one locus. Thus, it is not possible to average information from several independent loci. Keeping these reservations in mind, mtDNA still remains the most useful genetic marker for phylogeography and recent statistical developments, in particular, nested clade analysis (Templeton 1998), have provided sophisticated tools for discriminating between patterns of historical and ongoing gene flow and migration, range expansions and secondary contact between divergent evolutionary lineages.

Variation at mtDNA may be analysed in several different ways. The approach yielding maximum resolution is to sequence the mtDNA region of interest. An alternative is to conduct a more coarse-grained screening, which does not necessarily detect all variation present. This can be done using restriction enzymes, i.e. enzymes that recognise a specific sequence and then cut the DNA at a specific site. This results in a number of DNA fragments (Fig. 4.6) which can then be visualised using gel electrophoresis and stains for DNA such as ethidium bromide. Mutations may lead to gain or loss of the sites recognised by the different restriction enzymes and, consequently, different banding patterns for different haplotypes. Another useful technique for rapid screening of variation is single-strand conformation polymorphism (SSCP) and related methods (e.g. see Sunnucks et al. 2000). In SSCP the principle is to make the DNA single-stranded, electrophorese it through a gel and then visualise the DNA. Depending on sequence differences the DNA will 'curl' in different ways and migrate at different speeds, resulting in different bands on the gel.

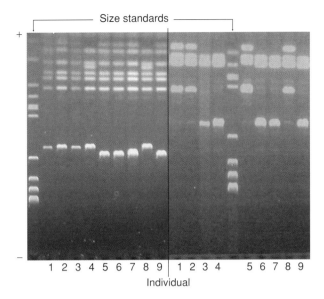

Fig. 4.6 Restriction fragment size variation among individuals detected in the mtDNA in the PCR amplified ND1 gene region by electrophoretic separation in agarose gels followed by ethidium bromide staining and visualisation under UV light. Left: variants observed following restriction digestion with the *HaeIII* enzyme. Right: variants observed following restriction digestion with the *RsaI* enzyme. (Image credit: E. Verspoor.)

4.4.3 Microsatellite DNA

Satellite DNA is non-coding nuclear DNA found throughout the genomes of eukaryotes. It consists of repeated sequences and is named according to the length of the repeat motifs. Minisatellites consist of relatively long repeat units. Minisatellite single-locus probes have previously been extensively used in Atlantic salmon (e.g. Prodöhl *et al.* 1995), but have now been largely replaced by microsatellites and will not be described further in this chapter. The special features of microsatellite DNA are summarised in Box. 4.3; an example of microsatellite variation observed in Atlantic salmon is shown in Fig. 4.7.

Since the early 1990s microsatellites have gradually replaced allozymes as the preferred marker for population studies. There are several reasons for this, with the following presumably being the most important. Microsatellite loci are typically short, in the range of approximately 80–400 base pairs (bp). This makes it easy to amplify the loci using PCR, and the amplified products can subsequently be analysed on either manual sequencing gels or systems for automated sequencing. Also, the much higher variability of microsatellites, as compared to allozymes, results in increased power for a number of applications, ranging from analysis of kinship and parentage assignment to assignment of individuals to populations and detection of population bottlenecks (reviewed by Luikart and England 1999).

Of course, despite the usefulness of microsatellite markers they are not without problems, and one of these, the presence of so-called null alleles (Pemberton *et al.* 1995) deserves special mention. Null alleles occur when mutations take place in the primer binding regions of the microsatellite locus, i.e. not in the microsatellite DNA itself. This may result in one of the two primers being unable to anneal satisfactorily to the target DNA, and the allele with the mutation

Box 4.3 Microsatellite DNA.

Microsatellites consist of a short sequence motif, such as 'TG', repeated a number of times, such as 'TGTGTGTGTGTG...' (see Jarne and Lagoda 1996). The sequence motif may consist of a single base, leading to a mononucleotide microsatellite, two bases leading to a dinucleotide microsatellite, three bases leading to a trinucleotide microsatellite, etc. Most microsatellites employed in population genetic studies consist of di-, tri- and tetranucleotide repeats. In addition to these categories microsatellites may be subdivided into three types: 'perfect' loci consisting of non-interrupted sequences of repeat units, such as $(TG)_{25}$, 'compound' microsatellites consisting of sequences of different repeat units, such as $(TG)_{10}(TC)_7(TG)_{12}$, and 'interrupted' microsatellites, consisting of repeat units interrupted by non-repetitive DNA, such as $(TG)_{10}ACATGATAC(TG)_{12}$. In practice, microsatellite DNA is analysed by first generating a huge number of copies of the locus of interest in each individual being analysed. This is done using polymerase chain reaction (PCR) (see Chapter 3). Next, the amplified microsatellite DNA is electrophoresed through a gel, where different alleles migrate at different speeds, depending on the length of the microsatellite allele (see Fig. B4.3). Equipment developed for automated sequencing of DNA can also be used for analysing microsatellites, and this has drastically facilitated analysis compared to previous use of 'manual' sequencing gels.

The mutational processes at microsatellite loci are quite distinct from those occurring in other types of DNA. First, mutation rate is very high, with values from 1 in 10 000 to 1 in 1000 bases per generation reported in different studies (Jarne and Lagoda 1996). This leads to high levels of variation, with numbers of alleles at individual loci often ranging between 10 and 20 or more. Second, mutation involves primarily insertion or deletion of one or a few repeat units, presumably resulting from slippage of the DNA polymerase during replication. It was initially suggested that mutation at microsatellite loci follows a strict stepwise mutation mode, involving insertion or deletion of a single repeat unit. Later studies have, however, demonstrated that mutations involving several repeat units also occur. The presently most favoured mutation model is the two-phase model by Di Rienzo *et al.* (1994), where most mutations involve insertion or deletion of a single repeat unit, but a small fraction of mutations involve several repeats.

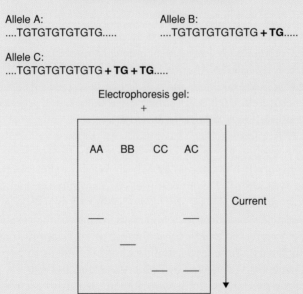

Fig. B4.3 Schematic representation of how genetic variation at a microsatellite DNA locus may be visualised using gel electrophoresis.

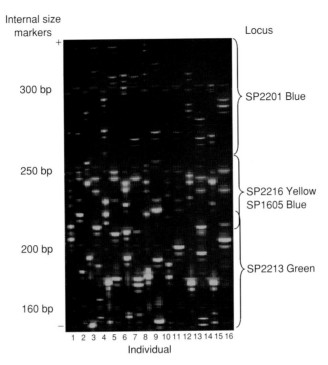

Fig. 4.7 Amplified fragment size variation among individuals for four microsatellite loci detected by PCR amplification and fluorescent labelling based on automated laser scanning of gels. For each locus a characteristic set of similarly sized fragments is produced, with homozygous individuals having one set of fragments and heterozygous individuals two sets of fragments. (Image credit: E. Verspoor.)

is consequently not PCR amplified. The presence of an amplified allele and a non-amplifying allele is therefore erroneously interpreted as a homozygote, and a null allele homozygote is interpreted as an individual in which for technical reasons the locus would not amplify.

4.4.4 Other types of molecular marker

Some other types of molecular marker should also be mentioned briefly. Analysis of coding nuclear genes is of increasing interest, particularly as this may allow for analysing genetic variation at ecologically important genes which are likely to be subject to selection. So far, major histocompatibility complex (MHC) genes have attracted most interest. MHC genes encode glycoproteins on the surface of T-cells. The glycoproteins enable T-cells to recognise and eliminate foreign antigens and MHC loci are thereby directly involved in the immune response. In Atlantic salmon it has been found that specific alleles at the MHC class IIB gene are involved in resistance against the disease furunculosis (Langefors *et al.* 2001). Also, MHC has been shown to be involved in mate choice in a range of species, including Atlantic salmon (Landry *et al.* 2001), presumably through olfactory recognition.

Single nucleotide polymorphisms, SNPs, are another type of marker likely to become of great importance in the future (Landegren *et al.* 1998). SNPs consist simply of single base substitutions in nuclear DNA. Such polymorphisms occur at high frequency in genomes, on average one SNP per 100–300 bp, and thus provide an almost unlimited source of molecular markers. Perhaps the most important perspectives are the potentials for analysing SNPs using microarray (chips) technology, whereby genome-wide screening of a huge number of SNPs may be conducted with relatively limited labour, provided the technique has become established and a high number of SNPs have been identified.

Finally, AFLP (amplified fragment length polymorphism) is a type of marker with considerable potential for population studies. Very briefly, AFLP is a combination of restriction enzyme analysis and PCR, and the markers are dominant, i.e. a specific band may represent both a homozygote and a heterozygote. The major advantage of the method is that a high number of loci ($\gg 100$) can be screened with a relatively minor effort, and the method has proven to be very useful in determining the population of origin of individuals using so-called assignment tests (Campbell *et al.* 2003). Readers interested in more information about the AFLP technique should consult Mueller and Wolfenbarger (1999).

4.5 Studying populations: issues and limitations

In the previous sections we have described the basics of population and quantitative genetics and given an overview of the most commonly used molecular markers, in particular those applied or likely to be applied to Atlantic salmon. We will now return to the biological issues that we listed in the introduction, and describe how genetic approaches can be used for resolving questions concerning the genetic structure of Atlantic salmon, and problems of importance for the conservation of populations. However, before we go into the details of empirical population genetic studies, a few sentences are needed to put the issues into a general scientific perspective.

4.5.1 Types of study and their limitations

Science is fundamentally about curiosity. Ultimately, the aim is to understand patterns and processes in nature, for their own sake and/or because it serves a practical, applied purpose. Hence, purely descriptive research rarely leads anywhere and the only question that is likely to arise is 'so what?' Good science, in contrast, is problem-oriented; it asks questions and tests hypotheses. In population genetics, including research on Atlantic salmon, there are basically three ways of targeting an issue. First, the issue may be treated theoretically, i.e. by developing theory possibly supported by mathematical models, which could subsequently be tested empirically. Second, the issue may be targeted empirically, by analysing populations and then using the observed results to test the initial hypothesis. Third, in order to test specific hypotheses it may be necessary to conduct experiments, either by manipulating natural populations or by conducting trials in semi-natural, experimental settings. The three approaches are certainly not mutually exclusive. On the contrary, it would often be highly beneficial to target the issue from all three angles. As an example, consider the problem of gene flow from domesticated salmonids into wild populations and its possible biological consequences. The problem has been addressed theoretically, e.g. by Mork (1991) and Lynch and O'Healy (2001), by considering cases where wild populations have become exposed to domesticated fish (e.g. Clifford *et al.* 1998; Hansen 2002), and by conducting experiments where wild and domesticated fish are brought together in a common, controlled environment (McGinnity *et al.* 1997; Fleming *et al.* 2000). The combined results of all three approaches contribute to a much deeper understanding of the problem than any single approach would do on its own. Thus, (1) we know from theory that gene flow from domesticated salmonids to wild populations is expected to be harmful, (2) it has been demonstrated by a controlled experiment, and (3) it is shown that the principles also apply in real life, i.e. in wild populations.

4.5.2 Mixed-stock analysis and assignment tests

Methods for assessing the origin of individual fish in fisheries have been a major topic in fisheries biology for decades, including studies on Atlantic salmon. The common solution to the problem of identifying the source population of individuals has been to implement extensive tagging and recapture programmes. However, genetic methods have several advantages over traditional tagging for identification of the origin of individuals. The primary reason is that all individuals in a population carry the genetic signature of that population, meaning that all individuals are marked – for life. Further, genetic methods avoid the costs associated with actual tagging, the loss of tags, and bias due to possible viability or catchability effects of the external tags. Basically there are two different ways of exploiting genetic information for assessing the origin of individuals: mixed-stock analysis and assignment tests.

Traditional mixed-stock analysis (MSA) estimates the proportion of each 'stock' (population) in a mixed sample of fish of unknown origin based on genetic information from potentially contributing baseline populations (Pella and Milner 1987). The method determines the relative contributions of baseline stocks with the highest probability of providing the observed multilocus genotypic frequencies in the mixed sample. Baseline genotype frequencies are obtained from the observed allele frequencies of the baseline stocks. Mixed-stock analysis is commonly applied when fisheries are targeted on mixed (feeding) aggregations of fish from several stocks. Baseline data are collected from potentially contributing stocks and their most likely proportions in the catch (mixed sample) is estimated. It is known that mixed harvesting of populations can lead to the extirpation of minor stocks through high harvest rates and unequal productivity of different stocks. Therefore, MSA can assist fishery managers to target fisheries in areas or during time periods where threatened stocks are not present. MSA has gained wide application for regulation of coastal fisheries of Pacific salmon, where fisheries are closed or opened according to stock proportions evaluated by real-time MSA.

The principle of assignment tests is to estimate the probability of encountering the genotype of an individual of unknown origin in a number of potential source populations (baseline populations). The individual is then assigned to the population where its genotype is most likely to occur. This means that, in contrast to mixed-stock analysis, each fish is treated individually. Therefore, assignment tests have their major application in cases where the origin of one or a few fish is required. Assignment methods have gained wide use in population and conservation biology for various applications, such as to discriminate among species, to investigate hybridisation among populations and to identify migrant and indigenous individuals within populations. However, probably the best-known applications are in fish forensics. An illustrative example of the latter is the study by Primmer *et al.* (2000). They identified a winning salmon in a fishing contest in a Finnish lake as being 600 times more likely to have originated from one of the regions that supply most of Finland's fish markets than from the lake in which the contest was held. Confronted with the evidence, the angler confessed to having purchased the salmon at a local fish shop!

The initial sampling of baseline populations is often the most crucial step of both procedures. It is of great importance that all potential source populations are sampled and that the samples are representative of the population, e.g. of sufficient size (see also Chapter 8). The power of these methods is critically dependent on the numbers and variability of the genetic markers used, so a large number of highly variable markers such as microsatellites are well suited for these analyses. Finally, however, the power is ultimately determined by the levels of genetic differentiation among populations. Consequently, these methods can only be applied when there is significant population structure.

4.5.3 Estimating effective population size and detecting population declines

Molecular markers are being applied increasingly for estimating demographic parameters within populations, i.e. by either directly estimating effective population size (N_e) or by detecting changes in effective population size.

As stated previously, N_e is nearly always considerably lower than the census population size due to differences in the number of males and females contributing to reproduction, variance in reproductive success and family size, and variance of N_e over several generations (Frankham 1995). Consequently, even if census population size estimates are available it may be difficult to assess N_e due to the number and complexity of factors affecting it. However, recent statistical developments for determining family structure in assemblages of individuals without parental information (Smith *et al.* 2001; Wang 2004) hold great promise for studying variance in reproductive success in wild salmonid populations (Hansen and Jensen 2005; Herbinger *et al.* 2006) and could thus be used for estimating N_e based on census population size estimates.

N_e can be estimated directly essentially by estimating random genetic drift using molecular markers. The so-called temporal method has proven particularly useful (e.g. Waples 1989; Jorde and Ryman 1995). The basic principle is to sample a population at two or more points in time separated by a specified number of generations. The smaller N_e is, the more random genetic drift will take place. Thus, based on the random changes in allele frequencies that have occurred during the interval it is possible to estimate the effective population size. Another approach is to estimate N_e from linkage or rather gametic phase disequilibrium between two loci (Hill 1981; Waples 1991). Here, the principle is that genetic drift causes gametic phase disequilibrium and, conversely, estimating gametic phase disequilibrium yields an estimate of genetic drift which again yields an estimate of N_e.

Instead of directly estimating N_e another approach is to test for indications of a significantly lowered N_e (e.g. Luikart *et al.* 1998). A simple way of doing this is to compare samples taken from the same population at different times and test whether there has been a loss of allelic diversity (e.g. Luikart *et al.* 1998). Other newer methods use various approaches for detecting shifts of allelic distributions within single samples that could indicate presence of a recent bottleneck (e.g. Cornuet and Luikart 1996; Garza and Williamson 2001).

4.5.4 Parentage assignment

Molecular markers can be used for genetic parentage assignment that in turn can provide detailed information about family structure, breeding strategies and reproductive success of the different individuals making up natural populations. Such information, which is often very difficult to obtain from observation studies, contributes to our understanding of several issues related to management and conservation of Atlantic salmon, and some examples are given here.

It is often of interest to find out what affects the expression and evolution of specific phenotypic traits. For instance, researchers have sought to elucidate the extent to which Atlantic salmon males showing different phenotypes, such as more or less red breeding coloration, smaller or larger body size, and earlier or later ascent to spawning sites, also have different reproductive success. To investigate this, genetic parentage studies can be carried out, either in experimental set-ups or under natural conditions (e.g. see Taggart *et al.* 2001). The principles of parentage assignment are described in Box 4.4.

A special issue in molecular parentage studies in salmonids concerns the biological significance of alternative reproductive morphs and spawning tactics. Genetic paternity analysis has been used to estimate the reproductive success of anadromous Atlantic salmon males in competition with 'sneak-spawning' mature male parr. Based on genetic paternity analyses it is possible to make direct estimates of the frequencies with which anadromous salmon and mature male parr sire the fry. These estimates first of all tell us something about the evolutionary processes that shape the complex salmonid life history. But knowledge of paternity patterns also has direct implication for our understanding of how genetic diversity is maintained within populations, which can be of special interest in a conservation genetic context. In a genetic paternity study of Atlantic salmon, Martinez *et al.* (2000) showed that, even though a few dominant anadromous males may be able to exclude all other anadromous males from spawning, sneak-spawning mature male parr sired considerable proportions of the offspring produced within a local site. This led to a much higher effective population size N_e compared to that if no mature male parr had been present. The authors concluded that the genetic input from the mature male parr can be crucial for the maintenance of genetic variability within small isolated populations of salmonids.

Molecular parentage analysis in salmonids has included evaluations of the reproductive performance of stocked individuals in restoration programmes (Letcher and King 2001), and examinations of the genetic consequences of interactions between wild and stocked individuals of hatchery origin. Based on estimates of the relative paternity contributions of males of wild v. hatchery origin Fleming *et al.* (2000) showed that anadromous males of hatchery origin had overall low reproductive success, relative to wild males. Based on this experiment it was estimated that the overall lifetime reproductive success of the hatchery fish was only 16% compared to that of wild fish.

Information from genetic parentage assignment in fish and other taxonomic groups has also been used in genetic 'mark–recapture' analyses to estimate the size of a contemporary breeding population (Jones and Avise 1997; Pearse *et al.* 2001; Fiumera *et al.* 2002). In brief, the approach in genetic 'mark–recapture' studies has been to sample offspring from a female, genotype them and identify the paternal genotypes that have contributed to her brood. Each genetic 'mark' is the first instance in which a male's genotype is observed ('captured') in a sample of eggs or offspring, and each 'recapture' is then represented by any later observation of that same paternal genotype in samples of offspring from other females. By applying these data in 'classical' mark–recapture analyses (e.g. King 1995), an estimate can then be generated of the total number of males breeding at a local site.

Box 4.4 Parentage assignment using genetic markers.

The principle of parentage assignment is to sample tissue non-invasively (e.g. adipose fin-clips) from all, or at least as many as possible, of the parental fish entering a given spawning site. After spawning, eggs or larvae are sampled from the redds and taken to the laboratory, where molecular genotyping is carried out for all putative parental fish and their offspring (eggs). Each offspring can then, based on its specific multilocus genotype, be assigned a maternal and a paternal genotype, using genetic exclusion procedures as shown in the example in Fig. B4.4. Based on the identification of the maternal and paternal genotypes assigned to each offspring, estimates can be made of the total number of offspring that were produced by each male and female present on the spawning site. For instance, it can then be evaluated whether males that vary with respect to specific phenotypes also show different levels of reproductive success. The answers to such questions have direct significance, not just at a level of the individual males, but also in the context of the evolutionary forces involved in population divergence, local adaptations and ultimately speciation.

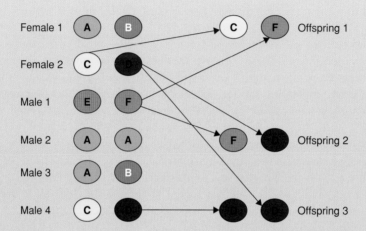

Fig. B4.4 Example of the procedure commonly used for genetic parentage assignment based on codominant molecular markers such as microsatellites. The genotypes for each of two female and four male putative parents are shown as coloured symbols (left panel), together with the genotypes of three offspring (right panel). Based on these genotypes, it can be seen that Female 1 does not share alleles with any of the offspring, and she is consequently excluded as a putative mother. If all parents have been sampled it can be deduced that Female 2 must be the mother of all three offspring. Given the assignation of the maternal allele in each offspring (**C** in Offspring 1, **D** in Offspring 2, and **D** in Offspring 3), it is clear that only the genotype of Male 1 (**F**) corresponds with those of Offspring 1 and Offspring 2, whereas only the genotype of Male 4 (**D**) corresponds with that of Offspring 3. In this hypothetical example, all offspring could be assigned parentage based on genotype information from one locus. Depending on the numbers of putative parents and the molecular resolution of the chosen genetic markers (i.e. how much allelic differentiation individuals display) it is normally necessary to analyse several loci in order to assign parentage unambiguously. Several computer programs have been developed to automate analysis of such genetic parentage data (see Wilson and Ferguson 2002 for a review of these).

4.5.5 Relatedness estimation

As with genetic parentage studies, relatedness estimation can be used in both empirical and applied studies of salmonid populations. Relatedness is a measure of the degree of genetic similarity between two individuals. This similarity can vary from unity (indicating that two individuals are genetically identical, as would be the case for identical twins and clones), over a progressively decreasing degree of relatedness (e.g. from full-sibs, half-sibs, cousins, to great-grand-offspring, etc.) to no relatedness (as would be the case for two individuals that share no detectable degree of ancestry or genetic similarity for any of a number of genetic markers). Relatedness can be estimated in several ways depending on what population level and composition one is looking at (e.g. within groups of sibs, or within a whole population largely consisting of unrelated individuals) and the associated sampling variance (Van de Casteele et al. 2001; Rousset 2002). All measures of relatedness represent the probability that two or more individuals share genes that originated from a common ancestor (this is also known as genes being *identical-by-descent*). Estimates of relatedness can be obtained either from path analysis of known pedigrees or via statistical methods using data from highly variable genetic markers, such as microsatellites. Pedigree data allow us to study the inheritance of specific phenotypic traits (e.g. Mousseau et al. 1998) and relatedness estimates are routinely used in quantitative genetics, but knowledge about relatedness may also be important from ecological and management perspectives. Small population sizes may lead to inbreeding depression, and in order to analyse the effects of inbreeding, relatedness patterns are often of interest.

Relatedness estimates can also be used to analyse dispersal behaviour associated with sex, age, or other phenotypic characteristics, by comparing relatedness between different groups of individuals. For instance, if females within a local site show closer genetic relatedness than males this indicates a higher propensity for males to disperse from their natal patch, than for females (for an example for fish see Knight et al. 1999). A specific dispersal pattern ultimately has implications for predictions about the re-establishment of extinct populations via natural recolonisation. Finally, relatedness estimates have been used to analyse the extent to which salmonids exhibit kin-biased behaviour and inbreeding avoidance. For instance, the extent to which spawning with close relatives is actively avoided has direct implications for the risk of inbreeding depression and extinction in fragmented or heavily exploited populations (Landry et al. 2001). Kin recognition and kin-biased behaviour have been demonstrated in both juveniles and adults in several experimental studies of salmonid species (review in Brown and Brown 1996). However, the extent to which kin-biased behaviour has a significant structuring effect on natural salmonid populations remains to be resolved.

4.6 Future perspectives: going beyond quantifying genetic differentiation and understanding local adaptation

Developments in both molecular biology and statistical genetics are proceeding at a very fast pace and will undoubtedly lead to even more improved research opportunities in population genetics. For instance, assignment of individuals to populations is likely to be facilitated both by new statistical developments and by the availability of techniques for screening large numbers of loci, such as SNPs. However, perhaps the most significant future development will be the ability to go beyond merely quantifying genetic differentiation and obtain a deeper understanding of the *biological* significance of the differentiation that we observe. As described in other chapters, Atlantic salmon along with many other salmonid fishes exhibit significant genetic differentiation among populations. The obvious question is then, what is the biological significance of this differentiation? Is it just a reflection of genetic differentiation due to random genetic drift and limited dispersal of individuals, but not reflecting any divergence at biologically important traits, or are populations likely to be adapted to local conditions? This question has become even more relevant with the advent of highly variable molecular markers, such as microsatellites, which make it possible to detect even very small genetic differences among populations which may be of questionable biological importance (e.g. Waples 1998).

It is possible to assess the potential for local adaptations in populations by considering information obtained from neutral genetic markers. Thus, by considering estimates of effective population size and gene flow and assuming realistic selection coefficients, the most likely scale and extent of local adaptation can be assessed (e.g. see Adkison 1995; Hansen *et al.* 2002). However, the most obvious way of addressing the issue of local adaptation is to try to demonstrate it in the wild, either by generally evaluating whether individuals from a population exhibit higher fitness in their 'home' environment than in a 'foreign' environment, or by more specifically identifying traits that result in higher fitness in the native environment. Spectacular examples of the first approach are found in the former Soviet Union, where huge transplantation experiments with Pacific salmon species have been conducted, in many cases demonstrating lower fitness of the transplanted populations in the 'foreign' environment (Altukhov *et al.* 2000). The other approach, aimed at identifying traits reflecting local adaptations, has been the most common. There are by now several examples available of differences among populations in traits that are assumed to represent local adaptations (e.g. reviewed by Taylor 1991; Palm and Ryman 1999). In many cases, however, definitive proof of local adaptation is lacking, as there are few studies demonstrating that differences in traits among populations are maintained by different selection pressures in different environments. A notable exception is a recent study by Koskinen *et al.* (2002). They used a combination of neutral microsatellite markers and analysis of quantitative traits and showed that differences in life-history traits among introduced grayling *Thymallus thymallus* populations, derived from a common source population, must be the result of strong local selection rather than drift. The study by Koskinen *et al.* (2002) represents an approach that is finding increased use in studies of local adaptation and demonstration of selection. This involves comparison of F_{ST} at neutral markers and Q_{ST}, a measure of differentiation among populations at quantitative traits, analogous to F_{ST} (reviewed by Merilä and Crnokrak 2001 and McKay and Latta 2002). In principle, if Q_{ST} is higher than F_{ST} then this would indicate local selection acting on the trait. However, there are many possible exceptions to this rule and the precise relationships between F_{ST} and Q_{ST} is currently being debated (e.g. Hendry 2002).

Yet another approach for demonstrating local selection that is expected to be of much future importance consists of directly analysing quantitative trait loci (QTLs). In practical terms this means analysis of polymorphic molecular markers, such as microsatellites, that are tightly linked to a polymorphic locus that has an effect on a quantitative trait. QTLs are currently being identified in a number of species of commercial and/or scientific interest, including salmonid species such as Atlantic salmon and rainbow trout (e.g. Perry *et al.* 2001). Finally, there is the possibility of directly analysing coding nuclear loci that are presumably subject to selection. Analysis of MHC genes is an obvious example that has been mentioned previously. Even though most traits involved in local adaptations are presumably quantitative traits, the developments in genomics will no doubt lead to the discovery of more single loci of significant ecological importance, and it will be of much interest to understand the function and distribution of different alleles in different environments.

There can be little doubt that one of the most important future research themes will be the geographical distribution and genetic architecture of local adaptation. This calls for a unification of disciplines that have previously evolved along somewhat different lines, such as molecular population genetics and quantitative genetics, and it will become important to have a broader knowledge of the fields and to be able to 'speak each other's language'.

4.7 Summary and conclusions

- Most living organisms, including Atlantic salmon, are organised into units, populations, which consist of individuals that are more likely to reproduce with each other than with individuals from other populations. Understanding the genetic population structure of a species is imperative for proper management and conservation, as populations are the basic units of recruitment and as individual populations may be adapted to the local environment.
- The evolutionary forces that shape the distribution of genetic variation are mutation, natural selection, random genetic drift and migration (gene flow). Random genetic drift depends on the crucial parameter of effective population size, a measure of the number of individuals that effectively pass their genes on to the next generation.
- Populations can be organised in different ways according to the patterns of gene flow, and various theoretical models have been described, including the island model, stepping-stone model and metapopulation models, where populations may go extinct followed by recolonisation of empty patches from other populations. Genetic differentiation among populations can be quantified by the parameters F_{ST} or G_{ST}.
- Population genetics focuses on variation due to segregation of alleles at single loci at the population level. However, most traits of interest in natural populations are complex quantitative traits that are influenced by many genes and by environmental factors, and this is the focus of the discipline of quantitative genetics. Heritability is a fundamental concept in quantitative genetics and measures the proportion of the total observed variation for a particular trait, in a given environment, which is explained by the additive effects of genes. Another fundamental concept is genotype by environment interaction, i.e. genotypes may perform differently in different environments. The disciplines of molecular population genetics and quantitative genetics are becoming increasingly integrated, e.g. for studying local adaptation.
- A number of different genetic markers have been used for characterising wild populations, including allozyme electrophoresis, mitochondrial DNA and microsatellite DNA. Microsatellite DNA is currently the most useful and widely applied type of marker, but new types of marker and screening techniques are continually being developed.
- The use of genetic markers has yielded unprecedented opportunities for population genetics and ecological research and for practical management of populations. Some of the applications that are most relevant for salmonid fish include mixed-stock analysis, where the contribution of individual populations to mixed fisheries can be estimated, assignment tests, where the most likely population of origin of an individual can be determined, estimation of effective population size and detection of recent population bottlenecks, parentage assignment, where the parents of an individual can be identified using genetic markers, and estimation of relatedness, where it can be determined if individuals are likely to be, e.g. full-sibs, half-sibs or unrelated.

- The developments in both molecular biology and in statistical genetics are proceeding at a very fast pace and will undoubtedly lead to even more improved research opportunities. The genetic basis of local adaptations is likely to be a central research theme and will require integration of molecular population genetics and quantitative genetics.

References

Adkison, M. (1995) Population differentiation in Pacific salmon: local adaptation, genetic drift or the environment? *Canadian Journal of Fisheries and Aquatic Sciences*, **52**: 2762–2777.

Altukhov, Y.P., Salmenkova, E.A. and Omelchenko, V.T. (2000) *Salmonid Fishes: Population biology, genetics and management*. Blackwell Science, Oxford.

Avise, J.C. (1994) *Molecular Markers, Natural History and Evolution*. Chapman & Hall, New York.

Birky, C.W. Jr, Fuerst, P. and Maruyama, T. (1989) Organelle gene diversity under migration, mutation and drift: equilibrium expectations, approach to equilibrium, effects of heteroplasmic cells and comparison to nuclear genes. *Genetics*, **121**: 613–627.

Brown, G.E. and Brown, J.A. (1996) Kin discrimination in salmonids. *Reviews in Fish Biology and Fisheries*, **6**: 201–219.

Campbell, D., Duchesne, P. and Bernatchez, L. (2003) AFLP utility for population assignment studies: analytical investigation and empirical comparison with microsatellites. *Molecular Ecology*, **12**: 1979–1991.

Clifford, S.L., McGinnity, P. and Ferguson, A. (1998) Genetic changes in Atlantic salmon (*Salmo salar*) populations of northwest Irish rivers resulting from escapes of adult farm salmon. *Canadian Journal of Fisheries and Aquatic Sciences*, **55**: 358–363.

Cornuet, J.M. and Luikart, G. (1996) Description and power analysis of two tests for detecting recent population bottlenecks from allele frequency data. *Genetics*, **144**: 2001–2014.

Dannewitz, J., Maes, G.E., Johansson, L., Wickström, H., Volckaert, F.A.M. and Järvi, T. (2005) Panmixia in the European eel: a matter of time . . . *Proceedings of the Royal Society, London, Series B: Biological Sciences*, **272**: 1129–1137.

Di Rienzo, A., Peterson, A.C., Garza, J.C., Valdès, A.M., Slatkin, M. and Freimer, N.B. (1994) Mutational processes of simple-sequence repeat loci in human populations. *Proceedings of the National Academy of Sciences USA*, **91**: 3166–3170.

Falconer, D.S. and Mackay, T.F.C. (1996) *Introduction to Quantitative Genetics* (4th edn). Longman Group, Harlow, UK.

Fiumera, A.C., Porter, B.A., Grossman, G.D. and Avise, J.A. (2002) Intensive genetic assessment of the mating system and reproductive success in a semi-closed population of the mottled sculpin, *Cottus bairdi*. *Molecular Ecology*, **11**: 2367–2377.

Fleming, I.A., Hindar, K., Mjolnerod, I.B., Jonsson, B., Balstad, T. and Lamberg, A. (2000) Lifetime success and interactions of farm salmon invading a native population. *Proceedings of the Royal Society of London Series B Biological Sciences*, **267**: 1517–1523.

Frankham, R. (1995) Effective population size/adult population size ratios in wildlife: a review. *Genetical Research, Cambridge*, **66**: 95–107.

Frankham, R., Ballou, J.D. and Briscoe, D.A. (2002) *Introduction to Conservation Genetics*. Cambridge University Press, Cambridge.

Franklin, I.R. (1980) Evolutionary change in small populations. In: M.E. Soulé and B.A. Wilcox (Ed.) *Conservation Biology: An evolutionary-ecological perspective*, pp. 135–150. Sinnauer Associates, Sunderland, MA.

Garza, J.C. and Williamson, E.G. (2001) Detection of reduction in population size using data from microsatellite loci. *Molecular Ecology*, **10**: 305–318.

Hansen, M.M. (2002) Estimating the long-term effects of stocking domesticated trout into wild brown trout (*Salmo trutta*) populations: an approach using microsatellite DNA analysis of historical and contemporary samples. *Molecular Ecology*, **11**: 1003–1015.

Hansen, M.M. and Jensen, L.F. (2005) Sibship within samples of brown trout (*Salmo trutta*) and implications for supportive breeding. *Conservation Genetics*, **6**: 297–305.

Hansen, M.M., Ruzzante, D.E., Nielsen, E.E., Bekkevold, D. and Mensberg, K.-L.D. (2002) Long-term effective population sizes, temporal stability of genetic composition and potential for local adaptation in anadromous brown trout (*Salmo trutta*) populations. *Molecular Ecology*, **11**: 2523–2535.

Hanski, I. and Gilpin, M. (1991) Metapopulation dynamics: brief history and conceptual domain. In: M. Gilpin and I. Hanski (Ed.) *Metapopulation Dynamics: Empirical and theoretical investigations*, pp. 3–16. Academic Press, London.

Hendry, A.P. (2002) QST > = ≠ < FST? *Trends in Ecology and Evolution*, **17**: 502.

Herbinger, C.M., O'Reilly, P.T. and Verspoor, E. (2006) Unravelling first generation pedigrees in wild endangered salmon populations using molecular genetic markers. *Molecular Ecology*, **15**: 2261–2275.

Hill, W.G. (1981) Estimation of effective population size from data on linkage disequilibrium. *Genetical Research*, **38**: 209–216.

Hill, W.G. and Zhang, X.-S. (2004) Genetic variation within and among animal populations. In: G. Simm, B. Villanueva, K.D. Sinclair and S. Townsend (Ed.) *Farm Animal Genetic Resources*, BSAS Occasional Publication No. 30, pp. 67–84. Nottingham University Press, Nottingham, UK.

Jarne, P. and Lagoda, P.J.L. (1996) Microsatellites, from molecules to populations and back. *Trends in Ecology and Evolution*, **11**: 424–429.

Jones, A.G. and Avise, J.C. (1997) Microsatellite analysis of maternity and the mating system in the Gulf pipefish *Syngnathus scovelli*, a species with male pregnancy and sex-role reversal. *Molecular Ecology*, **6**: 203–213.

Jorde, P.E. and Ryman, N. (1995) Temporal allele frequency change and estimation of effective size in populations with overlapping generations. *Genetics*, **139**: 1077–1090.

Kimura, M. (1983) *The Neutral Theory of Molecular Evolution*. Cambridge University Press, Cambridge.

Kimura, M. and Weiss, G.H. (1964) The stepping stone model of population structure and the decrease of genetic correlation with distance. *Genetics*, **49**: 561–576.

King, M. (1995) *Fisheries Biology: Assessment and management*. Fishing News Books, Oxford.

Knight, M.E., Van Oppen, M.J.H., Smith, H.L., Rico, C., Hewitt, G.M. and Turner, G.F. (1999) Evidence for male-biased dispersal in Lake Malawi cichlids from microsatellites. *Molecular Ecology*, **8**: 1521–1227.

Koskinen, M.T., Haugen, T.O. and Primmer, C.R. (2002) Contemporary Fisherian life-history evolution in small salmonid populations. *Nature*, **419**: 826–830.

Kruuk, L.E.B., Clutton-Brock, T.H., Slate, J., Pemberton, J.M., Brotherstone, S. and Guinness, F.E. (2000) Heritability of fitness in a wild mammal population. *Proceedings of the National Academy of Sciences, USA*, **97**: 698–703.

Landegren, U., Nilsson, M. and Kwok, P.Y. (1998) Reading bits of genetic information: methods for single-nucleotide polymorphism analysis. *Genome Research*, **8**: 769–776.

Landry, C., Garant, D., Duchesne, P. and Bernatchez, L. (2001) 'Good genes as heterozygosity': the major histocompatibility complex and mate choice in Atlantic salmon (*Salmo salar*). *Proceedings of the Royal Society of London, Series B Biological Sciences*, **268**: 1279–1285.

Langefors, A., Lohm, J., Grahn, M., Andersen, O. and von Schantz, T. (2001) Association between major histocompatibility complex class IIB alleles and resistance to *Aeromonas salmonicida* in Atlantic salmon. *Proceedings of the Royal Society of London, Series B Biological Sciences*, **268**: 479–485.

Letcher, B.H. and King, T.L. (2001) Parentage and grandparentage assignment with known and unknown matings: application to Connecticut River Atlantic salmon restoration. *Canadian Journal of Fisheries and Aquatic Sciences*, **58**: 1812–1821.

Luikart, G. and England, P.R. (1999) Statistical analysis of microsatellite DNA data. *Trends in Ecology and Evolution*, **14**: 253–256.

Luikart, G., Sherwin, W.B., Steele, B.M. and Allendorf, F.W. (1998) Usefulness of molecular markers for detecting population bottlenecks via monitoring genetic change. *Molecular Ecology*, **7**: 963–974.

Lynch, M. and O'Healy, M. (2001) Captive breeding and the genetic fitness of natural populations. *Conservation Genetics*, **2**: 363–378.

Martinez, J.L., Moran, P., Perez, J., De Gaudemar, B., Beall, E. and García-Vázquez, E. (2000) Multiple paternity increases effective size of southern Atlantic salmon populations. *Molecular Ecology*, **9**: 293–298.

McGinnity, P., Stone, C., Taggart, J.B., Cooke, D., Cotter, D., Hynes, R., McCamley, C., Cross, T. and Ferguson, A. (1997) Genetic impact of escaped farmed Atlantic salmon (*Salmo salar* L.) on native populations: use of DNA profiling to assess freshwater performance of wild, farmed, and hybrid progeny in a natural river environment. *ICES Journal of Marine Science*, **54**: 998–1008.

McKay, J.K. and Latta, R.G. (2002) Adaptive population divergence: markers, QTL and traits. *Trends in Ecology and Evolution*, **17**: 285–291.

Merilä, J. and Crnokrak, P. (2001) Comparison of genetic differentiation at marker loci and quantitative traits. *Journal of Evolutionary Biology*, **14**: 892–903.

Mork, J. (1991) One-generation effect of farmed fish immigration on the genetic differentiation of wild Atlantic salmon in Norway. *Aquaculture*, **98**: 267–276.

Mousseau, T.A., Ritland, K. and Heath, D.D. (1998) A novel method for estimating heritability using molecular markers. *Heredity*, **80**: 218–224.

Mueller, U.G. and Wolfenbarger, L.L. (1999) AFLP genotyping and fingerprinting. *Trends in Ecology and Evolution*, **14**: 389–394.

Nei, M. (1973) Analysis of gene diversity in subdivided populations. *Proceedings of the National Academy of Sciences, USA*, **70**: 3321–3323.

Nielsen, E.E., Hansen, M.M. and Loeschcke, V. (1999) Genetic variation in time and space: microsatellite analysis of extinct and extant populations of Atlantic salmon. *Evolution*, **53**: 261–268.

Palm, S. and Ryman, N. (1999) Genetic basis of phenotypic differences between transplanted stocks of brown trout. *Ecology of Freshwater Fish*, **8**: 169–180.

Pearse, D.E., Eckerman, C.M., Janzen, F.J. and Avise, J.A. (2001) A genetic analogue of 'mark-recapture' methods for estimating population size: an approach based on molecular parentage assessments. *Molecular Ecology*, **10**: 2711–2718.

Pella, J.J. and Milner, G.B. (1987) Use of genetic marks in stock composition analysis. In: N. Ryman and F. Utter (Ed.) *Population Genetics and Fishery Management*, pp. 247–276. University of Washington Press, Seattle, WA.

Pemberton, J.M., Slate, J., Bancroft, D.R. and Barrett, J.A. (1995) Non-amplifying alleles at microsatellite loci: a caution for parentage and population studies. *Molecular Ecology*, **4**: 249–252.

Perry, G.M.L., Danzmann, R.G., Ferguson, M.M. and Gibson, J.P. (2001) Quantitative trait loci for upper thermal tolerance in outbred strains of rainbow trout (*Oncorhynchus mykiss*). *Heredity*, **86**: 333–341.

Primmer, C.R., Koskinen, M.T. and Piironen, J. (2000) The one that did not get away: individual assignment using microsatellite data detects case of fishing competition fraud. *Proceedings of the Royal Society of London, Series B*, **267**: 1699–1704.

Prodöhl, P.A., Taggart, J.B. and Ferguson, A. (1995) A panel of minisatellite (VNTR) DNA locus specific probes for potential application to problems in salmonid aquaculture. *Aquaculture*, **137**: 87–97.

Rousset, F. (1997) Genetic differentiation and estimation of gene flow from *F*-statistics under isolation by distance. *Genetics*, **145**: 1219–1928.

Rousset, F. (2002) Inbreeding and relatedness coefficients: what do they measure? *Heredity*, **88**: 371–380.

Sakamoto, T., Danzmann, R.G., Gharbi, K., Howard, P., Ozaki, A., Khoo, S.K., Woram, R.A., Okamoto, N., Ferguson, M.M., Holm, L.E., Guyomard, R. and Hoyheim, B. (2000) A microsatellite linkage map of rainbow trout (*Oncorhynchus mykiss*) characterized by large sex-specific differences in recombination rates. *Genetics*, **155**: 1331–1345.

Smith, B.R., Herbinger, C.M. and Merry, H.R. (2001) Accurate partition of individuals into full-sib families from genetic data without parental information. *Genetics*, **158**: 1329–1338.

Ståhl, G. (1987) Genetic population structure of Atlantic salmon. In: N. Ryman and F. Utter (Ed.) *Population Genetics and Fishery Management*, pp. 121–140. University of Washington Press, Seattle, WA.

Sunnucks, P., Wilson, A.C.C., Beheregaray, L.B., Zenger, K., French, J. and Taylor, A.C. (2000) SSCP is not so difficult: the application and utility of single-stranded conformation polymorphism in evolutionary biology and molecular ecology. *Molecular Ecology*, **9**: 1699–1710.

Taggart, J.B., McLaren, I.S., Hay, D.W., Webb, J.H. and Youngson, A.F. (2001) Spawning success in Atlantic salmon (*Salmo salar* L.): a long-term DNA profiling-based study conducted in a natural stream. *Molecular Ecology*, **10**: 1047–1060.

Tave, D. (1993) *Genetics for Fish Hatchery Managers* (2nd edn). Van Nostrand Reinhold, New York.

Taylor, E.B. (1991) A review of local adaptation in Salmonidae, with particular reference to Pacific and Atlantic salmon. *Aquaculture*, **98**: 185–207.

Templeton, A.R. (1998) Nested clade analysis of phylogeographic data: testing hypotheses about gene flow and population history. *Molecular Ecology*, **7**: 381–397.

Tessier, N. and Bernatchez, L. (1999) Stability of population structure and genetic diversity across generations assessed by microsatellites among sympatric populations of landlocked Atlantic salmon (*Salmo salar* L.). *Molecular Ecology*, **8**: 169–179.

Utter, F.M., Aebersold, P. and Winans, G. (1987) Interpreting genetic variation detected by electrophoresis. In: N. Ryman and F. Utter (Ed.) *Population Genetics and Fishery Management*, pp. 21–46. University of Washington Press, Seattle, WA.

Van de Casteele, T., Galbusera, P. and Matthysen, E.A. (2001) Comparison of microsatellite-based pairwise relatedness estimators. *Molecular Ecology*, **10**: 1539–1549.

Verspoor, E. and Jordan, W.C. (1989) Genetic variation at the *Me-2* locus in the Atlantic salmon within and between rivers: evidence for its selective maintenance. *Journal of Fish Biology*, **35A**: 205–213.

Wang, J. (2004) Sibship reconstruction from genetic data with typing errors. *Genetics*, **166**: 1963–1979.

Waples, R.S. (1989) A generalized approach for estimating effective population size from temporal changes in allele frequency. *Genetics*, **121**: 379–391.

Waples, R.S. (1991) Genetic methods for estimating the effective size of cetacean populations. In: A.R. Hoelzel (Ed.) *Genetic Ecology of Whales and Dolphins, Report from the International Whaling Commission*, Special Issue 13, pp. 279–300.

Waples, R.S. (1998) Separating the wheat from the chaff: patterns of genetic differentiation in high gene flow species. *Journal of Heredity*, **89**: 438–450.

Wilson, A.J. and Ferguson, M.M. (2002) Molecular pedigree analysis in natural populations of fishes: approaches, applications, and practical considerations. *Canadian Journal of Fisheries and Aquatic Sciences*, **59**: 1696–1707.

Wirth, T. and Bernatchez, L. (2001) Genetic evidence against panmixia in the European eel. *Nature*, **409**: 1037–1040.

Wright, S. (1931) Evolution in Mendelian populations. *Genetics*, **16**: 97–159.

Part II
Population Genetics

5 Biodiversity and Population Structure

T. L. King[1], E. Verspoor[1], A. P. Spidle, R. Gross, R. B. Phillips, M-L. Koljonen, J. A. Sanchez and C. L. Morrison

North Atlantic rivers with populations of Atlantic salmon. Photos credit: clockwise from top left, 1, 3, 8, 9, 10, 11, 12 (E. Verspoor), 2, 4, 6, 7 (J. Webb), 5 (K.M. Ebert, Danish Anglers Association), central photo (R. Saunders, US NOAA Fisheries), globe image (GEBCO Digital Atlas (Centenary Edition), published on CD-ROM on behalf of the Intergovernmental Oceanographic Commission and the International Hydrographic Organization as part of the General Bathymetric Chart of the Oceans, British Oceanographic Data Centre, Liverpool, UK).

[1] Joint first authors.

The range of the Atlantic salmon spans thousands of kilometres across the North Atlantic. Throughout the species range, freshwater spawning habitat is highly fragmented such that the species is subdivided into large numbers of spatially disconnected groups of breeders. Between rivers, these groups show limited mixing as anadromous salmon home to natal river systems to spawn after their marine migrations and resident forms are isolated by their non-anadromous behaviour. Even within river systems anadromous and resident non-anadromous forms are generally isolated by physical barriers such as impassable falls, and where barriers are absent by the often discontinuous distribution of spawning and juvenile habitat. Collectively, these aspects of the salmon's biology give good reason to suspect that the species is partitioned into numerous distinct genetic populations. Yet, in apparent contradiction, the species shows little morphological differentiation. Studies of molecular variation indicate the latter characteristic masks a considerable level of evolutionary diversity and population structuring.

5.1 Introduction

Understanding of evolutionary genetic diversity and population structure in the Atlantic salmon has advanced significantly over the last 40 years, largely from research into the spatial and temporal distribution of molecular genetic variation. The work has delivered insights far beyond those gained from studies of variation in morphology and behaviour. The insights which have been gained from molecular genetic work are considered here.

Studies of molecular genetic variation have proven to be much more informative than studies of morphology and behaviour because the distribution of genetic variation is a direct consequence of the way in which a species is structured into populations, by the gene flow among populations, and by the extent of their adaptive divergence. Unlike a species' morphology and behaviour, its molecular genetic character does not change with developmental stage or environmental conditions. Therefore, using the framework of population genetics theory (Chapter 4), robust inferences can be drawn about population structuring from the distribution of molecular variation. Furthermore, the pattern of distribution is also highly informative about the historical processes underlying its evolution. The pattern will be different depending on the relative importance of genetic drift, gene flow, mutation and natural selection, and on the length of time during which salmon in different locations have been separated.

The studies carried out on the Atlantic salmon encompass a number of classes of molecular variation. These include chromosome structure (i.e. cytogenetics), protein variation (an indirect assessment of variation at nuclear protein coding genes), mitochondrial DNA (mtDNA) haplotype variation (reflecting the distribution of maternal lineages) and variation in nuclear DNA composition; much of this latter work, as discussed in Chapter 4, has been focused on the analysis of microsatellite loci. Recent reviews have been carried out of cytogenetic studies (Hartley 1987), studies of protein variation (Verspoor *et al.* 2005) and work on mtDNA (Gross *et al.* in prep.). In all cases, the work has focused largely on the analysis of the patterns of differentiation among and within river systems.

The loci represented by the molecular variation studied to date represent only a small part of the Atlantic salmon genome, and are a largely arbitrary selection. Marker choice has been dictated, for the most part, by available methodologies. That said, the variation at most of the nuclear loci (i.e. allozymes and microsatellites) studied to date appears to be unlinked (i.e. it assorts independently each generation). This suggests that they are widely distributed throughout the salmon's nuclear genome and the variation can serve as 'molecular markers' for the genome as a whole. This will also be the case for mtDNA variation studied in relation to the mitochondrial genome, as the molecule behaves as a single, maternally inherited, haploid locus due to the absence of recombination (Chapter 3).

The analysis of spatial and temporal patterns of molecular variation has provided insight into the genetic nature of the Atlantic salmon at three evolutionary levels. At the broadest level, insight has been gained into the evolutionary relationships among fish assigned to *Salmo salar* (Linnaeus), and fish placed in other taxa, an area of biology known as phylogenetics. In the case of Atlantic salmon, the main interest is in how the species is evolutionarily related to other members of the subfamily Salmoninae and the genus *Salmo*. This work has informed the evolutionary validity of taxonomic assignments of fish made on the basis of morphological criteria, and to establish the taxonomic validity and evolutionary distinctiveness of *S. salar* L. The focus of this chapter, however, is on evolutionary relationships among salmon from different parts of the range, within *S. salar*, and the structuring of those salmon into genetic populations. In other words, the main focus is on the nature and extent of intraspecific biodiversity.

Initially the chapter deals with the division of the species into evolutionary (i.e. phylogenetic) lineages, how these are distributed geographically, and how they developed historically. This aspect of the species' biology is referred to as its phylogeography, and evolves as a result of the historical reproductive isolation of salmon populations. Where gene flow is absent, populations become independent evolutionary lineages and diverge genetically due to genetic drift, genetic mutations and natural selection. Where independent sets of lineages come to occupy distinct geographical areas, they develop into distinct phylogeographic groups which diverge from other such groups in proportion to the time they are isolated from each other. Over time, older phylogeographic groups may be subdivided with more recently evolved, and less divergent, phylogeographic groups nested within them. Where populations have been isolated relatively recently, phylogeographic divergence may be limited and difficult to detect.

Within phylogeographic groups, species can be further structured into distinct genetic populations. A consideration of the nature of this substructuring is the focus of the second part of the chapter. The occurrence of population structuring in the absence of phylogeographic divergence is expected where gene flow, though not absent, is highly restricted or sufficiently sporadic that there is a subdivision of the species into breeding groups which are more or less distinct but, due to the gene flow, are not distinct evolutionary lineages. In such situations, divergence at effectively neutral loci (i.e. loci more affected by genetic drift than natural selection) will be limited or transitory, and populations which go extinct are likely to be re-established by migrants from other populations. Populations connected in this way are described as a metapopulation. Component populations of a metapopulation, though they may show only limited or transient phylogeographic divergence, may still have significant adaptive genetic divergence as a result of local variation in selective pressures (Chapter 7).

The consideration of the species at these three levels of genetic organisation reflects fundamental evolutionary and biological differences in the relationships of populations to each other, and it is critical that these differences in the relationships of salmon populations are taken into account in the species' management and conservation. However, it must also be recognised that there is a broader continuum of relatedness within and between populations (within species) than there is between species. Statistically significant differences may be detected between close intraspecific groups and they must be closely examined to determine whether or not the statistical significance of neutral genetic differentiation reflects true biological differentiation important to management decisions. Furthermore, as the historical geography of a species changes, so may the genetic nature of its populations, as patterns of gene flow and isolation are altered. Just as salient, these characteristics may be changed as a result of human activities, for example, habitat alteration or stocking activities. In general, such activities, in management time frames at least, can lead to the loss of intraspecific biodiversity.

5.2 Evolutionary relatedness to other salmonids

The Atlantic salmon is placed in the Salmoninae, one of many taxonomic subfamilies of fish defined by biologists. The fish in this family share many fundamental aspects of their biology, giving rise to the view that they represent an evolutionary lineage distinct from other fishes. However, most species assigned to this family still appear to be separated by millions of years of independent evolution.

Establishing the evolutionary relatedness of species within this family of fishes has proven problematic and, within the Salmonidae, some relationships still remain unresolved (Crespi and Fulton 2004). These difficulties probably arise for a number of reasons. The first is the group's rapid adaptive radiation following what appears to have been the tetraploidisation of their common ancestor 50–100 million years ago (Allendorf and Thorgaard 1984). Other reasons are the occurrence of interspecific hybridisation and introgression (Utter and Allendorf 1994) and, in many areas, the recent and rapid evolution of populations in habitat which has emerged following the retreat of Pleistocene glaciers. With regard to the latter factor, many areas have assemblages of sympatric morphotypes within the same species that show high though varying degrees of reproductive isolation (Phillips and Oakley 1997).

The family Salmonidae currently consists of three subfamilies. These are the Coregoninae (whitefishes and ciscoes), Thymallinae (graylings) and Salmoninae (lenoks, huchen, trouts, charrs and salmons), whose phylogenetic relationships are detailed in Box 5.1. The Salmoninae, within which the Atlantic salmon is placed, are viewed as being composed of between five and nine genera (reviewed in Phillips and Oakley 1997 and Crespi and Fulton 2004). One of these is *Salmo*, the genus within which the Atlantic salmon is placed along with the brown trout *S. trutta* L.

Brown trout and Atlantic salmon appear to be closely related based on both morphological and DNA sequence affinities. However, in contrast to the brown trout, the Atlantic salmon is seen as monophyletic (i.e. being a single evolutionary lineage), with only a single species *S. salar* defined. The brown trout is seen as polyphyletic with several distinct types of brown trout recognised, including the marbled trout, *S. marmorata*. These types, which are likely to be relatively recently diverged, are collectively viewed as part of the *Salmo trutta* complex (reviewed in Bernatchez 2001; Presa *et al.* 2002).

Rainbow and cutthroat trout endemic to the Pacific Northwest, as mentioned in Chapter 2, were originally placed in the genus *Salmo*, despite their having similarities with the Pacific salmon in both morphology and life history (Regen 1914). Further research has shown that they are part of the same evolutionary lineages as Pacific salmon. As such, in 1988, they were reclassified into the genus *Oncorhynchus* by the AFS-ASIH Committee on Names of Fishes (Smith and Stearley 1989).

On the other hand, recent DNA analyses (Phillips *et al.* 2000; Snoj *et al.* 2002; Crespi and Fulton 2004) indicate that *Acantholingua ohridana*, a salmonid endemic to Lake Ohrid in

> **Box 5.1** Phylogeny of the family Salmonidae.
>
> The family Salmonidae appears to have evolved from a common ancestor following genome duplication (Allendorf and Thorgaard 1984). This origin would account for why the group has approximately twice the amount of DNA per cell as other closely related families. A tetraploid ancestry (i.e. composed of four sets of chromosomes) would also account for the presence in most species in the family of duplicate allozyme loci. The initially tetraploid genome has now largely evolved in each species to the point where it again behaves as a diploid. However, at some loci there are residual inheritance patterns typical of a tetraploid genome, an effect known as pseudolinkage (Wilkins 1972a,b; Allendorf and Danzmann 1997) (see Chapter 3).
>
> The analysis of duplicated genes, such as the growth hormone genes *GH1* and *GH2*, shows that sequences from species in the subfamily Salmoninae, within the Salmonidae, belong to two evolutionary groups (Phillips *et al.* 2003). This is consistent with the gene duplication preceding the species radiation (Devlin 1993; Phillips *et al.* 2003), and suggests that the radiation of the Salmonidae may have been driven by adaptive benefits of the genome duplication event.
>
> The family Salmonidae is subdivided into three subfamilies: Coregoninae (whitefishes and ciscoes), Thymallinae (graylings) and Salmoninae (lenoks, huchen, trouts, charrs and salmons). The Salmoninae, within which the Atlantic salmon is placed, has between five and nine genera (reviewed in Phillips and Oakley 1997 and Crespi and Fulton 2004): *Brachymystax, Hucho, Parahucho, Salvelinus, Salmo, Oncorhynchus, Salvethymus, Salmothymus* and *Acantholingua*. The genus *Salvethymus* was erected for a single morphologically divergent endemic species in Lake El'gygytgyn in Russia. However, allozyme and cytogenetic data suggest a close relationship to Arctic charr and some have proposed that the species be placed in the genus *Salvelinus* (Glubokovsky and Frolov 1994). The genus *Parahucho* was also erected for a single species, *Hucho perryi*, which has been shown to be quite different from *Hucho hucho*. In this case, cytogenetic (Rab *et al.* 1994), allozyme (Osinov 1991) and DNA sequence variation all support this as a monotypic genus (Phillips and Oakley 1997).
>
> The genus *Acantholingua* contains a single species (*A. ohridana*) endemic to Lake Ohrid in Macedonia (Hadzisce 1961). However, Svetovidov (1975) placed it in the genus *Salmothymus*, which also included *S. obtusirostris*, a widely distributed species in the Adriatic drainage. Behnke (1965) suggested that *Salmothymus* should be a subgenus of *Salmo*. Because of its unusual and apparently primitive morphology, Stearley and Smith (1993) considered these species to be 'archaic trouts' which they placed in a basal position in the Salmoninae. However, analysis of both nuclear and mitochondrial data have shown that *A. ohridana* is closely related to brown trout, *Salmo trutta*, supporting its inclusion in the genus *Salmo* (Phillips *et al.* 2000; Crespi and Fulton 2004). Analysis of additional molecular data from *Salmothymus obtusirostris* produced a tree with these two taxa as sister species and brown trout on an adjacent branch, supporting their inclusion in the genus *Salmo* (Snoj *et al.* 2002).
>
> The evolutionary relationships of the genera in the Salmoninae remain uncertain. The molecular data supports *Brachymystax/Hucho* as the basal taxa in the subfamily. Morphological data supports *Salmo* and *Oncorhynchus* as sister taxa (Norden 1961), but most of the recent molecular data supports *Oncorhynchus* and *Salvelinus* as sister taxa (Buisine *et al.* 2002; Oakley and Phillips 1999; Crespi and Fulton 2004).

Macedonia, and *Sternopygus obtusirostris*, a widely distributed species in Adriatic drainages, should be placed in the genus *Salmo* with *S. salar* and *S. trutta* (see Box 5.1 for more detail). Thus molecular genetic analysis places Atlantic salmon in a higher-order evolutionary grouping, defined by the genus *Salmo*, with three other species (Fig. 5.1), with these new species in the genus most closely related to brown trout.

5.3 Phylogeographic diversity

Atlantic salmon are a distinct evolutionary lineage from other salmonid fishes. However, studies of molecular genetic variation also show significant intraspecific phylogenetic diversity within this lineage. This is not surprising. Nearctic and Palearctic fish species, such as the

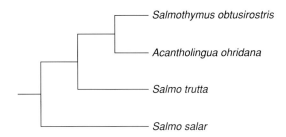

Fig. 5.1 Phylogenetic relationships within the designated genus *Salmo*. Based on Crespi and Fulton (2004).

Atlantic salmon, have a history of repeated range contractions and expansions during the Quaternary period, caused by the advance and retreat of glaciers and associated climatic shifts (Hewitt 1996, 1999). For such species, each range contraction would have been characterised by extinctions of northern populations, when the temperature decreased and ice advanced, and each northward expansion, as temperatures increased and glaciers receded, with the establishment of new populations by migrants from southern refugia (Taberlet *et al.* 1998). This process interacts with a species' basic biology, and historical geographical factors, to determine the nature and extent of modern stock structure in geographically isolated groups of intraspecific lineages (Bernatchez and Wilson 1998; Verspoor *et al.* 2002).

Holarctic rivers and lakes in previously glaciated areas have been inhabited by salmonid fishes for only about 8000–40 000 years, depending on when glaciers receded (Behnke 1972). For the Atlantic salmon, this represents a few thousand generations such that genetic differentiation between populations might, on first consideration, be expected to be minimal. Yet marked ecological specialisations and reproductive isolation can evolve in such time frames, particularly if there is physical isolation and differential selection across environments (Orr and Smith 1998). In Lake Victoria in Africa, for example, the last ~ 14 000 years has seen the evolution of cichlid fishes into hundreds of distinct morphological species (Seehausen 2002). Morphological diversification has not been detected in the Atlantic salmon (Claytor *et al.* 1991) but there is considerable diversity with regard to physiological, behavioural and life-history traits (Chapter 2), much of which is heritable (Chapter 7).

The molecular analyses which shed light on the intraspecific phylogenetics and phylogeography of the Atlantic salmon have emerged from historical work aimed at meeting a well-defined management need. In the 1960s, tagging studies showed that salmon from the Eastern and Western Atlantic Ocean migrated to communal feeding grounds off West Greenland in the Davis Strait where they were exploited in a high-seas drift net fishery. To regulate exploitation more effectively, managers needed insight into the respective contributions made by European and North American stocks to the catch. This led to research on both phenotypic and genetic methods for discriminating Eastern and Western Atlantic populations. Work on genetic markers has, over the years, encompassed the analysis of heritable variation in chromosomes, proteins, mitochondrial DNA and nuclear DNA.

For molecular variation to be informative about species' phylogeography, the differentiation must be predominantly due to stochastic evolutionary forces (i.e. by genetic drift rather than natural selection). While most of the molecular variation studied to date in Atlantic salmon has been generally assumed to be caused by genetic drift, there is evidence to suggest that not all detectable variation is selectively neutral. The allozyme locus *MEP-2** has provided compelling evidence that differentiation is influenced by natural selection (Verspoor *et al.* 2005; Chapter 7). Vasemägi *et al.* (2005a,b) showed that, along with several expressed sequence tag-linked microsatellites, some genomic microsatellite loci also deviated from neutral expectations. However, where it has been possible to test whether selection is important in the evolution of loci, drift and gene flow/isolation are indicated to be the primary driving forces behind differentiation (Verspoor 1994; Jordan *et al.* 1997).

The molecular evidence for phylogeographic divergence in the Atlantic salmon is considered in the sections below, first in relation to anadromous salmon on a range-wide scale between continents and then, separately, within the Eastern and the Western Atlantic regions. This is followed by a review of what is known about the phylogeographic divergence of anadromous and non-anadromous salmon. In the final part of this section, consideration is given to the historical processes responsible for the observed phylogeographical divergence patterns.

5.3.1 Range-wide

Studies of chromosome numbers and structure using light microscopy show major differences between European and North American salmon. Atlantic salmon from Eastern Atlantic drainages characteristically show a diploid number of 58 with 29 chromosome pairs and 74 chromosome arms. In contrast, in Atlantic salmon from the Western Atlantic, a diploid number of 54 with 27 chromosome pairs and 72 chromosome arms appears to be typical (Hartley 1987). In addition to this gross difference, salmon in these two parts of the species range are also divergent in the detailed structure of their chromosomes (Box 5.2). However, this work has often been ignored, in part because studies of this class of variation has been relatively restricted in their geographical coverage, largely because of the need for live animals to be brought to the laboratory to obtain the metaphase chromosomes required for karyotyping (Chapter 3). Unfortunately, collecting and transporting live fish to laboratories from remote locations is both costly and difficult.

More widely based evidence for the phylogeographic divergence of Atlantic salmon in the Eastern and Western Atlantic parts of the species range has been obtained from electrophoretic studies of molecular variation. The first of these were reported in the late 1960s and early 1970s and involved the analysis of variation in transferrin, a serum protein. Payne *et al.* (1971) analysed the distribution of frequencies for the four alleles resolved, using data for nearly 4500 Atlantic salmon from Britain, Ireland and Canada. Samples from European rivers were found to be either fixed for the widespread allele, Tf_1, or to have a further allele, Tf_2, present at a low frequency (< 0.1%). In contrast, North American samples were highly polymorphic for Tf_1 and a third allele, Tf_4, found only in North America, with the latter often more common; a further rare allele, Tf_3, was also found in North American salmon. The observed qualitative and quantitative differentiation led the authors to conclude there was little or no gene exchange between the two sides of the Atlantic, and that this had been the case for a very long time.

Studies of allozymes carried out later, in the 1980s and 1990s, indicated that the differentiation seen for transferrin was widespread across the genome. First, Cross (1981) reported significant differentiation at four of six polymorphic loci which he screened in salmon from three European and two North American rivers. This was supported by Ståhl (1987) who

Box 5.2 Chromosomal divergence of European and North American Atlantic salmon.

The karyotype typical of most Atlantic salmon from Eastern Atlantic Ocean drainages has a diploid number of 58 with 29 chromosome pairs and NF of 74 (Fig. B5.2a) (Hartley and Horne 1984). This karyotype shows 8 pairs of metacentrics (chromosomes with centromere centrally located), 14 pairs of large acrocentrics (chromosomes with centromere toward, but not at, the end) and 7 pairs of small acrocentrics. The largest metacentric pair has a prominent C band about 75% of the way down the arm, which probably marks a site of a tandem fusion. The 14 large acrocentric pairs all have one interstitial C band in the centre of the chromosome, suggesting that they are the result of a fusion between two acrocentric chromosomes (Ueda and Kobayashi 1990; Phillips and Rab 2001). The largest acrocentric chromosome pair has two interstitial C bands, suggesting it is the result of tandem fusions between three acrocentric chromosomes.

The karyotype typical of most Atlantic salmon from the Western Atlantic has a diploid number of 54 with 27 chromosome pairs and NF of 72 (Fig. B5.2b; Roberts 1970). There are 9 pairs of metacentrics, 13 pairs of large acrocentrics and 5 pairs of small acrocentrics. The largest acrocentric chromosome appears identical to that in the karyotype of the salmon from the Eastern Atlantic. The differences between the Eastern and Western Atlantic karyotypes of Atlantic salmon include both Robertsonian and tandem fusions/fissions. Chromosome pair #9 in the Western Atlantic karyotype appears to be derived from two acrocentrics in the Eastern Atlantic karyotype as a result of a Robertsonian fusion. This would reduce the diploid chromosome number from 58 to 56. There have been reports of fish with 56 chromosomes in both Europe and North America (Boothroyd 1959; Hartley and Horne 1984). To account for the diploid number of 54, we assume there must have been another two Robertsonian fusions. The largest chromosome pair (#1) in the Eastern Atlantic karyotype appears to be the result of a tandem fusion, which would reduce NF. However, the Eastern Atlantic karyotype has a larger NF than the Western Atlantic karyotype, suggesting a minimum of four tandem fusions in the latter. Thus, the minimum number of chromosomal changes would be eight between the two karyotypes.

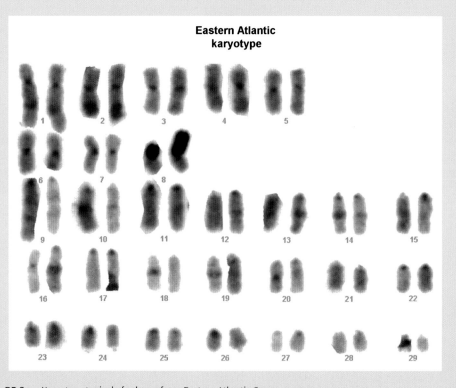

Fig. B5.2a Karyotype typical of salmon from Eastern Atlantic Ocean.

Box 5.2 (cont'd)

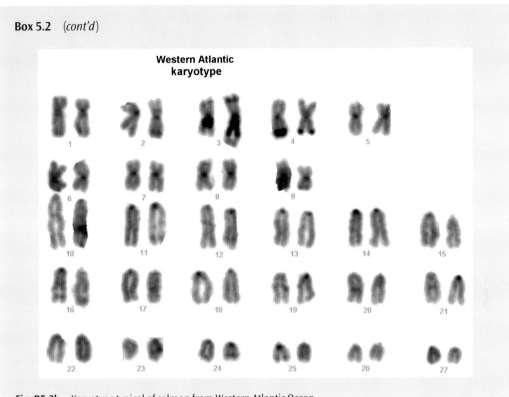

Fig. B5.2b Karyotype typical of salmon from Western Atlantic Ocean.

published results for a broader analysis of European populations and for variation in the tributaries of one of the primary Western Atlantic river systems, the Miramichi. Since then, populations from throughout both continental ranges of the salmon have been surveyed, encompassing more than 25 000 individuals from over 400 locations in > 200 river systems across the species distribution (Verspoor et al. 2005). In all cases, the substantive differentiation detected in the 1960s and 1970s has been confirmed (Verspoor 1988, 2005; Jordan et al. 1997).

The overall differentiation between the Eastern and Western Atlantic at allozyme loci has been recently evaluated using a set of representative rivers from across the species range (Fig. 5.2; Verspoor 2005), and shows intercontinental divergence relative to divergence among rivers within continents. On average, variation between continents ($F_{ST} = 0.330$) is over four times as great between Europe and North America compared to among rivers within North America ($F_{ST} = 0.076$), and almost twice as great compared to among rivers within Europe ($F_{ST} = 0.176$). This supports the view, first expressed by Payne et al. (1971), that there is a deep evolutionary divergence between populations on the two sides of the Atlantic.

At some allozyme loci, populations on one side of the Atlantic have allelic variation which is common or nearly fixed that is absent or rare in populations on the other side. For example, this occurs with *MDH-3,4** and *ME** (Verspoor et al. 2005), where North American populations have a common allele absent in Europe. At *EST-D**, North American and European populations show virtual fixation for alternate alleles. The allele typical of Western Atlantic

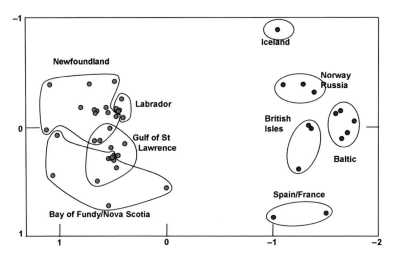

Fig. 5.2 A two-dimensional multidimensional scaling (MDS) plot of genetic distances (D_A) among Western and Eastern Atlantic Ocean populations of Atlantic salmon estimated derived from the analysis of 23 allozyme loci. Based on Verspoor (2005).

populations has only been found in Europe in the Kola Peninsula in northern Russia while the allele typical of the Eastern Atlantic is only found sporadically and at low frequency in North American populations on the island of Newfoundland (Makhrov *et al.* 2005; Verspoor, unpubl.).

The more recent but also extensive work on mtDNA variation supports deep phylogeographic division between salmon in the Eastern and Western Atlantic (Box 5.3). This work has focused largely on the analysis of base sequence variation in the *16S rRNA/ND1* gene region using PCR amplification and restriction enzyme digestion with four informative enzymes. To date, data are available for more than 8000 salmon from 148 Eastern Atlantic populations (Nielsen *et al.* 1996; Verspoor *et al.* 1999; King *et al.* 2000; Nilsson *et al.* 2001; Knox *et al.* 2002; Consuegra *et al.* 2002; Asplund *et al.* 2004; Hindar, unpubl.) and 3490 salmon from 126 Western Atlantic populations (King *et al.* 2000; Verspoor, unpubl.), and it has been synthesised and reviewed by Gross *et al.* (unpubl.). Work has also been carried out on variation in the ND5/6 (Nielsen *et al.* 1996) and D-loop regions (King *et al.* 2000; Nilsson *et al.* 2001).

Based on variation between the Eastern and Western Atlantic in the more extensively studied *16S RNA/ND1* haplotypes, continental differences account for most (44.5%) of the total genetic variance detected (Gross *et al.* unpubl.). In contrast, of the remaining variation, 22.0% is distributed between samples within regions within continents, and 33.5% occurs among individuals within populations. At the same time, the average level of variation, as measured by the number and diversity of the haplotypes observed, was significantly higher in Eastern Atlantic populations of salmon than in those in the Western Atlantic. The strong continental differentiation seen for *16S RNA/ND1* haplotypes is linked to the distribution of two deep phylogenetic groups (Box 5.3). Most haplotypes detected in North American salmon belong to one group and the vast majority of those found in European salmon belong to the other. The variant typical of North American salmon populations occurs in Europe but is restricted in its distribution there to the Kola Peninsula in Northern Russia where it occurs

Box 5.3 Continental divergence at mtDNA.

Existing work on mtDNA variation in the Atlantic salmon has recently been reviewed by Gross *et al.* (unpubl.). Early on, work focused largely on haplotype variation detected by restriction enzymes in the *16S rRNA/ND1* region of the mitochondrial genome (Nielsen *et al.* 1996; Nielsen 1998; Verspoor *et al.* 1999; King *et al.* 2000; Nilsson *et al.* 2001). However, some studies also assessed variation in other regions such as the *N3/4* and *N5/6* genes (Nilsson *et al.* 2001) and the D-loop (King *et al.* 2000). More recently, studies have also been extended to include surveys of populations using direct sequencing (Verspoor *et al.* 2005; King *et al.* unpubl.).

Analysis of the *16S rRNA/ND1* gene region using the restriction enzymes *Hae*III, *Ava*II, *Hinf*I and *Rsa*I has identified haplotype variation marking eleven common and four rare maternal lineages. Analysis of their genetic relatedness shows they cluster into two deeply divergent evolutionary groups, one containing two haplotypes and the other the remaining 13 (Fig. B5.3). These two evolutionary groups are marked by mutations detected by the restriction enzymes *Ava*II and *Hae*III (Verspoor *et al.* 1999; King *et al.* 2000; Nilsson *et al.* 2001). Nilsson *et al.* (2001) sequenced a 1224 bp segment of the *16S rRNA/ND1* region for seven restriction enzyme haplotypes in 37 fish from across the distribution. Observed variation was restricted to single base-pair substitutions at 16 (1.3%) nucleotide positions, with divergence between the closest haplotypes (BBBA and NNBA) in the two evolutionary groups of 0.82%. Verspoor *et al.* (2002, unpubl.) found variation at 23 sites for a 1330 bp fragment in the same region (1.7%), the increase due to a more detailed analysis of North American populations, with divergence between the two closest haplotypes in the two groups of 0.83%. Consuegra *et al.* (2002) rooted the haplotype network with the *16S rRNA/ND1* sequence of *Salmo trutta* and found it to be equidistant (71 substitutions) from the Eastern Atlantic haplotype BBBB and the Western Atlantic haplotype NNBA. BBBB, currently the most widely distributed haplotype in the Eastern Atlantic, was also present in Paleolithic samples suggesting that BBBB, not BBBA, is likely to be

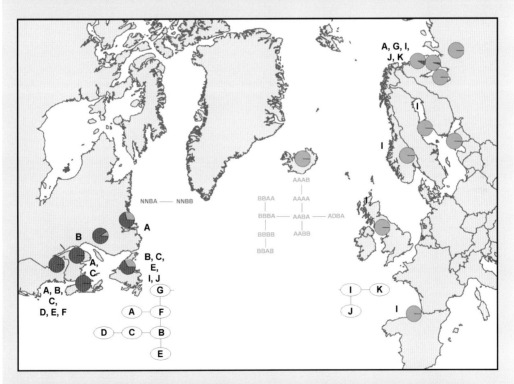

Fig. B5.3 Networks of mitochondrial DNA *16S RNA/ND1* (green and orange) and control region or D-loop (black) haplotypes. Each step in the network indicates a change of a single base. The two distinct evolutionary groups are separated by 9 and 14 substitutions (i.e. missing haplotypes), respectively, in relation to the two gene regions. The pie charts show the proportions of the two *16S RNA/ND1* groups found in different parts of the species North Atlantic range, while the letters show the occurrence in different regions of the D-loop haplotypes. Based on Gross *et al.* (unpubl.) and King *et al.* (unpubl.).

Box 5.3 (cont'd)

Table B5.3 Variable nucleotide positions along the 980-bp sequence in the Atlantic salmon mitochondrial DNA control region (D-loop). The numbering follows the sequence of Hurst *et al.* (1998) and GenBank accession number U12143. General location designations are: W = Western Atlantic; E = Eastern Atlantic; B = Baltic.

Haplotype	Sequence position																							No of Indiv.	Location
			1	3	3	4	4	5	5	5	5	7	7	9	9	9	9	9	9	9	9	9	9		
	2	4	7	0	9	4	7	0	3	6	7	0	1	4	4	6	6	6	7	7	7	7	8		
	0	3	6	3	0	9	3	4	0	3	1	3	1	2	3	4	5	6	1	2	7	8	0		
Hurst *et al.*	C	C	T	G	T	T	G	C	T	T	G	G	T	G	C	–	–	A	–	–	C	T	G		
A	T	C	T	G	T	T	G	C	T	–	A	G	C	A	C	C	T	A	C	A	–	–	A	12	W, E
B	T	C	T	G	T	T	G	C	T	–	G	A	C	A	C	C	T	A	C	A	–	–	A	24	W
C	T	C	T	G	T	T	G	C	T	–	G	A	C	A	C	C	T	G	C	A	–	–	A	5	W
D	T	T	T	G	T	T	G	C	T	–	G	A	C	A	C	C	T	G	C	A	–	–	A	1	W
E	T	C	T	G	T	T	G	C	T	–	T	A	C	A	C	C	T	A	C	A	–	–	A	2	W
F	T	C	T	G	T	T	G	C	T	–	G	G	C	A	C	C	T	A	C	A	–	–	A	2	W
G	T	C	T	G	T	T	G	C	T	–	G	G	T	A	C	C	T	A	C	A	–	–	A	1	E
I	C	C	T	A	C	C	T	T	C	–	G	G	T	A	A	–	–	A	–	–	C	T	G	22	E, W, B
J	C	C	T	A	T	C	T	T	C	–	G	G	T	A	A	–	–	A	–	–	C	T	G	1	E, W
K	C	C	C	A	C	C	T	T	C	–	G	G	T	A	A	–	–	A	–	–	C	T	G	1	B

the ancestral European haplotype and gives a revised estimate of divergence of 0.73% between the Eastern and Western Atlantic lineages, based on sequence data in Nilsson *et al.* (2001). Analysis of the distribution of evolutionary groups shows that almost all salmon in the Eastern Atlantic belong to the larger group of haplotypes and almost all salmon in the Western Atlantic are from the other. Two of the common Eastern haplotypes (BBBA and BBBB) occur sporadically in the Western Atlantic in Quebec, Labrador and, particularly, Newfoundland, while the main Western haplotype (NNBA) occurs sporadically and at low frequencies in the Kola Peninsula in northern Russia (Fig. B5.3).

The same continental divergence in mtDNA lineages has been revealed by the analysis of variation in the control region or D-loop (King *et al.* 2000), though the available population data is more limited. Kauppi *et al.* (1997) sequenced the 940 bp control region (D-loop) in nine salmon sampled from five Baltic populations and one individual from Newfoundland, Canada. No variation was found in Baltic salmon but ~ 1.5% sequence divergence was found between the Eastern and Western Atlantic fish. More recently, King *et al.* (in prep.) surveyed sequence variation in the control region among Western and Eastern Atlantic salmon. Ten haplotypes were identified which formed two distinct networks separated by more than 14 missing haplotypes (Fig. B5.3). Average sequence divergence between the two networks was also found to be ~ 1.5% (eight substitutions, six insertions/deletions; Table B5.3), similar to that found by Kauppi *et al.* (1997).

King *et al.* (2000, unpubl.) found five D-loop haplotypes (B–F) exclusive to Western Atlantic salmon, one (K) to samples from the Eastern Atlantic (Kola Peninsula); three (A, I, J) were shared. The largest network of seven haplotypes (A–G) consisted of the haplotypes found primarily in the Western Atlantic of which A, B and C occurred in all Western samples. In the other network of three haplotypes (I–K), haplotype I was the most common and found in all Eastern Atlantic collections. Haplotypes A and G, in the Western Atlantic network, were both present in salmon from the River Kachkovka in Russia (Fig. B5.3). Conversely, haplotypes I and J, in the Eastern Atlantic network, occurred in salmon from Newfoundland (Fig. 5.2). These results concur with the results of studies of *ND1* gene (see above) and lend further support to there having been historical gene exchange across the Atlantic. This accounts for the higher control region diversity in Newfoundland and on the Kola Peninsula, mirroring the results of *ND1* studies.

The greater haplotype diversity in North America compared to Europe seen in the control region (King *et al.* 2000) contrasts to what is seen for the restriction enzyme haplotype analysis of the *ND1* region. However, this is probably in part, at least, a sampling artefact. In a more detailed survey based on DNA sequencing, 10 haplotypes were resolved by sequencing of only part of the *ND1* gene in 165 salmon from the Bay of Fundy and Southern Uplands of Nova Scotia, of which only three were detectable by restriction enzymes (Verspoor *et al.* 2002); furthermore, a more extensive population survey resolved 39 haplotypes of which, again, only three were detectable by restriction enzymes. Thus, the greater D-loop variation is probably mirrored in the *16S rRNA/ND1* region as well.

at low frequencies. Conversely, two of the most common haplotypes in Europe also occur in North America, but only sporadically in Newfoundland and Labrador, in the north of the species' Western Atlantic range. The same continental divergence in mtDNA lineages is revealed by the analysis of variation in the control region (D-loop) of this molecule (King *et al.* 2000, unpubl.), though the data set for this region is more limited. Haplotype variation in the D-loop is associated with two distinct evolutionary groups (Box 5.3), one dominating in the Eastern Atlantic and the other in the Western Atlantic. The limited occurrence of the typical European lineage in the Western Atlantic Newfoundland, and of the typical North American lineage in the Kola region in Europe, also occurs in the *16S RNA/ND1* analysis.

The deep phylogeographic divergence of Eastern and Western Atlantic salmon is supported by more recent direct analyses of nuclear DNA. Populations on both sides of the Atlantic are virtually fixed for divergent sets of alleles at the *Ssa-A45/1* minisatellite locus (Taggart *et al.* 1995), as well as an insertion-deletion polymorphism in the flanking region of the *Spss1605* microsatellite locus (Verspoor, unpubl.). Many examples of divergence have been found with respect to microsatellite loci themselves (McConnell *et al.* 1995; King *et al.* 2001; Koljonen *et al.* 2002; Gilbey *et al.* 2005).

The general level of differentiation found at microsatellite loci is best illustrated by the recent, extensive range-wide survey of microsatellite DNA variation by King *et al.* (2001). In this study, a strong relationship was found between genetic and geographic distance (Fig. 5.3), with an overall correlation of $r = 0.93$ ($t = 17.3$; $P < 0.0001$). This compares to weaker but still highly significant correlation within the Eastern Atlantic of $r = 0.69$ ($t = 4.0$; $P = 0.0001$) and

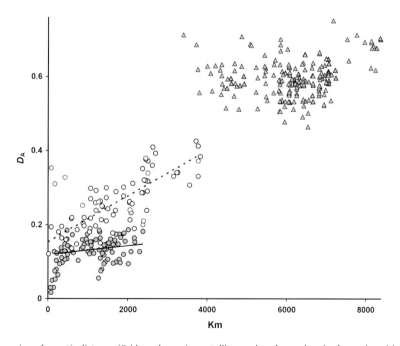

Fig. 5.3 Regression of genetic distance (D_A) based on microsatellite markers for each pair of samples with the geographic distance (km) separating sampling locations, estimated as the shortest ocean distance between river mouths. △ comparisons between continents, ○ comparisons within Eastern Atlantic populations, ● comparisons within Western Atlantic populations. Source: King *et al.* (2000).

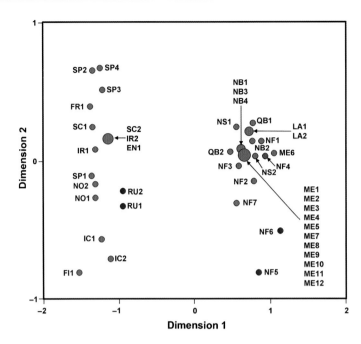

Fig. 5.4 Multidimensional scaling plot illustrating nonparametric distribution of pairwise genetic distances (D_A, Nei *et al.* 1983) based on 11 microsatellite loci between Atlantic salmon from throughout their range. Circles are scaled to indicate the number of indistinguishable populations found near the circle's centre. Red circles indicate presence of mtDNA haplotypes characteristic of the other side of the Atlantic from which the fish were collected. Based on King *et al.* (2001).

of $r = 0.43$ ($t = 3.3$; $P = 0.0008$) within the Western Atlantic. This strong association of differentiation with distance, both across the range and within each continental region, supports the view that historical gene flow has been a function of distance.

The overall pattern of divergence across microsatellite loci (Figs 5.4 and 5.5; King *et al.* 2001, unpubl.) supports the deep genetic divergence between Eastern and Western Atlantic populations indicated by allozymes and mtDNA. There is a close correspondence between patterns of genetic variation and geographical distributions of populations, and the estimate of genetic differentiation across continents for microsatellites ($F_{ST} = 0.274$) is similar to that found for allozymes (see above). The extent of continental divergence is such that variation at just four microsatellites is sufficient to allow native salmon to be assigned to continent of origin with > 99.99% accuracy (Chapter 9).

Molecular genetic studies provide widely based and compelling support for a deep phylogeographic division in the Atlantic salmon, between populations in the Eastern and Western Atlantic regions. This division represents a large and important part of the biodiversity which exists within the species, and should be protected by appropriate conservation and management legislation and policies. Mixing of the two groups of stocks, either deliberately by stocking or inadvertently through farm escapes, poses a serious threat to this biodiversity, and carries with it a serious risk of outbreeding depression and disruption of local adaptation (Chapters 11 and 12), in addition to the possibility of non-native disease transfer (see also Gross 1998). Thus, there should be a strong presumption against human-mediated movements of fish between the two regions.

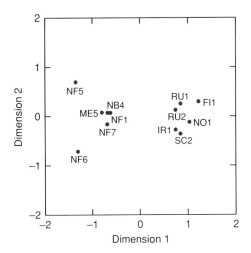

Fig. 5.5 Multidimensional scaling plot illustrating nonparametric distribution of pairwise genetic distances (D_A, Nei *et al.* 1983) calculated over 25 microsatellite loci from Atlantic salmon collections in Eastern and Western Atlantic Ocean. Based on King *et al.* (unpubl.).

Stock transfers have occurred in the past for stocking programmes (MacCrimmon and Gots 1979) and European aquaculture strains have recently been used in North America and North American strains in Europe. Despite the absence of obvious morphological differentiation (Claytor *et al.* 1991), according native populations in North America and Europe the subspecies status they deserve (Box 5.4) would go a long way toward helping develop appropriate legislation and protective measures. In the US, resources available for research and protection of taxa listed under the Endangered Species Act increase with taxonomic generality, where a listed species is eligible for greater resources than a listed subspecies, and a listed subspecies is eligible for greater resources than a designated distinct population segment. Formal taxonomic recognition of the evolutionary distinctiveness of salmon on the two sides of the Atlantic would also help protective legislation and management policies in Europe.

5.3.2 Eastern Atlantic

Molecular genetic studies show further, though shallower, phylogeographic structuring within the Eastern Atlantic. The structuring is inconsistent with low-level contemporary gene flow where differentiation would be expected to be a simple function of physical separation. Rather it strongly points to regional differentiation being historical and the result of a high degree of reproductive isolation of populations in different regions within this part of the species range.

Regional differentiation of Eastern Atlantic populations was first reported by Payne *et al.* (1971). He found a small but consistent discontinuity in transferrin allele frequency between rivers in the southern and northern parts of the British Isles. The existence of marked differences between the rivers Bandon and Shannon, two neighbouring Irish rivers on the regional boundary, argued that gene flow between the two regions was absent. Based on this, they proposed the existence of a 'Boreal' race in the north and 'Celtic' race in the south and suggested this reflected the pattern of postglacial colonisation of the British Isles.

Box 5.4 The case for subspecific designation.

Payne *et al.* (1971) was the first to propose that Atlantic salmon from Europe and North America be defined as separate subspecies based on the high level of genetic differentiation he found between the two continents at the serum transferrin protein locus (see section 5.2.1). However, at the time, this was not widely accepted. It was pointed out that a single locus, essentially a single trait, was insufficient as the foundation of such a taxonomic revision (Behnke 1972), and that the names proposed were invalid according to accepted rules of zoological nomenclature (Gruchy 1971). Furthermore, it was unclear whether the divergence at this locus was phylogenetic or due to regionally varying selection, as transferrin variants in other species show functional differences which could be subject to selection (Verspoor 1986). However, since Payne's initial proposal, a large body of genetic evidence has been amassed which now provides overwhelming and compelling support for populations on the two sides of the Atlantic being designated as distinct subspecies.

Initial allozyme studies, based on multiple protein-coding loci (Cross 1981; Ståhl 1987) found significant differentiation between Eastern and Western Atlantic salmon but less marked than found for serum transferrin. However, this was based on limited and biased geographical sampling and is not borne out by the larger body of allozyme data. At one locus, *EST-D**, most populations in the two regions appear to be fixed for alternate allelic variants (Verspoor *et al.* 2005). A more extensive review of variation in representative populations across the species range shows continental differentiation to be similar to that seen between recognised subspecies of cutthroat trout, *Oncorhynchus clarkii* (Verspoor *et al.* 2005). Furthermore, typical genetic distances (Nei's D) of 0.085 are well above the 0.05 threshold cited by Nei (1987) as typical of subspecies. This view is supported by observed karyotype differentiation, with the two geographic groups showing a minimum of eight chromosomal changes, including differences in chromosome numbers (Box 5.2). This divergence is only consistent with a high degree of reproductive isolation historically, and a lengthy time period of independent evolution. The observed divergence is also seen in designated subspecies of *O. clarkii* (Loudenslager and Gall 1980).

Divergence of mtDNA, which appears to be largely phylogenetic, is also highly marked between Eastern and Western Atlantic populations (Box 5.3). Populations in the Eastern and Western Atlantic, with the exception of two regions, are fixed for haplotypes from two different and deeply diverged evolutionary lineages. A similar phylogenetic division is seen with respect to microsatellite DNA, also generally assumed to be selectively neutral and phylogenetically informative (section 5.2.1; King *et al.* 2001). At some loci, as for mtDNA, genetic differences among many populations are fixed (e.g. Taggart *et al.* 1995; Gilbey *et al.* 2005), while at most, as with allozymes, there are overlapping allele distributions but major differences in allele frequencies (King *et al.* 2001).

All the classes of genetic variation examined show a major phylogeographic split between Eastern and Western Atlantic salmon populations, though completely fixed genetic differences between the two are lacking. Variation typical of one side of the Atlantic occur in some individuals in some populations in the same restricted parts of the species range on the other side, parts of the species range which were glaciated during the last Pleistocene. This is most parsimoniously explained as the result of limited gene flow between the two parts of the species range at the time the regions in question were being colonised following their deglaciation (section 5.2.5). As such the deep phylogenetic divergence between genetic types on the two sides of the Atlantic can only be reconciled with populations on the two sides of the Atlantic having been highly isolated for much of the time since they diverged, something the mtDNA divergence suggests occurred in the order of 600 000 years ago (Box 5.7).

These results therefore strongly support the contention of Payne *et al.* (1971) that the species *Salmo salar* should be split into Eastern (Old World) and Western (New World) subspecies. According to the International Code of Zoological Nomenclature, the Eastern Atlantic populations of Atlantic salmon constitute the nominate subspecies and should be designated *Salmo salar salar* Linnaeus 1758. The Western Atlantic populations found in North America, based on precedence, should be called *S. salar sebago* Girard 1854. This conforms to accepted naming practice but will be confusing due to the previous use of the name for non-anadromous forms. In contrast, the names suggested by Payne *et al.* (1971), *S. s. europaeus* and *S. s. americanus*, have the merit of clarity but must be rejected in accordance with accepted rules of nomenclature (Gruchy 1971).

Analysis of multilocus allozyme variation provides support for a north–south genetic differentiation in Britain and Ireland (Jordan *et al.* 2005). Their study examined variation in 76 river systems and found small but significant allele frequency differentiation within and among river systems, consistent with highly restricted gene flow at both spatial scales. A

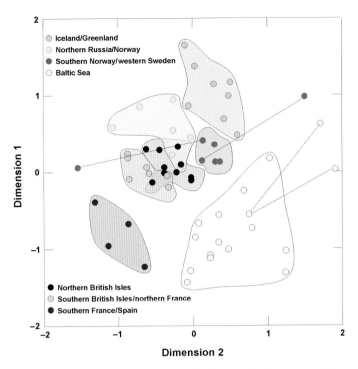

Fig. 5.6 Multidimensional scaling plot illustrating structured distribution of pairwise genetic distances (D_A, Nei *et al.* 1983) calculated over 12 polymorphic allozyme loci from Atlantic salmon collections in the Eastern Atlantic Ocean; collections of anadromous salmon (circles within defined regional clusters) and non-anadromous salmon from the same river are connected by lines. Source: Verspoor *et al.* (2005).

north–south cline in allele frequencies was detected at the *AAT-4** locus, as well as significant genetic differentiation between the 'Celtic' and 'Boreal' regions defined by Payne *et al.* (1971). This differentiation, however, is relatively small compared to regional variation across Europe as a whole (Verspoor *et al.* 2005; Fig. 5.6).

In Europe, the most striking phylogeographic division occurs between salmon in Baltic Sea and in Atlantic Ocean drainages (Fig. 5.6), something first noted by Ståhl (1987) based on allozyme work. This division is strongly supported by the distribution of mtDNA variation (Verspoor *et al.* 1999; Nilsson *et al.* 2001; Box 5.5), and by studies of microsatellite locus variation (Koljonen *et al.* 2002; Säisä *et al.* 2005; Verspoor, unpubl.). A recent synthesis of mtDNA data (Gross *et al.* unpubl.) from across the species range (see Box 5.5) shows that, within the Eastern Atlantic region, 28.3% of the genetic variance present in the Atlantic salmon was distributed between Atlantic and Baltic populations. In contrast, only 15.7% of the variance among the Atlantic rivers was due to differences between eight geographic regions (Fig. B5.5). In a study of regional microsatellite variation, Koljonen *et al.* (2002) reported genetic distances (D_A) of 0.34 between Baltic Sea and Atlantic populations in Europe, about half the distance they found between North American and European populations.

Phylogeographic substructuring is also apparent within the Baltic Sea. Koljonen *et al.* (1999), in a detailed analysis of allozyme variation at seven polymorphic loci in over 5000 salmon from 24 rivers in four countries, found a marked dichotomy between salmon from rivers in the south-east (Russia, Estonia, Latvia, southern Sweden) and the north-west (northern

Box 5.5   MtDNA variation in the Eastern Atlantic.

Over 7400 individuals from 143 sampling locations have been analysed for restriction enzyme detected variation in the *16S RNA/ND1* region, the most extensively studied part of the salmon's mtDNA. Of the 14 haplotypes resolved, four (AAAA, AABA, BBBB, BBBA: see Fig. B5.3) are common and widespread, and the remainder rare and found in only one or a few populations. AABA is the most common and widely distributed, found in 87% of populations (average frequency 0.392), closely followed by BBBB in 83% (average frequency 0.388). The main regional differences are with respect to haplotype frequency, with no regions showing any tendency to fixation for alternative haplotypes.

There is considerable variation with respect to the different regions of Europe (Fig. B5.5a). No apparent relationship is seen between genetic divergence and geographic separation (Fig. B5.5b). For example, the highly separated regions of Iberia and Iceland are closer than the spatially less separated regions of Karelia and Kola in the White Sea, or the Barents and White Seas rivers on the Kola Peninsula. As a result, aside from the grouping of Baltic populations, largely due to a low frequency of haplotypes other than AABA, no strong geographical clustering of regional population groups occurs. This strongly supports the view that the patterns are the product of historical factors rather than patterns of contemporary gene flow.

One of the common haplotypes, AABA, dominates in the Baltic Sea area (average frequency 0.816) whereas BBBB is the main haplotype in Atlantic drainages (average frequency 0.489), including those in the White and Barents Sea areas. BBBA occurs in most Atlantic populations but is at its highest frequency around the British Isles (average frequency 0.306) and along the Barents Sea coast of Scandinavia and the Kola Peninsula (average frequency 0.290). It is absent in both anadromous and resident (i.e. non-anadromous) populations in the Baltic Sea, and from populations in Karelia in the White Sea, in the eastern Barents Sea and in 90% of populations in Iceland.

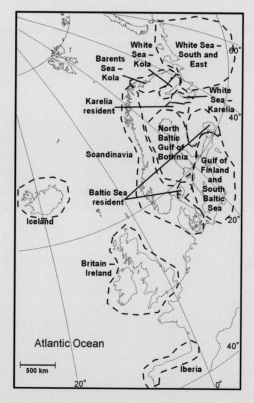

Fig. B5.5a   Regions referred to in text and Fig. B5.5b.

Box 5.5 (cont'd)

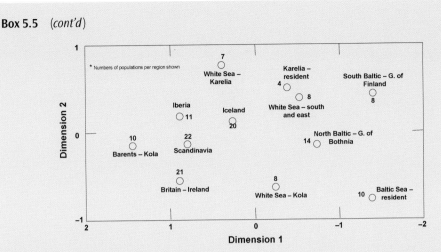

Fig. B5.5b A two-dimensional multidimensional scaling (MDS) plot of genetic distances (D_A) in Atlantic salmon among Eastern Atlantic Ocean regions estimated from regional *16S RNA/ND1* haplotype frequencies. Based on data in Gross *et al.* (unpubl.).

AAAA occurs sporadically, with its highest frequency in the northern Baltic Sea, in lakes Saimaa and Ladoga, around the British Isles and in Iceland. It is largely absent in the Barents and White Seas. NNBA, the haplotype typical of Western Atlantic populations of salmon, occurs in 14 populations on the Kola Peninsula along the Barents and White Sea coast (average frequency 0.085).

AAAA, AABA and BBBB occur in the Baltic Sea, but their distribution is not continuous. Gulf of Bothnia populations show all three haplotypes while those in the southern Baltic area (including Gulf of Finland, Gulf of Riga and the Baltic Main Basin) are fixed, or almost fixed, for AABA. AAAA is absent in southern Baltic and BBBB is very rare. Along the Swedish west coast, AABA occurs at high frequency in five rivers closest to the strait between Denmark and Sweden, while further north in Scandinavia the AABA haplotype almost disappears, and is replaced by BBBB. Landlocked, resident salmon in Lake Vänern in Sweden are fixed for the common Baltic haplotype AABA, and this haplotype also dominates in Russian Lake Onega. AAAA and AABA occur in Lake Ladoga while Lake Saimaa in Finland is almost fixed for AAAA.

Finland, northern Sweden). These population groups account for 7.5% of the total variation in allozyme loci (Koljonen *et al.* 1999) and 14.3% in mtDNA haplotype frequencies (Gross *et al.* unpubl.), and divergence among them with regard to microsatellite locus variation ($D_A = 0.27$) is only slightly smaller than that seen between Atlantic Ocean and Baltic Sea populations in Europe ($D_A = 0.34$). A recent, more detailed study of microsatellite variation in Baltic salmon revealed three clear groupings of populations, corresponding to the northern (Gulf of Bothnia), eastern (Gulf of Finland and eastern Baltic Main Basin) and southern regions (Western Baltic Main Basin), and the differences among these groups accounted for 5.6% of the total variation, which was much higher than that among population groups in the Eastern Atlantic Ocean (2.2%) from Ireland to the White Sea (Säisä *et al.* 2005).

The distribution of allozyme variation across Europe suggests that regional phylogeographic differentiation in the Eastern Atlantic may be widespread, beyond that already noted for the British Isles and in the Baltic. The recent broad-scale analysis of variation in 56 representative European rivers for 12 polymorphic allozyme loci (Fig. 5.6; Verspoor *et al.* 2005)

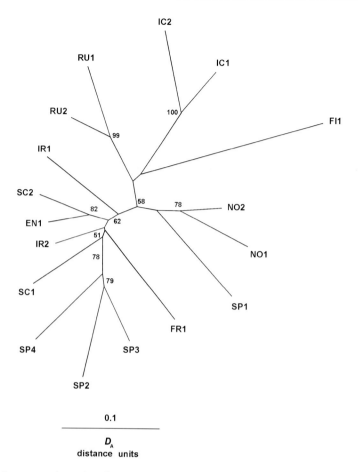

Fig. 5.7 Genetic distance tree (D_A, Nei et al. 1983) based on 11 microsatellite loci for Eastern Atlantic salmon collections. Numbers indicate bootstrap support for each node after 10 000 replicates. Absence of number indicates less than 50% bootstrap support for that node; abbreviations for populations refer to Table 5.2. Based on King et al. (2001).

showed populations clustering into regional groupings. Based on geographic divisions, in addition to the Baltic Sea region, these were Iceland/Greenland, northern Russia/northern Norway, southern Norway/western Sweden, the northern British Isles, the southern British Isles/northern France, and southern France/Spain. These groups show little overlap: the 'genetic map' (Fig. 5.6) for Europe, based on genetic distances, shows a remarkable correspondence to the geographic map of Europe.

Studies of microsatellite variation in salmon populations in the Eastern Atlantic have been less extensive. King et al. (2001, unpubl.) have identified three major significantly differentiated geographical groupings: Iceland, Baltic and Atlantic Europe (Fig. 5.7). There is also a strong geographical ordering of Atlantic drainage populations (including Iceland), as found for allozymes (Fig. 5.7), with significant differentiation between many of the tree branches. At one end of the tree are found all but one Spanish populations and at the other end rivers from Iceland. Next to the Spanish populations are rivers from France, Scotland, Ireland and England, followed by rivers from Norway and Russia, and then the Icelandic rivers. The existence of phylogenetic substructuring within northern Russia is supported by detailed regional work on allozyme, mtDNA and microsatellite variation (Makhrov et al. 2005; Tonteri et al. 2005).

The pattern of regional differentiation in European Atlantic salmon reflects an association of genetic divergence with geographic separation (= marine distances between river mouths; e.g. Fig. 5.3), seen for all classes of molecular markers. However, the physical separation of populations can explain only part of the observed differentiation as greater genetic differentiation is often seen over relatively short distances. A simple association with distance does not explain the marked differentiation seen between salmon in Baltic Sea and Eastern Atlantic Ocean drainages. With respect to allozymes, mtDNA and microsatellite variation, Baltic Sea populations are no closer to immediate neighbours in North Sea rivers on Sweden's west coast than they are to populations elsewhere. Indeed, they have their strongest affinity with Icelandic populations (e.g. Fig. 5.7). This is also seen for the *AAT-4*25* allele, the alternate allele at the locus in almost all Baltic populations. It is most frequent outside the Baltic in Iceland (Verspoor et al. 2005) as is the main Baltic mtDNA haplotype, AABA (Gross et al. unpubl.).

Regional genetic differentiation in Europe also occurs with respect to the level of variation as well as the type of variants present. With respect to allozymes, mtDNA and microsatellites, the lowest levels of heterozygosity seen in anadromous populations occur in the Baltic Sea region and the highest in Atlantic populations in the British Isles region and adjacent northern France (Fig. 5.8). The next highest levels are seen in populations in Scandinavia and the neighbouring Kola region of northern Russia. For allozymes and microsatellites, the lowest heterozygosity is found in non-anadromous populations. For mtDNA data, low diversity occurs in non-anadromous populations, but also in eastern White Sea populations and in the easternmost area of distribution range (e.g. River Pechora).

5.3.3 Western Atlantic

Regional differentiation is also present in the Western Atlantic. This was first detected by Möller (1970) in his studies of transferrin variation. Later, Payne (1974), who also looked at transferrin variation, but in a different set of populations, suggested that the differentiation was clinal (i.e. that the allele frequencies varied in a continuous rather than stepwise manner). However, a synthesis of the data from the two studies (Verspoor 1986) showed the variation was regional with frequencies of Tf_1, the variant for which most European populations are nearly fixed, higher in Newfoundland and southern Labrador than in more southerly populations. More recent work on allozyme (Verspoor 1988, 2005; Verspoor and Jordan 1989; Cordes et al. 2005), mtDNA (Verspoor et al. 2002, 2005) and microsatellite data (McConnell et al. 1997; King et al. 2001, unpubl.) suggests phylogeographic structuring is also likely to be widespread in the Western Atlantic part of the salmon's range.

A synthetic analysis of the data from allozyme studies (Verspoor et al. 2005) shows salmon in the Western Atlantic to be genetically clustered into a number of more or less distinct regional phylogeographic groups (Fig. 5.9). The groups are characterised by regional differences in population allele frequencies and the occurrence of regionally restricted variants. For a number of the allozyme loci, the allelic variants normally most common in Europe but less common in North America are at greatest frequency in Newfoundland, as was found for transferrin. The work suggests that regions such as Labrador, Newfoundland, the Gulf of

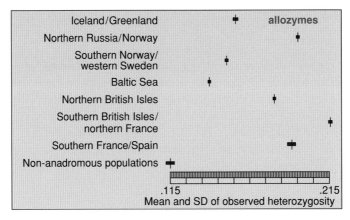
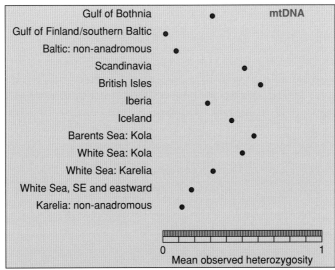
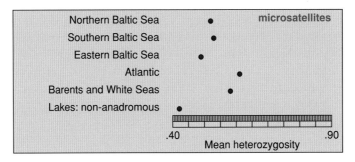

Fig. 5.8 Mean heterozygosity across loci for different regions and life-history groups in the Eastern Atlantic for allozymes (source: Verspoor *et al.* 2005), mtDNA (based on Gross *et al.* unpubl.), and microsatellites (based on Säisä *et al.* 2005).

St Lawrence, the Bay of Fundy and Maine represent distinct phylogeographic groups, though the exact geographical boundaries of these groups remain to be fully resolved. There is also an indication, from the restricted distribution of rare variants, that at least some of these groups are further subdivided (Verspoor 2005).

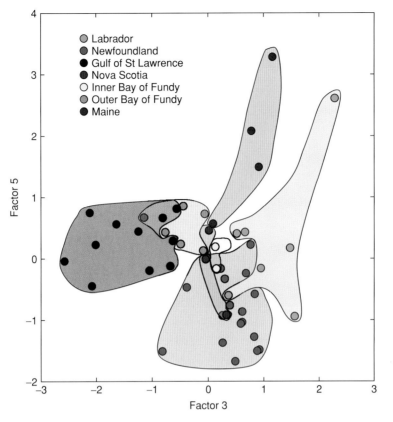

Fig. 5.9 Principal component analysis (PCA) plot, based on allele frequency differences observed for the main polymorphic allozyme loci, showing the structured distribution of genetic variation from Atlantic salmon collections in the Western Atlantic Ocean. Based on Verspoor et al. (2005).

The existence of regional phylogeographic structuring in the Western Atlantic is also supported by studies of mtDNA variation, though the work on this class of variation has been less extensive (Box 5.6). Overall, variation among populations in haplotype frequencies shows a weaker geographic structuring than in the Eastern Atlantic. Gross *et al.* (unpubl.) found only 10.7% of variance in haplotype frequencies to be due to differences between provinces, markedly less than seen among regions in Europe (see above). Most of the genetic variance was distributed among populations within provinces and within the samples (45.5% and 43.9%, respectively). The greatest nucleotide diversity in the Western Atlantic occurred in Newfoundland and Quebec, reflecting the presence of different phylogenetic lineages in these regions.

The more limited regional differentiation observed in the Western Atlantic, to some extent at least, reflects the fact that the analysis was based on political boundaries. Regional differences are more likely to be aligned to historical geographical units which would have controlled patterns of historical gene flow. This view is supported by detailed analyses of variation in some of the parts of the species' Western Atlantic range. For example, Verspoor *et al.* (2002) found a unique mtDNA lineage (i.e. haplotype) in a more extensive analysis of the *ND1* gene, detected by the restriction enzyme *Alu*I which occurs at high frequencies in

> **Box 5.6** MtDNA variation in the Western Atlantic.
>
> Much of the work carried out on mtDNA variation in the Western Atlantic has centred on restriction enzyme detected variation in the *16S RNA/ND1* region using the standard suite of restriction enzymes – *Ava*II, *Hae*III, *Hin*fI and *Rsa*I. These enzymes resolve four haplotypes in this part of the species range. The most commonly found, and the only ones found in most populations screened, are the two representatives of the less diverse of the two main evolutionary lineages resolved – NNBA and NNBB (Fig. B5.3). NNBB, exclusive to the Western Atlantic, was found only in Nova Scotia with the average frequency 0.131 (Verspoor, unpubl.). BBBB and BBBA, in the other lineage, whose haplotypes are typical of Eastern Atlantic populations, occur in both anadromous and non-anadromous populations in Labrador, Newfoundland and northern Quebec, with some populations fixed or nearly fixed for one or other of these Eastern Atlantic haplotypes (Verspoor, unpubl.). However, additional restriction enzymes allow further subtypes of NNBA to be discriminated. King *et al.* (2000) revealed two widely distributed and one rare subtype resolved by *Hae*I and *Hha*I, while Verspoor *et al.* (2002; unpubl.) found a further subtype using *Alu*I in seven populations from Nova Scotia and New Brunswick.
>
> Complete sequencing of a 710 bp sub-region of the *ND1* gene in 743 individuals from the Bay of Fundy and the Atlantic coast of Nova Scotia, resolved 32 variable sites which defined a total of 36 haplotypes, of which only two are detectable by restriction enzymes (Verspoor, unpubl.). This shows that restriction enzyme-based studies reveal only a small part of the mtDNA variation which exists within and among salmon populations.
>
> Genetic variation has also been resolved in the control region (D-loop) where two haplotypes identified, AAA and BBB, are common and widespread while the other three found are rare and confined to few or single populations (King *et al.* 2000). In the Gander River, Newfoundland, 15 fish from two tributaries had a haplotype, ABB, fixed in four of five Eastern Atlantic populations examined. The river also had the *ND1* BBBA haplotype, the second most common variant in Eastern Atlantic salmon. Haplotype ABB was also found in 12 salmon outside of Gander River (one Maine, 11 Canada) but, unexpectedly, possessed *ND1* variations typical of Western Atlantic populations.

populations associated with the Minas Basin of the inner Bay of Fundy, but is absent elsewhere in North America. A further haplotype variant, resolved by the restriction enzyme *Rsa*I, occurs in the southern uplands of Nova Scotia and is absent elsewhere in the species range (Verspoor *et al.* 2002, 2005). The restricted distribution of such variants strongly supports the view that they mark highly distinct phylogeographic groups, a view reinforced by the more limited work in the region on microsatellite variation. The latter shows that salmon in the rivers of the Minas Basin in the inner Bay of Fundy, such as the Stewiake, and of the Southern Uplands of Nova Scotia are distinct from each other and from other rivers in North America (McConnell *et al.* 1997; King *et al.* 2001; Verspoor, unpubl.).

Microsatellite data support the existence of widespread phylogeographic structuring as well. King *et al.* (2001) found that Atlantic salmon from both Eastern and Western Atlantic populations could be correctly assigned to country and/or province of origin an average of 82.7% of the time (Table 5.1). Recent work (Cornuet *et al.* 1999) indicates that a misclassification rate of 25% with 10 microsatellite loci is expected if there has been isolation over a period of 100–300 generations. Thus, the observed levels of divergence are consistent with the isolation of populations for a period much longer than this. Also, with microsatellites, only 81 of 1131 (7.2%) US and Canadian anadromous fish were misclassified to their country of origin. This was achieved despite introduction into the Penobscot River as late as the 1960s, of Quebec salmon from the St-Jean and the Saguenay rivers. This suggests that Maine salmon represent a group of well-adapted populations highly distinct from populations elsewhere in North America. The observed patterns of differentiation are further supported by allozyme analyses (Figs 5.9 and 5.10).

Table 5.1 Results of maximum-likelihood assignment tests. Samples from each political jurisdiction are lumped. For example, ME = ME1 + ME2 + ME3 + ME4; NB = NB1 + NB2; etc. The landlocked samples (ME13 and ME14) were, however, pooled into a separate category (MEL). The expectation of numbers correctly classified by chance alone were calculated assuming equal probability of membership in any single population. Probability of membership (p) in any grouping is assumed to be proportionate to the number of populations comprising that group. Specifically, for any single population $p = 1/13$; $p = 2/13$ for the United States grouping; $p = 5/13$ for the Canadian grouping; $p = 7/13$ for the North American grouping; and $p = 6/13$ for the European grouping.

Population	ME	MEL	NB	NS	QB	NF	LB	IC	NO	FN	SC	IR	SP	United States	Canada	North America	Europe	All
ME	571	5	36	9	1	12	5	0	0	0	0	0	0	576	63	639	0	639
MEL	1	90	0	0	0	1	0	0	0	0	0	0	0	91	1	92	0	92
NB	8	0	76	7	23	5	3	0	0	0	0	0	0	8	114	122	0	122
NS	5	1	8	88	6	2	0	0	0	0	0	0	0	6	104	110	0	110
QB	0	0	25	5	89	1	2	0	0	0	0	0	0	0	122	122	0	122
NF	5	0	7	5	4	71	1	0	0	0	0	0	0	5	88	93	0	93
LB	0	2	6	1	2	2	32	0	0	0	0	0	0	2	43	45	0	45
IC	0	0	0	0	0	0	0	95	1	0	0	0	0	0	0	0	96	96
NO	0	0	0	0	0	0	0	0	99	0	2	0	0	0	0	0	101	101
FN	0	0	0	0	0	0	0	0	0	61	0	0	0	0	0	0	61	61
SC	0	0	0	0	0	0	0	0	0	0	38	8	7	0	0	0	53	53
IR	0	0	0	0	0	0	0	1	0	0	11	48	4	0	0	0	64	64
SP	0	0	0	0	0	0	0	0	7	0	5	4	68	0	0	0	84	84
Sample size	639	92	122	110	122	93	45	96	101	61	53	64	84	731	492	1223	459	1682
Observed number correctly classified	571	90	76	88	89	71	32	95	99	61	38	48	68	667	471	1223	459	1682
Percent correctly classified	89.4	97.8	62.3	80.0	73.0	76.3	71.1	99.0	98.0	100.0	71.7	75.0	81.0	91.2	95.7	100.0	100.0	100.0
Expected number correctly classified	49.2	7.1	9.4	8.5	9.4	7.2	3.5	7.4	7.8	4.7	4.1	4.9	6.5	112.5	189.2	658.5	211.8	1682.0
χ^2	5540.2	971.6	472.9	747.7	675.4	569.8	235.3	1039.5	1071.3	675.7	282.3	376.9	586.1	2734.4	419.6	483.8	288.3	0.0

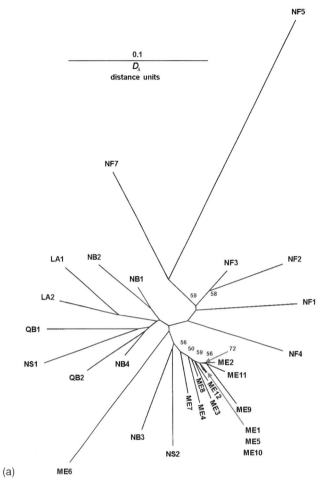

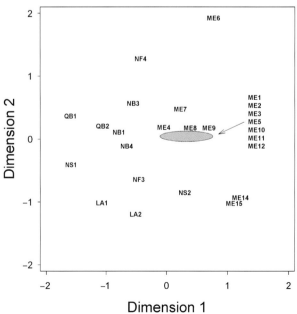

Fig. 5.10 Genetic distance (a) tree (D_A, Nei *et al.* 1983) and (b) multidimensional scaling plot based on 11 microsatellite loci for Western Atlantic salmon populations. Numbers on tree branches indicate bootstrap support for each node after 10 000 replicates. Absence of number indicates less than 50% bootstrap support for that node. Abbreviations for populations refer to Table 5.2. The land-locked Ouananiche Beck collection (NF-6) is not included in these graphics. Based on King *et al.* (2001).

In some cases stock transplants between rivers may have been successful and affected the observed differentiation among rivers. The St-Jean in Quebec was stocked with fish from the Miramichi and St Croix rivers in New Brunswick for 20 years, around 1900 (Fontaine *et al.* 1997), which may explain why the St-Jean River salmon showed a lower genetic distance and a lower than average assignment rate with these two New Brunswick rivers (King *et al.* 2001). However, there was also a relatively high rate of misclassification (32.4%) between the two Quebec rivers examined. As such, the differences in the rate of misclassification probably reflect historical levels of regional differentiation. In the support of the latter view, allozyme studies place Quebec and New Brunswick rivers, flowing into the Gulf of St Lawrence, into a distinct phylogeographic group from salmon rivers in Maine (Verspoor *et al.* 2005).

The distinctness of Maine's salmon from other populations in North America is further supported by the work of Spidle *et al.* (2003). Multidimensional scaling of genetic distance shows, with the exception of Cove Brook, a tight clustering of river populations within Maine. The study indicates the Kennebec and Penobscot rivers should be included with populations in the other Maine rivers accorded 'distinct population segment' (DPS) status for protection under the US Endangered Species Act (i.e. wild Atlantic salmon populations found in the Dennys, East Machias, Machias, Pleasant, Narraguagus, Ducktrap and Sheepscot rivers, and in Cove Brook). In the analysis, Cove Brook and the landlocked salmon samples are distinct from Maine's anadromous populations but cluster together with the other Maine populations separate from Canadian populations of anadromous salmon (Fig. 5.11). Multilocus assignment tests show high precision in allocation of fish to country of origin and a homogeneously high assignment rate among the DPS and Penobscot rivers compared to other sets of rivers (Table 5.2). However, the assignment tests do not show as much similarity between Atlantic salmon in the Kennebec and other Maine anadromous salmon as is indicated by the analysis of genetic distance. With this exception, the analyses support a shared ancestry of all Maine salmon, distinct from salmon populations in Canada which have been examined.

Taken together, the available molecular evidence makes a compelling case for a high degree of phylogeographic structuring of the Atlantic salmon in the Western, as well as the Eastern, Atlantic. However, as with the Eastern Atlantic, the full extent of this structuring remains to be resolved.

5.3.4 Resident (non-anadromous) salmon

Resident or non-anadromous salmon are a common component of the freshwater fauna of river systems, particularly in the Western Atlantic and in European Russia (Chapter 2). Resident and anadromous populations of Atlantic salmon, prior to 1947, were seen as distinct taxa and assigned to *S. s. salar* and *S. s. sebago*, respectively. However, this distinction was rejected by Wilder (1947) as he was unable to demonstrate consistent morphological differences between the two types, a view which is now widely accepted. However, the concept of a non-anadromous subspecies of landlocked Atlantic salmon persists in the literature with continued reference to non-anadromous forms as *S. s. sebago* (e.g. Vuorinen 1982; Ståhl 1987; Pursianinen *et al.* 1998).

The actual ecological and evolutionary relationship between resident and anadromous Atlantic salmon work appears to be complex. Non-anadromous and anadromous populations, even when they occur in the same river system, are found to be genetically differentiated (Vuorinen and Berg 1989; Verspoor 1994; King *et al.* 2000, 2001); this is the case even when

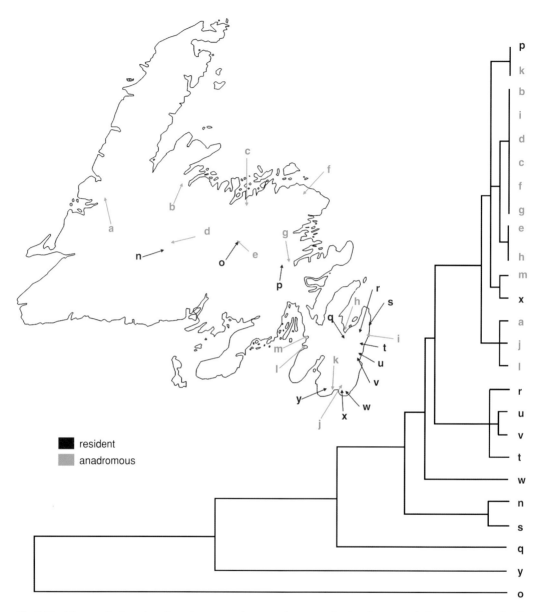

Fig. 5.11 The genetic clustering of anadromous and non-anadromous Atlantic salmon populations in Newfoundland based on differentiation at 21 allozyme loci. Based on Verspoor (1994).

they are found in the same habitat (Verspoor and Cole 1989; Verspoor 2005; see Box 5.8). The species shows both morphometric and meristic variation, but it is unlikely to be congruent with studied genetic variations (Claytor and Verspoor 1991), even among sympatric salmon populations. Historical factors affecting these phenotypic and genotypic characters are likely to have led to their independent evolution.

Verspoor (1994) examined the pattern of allozyme differentiation at 23 loci in 12 geographically paired sets of anadromous and non-anadromous salmon populations on the

Table 5.2 Atlantic salmon populations assayed for the range-wide comparison of microsatellite DNA variation referred to in Figs 5.4, 5.5, 5.7, 5.10, 5.11, 5.12 and 5.14.

Country	River	Year	Sample ID	N
Western Atlantic				
USA – Maine	Narraguagus	1994–98	ME1	642
	East Machias	1994–98	ME2	321
	Dennys	1993–98	ME3	264
	Ducktrap	1994–99	ME4	123
	Penobscot	1998–00	ME5	572
	Cove	1994–99	ME6	102
	Kend	1997–99	ME7	111
	Bond	1994–96	ME8	106
	Togus	1994–96	ME9	81
	Machias	1994–98	ME10	572
	Pleasant	1995–2000	ME11	483
	Sheepscot	1994–98	ME12	270
	Sebago Lake (resident)	1995–98	ME13	50
	Grand Lake (resident)	1995–98	ME14	50
			Maine total =	3747
Canada	Gander – Soulis Brook	1994	NF1	30
	Gander – Jonathan's Brook	1994	NF2	16
	Gander – Lower	1995	NF3	63
	Conne	1994	NF4	30
	Trepassey		NF5	23
	Ouananiche Beck		NF6	40
	Bristol Cove		NF7	45
	St John – Nashwaak	1996	NB1	66
	St John – Hatchery	1994	NB2	23
	St Croix	1996	NB3	57
	Miramichi	1999	NB4	56
	Michaels	1997	LA1	28
	Sand Hill	1997	LA2	16
	Saguenay		QB1	58
	Saint Jean	1999	QB2	63
	Stewiacke	1994	NS1	55
	Gold	1994–98	NS2	52
			Canada total =	721
Eastern Atlantic				
Norway	Vosso	1996	NO1	44
	Lone	1996	NO2	57
Ireland	Spaddagh	1994	IR1	29
	Blackwater	1991	IR2	34
Iceland	Ellidaar	1998	IC1	50
	Vesturdalsa	1998	IC2	46
Spain	Eo	1995	SP1	26
	Esva	1995	SP2	18
	Bidasoa	1994	SP3	26
	Sella	1994	SP4	14
Scotland	Shin	1997	SC1	24
	Nith	1997	SC2	29

Table 5.2 (cont'd)

Country	River	Year	Sample ID	N
Russia	Pecha		RU1	15
	Kachkovka		RU2	50
England	Hodder		EN1	47
France	Scorff		FR1	25
Baltic				
Finland	Tornionjoki	1998	FI1	61
			Europe total =	573
			Grand total =	5041

island of Newfoundland. No systematic differences in the genetic character of the two population types were found, consistent with each having evolved independently from anadromous ancestors. The two population types in Newfoundland do not form two separate clusters (Fig. 5.11). Some regional clustering is seen, particularly for the anadromous populations; however, there is no tendency for anadromous and resident populations from the same area to cluster together. Rather the resident populations show a much higher level of differentiation, even though they derived from the same geographical area. Indeed, across the five polymorphic loci found, variance among non-anadromous populations was over 10 times greater than for anadromous populations ($F_{ST} = 0.375$ vs $F_{ST} = 0.033$, respectively). Greater divergence of non-anadromous populations is also seen with regard to allozymes in European populations (Fig. 5.8; Ståhl 1987; Tonteri et al. 2005; Verspoor et al. 2005).

The same conclusion has been reached by King et al. (2001) for resident populations of Atlantic salmon in Maine, based on genetic studies of microsatellite variation. They found resident salmon from Maine had no unique genetic variation compared to anadromous populations (King et al. 2000, 2001). This finding suggests that non-anadromous salmon which colonised North America were not a monophyletic group derived from a spatially distinct glacial refuge relative to the region's anadromous fish. Furthermore, the absence of unique alleles at the loci examined in the resident populations of Maine is consistent with the divergence of non-anadromous populations as a result of stochastic processes (e.g. genetic drift) after colonisation of this part of the species range following deglaciation.

Other considerations point to the importance of stochastic processes such as genetic drift in the observed differentiation between resident and anadromous populations at the loci studied. Within a river system, genetic differentiation among the two types of populations often exceeds that seen between anadromous populations separated by hundreds or even thousands of kilometres. For example, the differentiation between the sympatric populations of the two forms in Little Gull Lake, Newfoundland, at allozyme loci, is as great as between anadromous populations on either side of the Atlantic (Verspoor and Cole 1989). Greater divergence among resident populations within a river than among anadromous populations in different river systems has been found by Tessier and Bernatchez (2000) in the Saguenay River in Quebec, with regard to microsatellite loci, as well as at allozyme loci in populations of the two forms in European rivers (Verspoor et al. 2005). This is again consistent with differentiation driven largely by stochastic processes since genetic drift would be expected to be greater in the more historically isolated and generally smaller resident populations than in populations of anadromous salmon.

The evidence clearly supports the polyphyletic origin of non-anadromous populations of Atlantic salmon in different river systems. However, this raises the question of whether populations of non-anadromous salmon within different river systems represent distinct phylogenetic groups. If populations have not had gene flow between them for thousands of years and have evolved a high degree of genetic divergence, then they represent distinct phylogeographic entities and should be recognised as such. This view is supported by the high degree of genetic isolation shown by non-anadromous populations sympatric with anadromous forms such as seen in Little Gull Lake, Newfoundland. In addition to the life-history and genetic divergence from a sympatric anadromous population there is also significant phenotypic divergence (e.g. Claytor and Verspoor 1991) and divergence in spawning behaviour (Verspoor and Cole 2005). Furthermore, the levels of genetic divergence observed between anadromous and non-anadromous forms in the same watershed are often greater than those seen between the anadromous populations of the Baltic and Atlantic Seas (Verspoor 1994, unpubl.; Verspoor *et al.* 2005), groups which are widely recognised as distinct phylogeographic groups.

Whether non-anadromous populations in the same river system will always be distinct phylogeographic groups is less clear. The non-anadromous forms of the River Neva–Lake Ladoga–Lake Onega region studied by Säisä *et al.* (2005) suggest that populations within river systems may have a common origin. However, if populations within a river are isolated from each other by becoming landlocked in different tributaries and evolved genetic differences, they would represent distinct phylogeographic groups. For example, populations NF6 and NF7 (Table 5.2) are reproductively isolated populations of non-anadromous salmon from the same river system in Newfoundland. The level of microsatellite divergence (Fig. 5.6) seen for these two populations is as great as between any two anadromous populations in North America. The same picture is seen with respect to allozyme and mtDNA variation within this river (Verspoor, unpubl.). However, there is evidence from northern Russia that in many cases non-anadromous populations found in the same river system will share a common evolutionary origin (Tonteri *et al.* 2005).

Collectively this work suggests an extensive and complex phylogeographic structuring of the species with respect to non-anadromous forms, though this requires further investigation. It is clear that the genetic diversity among populations of non-anadromous forms, both among and within river systems, represents a large and important part of the genetic diversity present in the species. It is, unfortunately, one which is all too often ignored, particularly in North America, due to its relatively limited commercial and recreational interest.

5.3.5 *Historical origins*

The observed patterns and levels of regional genetic differentiation do not conform to a simple model of increasing genetic differentiation with increasing physical separation which would be expected if genetic differentiation among regions was simply a function of levels of contemporary gene flow. Rather the patterns are most consistent with populations in different regions having been isolated for considerable periods of time with little or no gene flow between them. As such, the genetic differences reflect regionally distinct evolutionary lineages with differentiation being phylogeographic. However, in most cases the divergence of the

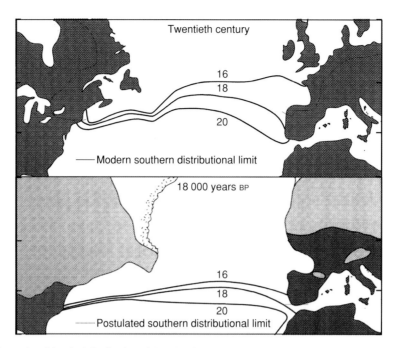

Fig. 5.12 The modern historical distribution of the Atlantic salmon relative to the southern limit of the Pleistocene ice sheets at the time of the last glacial maximum (LGM), 18 000 years BP, based on summer sea surface temperature (SST) isotherms for two periods which bracket the modern southern limit of the species (shown). Taken from Wilson *et al.* (2000).

phylogeographic groups is too recent for them to have evolved fixed genetic differences as seen at a high proportion of loci at the species level.

Much if not all the molecular variation appears to be predominantly the result of stochastic processes such as genetic drift rather than selection, and a number of different historical influences probably lie behind the observed phylogeographic partitioning. The most important of these is likely to be historical distribution and patterns of mixing of Atlantic salmon throughout the Pleistocene and the recent pattern of postglacial colonisation of the species' modern range. Most of the modern historical range of the Atlantic salmon is located in regions covered by Pleistocene ice sheets (Fig. 5.12).

In North America, the species' entire modern range lies within the limit of the last glacial maximum (LGM), 18 000 years BP. In Europe, the only parts not glaciated occur in west central and south-western areas such as Iberia. In the latter area, Atlantic salmon vertebrae have been found dating from 16 000 to 40 000 years BP (Consuegra *et al.* 2002), indicating that this area, well to the south of the main ice sheets, was a part of the species' Pleistocene range. However, almost certainly it occurred further north in non-glacial rivers within kilometres of the ice, at least along the Atlantic seaboard, as the species does now in both West Greenland and Iceland (MacCrimmon and Gots 1979). The conditions today at the ice edge in these more northerly regions are likely to be more severe than those at the southern edge of the Pleistocene ice sheets, given the lower solar energy input further north.

At times of glacial advances, the Atlantic salmon's range expanded southward, though not to the same extent that its northern limit moved south. Climate models at the time of the LGM indicate that climate change south of the species' current southern limit was much less marked than in the north (Fig. 5.12), such that the species is unlikely to have occurred south of the Iberian Peninsula in Europe. In the Western Atlantic, the distribution of salmon during the Pleistocene was probably even more constrained, as the southern limit of the ice sheet extended further south than in Europe and the climatic gradient south of this was even steeper (Fig. 5.12).

Atlantic salmon may also have occurred in regions near to the ice which are now submerged. During the Pleistocene, sea level was lower and presently submerged areas to the east of the ice sheets in North America, such as the unglaciated Georges Bank marine area, are likely to have had rivers with salmon populations which would have served as refugia (Verspoor *et al.* 2002). Such areas also probably existed to the south-west of the modern Irish coastline (Bowen *et al.* 2002). As the ice sheets advanced and retreated over the Pleistocene period, a scenario of changing distribution is likely to have repeated itself to varying degrees a number of times, of which the Weichselian/Wisconsin glacial cycle is the last one. There were a number of glacial advances and retreats within this last glacial period (Andersen and Borns 1997).

The most marked phylogeographic division in the Atlantic salmon, between populations on the two sides of the Atlantic, shows an almost total geographic separation of two deeply diverged phylogenetic lineages. The level of divergence, and its manifestation with regard to individual DNA loci as well as chromosome structure, suggests that the divergence of these two lineages is very old. How old can be roughly estimated from the level of mtDNA divergence between the two lineages (Box 5.7). This comparison suggests that they split from a common ancestor ~ 600 000–700 000 years BP. This places their split just before the start of the longest Pleistocene glacial period of the last 2.5 million years, a period of ~ 200 000 years of largely uninterrupted cold (Andersen and Borns 1997). Certainly, divergence would have been strongly promoted by prolonged isolation in separate glacial refugia on different sides of the Atlantic during this period of time. Within the main 'European' mtDNA lineage, divergence between its two component sub-lineages is 10% of that seen between the two main lineages, placing their evolution during the last glacial period within the last 115 000 years BP and probably prior to the LGM 18 000 years BP.

Isolation of salmon in the Eastern and Western Atlantic for much of the last 600 000–700 000 years would explain the strong association of each region with a highly diverged genetic lineage. The duration of the isolation also explains the absence of intermediate haplotypes between the two mtDNA lineages and the virtual fixation of differing sets of alleles at a number of nuclear marker loci. If one assumes a more or less constant rate of molecular evolution for effectively neutral variation, the much greater divergence seen between as compared to within continents is expected. We know that populations within continents, in most cases, have been isolated for less than 15 000 years (i.e. since their establishment following the deglaciation of the watersheds in which they now occur).

The lack of completely disjunct distributions of the two phylogeographic lineages which mark the two sides of the Atlantic is most likely to be the result of secondary contact and gene flow between the two phylogeographic groups early on in the current postglacial period (Mahkrov *et al.* 2005; Verspoor *et al.* 2005). In some more northerly regions, variants typical of North America are seen in Europe (Kola Peninsula), and variants typical of Europe are seen in North America (Newfoundland and Labrador). Yet native populations in the southern parts of Europe and North America, the areas most closely associated with refugial areas on the two continents, are fixed for haplotypes associated with one or the other of the deeply diverged mtDNA lineages as well as, in many cases, for different sets of allelic variants at some nuclear

Box 5.7 A molecular clock for timing phylogenetic divergence.

Molecular divergence can be used to provide rough estimates of the time groups of populations within a species split from a common ancestor, particularly where levels of divergence among intraspecific phylogenetic groups are high. The use of molecular divergence as a clock, however, requires considerable caution (Avise 2004) and that a number of conditions be met.

The first condition is that the rate of mutational substitution in the molecule used is fast enough to generate a significant number of differences in the time frame involved but slow enough such that mutations are unlikely to be 'overwritten', i.e. that multiple mutations occur at the same location. The second is that the divergence rate can be calibrated independently. This can be done by using observed levels of divergence for the locus, or similar locus, in the same or a closely related taxon and estimates of divergence time of phylogenetic groups within the taxon based on independent estimates of geological events known to be associated with their divergence.

Different types of DNA sequences (or proteins), or even different parts of the same gene (or protein), can evolve at markedly different rates, such that it is important to use similar genes for calibration to those being queried. The rate of evolution for a given stretch of DNA can also be expected to vary across major taxonomic groups. However, if these conditions can be satisfied, a rough estimate can be made of the expected percentage change per million years for the region of DNA in question.

The concept of a molecular clock fits with Kimura's theory of neutrality, as the rate of neutral evolution of DNA is the same as the mutation rate for neutral alleles; this assumes that variation at the locus in question is neutral to the effects of selection. It also assumes that mutations at the locus occur at similar rate in the taxon in question as in the taxon used for the calibration. If so, the difference between the sequences of segments (or protein) in two lineages or groups is proportional to the time since they diverged from a common ancestor (also known as the coalescence time). It is possible to use divergence at a locus as a molecular clock if a stretch of DNA is subject to selection provided the selection is more or less uniform across the time frame of divergence of the DNA used for calibration and the time frame of the sequence whose divergence time is being estimated.

A conventional calibration for the overall rate of evolutionary change in animal mtDNA is about 2% sequence divergence per million years between pairs of lineages separated for less than 10 million years; beyond 15–20 million years, mtDNA sequence divergence begins to plateau, presumably as the genome becomes saturated with substitutions at variable sites. This is the normal rate assumed for animals for the mtDNA molecule as a whole. However, the rate of divergence is known to vary among genes (Avise 2004), with the *16S rRNA* region generally evolving slower and the D-loop control region faster than other parts of the molecule, something supported by the limited evidence from sequencing the overall Atlantic salmon mtDNA genome (Gross *et al.* unpubl.). The data generally suggest that for enzyme coding genes such as *COI*, *ND2* and *ND1*, the rate of evolution is quite similar. Bermingham *et al.* (1997) found divergence rates of 1.2% for the *COI* gene and 1.3% per million years for the *ND2* and *ATPase6* genes in a group of coral reef fishes. This is based on divergence of closely related fish species on either side of the Panamanian Isthmus and estimates of the timing of the joining of North and South America 2.9–3.5 million year (Ma), based on geological considerations.

Nilsson *et al.* (2001) found a maximum of 0.73% (9 base substitutions) divergence in a 1224 bp segment of the *16S rRNA/ND1* between the two most closely related haplotypes in the Eastern and Western Atlantic lineages. For a slightly larger 1330 bp region, an estimate of 0.83 (14 base substitutions) was obtained Verspoor, unpubl.). Using the *COI* and *ND2* gene divergence rate cited above, this places the estimated time of divergence of these two lineages from a common ancestor around 600 000–700 000 years ago. Observed divergence between the two sublineages within the main European lineage is one base substitution or ~ 0.1% (Nilsson *et al.* 2001; Verspoor, unpubl.) suggesting they split in the last 100 000 years BP, though, given this is based on a single base substitution, the errors around this estimate are large. In contrast, divergence between Atlantic salmon and brown trout is 5.4 % (Verspoor, unpubl.), placing the split of these two species from their common ancestor at around 4–5 million years BP.

Recent work on D-loop sequence variation (Table B5.1) shows divergence between the two main evolutionary lineages of 1.4%. This is in accord with values for a 940 bp section of the control region of ~ 1.5% (Kauppi *et al.* 1997). This is greater than for the *16S RNA/ND1* gene region, as expected. However, no estimate of time of divergence can be derived using this regime as there is no suitable published calibration for rates of change in this part of the mtDNA in teleost fishes.

Microsatellite loci, in principle, have properties that make them suitable for dating evolutionary events. In particular they have a relatively high mutation rate, such that a reasonable number of mutational events can occur within the short time of modern human evolution. However, microsatellites have yet to be applied rigorously to estimating divergence times among the different Atlantic salmon phylogenetic lineages observed. Given the potential for homoplasy (i.e. alleles that are identical in state but not by descent) in this class of marker, a large number of such loci would need to be used to infer divergence times within phylogenetic and phylogeographic relationships.

allozyme and satellite DNA loci. This distribution is probably the result of a short period of limited, spatially restricted gene flow between the two continents following the LGM at the time of the postglacial establishment of populations in these regions. The distribution cannot be attributed to stocking. Unlike some more southerly parts of the species range, no transatlantic stocking has been carried out in these regions (MacCrimmon and Gots 1979).

Variants from both of the deeply diverged phylogenetic lineages are unlikely to have been present in refugial populations on both sides of the Atlantic, given their absence in more southerly populations; the latter might be expected to be most similar genetically to refugial populations. Variants typical of the other side of the Atlantic, found at numerous different loci within each continent, do not occur in the same individuals or, in most cases, in the same populations, only in the same general region. If contemporary gene flow was involved the variants at the different loci should be found in the same individuals in the same populations, and if it was relatively recent, at least in the same populations, as well as in the same region. On the other hand, only a regional association is expected if gene flow is old and followed by a substantial period without any gene flow. As 'European' variants occur in some Newfoundland and Labrador non-anadromous populations (probably established and isolated from other salmon ~ 10 000 years BP) the gene flow is likely to pre-date this time.

The interaction of the species' biological character with historical geographical factors likely accounts for the observed patterns of distribution of genetic variation within each continental area. The range of the salmon in North America, both today (Chapter 2) and historically (Fig. 5.13), will have been smaller and less geographically structured than in Europe; the modern historical range of the salmon in Europe is roughly three times larger than in North America. The latter alone could account for the higher overall genetic differentiation seen among Eastern

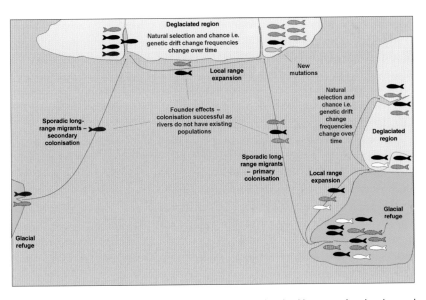

Fig. 5.13 A schematic summary of the historical processes likely to be involved in generating the observed patterns of genetic differentiation among Atlantic salmon populations. The pattern of variation arises because, once established, gene flow becomes highly restricted by inherent homing behaviour and to local population adaptation which leads to the relative maladaptation of migrants compared to native fish. The latter would make it unlikely that those fish that do stray will make a substantive contribution to future generations.

compared to Western Atlantic populations as well as the higher diversity of alleles and greater heterozygosity (King *et al.* 2001). It would also explain the lower mtDNA haplotype diversity resolved by restriction enzyme analysis in the Western Atlantic (King *et al.* 2000) relative to Europe. However, lower levels of diversity in the Western Atlantic may be due to population founding from a relatively small number of individuals when established by migrants from Europe. The Eastern Atlantic Ocean is the most likely location of the common ancestor of the two continental groups, given other members of the genus *Salmo* are endemic to Europe.

In contrast to continental divergence, phylogeographic divergence within continents does not appear to be characterised by different regions having different variants. Rather divergence involves differences in variant frequencies, the absence of variants in some regions and the presence of rare variants which are absent elsewhere (e.g. Säisä *et al.* 2005; Verspoor *et al.* 2005; Gross *et al.* unpubl.), as expected for more recent phylogeographic divergence. The fact that most populations occur in regions covered by ice at the time of the LGM, the observed phylogeographic differentiation within continents can be assumed to be less than 18 000 years old and, in many cases, less than 10 000 years old. This falls within the estimated evolutionary divergence times between the two mtDNA haplotypes within the Eastern Atlantic lineage (within the last 100 000 years; Box 5.6).

The lack of distinct regional sets of alleles suggests that, if there was more than one refugial area on each side of the Atlantic, these refuges were isolated for sufficiently brief periods of time to preclude the evolution of distinct sets of variants. Ice sheets were at their maximum extent for a few relatively short periods during the last 125 000 years since the last interglacial, at least in Europe, lasting less than 5000 years each (Boulton *et al.* 2001). Throughout the last glacial period, with a continually changing distribution of ice and range expansion and contraction, a considerable amount of historical mixing within each continental area occurred.

During each glacial cycle, the genetic character of local and regional groups of populations would have been conditioned by a combination of processes (Fig. 5.13). This most likely involved local range expansion from refugia into new adjacent habitat, combined with less frequent long-range migration to distant deglaciated regions. In the latter case, newly established populations would then serve as the main source of migrants for local range expansion, giving rise to the regional phylogeographic groups seen today. Insight into the processes of local range expansion at the edge of retreating glaciers is being gained in respect of studies of Pacific salmon in Alaska (e.g. Milner *et al.* 2000).

The distribution of molecular variation suggests a number of different refugial areas existed at the time of the LGM, in both Europe (Koljonen *et al.* 1999) and North America (Verspoor *et al.* 2002). Furthermore, in the Eastern Atlantic, the previously glaciated parts of the Atlantic salmon's modern range have been differentially colonised from these refugial areas. Overlaid on this are founder effects and genetic drift associated with the establishment of the first populations in many regions, as well as new local mutations. Combined with highly restricted historical gene flow, these stochastic (i.e. non-selective) processes would account for the restricted regional distribution of many rare mtDNA (Verspoor *et al.* 2002) and allozymes (Verspoor 2005) variants.

A number of authors over the years (Power 1958; Behnke 1972; Berg 1985) have proposed that non-anadromous populations of Atlantic salmon evolved independently in most of the river systems from anadromous colonisers. This view is based on the absence of consistent morphological differentiation of the two forms (Wilder 1947), the distribution of non-anadromous forms and their resident habit, and the observation that anadromous forms which are 'landlocked' are able to mature without going to sea. This view is supported by the molecular evidence.

Non-anadromous forms are likely to have evolved in one of two ways. The first is where early anadromous colonisers were landlocked by physical changes in river systems associated with postglacial isostatic rebound of the landscape, such that migration to sea and returning to spawn was precluded. The second possibility is that some early anadromous populations were 'landlocked' during a temporary cold phase when, while freshwater conditions were still suitable for salmon, lethally cold sea temperatures stopped the marine migration. This situation is seen today in some years in Ungava Bay (Power 1958). Berg (1985) suggested 'landlocking' by temperatures is likely to have occurred during the Younger Dryas period starting 11 000 years BP, when for 1000 years glaciers readvanced and the North Atlantic became much colder. As a result of this isolation, many if not all affected populations would appear to have been under strong selection for genetic types which did not migrate, giving rise to the non-anadromous forms observed today. These forms are inherently non-anadromous as, even for those populations which now could undergo a marine migration, they do not do so (e.g. Little Gull Lake, Newfoundland: Verspoor and Cole 1989, 2005).

5.4 Regional and local population structure

The full extent of phylogeographic structuring among Atlantic salmon populations within the Eastern and Western Atlantic regions remains to be established. Phylogeographic groups which have recently diverged relatively and maintained large effective population sizes will show only low levels of divergence. In such cases, differentiating phylogeographic divergence from the divergence of populations within metapopulations becomes difficult. However, regional and local differentiation detected using molecular markers clearly show the species to be highly structured into distinct genetic populations both among and within rivers.

Understanding population structuring, whether or not it is associated with phylogeographic divergence, is critical to the development of effective conservation and management policy and programmes. Genetic populations are the fundamental units responsible for the local character and abundance of Atlantic salmon. These units need to be the focus of management, as management challenges, with regard to exploitation, habitat alteration, and so on, are often likely to vary among populations or population assemblages.

Three aspects of population structuring are important. The first is the number of populations in a region or river and their spatial boundaries. The second is the extent of gene flow which occurs, or could potentially occur, among them. The latter relates to whether Atlantic salmon populations in specific circumstances are organised into metapopulations. As mentioned previously, these are groups of local populations linked to varying degrees over time by low-level or sporadic gene flow and within which, if populations become extinct, they are re-established by migrants from the other populations in the group (Adkison 1995). These two aspects are dealt with here. The third aspect of population structuring, adaptive differentiation, is considered in Chapter 7.

Insight into the structuring of the Atlantic salmon into genetic populations and metapopulations has been gained largely from studies of allozymes and microsatellite DNA loci. Mitochondrial DNA, a molecule expected to show higher levels of variation between populations due to its maternal, haploid inheritance and high mutation rate (Chapter 4), has so far proven to be less informative. Allozyme analyses have provided significant insight in many situations but the low level of allelic variation has limited their usefulness (Elo 1993; Verspoor *et al.* 2005). More informative has been recent work based on microsatellite loci. This is because of the number of microsatellite loci which can be screened and the high levels of allelic diversity most such loci display; most microsatellite loci show ten or more allelic variants, rather than just one or two as seen in allozymes.

The high level of polymorphism resolved for microsatellite loci allows these loci to be used both for the characterisation of genetic differences based on the distribution of diversity within and among samples, as for allozymes, and the analysis of the relatedness of individuals, a level of resolution not possible with allozymes. The analysis of relatedness can be used to study population structure and, as detailed in Chapter 8, to assign an individual to the region, river or population of origin.

5.4.1 Spatial scale and boundaries

Studies using physical tags have long indicated that anadromous salmon in different river systems belong to different genetic populations (Stabell 1984). However, the work has been limited to relatively few river systems. Furthermore, radio tracking shows that some salmon stray temporarily into rivers adjacent to the one in which they eventually spawn. Moreover, salmon that do stray permanently may not spawn successfully. Thus, estimates of straying probably overestimate genetic exchange.

Studies across the species range indicate levels of genetic exchange are generally low, consistent with widespread structuring of anadromous Atlantic salmon into discrete genetic populations both among and within rivers. Observed differences in the frequencies of genetic variants between samples from different rivers or tributaries are incompatible with their derivation from the same population. This has been found with respect to all classes of genetic variation studied (e.g. allozymes, Verspoor *et al.* 2005; mtDNA, Gross *et al.* in prep., microsatellites, King *et al.* 2001). The same conclusion is reached for non-anadromous salmon in different river systems, though levels of differentiation are generally much higher (Verspoor 1994; Verspoor *et al.* 2005); this is consistent with the fact that non-anadromous populations in different river systems will not have the scope for physical mixing due to their behavioural, and often physical, isolation.

Within river systems, the nature of population structuring is potentially more complex. In river systems with non-anadromous forms, those above impassable falls will be physically isolated and encompass one or more genetically distinct populations. Allozyme studies show resident salmon of the Upper Namsen, above an impassable falls, are highly differentiated from anadromous salmon in the lower river (e.g. Vuorinen and Berg 1989). In the headwaters of the Saguenay River in Quebec, genetic studies show there to be four distinct 'landlocked' populations which contribute to the mixed fishery in Lac St-Jean but which home to four distinct spawning tributaries in rivers entering the lake (Tessier *et al.* 1997). This is also seen where there are no physical barriers to mixing (Verspoor *et al.* 2005), even when non-anadromous and anadromous forms cohabit. In Little Gull Lake on the Gander River in Newfoundland, the two cohabiting forms belong to genetically distinct populations with no gene flow between them (Verspoor and Cole 1989). In this particular case, it appears to be due to the non-anadromous population spawning in the lake while the anadromous form spawns in the inlet and outlet streams (Box 5.8).

Box 5.8 Defining the spatial boundaries: a case study.

Molecular genetic variation can, in principle at least, be used to define the boundaries of populations within river systems. This becomes possible where sufficient genetic differences can be resolved to allow the population of origin of individuals to be assigned with a high degree of certainty. Levels of genetic differentiation among populations within rivers identified for allozymes and mtDNA variation has generally been insufficient for this purpose. Such analyses can be expected to become more common in the future using analyses of microsatellite variation, where the use of large numbers of differentiated loci gives the discriminatory power needed to assign individuals to their population of origin with much higher certainty than was previously possible (King *et al.* 2001; Spidle *et al.* 2001; Chapter 9).

One situation where it has been possible to use allozyme variation to define the spatial boundaries of a population is with regard to the non-anadromous salmon of Little Gull Lake, in the Gander River system on the Island of Newfoundland. In Little Gull Lake, allozyme variation serves as a good marker of the non-anadromous and anadromous populations of Atlantic salmon cohabiting in the lake (Verspoor and Cole 1989, 2005; Fig. B5.8). Non-anadromous individuals have combined *AAT-4*/MDH-3,4** genotypes which are very rare in the anadromous populations and vice versa; the three types characterising > 97% of non-anadromous salmon represent only 6.5% of anadromous fish. Using these genetic markers, small-scale spatial sampling of juvenile salmon in the inlet and outlet streams plus two upstream lakes was able to show that the non-anadromous population was largely, if not exclusively, restricted in distribution to the lake. Only a small proportion of individuals sampled in the upper of the two headwater lakes had genotypes consistent with being from the non-anadromous population. This analysis, in combination with the nature of observed morphological differences (Claytor and Verspoor 1991), suggests that the non-anadromous population is lake spawning and behaviourally constrained to remain in the lake, despite the fact that movement into the rivers is physically possible.

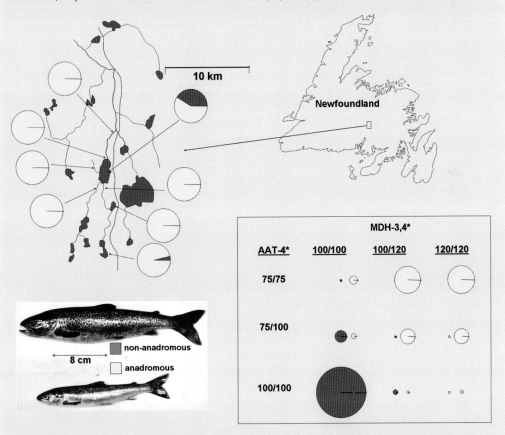

Fig. B5.8 Schematic depicting the delineation of anadromous and resident Atlantic salmon inhabiting Little Gull Lake in the Gander River system, Newfoundland, by the two-locus *AAT-4*/MDH-3,4** genotypes.

Tagging studies have suggested that population structuring of anadromous stocks within many river systems also occurs. For example, in two tributaries of the River Dee in north-east Scotland, all migrating smolts have been tagged for the last 15 years but none have been recaptured in a different tributary (A. Youngson, pers. comm.). Yet the two tributaries are separated by 30 km of the river's main stem which contains continuous juvenile and adult spawning habitat. Consistent with the tagging data, fish in the two tributaries show significant genetic differentiation for allozyme, mtDNA and minisatellite markers (Verspoor 1995), as do most of the salmon inhabiting tributaries in the system (Verspoor, unpubl.). Such differences have been found in most larger river systems which have been studied (e.g. Verspoor et al. 2005), including the Penobscot River in Maine which has been shown to support three distinct populations (two tributaries and the main stem) (Spidle et al. 2001). Only in smaller river systems does structuring appear to be absent. For example, a detailed study of the River Polla in northern Scotland, a small river with only 7 km of salmon habitat confined to the main stem of the river, found no evidence for more than one genetic population (Verspoor, unpubl.).

That many river systems contain multiple genetic populations of salmon is now widely accepted. What remains unclear, in cases where there is no obvious habitat fragmentation and separation of populations, are the spatial boundaries of populations and how these are defined. This aspect of population structure remains to be resolved and represents the next task to be addressed by molecular genetic studies. Without this knowledge, it will be difficult to implement a truly population-specific approach to the management and conservation of the species, which will be needed if management issues vary significantly in different parts of a river system. This will be the case, for example, where the timing of the return of adults from different populations varies and they are differentially exploited, or the disturbance of fish habitat is localised. Knowledge of the spatial boundaries of populations is also needed when collecting broodstock as part of supportive breeding programmes, to avoid population mixing and potential problems with outbreeding depression (see Chapter 7).

5.4.2 Metapopulation structure and gene flow

The evolution of distinct phylogeographic groups of populations requires a prolonged historical absence of consistent gene flow. However, interrelated populations may exist where gene flow is generally highly constrained or is sporadic and infrequent. Observations of natural recolonisation of rivers and tributaries which have lost their salmon stocks, due to dams or weirs, suggest that migration can occur when habitat is vacant. A recent genetic study suggests that recolonisation can also involve migration from one river system to another when rivers are relatively close to each other (e.g. Vasemägi et al. 2001). Migration may also occur among river systems where populations in rivers have been severely reduced (Consuegra et al. 2005). Gene flow among some populations is conditional depending on the status of individual populations. Groups of populations so linked, where population declines or extinctions occur sporadically and are countered by gene flow, constitute a higher-order unit of genetic and evolutionary organisation, the metapopulation.

Understanding of metapopulation structuring in Atlantic salmon is as yet limited. Recent research suggests that the salmon populations in Maine are operating as a metapopulation distinct from populations elsewhere in North America. This is supported by the genetic coherence of Atlantic salmon populations within Maine relative to other nearby North American populations (Fig. 5.14), implying that habitat conditions and linkages have resulted in a historical connectedness by migration, gene flow and population re-establishment. It is also

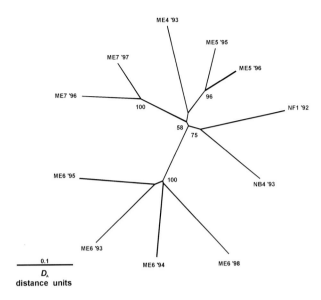

Fig. 5.14 Genetic distance tree (D_A, Nei et al. 1983) based on 12 loci illustrating population structure within a large river system in Maine. Numbers indicate bootstrap support for each node after 10 000 replicates. Absence of number indicates less than 50% bootstrap support for that node; abbreviations for populations refer to Table 5.2. After Spidle et al. (2001).

supported by observed natural population restoration. Atlantic salmon were effectively extirpated from several rivers in the state of Maine during the Industrial Revolution yet at least one, the Kennebec River, appears to have been repopulated with fish of Maine origin in the absence of any restoration efforts (Spidle et al. 2003; NRC 2004).

It is likely that the Maine metapopulation has, in part, been artificially maintained. Salmon of Machias and Narraguagus origin were artificially reared in order to enhance the population of the Penobscot River drainage, and then the enhanced Penobscot population was stocked throughout the state. However, propagation in Maine waters of fish imported from Canada has apparently been unsuccessful based on the lack of genetic affinity between the two population groups (King et al. 2001; NRC 2002, 2004; Spidle et al. 2003), suggesting specific adaptation to native habitat allowed fish of Maine origin to survive even when large numbers of Canadian fish were being released into these rivers (Baum 1997; Spidle et al. 2003). The success of native Maine fish, in the face of reduced population size and increased competition from stocked non-native fish, emphasises the potential for both local adaptation and demographic structure (Chapter 7; Ardren and Kapuscinski 2003) enabling populations to survive below levels commonly thought possible. That native fish were able to repopulate habitat from which they had been extirpated (e.g. Kennebec River) suggests that the Maine salmon metapopulation is a viable unit of genetic diversity and that the metapopulation is an important consideration in species management. This type of population structuring may be an important mechanism for dealing with the potential pitfalls of small population sizes (see Chapter 8).

Though few studies have been done, it seems metapopulation structuring is likely to be widespread in the Atlantic salmon. Metapopulations most likely exist wherever groups of either anadromous or non-anadromous populations occupy spatially connected or neighbouring habitat patches, among which migration and gene flow is possible and likely. Indeed, metapopulations are likely to be a part of the evolutionary dynamic of salmon populations and important to the maintenance of local populations. They are likely to be biologically important in so far as they increase the long-term effective size of individual constituent populations, reducing inbreeding and loss of genetic variation (see Chapters 4 and 9), and lead to the re-establishment of extinct populations.

5.5 Overview

Biodiversity and population structuring in the Atlantic salmon has been reviewed at a number of levels and a potted synthesis of the insight which has been gained from molecular genetic work over the last 40 years is given in Fig. 5.15. The levels considered are part of a continuous spectrum of potential evolutionary divergence and genetic isolation.

Regular low-level or sporadic gene flow within metapopulations will preclude the development of long-term phylogeographic divergence among constituent populations. However, with increasing genetic isolation, populations or groups of populations will diverge. As such the metapopulations, between which gene flow is absent, can be viewed as the most basic phylogeographic units, among which divergence will be conditioned by their size and degree of isolation. With the advance and retreat of glaciers over much of the salmon's range, metapopulation and phylogeographic structure is likely to have been lost and reconstructed repeatedly, newly evolving with each range expansion, with one exception. This is the subdivision of the species into distinct Eastern and Western Atlantic phylogenetic groups, which appears to have been sustained for more than 500 000 years and numerous cycles of glaciation. The remaining structuring seen is likely to be less than 18 000 years old.

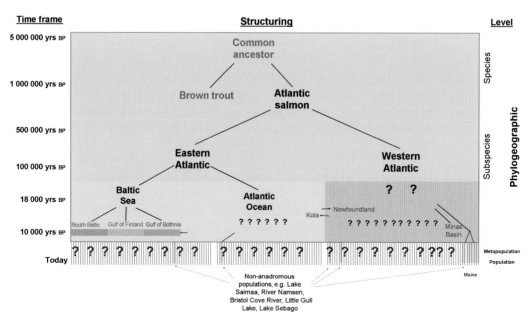

Fig. 5.15 Diagrammatic overview of hierarchical population structuring detected in the Atlantic salmon using molecular genetic markers. ? indicates aspects of structuring which remain to be resolved.

Phylogeographic divergence has important implications for conservation and management. It can be expected to be associated with significant adaptive differentiation (Chapter 7; Avise 2004). For example, salmon from the Baltic Sea region are resistant to the endemic monogenean skin parasite *Gyrodactylus salaris*, while those from European Atlantic drainages, where the parasite is non-native, are susceptible (Bakke *et al.* 2004). Given their greater divergence, the adaptive differentiation of Eastern and Western Atlantic phylogeographic groups is also likely to be greater. Though few studies have been done to examine this question, recent work (Cauwelier *et al.* unpubl.) shows that second-generation backcrosses of hybrid offspring to parent populations is associated with severely reduced viability of offspring. This is seen, though to a much smaller degree, where different populations from the same continental phylogeographic group are crossed (McGinnity *et al.* 2004). Indeed, there is evidence for adaptive differentiation among anadromous populations in different tributaries of the same river (Chapter 7), where phylogeographic divergence is limited and some gene flow within a metapopulation structure is likely. As such, it is essential that population structuring be considered in the development of conservation and management policy and programmes (Youngson *et al.* 2003).

5.6 Summary and conclusions

- Homing to natal rivers to spawn, a fragmented distribution of spawning and juvenile habitat, and a capacity for local adaptation provide the necessary prerequisites for population structuring in the Atlantic salmon.
- Atlantic salmon in the Eastern and Western Atlantic belong to two deeply divergent phylogeographic groups, between which there has only been limited gene flow, probably for more than 500 000 years.
- The Eastern Atlantic phylogeographic group is divided into two less divergent, more recently evolved, phylogeographic subgroups, one associated with the Atlantic Ocean and the other with Baltic Sea drainages; each of these is in turn further evolutionarily divided.
- Phylogeographic structuring is also apparent in the Western Atlantic, though this appears to be less extensive, but small-scale regional groups have been defined in areas such as the inner Bay of Fundy.
- Non-anadromous populations appear to have evolved independently in most river systems and regions from anadromous colonisers due to temporary or permanent historical 'landlocking'.
- The phylogeographic structuring in the Atlantic salmon arises from the historical distribution of the species during the Pleistocene glaciations, the pattern of postglacial colonisation, and historical environmental conditions.
- Contemporary gene flow among populations is highly restricted, even among tributaries within rivers.
- Limited and sporadic gene flow appears to occur among anadromous populations within regions or within rivers, linking them into metapopulation groups.
- The distinct evolutionary histories of populations in the Eastern and Western Atlantic justify their designation as distinct subspecies and they should be managed as distinct biodiversity components of the species (i.e. as distinct population segments within the framework of the US Endangered Species Act).

- Observed population structure has significant implications for conservation efforts throughout the species range, particularly for the peripheral populations; most important is a presumption against the introduction of cultured strains or stocks of Atlantic salmon from one continent into the other continent.

5.7 Management recommendations

- Assume that salmon stocks in different rivers belong to discrete genetic populations and that most river systems contain multiple genetic populations.
- The fundamental unit of management for Atlantic salmon is that population and management should take into account the potentially complex population structuring among and within rivers.
- Investigate population structuring within rivers being managed either by detailed physical tagging and behavioural monitoring, or by molecular genetic analysis.
- Abundant information obtained from throughout the range of Atlantic salmon for allozymes, microsatellite DNA and mitochondrial DNA indicate that deep phylogeographic differences exist between Eastern and Western Atlantic populations. This information strongly suggests that these distinct forms should be recognised by management agencies as the distinct subspecies, *Salmo salar salar* and *Salmo salar sebago*, respectively.
- Stop all transfer locations, and eliminate existing stocks, of European-origin salmon to North America and North American-origin salmon to Europe.
- Consider a policy of non-transfer of salmon from one river system to another and employ caution in moving salmon among locations within river systems.
- Contemporary molecular tools allow for population- or river-specific captive broodstock management; efforts should be made to mimic allelic and haplotype diversity present in recipient populations in such programmes.

Acknowledgements

The authors thank the numerous researchers throughout the world for providing Atlantic salmon tissue and DNA samples. TLK wishes to thank the US Geological Survey's Leetown Science Center, the US Department of Commerce's NOAA Fisheries and the US Fish and Wildlife Service for direct or in-kind funding. Anne Henderson provided valuable editorial assistance. EV wishes to thank all those in Fisheries and Oceans Canada for their support over the years with the work on Canadian Atlantic salmon populations.

Further reading

Avise, J.C. (2000) *Phylogeography: The history and formation of species*. Harvard University Press, Cambridge, MA.
Avise, J.C. (2004) *Molecular Markers: Natural history and evolution* (2nd edn). Sinauer Press, Sunderland, MA.
Cavalli-Sforza, L.L. (2000) *Genes, Peoples, and Languages*. North Point Press/Farrar, Straus & Giroux, New York.

NRC (National Research Council) (1996) *Upstream: Salmon and society in the Pacific Northwest*. National Academy Press, Washington, DC.
NRC (National Research Council) (2002) *Genetic Status of Atlantic Salmon in Maine: Interim Report from the Committee on Atlantic Salmon in Maine*. National Academy Press, Washington, DC.
NRC (National Research Council) (2004) *Atlantic Salmon in Maine*. National Academy Press, Washington, DC.

References

Adkison, M.D. (1995) Population differentiation in Pacific salmon: local adaptation, genetic drift, or the environment? *Canadian Journal of Fisheries and Aquatic Sciences*, **52**: 2762–2777.
Allendorf, F.W. and Thorgaard, G.H. (1984) Tetraploidy and the evolution of salmonid fishes. In: B.J. Turner (Ed.) *Evolutionary Genetics of Fishes*, pp. 1–53. Plenum Publishing Company, New York.
Allendorf, F.W. and Danzmann, R.G. (1997) Secondary tetrasomic segregation of MDH-B and preferential pairing of homeologues in rainbow trout. *Genetics*, **145**: 1083–1092.
Andersen, B.G. and Borns, H.W. Jr (1997) *The Ice Age World*. Scandinavian University Press, Oslo.
Ardren, W.A. and Kapuscinski, A.R. (2003) Demographic and genetic estimates of effective population size (N_e) reveals genetic compensation in steelhead trout. *Molecular Ecology*, **12**: 35–49.
Asplund, T., Veselov, A., Primmer, G., Bakhmet, I., Potutkin, A., Titov, S., Zubchenko, A., Studenov, I., Kaluzchin, S. and Lumme, J. (2004) Geographical structure and postglacial history of mtDNA haplotype variation in Atlantic salmon (*Salmo salar* L.) among rivers of the White and Barents Sea basins. *Annales Zoologici Fennici*, **41**: 465–475.
Avise, J.C. (2004) *Molecular Markers: Natural history and evolution* (2nd edn). Sinauer Press, Sunderland, MA.
Bakke, T.A., Harris, P.D., Hansen, H., Cable, J., and Hansen, L.P. (2004) Susceptibility of Baltic and East Atlantic salmon *Salmo salar* stocks to *Gyrodactylus salaris* (Monogenea). *Diseases of Aquatic Organisms*, **58**: 171–177.
Baum, E. (1997) *Maine Atlantic Salmon: A national treasure*. Atlantic Salmon Unlimited, Hermon, ME.
Behnke, R.J. (1965) A Systematic Study of the Family Salmonidae with Special Reference to the Genus *Salmo*. PhD thesis, University of California, Berkeley, CA.
Behnke, R.J. (1972) The systematics of salmonid fishes of recently glaciated lakes. *Journal of Fisheries Research Board of Canada*, **29**: 639–671.
Berg, O.K. (1985) The formation of non-anadromous populations of Atlantic salmon, *Salmo salar* L., in Europe. *Journal of Fish Biology*, **27**: 805–815.
Bermingham, E., McCafferty, S.S. and Martin, A.P. (1997) Fish biogeography and molecular clocks: perspectives from the Panamanian Isthmus. In: T.D. Kocher and C.A. Stepien (Ed.) *Molecular Systematics of Fishes*, chapter 8. Academic Press, New York.
Bernatchez, L. (2001) The evolutionary history of brown trout (*Salmo trutta* L.) inferred from phylogeographic, nested clade, and mismatch analyses of mitochondrial DNA variation. *Evolution*, **55**: 351–379.
Bernatchez, L. and Wilson, C.C. (1998) Comparative phylogeography of nearctic and palearctic fishes. *Molecular Ecology*, **7**: 431–452.
Boothroyd, E.R. (1959) Chromosome studies on three Canadian populations of Atlantic salmon, *Salmo salar* L. *Canadian Journal of Genetics and Cytology*, **1**: 161–172.
Boulton, G.S., Dongelmans, P., Punkari, M. and Broadgate, M. (2001) Paleoglaciology of an ice sheet through a glacial cycle: the European ice sheet through the Weichselian. *Quaternary Science Reviews*, **20**: 591–625.
Bowen, D.Q., Philipps, F.M., McCabe, A.M., Knutz, P.C. and Sykes, G.A. (2002) New data for the last glacial maximum in Great Britain and Ireland. *Quaternary Science Reviews*, **21**: 89–101.

Buisine, N., Trichet, V. and Wolff, J. (2002) Complex evolution of vitellogenin genes in salmonid fishes. *Molecular Genetics and Genomics*, **268**: 535–542.

Claytor, R.R. and Verspoor, E. (1991) Discordant phenotypic variation in sympatric resident and anadromous Atlantic salmon (*Salmo salar*) populations. *Canadian Journal of Zoology*, **69**: 2846–2852.

Claytor, R.R., MacCrimmon, H.R. and Gots, B.L. (1991) Continental and ecological variance components of European and North American Atlantic salmon, *Salmo salar* phenotypes. *Biological Journal of the Linnean Society*, **44**: 203–229.

Consuegra, S., Garcia de Leaniz, C., Serdio, A., Gonzalez Morales, M., Strauss, L.G., Knox, D. and Verspoor, E. (2002) Mitochondrial DNA variation in Pleistocene and modern Atlantic salmon from the Iberian glacial refugium. *Molecular Ecology*, **11**: 2037–2048.

Consuegra, S., Knox, D., Verspoor, E. and Garcia de Leaniz, C. (2005) Asymmetric gene flow and the evolutionary maintenance of genetic diversity in small, peripheral Atlantic salmon populations. *Conservation Genetics*, **6**: 823–842.

Cordes, J.F., Perkins, D.L., Kincaid, H.L. and May, B. (2005) Genetic analysis of fish genomes and populations: allozyme variation within and among Atlantic salmon from Downeast Rivers of Main. *Journal of Fish Biology*, **67** (Supplement A): 104–117.

Cornuet, J-M., Piry, S., Luikart, G., Estoup, A. and Solignac, M. (1999) New methods employing multilocus genotypes to select or exclude populations as origins of individuals. *Genetics*, **153**: 1989–2000.

Crespi, B.J. and Fulton, M.J. (2004) Molecular systematics of Salmonidae: combined nuclear data yields a robust phylogeny. *Molecular Phylogenetics and Evolution*, **31**: 658–679.

Cross, T.F. (1981) Section E: Biochemical genetics. *Salmon Research Trust of Ireland, Annual Report*, **26**: 49–51.

Devlin, R.H. (1993) Sequence of sockeye type 1 and type 2 growth hormone genes and the relationships of rainbow trout with Atlantic and Pacific salmon. *Canadian Journal of Fisheries and Aquatic Sciences*, **50**: 1738–1748.

Elo, K. (1993) Gene flow and conservation of genetic variation in anadromous Atlantic salmon (*Salmo salar*). *Hereditas*, **119**: 149–159.

Fontaine, P-M., Dodson, J.J., Bernatchez, L. and Slettan, A. (1997) A genetic test of metapopulation structure in Atlantic salmon (*Salmo salar*) using microsatellites. *Canadian Journal of Fisheries and Aquatic Sciences*, **54**: 2434–2442.

Gilbey, J., Knox, D., O'Sullivan, M. and Verspoor, E. (2005) Novel DNA markers for rapid, accurate, and cost-effective discrimination of the continental origin of Atlantic salmon (*Salmo salar* L.). *ICES Journal of Marine Science*, **62**: 1609–1616.

Glubokovsky, M.K. and Frolov, S.V. (1994) Phylogenetic relations and the systematics of chars of Lake El'gygytgyn. *Journal of Ichthyology*, **34**: 128–147.

Gross, M.R. (1998) One species with two biologies: Atlantic salmon (*Salmo salar*) in the wild and in aquaculture. *Canadian Journal of Fisheries and Aquatic Science*, **55** (Supplement): 131–144.

Gross, R., Consuegra, S., King, T., Lumme, J., Nilsson J. and Verspoor E. **In Preparation.** Mitochondrial DNA variation in the Atlantic salmon (*Salmo salar* L.): implications for intraspecific population structuring and adaptive population differentiation.

Gruchy, C.G. (1971) Salmon nomenclature. *Nature*, **234**: 360.

Hadzisce, S. (1961) Zur Kentnis des *Salmothymus orhidanus* (Steindachner) (Pisces, Salmonidae). *International Vereinigung theoretische angewandte Limnology Verhandlungen*, **14**: 785–791.

Hartley, S.E. (1987) The chromosomes of salmonid fishes. *Biological Reviews*, **62**: 197–214.

Hartley, S.E. and Horne, T.M. (1984) Chromosome polymorphism and constitutive heterochromatin in Atlantic salmon, *Salmo salar*. *Chromosoma*, **89**: 377–380.

Hewitt, G.M. (1996) Some genetic consequences of ice ages, and their role in divergence and speciation. *Biological Journal of the Linnean Society*, **58**: 247–276.

Hewitt, G.M. (1999) Post-glacial re-colonization of European biota. *Biological Journal of the Linnean Society*, **68**: 87–112.

Hurst, C.D., Bartlett, S.E., Bruce, I.J. and Davidson, W.S. (1998) The complete nucleotide sequence of the mitochondrial DNA of the Atlantic salmon, *Salmo salar*. GenBank Accession number U12143. NCBI, Bethesda, MD.

Jordan, W.C., Verspoor, E. and Youngson, A.F. (1997) The effect of natural selection on estimates of genetic divergence among populations of the Atlantic salmon (*Salmo salar* L.). *Journal of Fish Biology*, **51**: 546–560.

Jordan, W.C., Cross, T.F., Crozier, W.W., Ferguson, A., Galvin, P., Hurrell, R.H., McElligott, E.A., McGinnity, P., Martin, S.A.M., Moffett, I.J.J., Price, D.J., Youngson, A.F. and Verspoor, E. (2005) Allozyme variation in Atlantic salmon from the British Isles: associations with geography and the environment. *Journal of Fish Biology*, **67** (Supplement A): 146–168.

Kauppi, R., Kvist, L., Ruokonen, M., Soivio, A. and Lumme, J. (1997) Lack of variation in mitochondrial DNA of the Baltic salmon (*Salmo salar*) indicates a bottleneck during or long before postglacial recolonization of the Baltic Sea. *Oulanka Reports*, **17**: 19–23.

King, T.L., Spidle, A.P., Eackles, M.S., Lubinski, B.A. and Schill, W.B. (2000) Mitochondrial DNA diversity in North American and European Atlantic salmon with emphasis on the Downeast rivers of Maine. *Journal of Fish Biology*, **57**: 614–630.

King, T.L., Kalinowski, S.T., Schill, W.B., Spidle, A.P. and Lubinski, B.A. (2001) Population structure of Atlantic salmon (*Salmo salar* L.): a range-wide perspective from microsatellite DNA variation. *Molecular Ecology*, **10**: 807–821.

Knox, D., Lehmann, K., Reddin, D.G. and Verspoor, E. (2002) Genotyping of archival Atlantic salmon scales from northern Quebec and West Greenland using novel PCR primers for degraded mtDNA. *Journal of Fish Biology*, **60**: 266–270.

Koljonen, M-L., Jansson, H., Paaver, T., Vasin, O. and Koskiniemi, J. (1999) Phylogeographic lineages and differentiation pattern of Atlantic salmon (*Salmo salar*) in the Baltic Sea with management implications. *Canadian Journal of Fisheries and Aquatic Sciences*, **56**: 1766–1780.

Koljonen, M-L., Tahtinen, J., Saisa, M. and Koskiniemi, J. (2002) Maintenance of genetic diversity of Atlantic salmon (*Salmo salar*) by captive breeding programmes and the geographic distribution of microsatellite variation. *Aquaculture*, **212**: 69–92.

Loudenslager, E.J. and Gall, G.A.E. (1980) Geographic patterns of protein variation and subspeciation in cutthroat trout, *Salmo clarki*. *Systematic Zoology*, **29**: 27–42.

Makhrov, A.A., Verspoor, E., Artamonova, V.S. and O'Sullivan, M. (2005) Atlantic salmon colonization of the Russian Arctic coast: pioneers from North America. *Journal of Fish Biology*, **67** (Supplement A): 68–79.

MacCrimmon, H.R. and Gots, B.L. (1979) World distribution of Atlantic salmon, *Salmo salar*. *Journal of the Fisheries Research Board of Canada*, **36**: 422–457.

McConnell, S.K., O'Reilly, P.T., Hamilton, L., Wright, J.M. and Bentzen, P. (1995) Polymorphic microsatellite loci from Atlantic salmon (*Salmo salar*): genetic differentiation of North American and European populations. *Canadian Journal of Fisheries and Aquatic Sciences*, **52**: 1863–1872.

McConnell, S.K., Ruzzante, D.E., O'Reilly, P.T., Hamilton, L. and Wright, J.M. (1997) Microsatellite loci reveal highly significant genetic differentiation among Atlantic salmon (*Salmo salar* L.) stocks from the east coast of Canada. *Molecular Ecology*, **6**: 1075–1089.

McGinnity, P., Prodöhl, P., Ó Maoiléidigh, N., Hynes, R., Cotter, D., Baker, N., O'Hea, B. and Ferguson, A. (2004) Differential lifetime success and performance of native and non-native Atlantic salmon examined under communal natural conditions. *Journal of Fish Biology*, **65** (Supplement A): 173–187.

Milner, A.M., Knudsen, E.E., Soiseth, C., Robertson, A.L., Schell, D., Phillips, I.T. and Magnusson, K. (2000) Colonization and development of stream communities across a 200-year gradient in Glacier Bay National Park, Alaska, USA. *Canadian Journal of Fisheries and Aquatic Sciences*, **57**: 2319–2335.

Möller, D. (1970) Transferrin polymorphism in Atlantic salmon (*Salmo salar*). *Journal of Fisheries Research Board of Canada*, **27**: 1617–1625.

Nei, M. (1987) *Molecular Evolutionary Genetics*. Columbia University Press, New York.

Nei, M., Tajima, F. and Tateno, Y. (1983) Accuracy of estimated phylogenetic trees from molecular data. *Journal of Molecular Evolution*, **19**: 153–170.

Nielsen, J.L. (1998) Population genetics and the conservation and management of Atlantic salmon (*Salmo salar*). *Canadian Journal of Fisheries and Aquatic Sciences*, **55** (Supplement 1): 145–152.

Nielsen, E.E., Hansen, M.M. and Loeschcke, V. (1996) Genetic structure of European populations of *Salmo salar* L. (Atlantic salmon) inferred from mitochondrial DNA. *Heredity*, **77**: 351–358.

Nilsson, J., Gross, R., Asplund, T., Dove, O., Jansson, H., Kelloniemi, J., Kohlmann, K., Loytynoja, A., Nielsen, E.E., Paaver, T., Primmer, C.R., Titov, S., Vasemägi, A., Veselov, A., Ost, T. and Lumme, J. (2001) Matrilinear phylogeography of Atlantic salmon (*Salmo salar* L.) in Europe and postglacial colonization of the Baltic Sea area. *Molecular Ecology*, **10**: 89–102.

Norden, C.R. (1961) Comparative osteology of representative salmonid fishes, with particular reference to the grayling (*Thymallus arcticus*) and its phylogeny. *Journal of Fisheries Research Board of Canada*, **8**: 679–701.

NRC (National Research Council) (2002) *Genetic Status of Atlantic Salmon in Maine: Interim Report from the Committee on Atlantic Salmon in Maine*. National Academy Press, Washington, DC.

NRC (National Research Council) (2004) *Atlantic Salmon in Maine*. National Academy Press, Washington, DC.

Oakley, T.H. and Phillips, R.B. (1999) Phylogeny of Salmoninae fish based on growth hormone introns: Atlantic (*Salmo*) and Pacific (*Oncorhynchus*) salmon are not sister taxa. *Molecular Phylogenetics and Evolution*, **11**: 381–393.

Orr, M. and Smith, T.B. (1998) Ecology and speciation. *Trends in Ecology and Evolution*, **13**: 502–506.

Osinov, A. (1991) Genetic divergence and phylogenetic relationships between lenoks of genus *Brachymystax* and huchens of genera *Hucho* and *Parahucho*. *Genetika*, **27**: 2127–2135.

Payne, R.H. (1974) Transferrin variation in North American populations of the Atlantic salmon, *Salmo salar*. *Journal of the Fisheries Research Board of Canada*, **31**: 1037–1041.

Payne, R.H., Child, A.R. and Forrest, A. (1971) Geographical variation in the Atlantic salmon. *Nature*, **231**: 250–252.

Phillips, R.B. and Oakley, T.H. (1997) Phylogenetic relationships among the Salmoninae based on nuclear and mitochondrial DNA sequences. In: T.D. Kocher and C.A. Stepien (Ed.) *Molecular Systematics of Fishes*, pp. 145–162. Academic Press, New York.

Phillips, R.B. and Rab, P. (2001) Chromosome evolution in the Salmonidae (Pisces): an update. *Biological Reviews*, **76**: 1–25.

Phillips, R.B., Matsuoka, M.P. and Reed, K.M. (2000) Phylogenetic analysis of mitochondrial and nuclear sequences supports inclusion of *Acantholingua ohridana* in *Salmo*. *Copeia*, **2000**: 546–550.

Phillips, R.B., Matsuoka, M.P., Konkol, N.R. and McKay, S.J. (2003) Molecular systematics and evolution of the growth hormone introns in the Salmoninae. *Environmental Biology of Fishes*, **69**: 433–440.

Power, G. (1958) The evolution of the freshwater races of the Atlantic salmon (*Salmo salar* L.) in eastern North America. *Arctic*, **11**: 86–91.

Presa, P., Pardo, B.G., Martinez, P. and Bernatchez, L. (2002) Phylogeographic congruence between mtDNA and rDNA ITS markers in brown trout. *Molecular Biology and Evolution*, **19**: 2161–2175.

Pursiainen, M., Makkonen, J. and Piironen, J. (1998) Maintenance and exploitation of landlocked salmon, *Salmo salar* m. *sebago*, in the Vuoksi watercourse. In: J.G. Cowx (Ed.) *Stocking and Introduction of Fish*, 46–58. Blackwell Science, Oxford.

Rab, P., Slechta, V. and Flajshans, M. (1994) Cytogenetics, cytotaxonomy, and biochemical genetics of Huchonine salmonids. *Folia Zoology*, **43**: 97–107.

Regen, C.T. (1914) The systematic arrangement of the fishes of the family Salmonidae. *Annual Magazine of Natural History*, **13**: 405–408.

Roberts, F.L. (1970) Atlantic salmon (*Salmo salar*) chromosomes and speciation. *Transactions of the American Fisheries Society*, **99**: 105–111.

Säisä, M., Koljonen, M-L., Gross, R., Tähtinen, J., Koskiniemi, J., Vasemägi, A., Nilsson, J., Paaver, T. and Ojala, M. (2005) The genetic structure and postglacial colonization of Atlantic salmon in the Baltic Sea area based on microsatellite loci. *Canadian Journal of Fisheries and Aquatic Science*, **62**: 1887–1904.

Seehausen, O. (2002) Patterns in fish radiation are compatible with Pleistocene desiccation of Lake Victoria and 14 600 year history for its cichlid species flock. *Proceedings of the Royal Society of London, Series B Biological Sciences*, **269**: 491–497.

Smith, G.R. and Stearley, R.F. (1989) The classification and scientific names of rainbow and cutthroat trouts. *Fisheries*, **14**: 4–10.

Snoj, A., Melkic, E., Susnik, S. and Muhamedagic, S. (2002) DNA phylogeny supports revised classification of *Salmothymus obtusirostris*. *Biological Journal of the Linnean Society*, **77**: 399–411.

Spidle, A.P., Schill, W.B., Lubinski, B.A. and King, T.L. (2001) Fine-scale population structure in Atlantic salmon from Maine's Penobscot River drainage. *Conservation Genetics*, **2**: 11–24.

Spidle, A.P., Kalinowski, S.T., Lubinski, B.A., Perkins, D.L., Beland, K.F., Kocik, J.F. and King, T.L. (2003) Population structure of Atlantic salmon in Maine with reference to populations from Atlantic Canada. *Transactions of the American Fisheries Society*, **132**: 196–209.

Stabell, O.B. (1984) Homing and olfaction in salmonids: a critical review with special reference to the Atlantic salmon. *Biological Reviews*, **59**: 333–388.

Ståhl, G. (1987) Genetic population structure of Atlantic salmon. In: N. Ryman and F. Utter (Ed.) *Population Genetics and Fishery Management*, pp. 121–140. Washington Sea Grant, Seattle, WA.

Stearley, R.F. and Smith, G.R. (1993) Phylogeny of the Pacific trouts and salmons (*Oncorhynchus*) and genera of the family Salmonidae. *Transactions of the American Fisheries Society*, **122**: 1–33.

Svetovidov, A. (1975) Comparative osteological study of the Balkan endemic genus *Salmothymus* in relation to its classification. (in Russian) *Zoology Zhurnal*, **54**: 1174–1190.

Taberlet, P., Fumagalli, L., Wust-Saucy, A.G. and Cossons, J.F. (1998) Comparative phylogeography and postglacial colonization routes in Europe. *Molecular Ecology*, **7**: 453–464.

Taggart, J.B., Verspoor, E., Galvin, P.T., Moran, P. and Ferguson, A. (1995) A minisatellite marker for discriminating between European and North American Atlantic salmon (*Salmo salar*). *Canadian Journal of Fisheries and Aquatic Sciences*, **52**: 2305–2311.

Tessier, N. and Bernatchez, L. (2000) A genetic assessment of single versus double origin of landlocked Atlantic salmon (*Salmo salar*) from Lake Saint-Jean, Quebec, Canada. *Canadian Journal of Fisheries and Aquatic Sciences*, **57**: 797–804.

Tessier, N., Bernatchez, L. and Wright, J.M. (1997) Population structure and impact of supportive breeding inferred from mitochondrial and microsatellite DNA analyses in land-locked Atlantic salmon *Salmo salar* L. *Molecular Ecology*, **6**: 735–750.

Tonteri, A., Titov, S., Veselov, A., Zubchenko, A., Koskinen, M.T., Lesbarrères, D., Kaluzchin, S., Bakhmet, I., Lumme, J. and Primmer, G.R. (2005) Phylogeography of anadromous and non-anadromous Atlantic salmon (*Salmo salar*) from northern Europe. *Annales Zoologici Fennici*, **45**: 1–22.

Ueda, T. and Kobayshi, J. (1990) Karyotype differentiation of Atlantic salmon, *Salmo salar*: especially the sequential karyotype change. *La Kromosomo-II*, **58**: 1967–1972.

Utter, F.M. and Allendorf, F.W. (1994) Phylogenetic relationships among species of *Oncorhynchus*: a consensus view. *Conservation Biology*, **8**: 864–867.

Vasemägi, A., Gross, R., Paaver, T., Kangur, M., Nilsson, J. and Eriksson, L.-O. (2001) Identification of the origin of an Atlantic salmon (*Salmo salar* L.) population in a recently recolonized river in the Baltic Sea. *Molecular Ecology*, **10**: 2877–2882.

Vasemägi, A., Gross, R., Paaver, T., Koljonen, M-L., Säisä, M. and Nilsson, J. (2005a) Analysis of gene associated tandem repeat markers in Atlantic salmon (*Salmo salar* L.) populations: implications for restoration and conservation in the Baltic Sea. *Conservation Genetics*, **6**: 385–397.

Vasemägi, A., Nilsson, J. and Primmer, C.R. (2005b) Expressed sequence tag-linked microsatellites as a source of gene-associated polymorphisms for detecting signatures of divergent selection in Atlantic salmon (*Salmo salar* L.). *Molecular Biology and Evolution*, **22**: 1067–1076.

Verspoor, E. (1986) Spatial correlations of transferrin allele frequencies in Atlantic salmon (*Salmo salar*) populations from North America. *Canadian Journal of Fisheries and Aquatic Sciences*, **43**: 1074–1078.

Verspoor, E. (1988) Reduced genetic variability in first-generation hatchery populations of Atlantic salmon (*Salmo salar*). *Canadian Journal of Fisheries and Aquatic Sciences*, **45**: 1686–1690.

Verspoor, E. (1994) The evolution of genetic divergence at protein coding loci among anadromous and nonanadromous populations of Atlantic salmon *Salmo salar*. In: A. Beaumont (Ed.) *Genetics and Evolution of Aquatic Organisms*, pp. 52–67. Chapman & Hall, London.

Verspoor, E. (1995) Population structure: what genetics tells us. In: *Salmon in the Dee Catchment: The scientific basis for management*. Atlantic Salmon Trust, Moulin, Pitlochry, Scotland.

Verspoor, E. (2005) Regional differentiation of North American Atlantic salmon (*Salmo salar*) at allozyme loci. *Journal of Fish Biology*, **67** (Supplement A): 80–103.

Verspoor, E. and Cole, L.J. (1989) Genetically distinct sympatric populations of resident and anadromous Atlantic salmon, *Salmo salar*. *Canadian Journal of Zoology*, **67**: 1453–1461.

Verspoor, E. and Cole, L.J. (2005) Genetic evidence for lacustrine spawning of the non-anadromous Atlantic salmon population of Little Gull Lake, Newfoundland. *Journal of Fish Biology*, **67** (Supplement A): 200–205.

Verspoor, E. and Jordan, W.C. (1989) Genetic variation at the ME-2 locus in the Atlantic salmon within and between populations: evidence for its selective maintenance. *Journal of Fish Biology*, **35** (Supplement A): 205–213.

Verspoor, E., McCarthy, E.M. and Knox, D. (1999) The phylogeography of European Atlantic salmon (*Salmo salar* L.) based on RFLP analysis of the ND1/16sRNA region of the mtDNA. *Biological Journal of the Linnaean Society*, **68**: 129–146.

Verspoor, E., O'Sullivan, M., Arnold, A.L., Knox, D. and Amiro, P.G. (2002) Restricted matrilineal gene flow and regional differentiation among Atlantic salmon (*Salmo salar* L.) populations within the Bay of Fundy, eastern Canada. *Heredity*, **89**: 465–472.

Verspoor, E., Beardmore, J.A., Consuegra, S., Garcia de Leaniz, C., Hindar, K., Jordan, W.C., Koljonen, M-L., Mahkrov, A., Paava, T., Sánchez, J.A., Skaala, O., Titov, S. and Cross, T.F. (2005) Population structure in the Atlantic salmon: insights from 40 years of research into genetic protein variation. *Journal of Fish Biology*, **67** (Supplement A): 3–54.

Vuorinen, J. (1982) Little genetic variation in the Finnish Lake salmon, *Salmo salar sebago* (Girard). *Heriditas*, **97**: 189–192.

Vuorinen, J.A. and Berg, O.K. (1989) Genetic divergence of anadromous and nonanadromous Atlantic salmon (*Salmo salar*) in the River Namsen, Norway. *Canadian Journal of Fisheries and Aquatic Sciences*, **46**: 406–409.

Wilder, D.G. (1947) A comparative study of the Atlantic salmon, *Salmo salar* Linnaeus, and the lake salmon, *Salmo salar sebago* (Girard). *Canadian Journal of Research*, D, **25**: 175–189.

Wilkins, N.P. (1972a) Biochemical genetics of the Atlantic salmon *Salmo salar* L. I. A review of recent studies. *Journal of Fisheries Biology*, **4**: 487–504.

Wilkins, N.P. (1972b) Biochemical genetics of the Atlantic salmon *Salmo salar* L. II. The significance of recent studies and their application in population identification. *Journal of Fisheries Biology*, **4**: 505–517.

Wilson, R.C.L., Drury, S.A. and Chapman, J.L. (2000) The Great Ice Age: climate change and life. Routledge, London.

Youngson, A.F., Jordan, W.C., Verspoor, E., Cross, T. and Ferguson, A. (2003) Management of salmonid fisheries in the British Isles: towards a practical approach based on population genetics. *Fisheries Research*, **62**: 193–209.

6 Mating System and Social Structure

W. C. Jordan, I. A. Fleming and D. Garant

Upper: classical Atlantic salmon spawning habitat showing redd structure (inset: spawning salmon). (Photo credit: J. Webb and inset, D. Pugh, US Geological Survey.) Lower: spawning Atlantic salmon. (Photo credit: G. van Ryckevorsel.)

The mating system of Atlantic salmon, *Salmo salar*, has been the subject of many studies, based principally on behavioural observation (reviewed by Fleming 1996, 1998; Fleming and Reynolds 2004). However, the use of molecular genetic markers in such studies and their extension to an examination of any social structure within populations resulting from the mating system is still very much in its infancy. While there remains tremendous scope for further work, a body of recent research integrating genetic data with behavioural observations has produced novel insights into some aspects of the mating system and social structure in the species, through both controlled experiments and field studies of natural populations. It is thus timely now to review these studies and their conclusions, to consider the current implications for Atlantic salmon management and to recommend directions for future work.

6.1 Introduction

6.1.1 Definitions, approach and organisation

The amount and distribution of genetic variation within a population depends on its mating system and social structure. There have been many definitions of 'mating system' (e.g. Emlen and Oring 1977; Andersson 1994; Shuster and Wade 2003; Fleming and Reynolds 2004), but most include aspects of (1) the numbers of mates obtained by individuals of both sexes, (2) the manner in which mates are obtained through male–male or female–female competition for mates and resources, (3) courtship, (4) mate choice and sexual selection and (5) level of parental care. Post-mating mechanisms such as sperm competition and cryptic female choice (preferential use of sperm by female or egg), which further determine the number of successful fertilisations per individual, might also be included in a broad definition of mating system. In practice, it may only be possible to determine the outcome of a mating system (in terms of mean and variance in reproductive success) after these mechanisms have acted. By contrast, clear definitions of 'social structure' outside the human social science literature are fewer, but here we define social structure as the pattern of relationships among individuals within a population. These relationships may be described in terms of behaviour (e.g. dominance, aggression, spacing, cooperation) or genetics (level of relatedness), and a great deal of the study of social evolution concerns the fitness outcomes of the interaction between behavioural and genetic relationships (Hamilton 1964a,b; Trivers 1985; Bourke and Franks 1995; Frank 1998). For any species, both mating system and social structure are constrained to some degree by its evolutionary history, but both may also vary across populations according to local ecological conditions.

In this chapter we intend to focus on how molecular genetic techniques, usually in combination with behavioural observations, have illuminated our understanding of aspects of mating system and social structure in Atlantic salmon populations and to what extent they may be able to yield further insight into these topics. Studies based on behavioural observations alone are cited, where pertinent, but for a fuller description of these studies the reader is directed to reviews by Fleming (1996, 1998) and Fleming and Reynolds (2004). We introduce the topic of effective population size (N_e) and describe how an understanding of mating system can aid in estimation of N_e in Atlantic salmon populations and in prediction of how N_e might change in response to changing environmental conditions. We then discuss how a number of studies have provided information on the variance in individual reproductive success and patterns of mate choice under experimental conditions where elements of the mating system are controlled for or absent. These are compared with other studies (currently limited in number) that have shown us something about the same variables (including interspecific hybridisation) under conditions where all (or most) elements of the mating system interact. We go on to link the social structure of juvenile Atlantic salmon with the mating system and compare expectations on the strength and direction of associations among relatedness, behavioural interaction and individual fitness components from laboratory studies with results from the field. Initially, however, we present some background essential to the understanding of the methods used in parentage assignment and relatedness estimation from molecular markers.

6.1.2 Genetic markers in the analysis of mating system and social structure

In recent years, the development of highly variable genetic markers has added a new dimension to the study of mating systems and social structure in a wide range of species, often revealing a higher degree of complexity in animal populations than that apparent from behavioural observations alone (DeWoody and Avise 2001; Gibbs and Weatherhead 2001; Ross 2001; Griffith *et al.* 2002). Essentially, analysis of genetic markers allows the estimation of levels of relatedness among individuals, particularly the identification of parent–offspring relationships. A molecular marker approach is most powerful when allied to behavioural observation, but it is especially useful for species inhabiting environments such as the aquatic environment where behavioural observation is often constrained. Further difficulties in assigning parentage from behavioural observations occur for species where fertilisation is external with many candidate parents involved in a mating and/or little parental care is provided, as is the case with Atlantic salmon.

Although allozyme markers have been used in the past to infer parentage and assess relatedness (e.g. Blakey 1994; Peuhkuri and Seppa 1998; Bulmer *et al.* 2001), in most species the low levels of variability at these loci give them limited usefulness. Thus, it was only with the isolation of hypervariable codominant genetic markers (i.e. at which heterozygotes can be distinguished from homozygotes), such as mini- and microsatellite loci, that studies have been able to obtain reliable relatedness estimates and assign parentage with a high degree of statistical confidence (see Box 6.1). In the near future, however, dominant markers such as amplified fragment length polymorphisms (AFLPs) (Vos *et al.* 1995) may also become increasingly useful for such analyses, as statistical methods for dealing with such data are developed (Hardy 2003; Wang 2004a; Ritland 2005). Additionally, as the number of single nucleotide polymorphisms (SNPs) identified in non-model organisms increases, it is possible that these markers will be more commonly used (Glaubitz *et al.* 2003).

Obviously, estimation of relatedness should be both accurate and precise. In the particular case of parentage assignment, the power of the analysis is often expressed as an exclusion probability. The exclusion probability is generally defined as that of a randomly chosen individual or parental pair being genetically excluded as the parent(s) of a randomly chosen offspring (Villanueva *et al.* 2002), but this may be modified if some relationships are known a priori (e.g. maternal-offspring). Exclusion probabilities increase as the number of loci used and the number of alleles per locus increase (they are also increased if the frequencies of the alleles at a locus are even) (Bernatchez and Duchesne 2000; Villanueva *et al.* 2002). However, researchers are also often interested in the minimum number of loci required to give reasonable exclusion probabilities. Bernatchez and Duchesne (2000) and Villanueva *et al.* (2002) provide estimates of the number of loci needed under conditions with varying number of alleles, different allele

> **Box 6.1** Assigning parentage and estimating effective population size using molecular markers.
>
> **Parentage**
>
> Along with developments in molecular biology techniques that now allow routine analysis of highly polymorphic DNA markers have come parallel advances in statistical and computational methods for assigning parentage and assessing the confidence of that assignment. In early studies, most of which were designed to assign paternity under the assumption that maternity was known, a mismatch between the genotype of offspring and candidate father was considered grounds for exclusion from paternity assignment (e.g. Morin *et al.* 1994). However, it was quickly acknowledged that reliance on genotype mismatch(es) and exclusion was not a strong basis for paternity/parentage assignment as a number of males may be non-excluded due to high levels of relatedness within the population. Also, exclusion-based methods are very sensitive to errors in genotyping. Methods of parentage assignment based on likelihood are now much more widely used. The likelihood of any individual (Marshall *et al.* 1998) or male/female pair (Duchesne *et al.* 2002) being assigned as parent(s) given their genotype(s) and that of an offspring is calculated. Putative parents can then be accepted or rejected at a chosen probability level (usually > 90%). Such methods can also build in tolerance of genotyping errors (San Cristobal and Chevalet 1997). A potentially more powerful method of parentage assignment entails first grouping offspring into sibships and then assigning parents or parental pairs to those sibships (Wang 2004b). An alternative approach allows other biological data (such as behavioural observations) to be incorporated into a Bayesian framework for the calculation of expected parentage under explicit models of multiple mating (Neff *et al.* 2001).
>
> **Effective population size**
>
> Most methods of estimating effective population size use the temporal method (Nei and Tajima 1981; Waples 1989). In this method, samples are taken from a population at two or more time points and the allele frequencies of (usually) multiple loci are estimated. As genetic drift is higher in small populations, they display greater fluctuations in allele frequency over time and, using standard population genetic models, the degree of allele frequency fluctuation can be used to estimate the effective population size over the time interval concerned. Important developments of the temporal method have included modifications to allow for overlapping generations (Jorde and Ryman 1995; Waples 2005) and introduction of likelihood-based methods which use all the available information on allele frequencies instead of summary statistics such as means and variances (Berthier *et al.* 2002). Recently, a method that jointly estimates gene flow and effective population size (Wang and Whitlock 2003) allows relaxation of the assumption that populations are completely reproductively isolated from each other, a situation rarely found in nature.

frequency distributions and varying number of potential parents. As an example, when all alleles at each locus are at the same frequency, only six loci with more than two alleles at each would yield an exclusion probability ≥ 0.99 (Villanueva *et al.* 2002). Also, characteristics of the markers screened (e.g. number of alleles and their distribution) can be used to predict the potential to detect more than one reproductive partner for a given individual in cases where multiple mating occurs (Neff and Pitcher 2002). Many of the conditions that influence the accuracy of parentage assignment affect pairwise relatedness measures in a similar manner (Wang 2002).

6.2 Mating system

6.2.1 Effective population size

An understanding of mating system, and its effect on population dynamics, is important because of its effect on the genetically effective size of a population, more commonly known as simply effective population size, or N_e. While there are numerous definitions of N_e, in all it is inversely proportional to the rate of change in some measure of genetic variation (e.g. heterozygosity, allelic richness, rate of inbreeding or the level of fluctuation in allele frequencies) (Neigel 1996). That is, populations with a high N_e have a low rate of change in genetic variation. When migration and mutation rates are low, as is often the case, variation is lost over generations due to random sampling effects which alter the genetic composition of the population.

Therefore, from a conservation genetics perspective N_e is an important population parameter. By definition, maintaining high N_e reduces the rate of loss of genetic variation in a population and has effects both short and long term. In the short term, high levels of genetic variability may be associated with increased individual fitness (Wang et al. 2002a) and population persistence (Sacherri et al. 1998; Reed and Frankham 2003). Moreover, in populations with small N_e the probability of mating with a related individual (inbreeding) is greater, increasing the potential for inbreeding depression (Wang et al. 2002b). In the longer term, genetic variation is necessary for a population to be able to adapt to changes in environmental conditions.

In most situations, N_e is not the same as the number of potentially reproductive individuals in the population (the census population size, N_c) due to factors such as unequal sex ratio, fluctuation in population size over time, and variance in reproductive success among individuals (Frankham 1995). These factors interact in a complex manner, but can have such a large cumulative effect that it has been estimated that in natural populations N_e is often only ~ 10% of the value of N_c (Frankham 1995). All aspects of the mating system, including its effect on population dynamics, may influence N_e, but some more so than others, depending on the biology of the species in question.

Estimating N_e and comparing it to N_c could potentially be done for Atlantic salmon populations. However, most methods for estimating N_e involve using changes in allele frequencies over a number of loci within a population over time to infer the average N_e over that period (known as the temporal method) (Waples 1989, 2005; Wang 2005). Applying the temporal method to species with overlapping generations is problematic (Wang 2005) and, although the method has been modified to take overlapping generations into account (Jorde and Ryman 1995; Waples 2005), specific sampling schemes are required for estimates to be valid. Moreover, such estimates usually relate to N_e in previous generations (Waples 2005) and are not valid when environmental conditions and population demography change. So, while it may be possible to produce a number that could be called N_e for a very limited number of populations of Atlantic salmon, what would those numbers really tell us?

Better, we suggest, would be to understand how much the individual elements of the Atlantic salmon mating system influence N_e. Such an understanding would allow use of information that is often more easily estimated from populations, such as census size, sex ratio and age structure, to estimate N_e. Importantly, we may be able to predict how changes in those population parameters might change N_e. For example, with a knowledge of how the relative numbers of male parr to anadromous males at a spawning or in a population affect individual male reproductive success, we might be able to quantify the effect on N_e of the removal of anadromous males through, for example, fishing.

6.2.2 Factors affecting the variance in reproductive success of male alternative reproductive tactics

Variation in Atlantic salmon life histories produces breeding populations with large differences in individual body size with a resultant potential for asymmetries in reproductive success

Fig. 6.1 Mature parr and 4+ MSW fish. (Photo credit: J. Webb.)

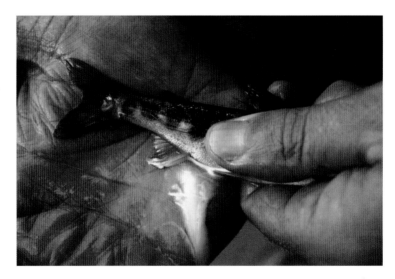

Fig. 6.2 Milt running from a mature male parr. (Photo credit: D. Hay.)

across individuals and age-at-maturity groups (Fleming 1996, 1998). The most marked size differences in alternative life-history types occur between mature male parr and anadromous males (Fig. 6.1), and the consequences of those differences for reproductive success have been the subject of much investigation. Although male parr have long been known to produce viable sperm and to be found in the vicinity of spawning adult fish (Jones and King 1950; Jones 1959) (Fig. 6.2), the extent to which male parr are successful in natural reproduction was not known until recently. Using allozyme polymorphisms as genetic markers, early studies (conducted in the late 1980s) confirmed Jones' (1959) earlier findings that male parr could fertilise eggs in spawning competition with adult males under experimental (Hutchings and Myers 1988) and semi-natural (Jordan and Youngson 1992) conditions. However, given the limited polymorphism of the allozyme loci used, reproductive success of mature male parr was estimated as a group only and not on an individual parr basis. The results from the two

Table 6.1 Results from experimental studies on the reproductive success of Atlantic salmon mature male parr using DNA markers. In some cases values are estimated from figures in original papers and are therefore approximate.

Reference (replicate unit)	No. of anadromous males	No. of anadromous females	No. of mature male parr	Overall parr* reproductive success	Individual parr** reproductive success
Morán et al. (1996) (experimental treatment)	1	1	1	24.73	—
	1	1	6	89.36	—
	1	1	12	39.36	3.5 (1.05–7.40)
Thomaz et al. (1997) (egg pocket)	1	1	1	26 (8–40)	26
	1	1	3	30	10
	1	1	6	27.5 (4–70)	5 (0–10.0)
	1	1	12	40	4
Martinez et al. (2000) (redd)	7	6	Unknown	65.1 (16.7–85.7)	Up to 46.7
García-Vázquez et al. (2001) (experimental treatment)	1	1	4	85.7	—
	2	1	5	42.5	—
	2	1	5	67.9	—
	3	1	7	77.3	—
Jones & Hutchings (2001) (experimental treatment)	1	1	5	4.7	(0–4)
	1	1	10	40.5 (25.9–55.0)	(0–25)
	1	1	20	29.9	(0–6)
	1	2	7	75.8	(0–30)
	1	2	10–23	52.5	(0–28)
	1–2	1	7	50.0	(0–19)
	0	1	5	100.0	(1–50)
Jones & Hutchings (2002) (egg nest)	4	4	20	37.1 (0–100)	1.9 (0–12.7)
	4	4	20	23.0 (0–100)	1.2 (0–13.3)
Garant et al. (2002) (egg nest)	1	1	23	15	2.5 (0–8)
	2	1	12	12	2.0 (0–8)
	6	1	7	6	1.0 (0–6)
Garant et al. (2003a) (egg nest)	12	12	20	24 (0–100)	—

* Mean proportion (%) of eggs fertilised by mature male parr as a group per replicate unit (range across replicates).
** Mean proportion (%) of eggs fertilised by individual mature male parr per replicate unit (range across individuals).

studies were consistent: as a group, male parr were shown to fertilise between 5–23% (Hutchings and Myers 1988) and 1–38% (Jordan and Youngson 1992) of the eggs in a redd (defined as a series of egg nests produced sequentially by a single female).

More recent studies utilising more informative markers have been able to estimate the reproductive success of male parr on both group and individual levels. Most experimental studies (summarised in Table 6.1) have involved a single anadromous female and male pair and variable numbers of mature male parr contained within artificial spawning enclosures. The results from different studies are highly variable, possibly due to the fact that studies varied in the level of analysis (egg nest, redd or over the entire experimental treatment), but some generalisations can still be made. Mature male parr as a group usually contributed to fertilisation of each egg nest (but see Jones and Hutchings 2002; Garant et al. 2003a; Weir et al.

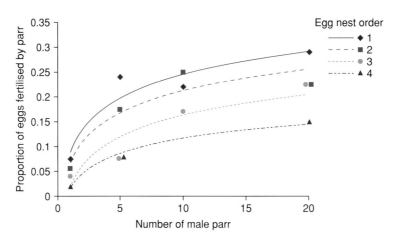

Fig. 6.3 Observed proportions of eggs per egg nest fertilised by male Atlantic salmon parr, competing with a single anadromous male as a function of the number of male parr present per mating. Source: Hutchings and Myers (1988).

2005), but where multiple parr were present often one or more of them did not contribute or the contribution was low and could not be detected in the sample size analysed. The group contribution of male parr tended to increase with number of parr present, but in general anadromous males fertilised the majority of eggs in a nest or redd (Table 6.1). In some cases, male parr success was observed to be inversely proportional to the order of nest construction within the redd, suggesting sperm depletion in parr during spawning (Hutchings and Myers 1988; Thomaz et al. 1997) (Fig. 6.3). Seasonal declines in parr participation have also been observed in experimental studies (Weir et al. 2005). However, whether such effects are found in nature, where male parr are generally in excess and turnover of male parr participating in each egg nest is possible, or they are an artefact of the experimental designs, where the number of parr was limited and fixed (~ 1–30 per enclosure) for the duration of redd construction, remains to be resolved.

On an individual basis, male parr tend to have a much lower success than anadromous males, but this may not always be true at either the egg nest or redd level (Martinez et al. 2000; Jones and Hutchings 2002). Indeed, no anadromous male contribution was detected in some egg nests in the Jones and Hutchings (2002) study. While total parr contribution increases as the number of parr present at spawning increases (Fig. 6.3), it appears that the mean success of each male parr decreases (Hutchings and Myers 1988; Thomaz et al. 1997; Garant et al. 2003a). Within parr, body size has been suggested to influence position in a dominance hierarchy and proximity to the redd (Myers and Hutchings 1987) and there is some evidence that parr reproductive success is also related to size, expressed in terms of proportion of eggs fertilised per egg pocket (Thomaz et al. 1997) (Fig. 6.4a). In contrast, others have found little or no evidence for parr fertilisation success to be size-dependent (Jones and Hutchings 2001, 2002; Garant et al. 2003a). Parr size, however, seems to influence both the probability of achieving any fertilisations at all and the number of nests in which a parr is successful (Jones and Hutchings 2001, 2002; Garant et al. 2002) (Fig. 6.4b). Thus, the actual role of parr size in reproductive success remains somewhat equivocal.

The reproductive success of mature male parr compared to anadromous males may be much greater than might be expected on the basis of, for example, their relative body masses.

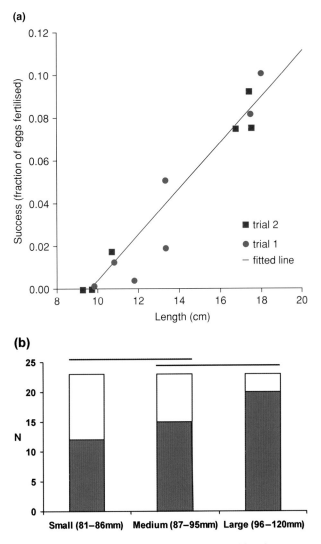

Fig. 6.4 Relationship between parr size and reproductive success measured (a) as the percentage of eggs fertilised in two trials (the regression of reproductive success on body length is highly significant) (source Thomaz *et al.* 1997) and (b) as the number of parr of different size classes that did (■) and did not (□) successfully fertilise at least one egg over a number of trials. Lines connect size classes that did not differ significantly (source: Jones and Hutchings 2001).

The question then becomes: what factors could be promoting male parr reproductive success? While providing a single answer might be too simplistic, several aspects of parr reproductive biology are likely to increase their success. First, from the behavioural point of view, as parr position themselves below anadromous fish during spawning, they are often closer than the anadromous male to the female vent and to the eggs and this proximity may result in higher fertilisation potential. Second, sperm quality is very likely to be a major factor influencing parr success. In absolute terms, anadromous males have larger testes and much greater ejaculate volumes than parr. However, mature male parr not only invest relatively more in gonadal growth than anadromous males, they also produce a relatively greater ejaculate volume with a higher concentration of sperm (of which more are motile) than adult males (Kazakov 1981; Gage *et al.* 1995; Vladic and Järvi 2001). Thus, although there are no apparent differences in sperm morphology between adults and parr (Gage *et al.* 1995; Vladic *et al.* 2002), higher levels of adenosine triphosphate (ATP) (Vladic and Järvi 2001) may produce the greater motility levels, ejaculate longevity (Gage *et al.* 1995) and enhanced fertilisation capacity of parr sperm (Vladic and Järvi 2001; Vladic *et al.* 2002). However, the relevance of a recent report that relative sperm velocity is the primary determinant of fertilisation success during sperm competition in Atlantic salmon (Gage *et al.* 2004) for the relative reproductive success of parr to anadromous males remains unclear, as the two male types appear not to differ in sperm velocity (Vladic and Järvi 2001).

Generally, experimental studies have therefore demonstrated that mature male Atlantic salmon parr can be successful in reproduction and may even be more successful than anadromous males under certain circumstances. As a group, male parr therefore have the potential to significantly increase N_e, although the extent to which they do so depends critically on the mean and variance in reproductive success among male parr in natural populations. The studies described above suggest that both the mean and the variance may be low relative to anadromous males. On this basis, Jones and Hutchings (2002) suggested that variance in reproductive success among anadromous males may have a greater overall influence on N_e than that of male parr. However, Jones and Hutchings (2002) did concede that male parr may have a more important role in populations where (or in years when) anadromous males are relatively scarce, as suggested by other authors (e.g. Morán and García-Vázquez 1998, Martinez *et al.* 2000; García-Vázquez *et al.* 2001). In a species with declining numbers of returning anadromous fish, the reproductive success of mature male parr may play an increasingly important role.

Reproductive success of male parr may also have negative consequences. Although there is evidence that adult Atlantic salmon of farmed origin have lower reproductive success than wild fish (Fleming *et al.* 1996, 2000; Weir *et al.* 2005), the success of mature male parr of farmed origin has been reported to be higher (Garant *et al.* 2003a) or at least equivalent to that of parr of wild origin. Thus, parr may be responsible for significant gene flow in the second and subsequent generations of invasion by farm fish into native populations, hastening the threat to their genetic integrity. Additionally, it appears that the redds of anadromous females of hatchery origin can have higher levels of contribution by secondary males (presumably parr) than those of wild females (Thompson *et al.* 1998). In such a situation parr can also act as means for the introgression of farm (or hatchery-reared) salmon traits into wild populations.

Alternative reproductive tactics are also deployed by anadromous males. However, the relative reproductive success of alternative anadromous male life-history types has received less attention than that of male parr and a limited number of situations in which multiple anadromous males have competed for reproductive success have been investigated, in some cases in the absence of mature male parr. Multiple anadromous male paternity in redds has been observed, but at a low level (Martinez *et al.* 2000; Jones and Hutchings 2002). For example, Martinez *et al.* (2000) found a contribution from a second anadromous male in only one of seven redds constructed in the presence of four competing anadromous males with mature male parr present, and in zero out of six redds with three anadromous males and parr absent. As anadromous males attack and chase male parr, it is possible that the presence of male parr deflects the attention of the dominant anadromous male and allows a second anadromous male to become involved in spawning; this, however, remains to be investigated. Using a combination of genetic analysis and detailed behavioural observation, Mjølnerød et al. (1998) found that timing of ejaculation, level of sperm depletion and behavioural dominance associated with male size all played a role in the reproductive success of anadromous males when multiple males participated in a spawning (in situations where parr were absent). However, in all ten egg nests analysed one male dominated fertilisation (> 80% success).

The experiments in laboratories and under semi-natural conditions described above have therefore elucidated various factors and mechanisms affecting mean and variance in reproductive success in male Atlantic salmon of different life-history types and have suggested variability in relative reproductive success across egg nests and redds. However, while these studies allow for control of many variables they suffer from the criticism that they may not reflect natural conditions. Nevertheless, two recent extensive studies on natural systems have been carried out, and both have contributed much to our understanding of the mating system in Atlantic salmon and its consequences.

6.2.3 Reproductive success estimates and mate choice under natural conditions

Reproductive success

Taggart et al. (2001) investigated the breeding system over 3 years within the Girnock Burn, a tributary of the Aberdeenshire Dee in Scotland. Fixed trap facilities on this stream allow capture and sampling (tissue and scales) of adult fish ascending the stream to spawn. In 3 years of study, spawning was carefully observed, redds identified and accurately mapped. Before hatching in the following spring, egg nests were excavated and 10–20 progeny sampled per redd. Both adult and progeny samples were then subjected to single-locus minisatellite profiling at up to ten loci to establish parentage.

In a similar study, Garant et al. (2001, 2003b, 2005) captured 41 adult male and 35 adult female Atlantic salmon in a branch of the Ste-Marguerite River in Quebec, Canada. These fish were tissue sampled and transferred to a 19 km-long upstream section of the same tributary that is isolated by two impassable waterfalls – the study stretch of the river had not previously been occupied by Atlantic salmon. Progeny were subsequently sampled by electrofishing in late summer of the following years. Again, adults and progeny were genotyped to establish parentage, in this case at five microsatellite loci. While similar in many respects, essential differences between the Taggart et al. and Garant et al. studies were that mature male parr were absent from the study site on the Ste-Marguerite River and that reproductive success was assessed in eyed ova (Taggart et al. 2001) or parr at the end of the summer after hatching (Garant et al. 2001, 2003b, 2005).

Both studies revealed a relatively high level of multiple mating and variance in reproductive success. In the Ste-Marguerite study the estimated average number of mates was 7.5 (range 2–18) per female and 6.4 (range 1–16) for males. Similarly, in the Girnock Burn more than 50% of both males and females contributed to more than one redd in almost all years (Fig. 6.5), but the maximum number of redds in which a single individual's contribution was detected was lower than that for the Ste-Marguerite study (six for females and seven for males). In 12% of redds the contribution of more than one anadromous male could be detected. Mature male parr were successful in spawning in at least 91% of redds over the course of the Girnock Burn study with average yearly contributions of 40–50% per redd. Thus the

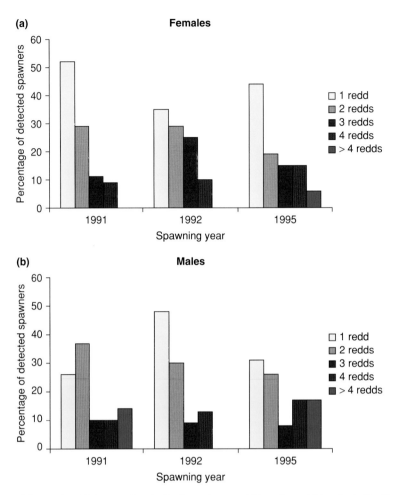

Fig. 6.5 Extent of detected multiple spawnings for both anadromous (a) females and (b) males in the Girnock Burn over three spawning seasons. Source: Taggart et al. (2001).

overall reproductive success of mature male parr per redd in nature agrees well with the average of 45% from the experiments summarised in Table 6.1.

Multiple mating appears to have fitness benefits in Atlantic salmon as reproductive success (assessed as number of offspring sampled in the summer after emergence) was positively correlated with the number of mates for both males and females in the Ste-Marguerite study. This increased reproductive success may be a reflection of having eggs distributed over multiple, often widely dispersed, redds with less chance of high mortality due to poor redd positioning (Taggart et al. 2001) and local density-dependent effects (Einum and Nislow 2005) and/or an inherent property of having a genetically more heterogeneous group of offspring. Body size in males and females was also correlated with reproductive success and number of mates when repeat-year spawners (which are large but had relatively poor success) were removed from the analyses (Fig. 6.6). A similar relationship between size of both males and females and number of matings was also detected in the Girnock Burn (Taggart et al. 2001). Grilse had significantly lower reproductive success than multi-sea-winter fish (Garant et al. 2003b).

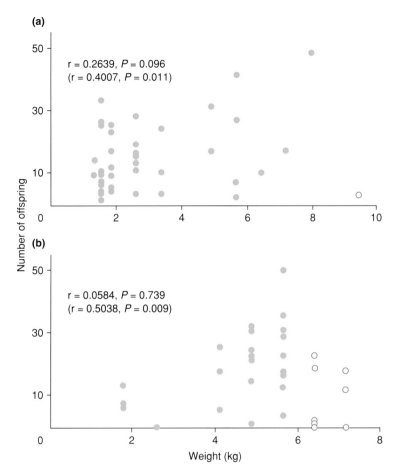

Fig. 6.6 Relationship between body size and reproductive success for anadromous (a) males and (b) females. The results of Pearson's correlation analysis are given both including and (in parentheses) excluding repeat spawners (indicated by open circles). Source: Garant *et al.* (2001).

Mate choice

Body size appears to be a component of number of mates and reproductive success among anadromous males and females (Garant *et al.* 2001; Taggart *et al.* 2001). Both males and females appear to be choosing large mates, with little evidence for size-assortative mating – a weak, but significant effect was found in one of the three years of study by Taggart *et al.* (2001).

Mate choice based on odours determined by genotype at the major histocompatibility complex (MHC) has been demonstrated for a number of vertebrates (reviewed by Jordan and Bruford 1998 and Bernatchez and Landry 2003). MHC genes code for proteins that have an important role in the immune system and exceptionally high individual MHC heterozygosity has been observed in a range of species – presumably to confer resistance to as wide a range of diseases as possible – and appears positively correlated with a range of fitness-related traits (reviewed by Apanius *et al.* 1997; Bernatchez and Landry 2003). At least part of this heterozygosity may result from MHC-disassortative mate choice: i.e. individuals choose mates that have an MHC genotype that is dissimilar to their own in order to produce progeny that are highly heterozygous either at the genome-wide level or specifically at the MHC. Juvenile salmonids are able to discriminate between MHC-identical and MHC-dissimilar siblings on the basis of odour cues (Olsén *et al.* 1998, 2002), indicating the potential for an MHC-based mate choice system. In this context, Landry *et al.* (2001) compared the pattern of mate choice with MHC genotype in the experimental population of the Garant *et al.* (2001, 2003b, 2005) studies, and found a significant tendency for MHC-dissimilar mating (Fig. 6.7a). Given that the offspring were sampled at the end of the first summer, after natural selection has been observed to act on MHC genotype distributions (De Eyto *et al.*), the apparent strength of MHC-based mate choice may have been overestimated. However, given the level and mode of natural selection so far observed on MHC loci in Atlantic salmon (De Eyto *et al.*), it is unlikely that selection alone explains the lower-than-expected incidence of MHC-similar matings. There was no evidence for outbreeding across the genome (as assessed at a number of microsatellite loci) (Fig. 6.7b), suggesting that the observed mate choice has evolved to promote diversity specifically at MHC genes.

This lack of genome-wide outbreeding is perhaps surprising given the potential fitness benefits of increased genetic diversity in offspring from outbred matings (Wang *et al.* 2002a,b). While there may be some traits (Primmer *et al.* 2003) and types of genetic marker (Borrell *et al.* 2004; Pineda *et al.* 2003) that do not conform to the expected positive relationship between heterozygosity and components of fitness, many studies have reported associations between individual genetic diversity and components of fitness (e.g. levels of fluctuating asymmetry, Borrell *et al.* 2004; time of first feeding and age at smolting, McCarthy *et al.* 2003; Pineda *et al.* 2003; Borrell *et al.* 2004; growth rate, Blanco *et al.* 1998; competitive foraging ability, Primmer *et al.* 2003; aggression, Tiira *et al.* 2003).

This apparent paradox – a lack of inbreeding avoidance in a species in which homing behaviour to the natal stream make the chances of inbreeding high – may be resolved to some extent by the observation of high levels of multiple mating in the Girnock and Ste-Marguerite studies which can produce genetically more heterogeneous offspring than single mating (Garant *et al.* 2001; Taggart *et al.* 2001). Moreover, it appears that more outbred males are more successful in reproduction (Fig. 6.8), which may be due to mate choice by females or the outcome of male–male competition. Female mate choice in salmonids appears to be heavily constrained by male–male competition (Fleming and Reynolds 2004). Regardless of the mechanism, this effect produces less inbred offspring and increases female reproductive success (Fig. 6.9).

6.2.4 Hybridisation

Interspecific hybridisation may be viewed as a breakdown in mate choice, and naturally occurring hybridisation between Atlantic salmon and brown trout *Salmo trutta* has been documented across most areas of overlap between the species ranges using a wide choice of molecular marker types (Table 6.2). Furthermore, by combining data from different molecular markers, especially those from nuclear and mitochondrial DNA, it is often possible to determine the direction of hybridisation as nuclear markers are inherited from both species (and therefore identify the individual as a hybrid) but mitochondrial DNA is inherited from the female parent only.

Under normal circumstances, it appears that temporal and spatial differences in spawning maintain reproductive isolation between the species (Heggberget *et al.* 1988). Numerous

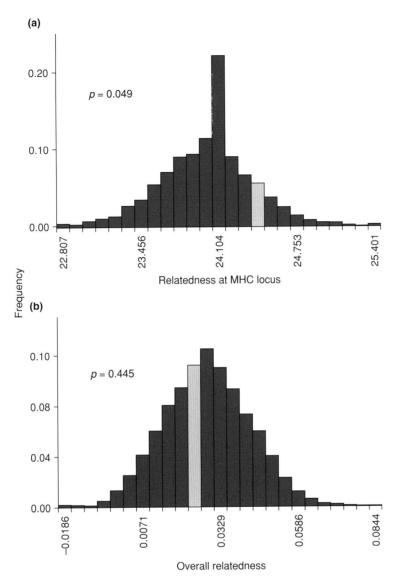

Fig. 6.7 Distributions of relatedness from simulated random matings between anadromous Atlantic salmon in the Ste-Marguerite study (see text). The grey bars indicate the observed level of relatedness between actual mates (a) specifically at an MHC locus and (b) across the genome as a whole assessed from five microsatellite loci. The observed relatedness at the MHC locus was significantly lower than if mating had been at random (the scale used here is inverse), while observed overall relatedness was not significantly outside the random of simulated random values. Source: Landry *et al.* (2001).

explanations for the breakdown of that reproductive isolation have been proposed. These include introductions of hatchery-reared fish (often non-native), either through fish farm escapes (Youngson *et al.* 1993; Hindar and Balstad 1994; Matthews *et al.* 2000) or stocking programmes (García de Leániz and Verspoor 1989; Jansson and Öst 1997). Hybridisation seems particularly frequent in those situations where one or both species are introduced

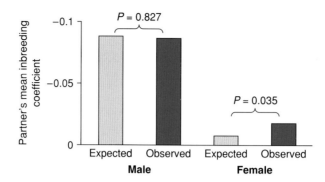

Fig. 6.8 Expected (population mean) and observed (mean of all partners) mean inbreeding coefficient for male and female Atlantic salmon partners. Here the inbreeding coefficient used is internal relatedness. Source: Garant *et al.* (2005).

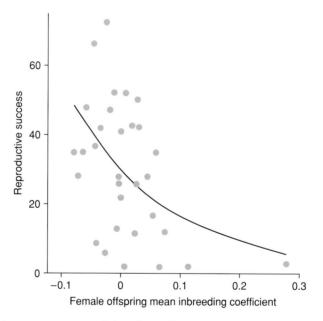

Fig. 6.9 Relationship between average mean inbreeding coefficient of offspring and reproductive success of each anadromous female. The relationship is significant even if the individual with the highest progeny inbreeding coefficient is excluded. Source: Garant *et al.* (2005).

outside their natural range (Verspoor 1988; Gephard *et al.* 2000; Ayllon *et al.* 2004). A low abundance of one species relative to the other (Crozier 1984), or similar body mass of brown trout and Atlantic salmon in some situations (Jordan and Verspoor 1993), may also promote hybridisation. Extreme environmental conditions or disturbance may increase competition for spawning grounds and consequent levels of hybridisation (García de Leániz and Verspoor 1989; Jansson and Öst 1997).

Most authors have suggested that, in addition to situation-specific factors, mature male Atlantic salmon parr may play an important role in interspecific matings. Such a role could potentially explain those situations where hybridisation is unidirectional and brown trout are the female parent (Table 6.2), but both hybridisation exclusively in the other direction and

Table 6.2 Reports of natural hybridisation between Atlantic salmon and brown trout; updated from Jordan and Verspoor (1993).

Reference	Country/area	Morphological type and life-history stage	Mean frequency (%)	Range (%)	Direction of hybridisation (parental female)
Payne et al. (1972)	British Isles	'salmon' adults	0.4	—	—
Solomon & Child (1978)	British Isles	'salmon' adults	0.3	—	—
Beland & Roberts (1981)	Stewiake River, Nova Scotia	salmonid parr	0.2	—	—
Taggart et al. (1981)	Ireland	'trout' parr	0.003	—	—
Fleming (1983)	British Isles	'trout' parr	0.1	0–4.2	—
Crozier (1984)	Lough Neagh system, N. Ireland	'trout' parr	0.4	0–5.9	—
Verspoor (1988)	Newfoundland	'salmon' parr, smolts and adults	0.9	0–11.8	—
García de Leániz & Verspoor (1989)	Spain	'salmon' parr	2.3	0–7.7	—
Hurrell & Price (1991)	S. W. England	salmonid parr	1.4	0–3.8	—
Jansson et al. (1991)	River Gronan, Sweden	salmonid parr	13.2	2.5–22.8	—
McGowan & Davidson (1992)	Newfoundland	salmonid parr	4.7	0–18.8	brown trout
Youngson et al. (1992)	Scotland	suspected hybrid adult females	—	—	3/3 Atlantic salmon
Jordan & Verspoor (1993)	British Isles	'salmon' parr	1.0	0–14.6	—
Hindar & Balstad (1994)	Norway	salmonid parr 1980–86 1987–92	0.2 0.9	0–2.0 0–8.0	— —
Elo et al. (1995)	Rivers Teno and Näätämö, Norway and Finland	salmonid parr and adults	0.1	0–1.6	—
Hartley (1996)	River Leven, England	'salmon' parr	18.8	—	9/10 brown trout 1/10 Atlantic salmon
Jansson & Öst (1997)	River Dalälven, Sweden	salmonid parr " adults	41.5 1.6	35.2–66.7 0.1–3.1	brown trout
Matthews et al. (2000)	Western Ireland	salmonid parr	1.2	0–3.1	Atlantic salmon
García-Vázquez et al. (2001)	S. Europe	salmonid parr and adults	2.5	0.9–3.2	Atlantic salmon
Taggart et al. (2001)	Girnock Burn, Scotland	eyed ova (1991 cohort)	0.01	—	—
Ayllon et al. (2004)	Kergulen Islands	salmonid parr and adults	6.2	4.2–28.6	brown trout

Fig. 6.10 Two anadromous adult hybrid salmon caught in the River Don, north-east Scotland, on their return from the sea. (Photo credit: D. Hay.)

bidirectional hybridisation have also been observed (Table 6.2). Aggression of brown trout towards mature male Atlantic salmon parr and a reported lower survival rate of hybrid progeny may also argue against a universal role for Atlantic salmon parr in hybridisation events (García-Vázquez *et al.* 2002).

In general, interspecific hybrids display reduced survival and fecundity, which act as post-mating isolating mechanisms reinforcing species barriers. Little is known of the survival of Atlantic salmon/brown trout hybrids relative to non-hybrids in the wild, but naturally occurring adult migratory hybrids have been identified (Table 6.2, Fig. 6.10). While in some situations hybrids can produce viable backcross offspring (García-Vázquez *et al.* 2003), usually they are functionally infertile (Youngson *et al.* 1992; Galbreath and Thorgaard 1995). When they are produced under experimental conditions backcross individuals are generally triploid and therefore infertile, limiting the possibility of introgression of genes from one species to the other (Verspoor and Hammar 1991).

6.3 Social structure

The mating system of the Atlantic salmon has a profound influence on the social structure of early juvenile populations. The construction of a closely spaced series of egg nests to form a redd by a female means that closely related fish (full- and half-sibling) emerge in near proximity to one another. However, depending on the distribution of spawning gravel within a river or stream, redds from different females, mating with the same or different males, may also be in close proximity. Extensive multiple mating by both sexes has the result that half-sibling fish often also occur in different redds, which may be widely separated from one another. Patterns of relatedness in natural populations of Atlantic salmon are therefore heavily dependent on the distribution of redds and matings, while the consequences of those patterns of relatedness are crucially affected by the ability to recognise and differentially respond to kin. Molecular markers can have a key role in describing those patterns of relatedness and how they might change during juvenile life.

6.3.1 Kin recognition and kin-biased behaviour

From laboratory studies it appears that salmonids have well-developed kin recognition abilities (reviewed in Brown and Brown 1996a). In Atlantic salmon specifically, parr can discriminate kin from non-kin using waterborne sensory cues (Brown and Brown 1992), including urine (Moore *et al.* 1994). Although not directly demonstrated in Atlantic salmon, Arctic charr (*Salvelinus alpinus*) can use odour cues related to MHC genotype in kin discrimination, in common with many other vertebrate species (Olsén *et al.* 1998, 2002). This is consistent with a possible role for MHC genotype in mate choice described above.

This ability to recognise kin appears to have fitness benefits as, in controlled settings, kin groups of Atlantic salmon parr show shorter nearest neighbour distances, lower levels of aggression and greater foraging opportunities (Brown and Brown 1993, 1996b), resulting in higher and less variable individual weight gain in kin groups than non-kin groups (Brown and Brown 1996b). In particular, subordinate individuals benefit from a greater tolerance by dominant kin (Brown and Brown 1996b; Griffiths and Armstrong 2002). However, in non-kin groups where genetic diversity is high, competition among individuals may be lower due to different habitat use by different genotypes (Griffiths and Armstrong 2001) and there is evidence that under conditions where there is a cost to aggregation, such as in winter refuges, juvenile salmon avoid kin in order to avoid sharing these costs with close relatives (Griffiths *et al.* 2003). Despite these caveats, it appears that an ability to recognise kin has fitness benefits in a wide range of fish species and may promote kin aggregations (Ward and Hart 2003).

6.3.2 Patterns of relatedness in nature and fitness

Given that Atlantic salmon, and salmonids in general, have well-developed kin recognition capabilities and that laboratory studies suggest that there are often benefits from living in aggregations of kin, is there much evidence that such kin groups actively occur in nature? It appears not, at least for Atlantic salmon.

Immediately after emergence, fry may be found in highly related groups, but these groups appear to breakdown as fry disperse, and any kin-related structure may weaken further as parr move to different habitats (Fontaine and Dodson 1999; Webb *et al.* 2001). Moreover, as multiple mating seems to be common in Atlantic salmon, fry from a single redd may commonly represent many half-sibling groups. Thus, while Mjølnerød *et al.* (1999) found that there was a significant negative relationship between genetic similarity and geographic distance between pairs of juvenile fish in a 300 m stretch of river (even in parr that were 3 years old), suggesting some degree of kin aggregation, genetic similarity explained only 0.9% of the variability in spatial distribution of fry and parr. An intriguing sex-related difference in the relationship between genetic similarity and geographic distance in parr appeared to result from a small number of closely spaced and relatively closely related mature male parr within the sample, perhaps associated with aggregation on spawning grounds (Mjølnerød *et al.* 1999). However, this observation requires further investigation and confirmation.

Fontaine and Dodson (1999) found no significant relationship between pairwise relatedness and geographic distance between pairs of fish in a series of 5 m × 20 m sections of stream. Furthermore, using mapped juvenile territories and identification of full- and half-sibling relationships, they found that very few highly related individuals held neighbouring territories

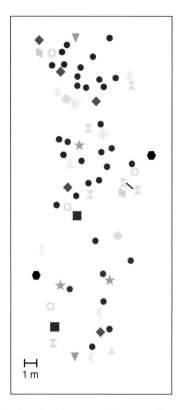 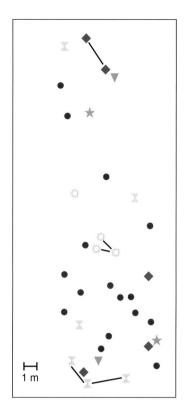

Fig. 6.11 Distribution of relatedness within two sections of Atlantic salmon fry habitat. Individuals from the same family are indicated by the same symbol within each section and lines connect siblings that are nearest neighbours. Closed circles represent individuals with no sibling identified within that section. Source: Fontaine and Dodson (1999).

(Fontaine and Dodson 1999) (Fig. 6.11) and, in contrast to laboratory studies, no evidence for growth benefits in individuals with neighbouring kin.

While there is thus little evidence of active kin aggregation in juvenile Atlantic salmon in nature, there is a possibility for kin groups to exist within migrating (both as smolts and adults) and marine feeding groups, analogous to the situation for mature male parr on spawning grounds (Mjølnerød et al. 1999). To date, no genetically based studies have investigated these potential phenomena, although results from physical tagging of Atlantic salmon smolt groups (Olsén et al. 2004) and microsatellite analysis of migratory brook charr schools (Fraser et al. 2005) suggest that there may be variation in patterns of relatedness within salmonid groups across life-history stages.

6.4 Summary and conclusions

It may be that we are still some way off being able to estimate N_e in Atlantic salmon populations through extrapolation from a knowledge of basic population variables (census size, sex ratio, age structure, and so on), but considerable insights into the relative importance of various factors have been made from the application of genetic marker techniques. The same techniques are also beginning to improve understanding of patterns of relatedness among parr in natural populations.

- The limited number of studies on the mating system of Atlantic salmon using genetic marker techniques has revealed an unforeseen complexity, especially in the level of multiple mating and the reproductive success of mature male parr. However, further long-term studies integrating both behavioural observations and genetic analyses are required to assess the level of individual, temporal and inter-population variability in mating strategies adopted.
- Experimental studies have provided much insight into the factors affecting the reproductive success of mature male parr, and suggest that overall success of parr may be high but both the mean and variance of individual parr success may be relatively low compared to that of anadromous males. Studies of natural situations confirm the high overall success of male parr, but future work is required to determine the mean and variance of individual male success in natural situations. Only then can the influence of mature male parr on the effective size of natural populations be assessed.
- Multiple mating appears common, with fitness benefits apparent in the higher survival of offspring from those parents with more mates and lower inbreeding levels of female's offspring. Whether this is a direct genetic benefit as a result of increased offspring diversity or a benefit of bet-hedging by spreading reproductive effort over a number of redds remains to be established.
- Variance in reproductive success is high within anadromous fish (for both males and females). Reproductive success is affected by body size, with multi-sea-winter fish more successful than grilse. However, repeat spawners appear to have lower success than their size would predict.
- There appears to be little assortative mating on the basis of body size, but there may be disassortative mating on the basis of MHC genotype and selection of males with low levels of inbreeding by females. However, work is needed to disentangle the potential roles of male and female mate choice from one another, as well as from the effects of male–male competition.
- While hybridisation between Atlantic salmon and brown trout can be common and occur locally at high frequency, reduced fecundity (and possibly viability) of hybrids suggest that introgression of genes from one species to the other is low.
- Social structure in natural populations appears to be weak despite potential kin-selected benefits and evidence for kin-biased behaviours in laboratory studies. Clumping of redds representing different matings, extensive dispersal from redds by fry (which are often half-siblings initially) and movement into different habitats as parr grow may explain the weak social structure in nature. However, little is known of kin structure in migrating and marine feeding groups of Atlantic salmon.

6.5 Management recommendations

- The importance of mature male parr in increasing effective population size, especially in small or declining populations, suggests that any measures of population size in Atlantic salmon should include estimates of numbers of mature male parr (conditioned by estimates of their relative reproductive success).

- Similarly, mature male parr may be important in speeding introgression of genes from one population into another (e.g. farmed and hatchery into wild populations) and should be considered in risk assessment and conservation management.
- The potential direct genetic effects of multiple mating should be recognised and breeding systems for producing fish for supplemental stocking should be designed accordingly.
- While introgression between brown trout and Atlantic salmon may be limited, the conditions which appear to promote hybridisation between the species should be avoided as high frequencies of hybrids may reduce population fitness.
- The apparent lack of kin aggregations among parr and the lack of convincing demonstration of kin-selected benefits in juvenile life in natural populations suggest that there is little need for maintaining sibling groups together when (re-)stocking rivers with parr. However, the kin structure of groups of smolts, marine feeding groups and returning adults may have to be taken into account when designing smolt release programmes and strategies for oceanic, coastal and riverine harvesting.

References

Andersson, M. (1994) *Sexual Selection*. Princeton University Press, Princeton.

Apanius, V., Penn, D., Slev, P.R., Ruff, L.R. and Potts, W.K. (1997) The nature of selection on the major histocompatibility complex. *Critical Reviews in Immunology*, **17**: 179–224.

Ayllon, F., Martinez, J.L., Davaine, P., Beall, E., García-Vázquez, E. (2004) Interspecific hybridization between Atlantic salmon and brown trout introduced in the subantarctic Kerguelen Islands. *Aquaculture*, **230**: 81–88.

Beland, K.F. and Roberts, F.L. (1981) Evidence of *Salmo salar* × *Salmo trutta* hybridization in a north American river. *Canadian Journal of Fisheries and Aquatic Sciences*, **38**: 552–554.

Bernatchez, L. and Duchesne P. (2000) Individual-based genotype analysis in studies of parentage and population assignment: how many loci, how many alleles? *Canadian Journal of Fisheries and Aquatic Sciences*, **57**: 1–12.

Bernatchez, L. and Landry, C. (2003) MHC studies in nonmodel vertebrates: what have we learned about natural selection in 15 years? *Journal of Evolutionary Biology*, **16**: 363–377.

Berthier, P., Beaumont, M.A., Cornuet, J-M. and Luikart, G. (2002) Likelihood-based estimation of effective population size using temporal changes in allele frequencies: a genealogical approach. *Genetics*, **160**: 741–751.

Blakey, J.K. (1994) Genetic evidence for extra-pair fertilizations in a monogamous passerine, the great tit *Parus major*. *Ibis*, **136**: 457–462.

Blanco, G., Presa, P., Vázquez, E. and Sanchez, J.A. (1998) Allozyme heterozygosity and development in Atlantic salmon, *Salmo salar*. *Fish Physiology and Biochemistry*, **19**: 163–169.

Borrell, Y.J., Pineda, H., McCarthy, I., Vázquez, E., Sanchez, J.A. and Lizana, G.B. (2004) Correlations between fitness and heterozygosity at allozyme and microsatellite loci in the Atlantic salmon, *Salmo salar* L. *Heredity*, **92**: 585–593.

Bourke, A.F.G. and Franks, N.R. (1995) *Social Evolution in Ants*. Princeton University Press, Princeton, NJ.

Brown, G.E. and Brown, J.A. (1992) Do rainbow trout and Atlantic salmon discriminate kin? *Canadian Journal of Zoology*, **70**: 1636–1640.

Brown, G.E. and Brown, J.A. (1993) Social dynamics in salmonid fishes: do kin make better neighbours? *Animal Behaviour*, **45**: 863–871.

Brown, G.E. and Brown, J.A. (1996a) Kin discrimination in salmonids. *Reviews in Fish Biology and Fisheries*, **6**: 201–219.

Brown, G.E. and Brown, J.A. (1996b) Does kin-biased territorial behaviour increase kin-biased foraging in juvenile salmonids? *Behavioural Ecology*, 7: 24–29.

Bulmer, M.S., Adams, E.S. and Traniello, J.F.A. (2001) Variation in colony structure in the subterranean termite *Reticulitermes flavipes*. *Behavioral Ecology and Sociobiology*, 49: 236–243.

Crozier, W.W. (1984) Electrophoretic identification and comparative examination of naturally occurring F1 hybrids between brown trout (*Salmo trutta* L.) and Atlantic salmon (*S. salar* L.) *Comparative Biochemistry and Physiology*, 78B: 785–790.

De Eyto, E., McGinnity, P., Consuegra, S., Coughlan, J., Tufto, J., Farrell, K., Jordan, W.C., Cross, T., Hindar, K., Megens, H-J. and Stet, R. (in press) Natural selection acts on MHC variability in wild Atlantic salmon. *Proceedings of the Royal Society of London, Series B*.

DeWoody, J.A. and Avise, J.C. (2001) Genetic perspectives on the natural history of fish mating systems. *Journal of Heredity*, 92: 167–172.

Duchesne, P., Godbout, M-H. and Bernatchez, L. (2002) PAPA (package for the analysis of parental allocation): a computer program for simulated and real parental allocation. *Molecular Ecology Notes*, 2: 191–193.

Einum, S. and Nislow, N.H. (2005) Local-scale density-dependent survival of mobile organisms in continuous habitats: an experimental test using Atlantic salmon. *Oecologia*, 143: 203–210.

Elo, K., Erkinaro, J., Vuorinen, J.A. and Niemelä, E. (1995) Hybridisation between Atlantic salmon (*Salmo salar*) and brown trout (*S. trutta*) in the Teno and Näätämö river systems, northernmost Europe. *Nordic Journal of Freshwater Research*, 70: 56–61.

Emlen, S.T. and Oring, L.W. (1977) Ecology, sexual selection, and the evolution of mating systems. *Science*, 197: 215–223.

Fleming, C.C. (1983) Population Biology of Anadromous Brown Trout (*Salmo trutta* L.) in Ireland and Britain. PhD Thesis, Queen's University, Belfast.

Fleming, I.A. (1996) Reproductive strategies of Atlantic salmon: ecology and evolution. *Reviews in Fish Biology and Fisheries*, 6: 379–416.

Fleming, I.A. (1998) Pattern and variability in the breeding system of Atlantic salmon (*Salmo salar*), with comparisons to other salmonids. *Canadian Journal of Fisheries and Aquatic Sciences*, 55: 59–76.

Fleming, I.A. and Reynolds, J.D. (2004) Salmonid breeding systems. In: A.P. Hendry and S.C. Stearns (Ed.) *Evolution Illuminated: Salmon and their relatives*, pp. 264–294. Oxford University Press, Oxford.

Fleming, I.A., Jonsson, B., Gross, M.R. and Lamberg, A. 1996. An experimental study of the reproductive behaviour and success of farmed and wild Atlantic salmon (*Salmo salar*). *Journal of Applied Ecology*, 33: 893–905.

Fleming, I.A., Hindar, K., Mjølnerød, I.B., Jonsson, B., Balstad, T. and Lamberg, A. (2000) Lifetime success and interactions of farm salmon invading a native population. *Proceedings of the Royal Society of London, Series B*, 267: 1517–1523.

Fontaine, P.M. and Dodson, J.J. (1999) An analysis of the distribution of juvenile Atlantic salmon (*Salmo salar*) in nature as a function of relatedness using microsatellites. *Molecular Ecology*, 8: 189–198.

Frank, S.A. (1998) *Foundations of Social Evolution*. Princeton University Press, Princeton, NJ.

Frankham, R. (1995) Effective population size/adult population size ratios in wildlife: a review. *Genetical Research*, 66: 95–107.

Fraser, D.J., Duchesne, P. and Bernatchez, L. (2005) Migratory charr schools exhibit population and kin associations beyond juvenile stages. *Molecular Ecology*, 14: 31–33.

Gage, M.J.G., Stockley, P. and Parker, G.A. (1995) Effects of alternative male mating strategies on characteristics of sperm production in the Atlantic salmon (*Salmo salar*): theoretical and empirical investigations. *Philosophical Transactions of the Royal Society of London, Series B*, 350: 391–399.

Gage, M.J.G., Macfarlane, C.P., Yeates, S., Ward, R.G., Searle, J.B. and Parker, G.A. (2004) Spermatozoal traits and sperm competition in Atlantic salmon: relative sperm velocity is the primary determinant of fertilization success. *Current Biology*, 14: 44–47.

Galbreath, P.F. and Thorgaard, G.H. (1995) Sexual maturation and fertility of diploid and triploid Atlantic salmon × brown trout hybrids. *Aquaculture*, **137**: 299–311.

Garant, D., Dodson, J.J. and Bernatchez, L. (2001) A genetic evaluation of mating system and determinants of individual reproductive success in Atlantic salmon (*Salmo salar*). *Journal of Heredity*, **92**: 137–145.

Garant, D., Fontaine, P-M., Good, S.P., Dodson, J.J. and Bernatchez, L. (2002) The influence of male parental identity on growth and survival of offspring in Atlantic salmon (*Salmo salar*). *Evolutionary Ecology Research*, **4**: 537–549.

Garant, D., Fleming, I.A., Einum, S. and Bernatchez, L. (2003a) Alternative male life-history tactics as potential vehicles for speeding introgression of farm salmon traits into wild populations. *Ecology Letters*, **6**: 541–549.

Garant, D., Dodson, J.J. and Bernatchez, L. (2003b) Differential reproductive success and heritability of alternative reproductive tactics in wild Atlantic salmon (*Salmo salar* L.). *Evolution*, **57**: 1133–1141.

Garant, D., Dodson, J.J. and Bernatchez, L. (2005) Offspring genetic diversity increases fitness of female Atlantic salmon (*Salmo salar*). *Behavioural Ecology and Sociobiology*, **57**: 240–244.

García de Leániz, C. and Verspoor, E. (1989) Natural hybridization between Atlantic salmon, *Salmo salar*, and brown trout, *Salmo trutta*, in northern Spain. *Journal of Fish Biology*, **34**: 41–46.

García-Vázquez, E., Morán, P., Martinez, J.L., Perez, J., de Gaudemar, B. and Beall, E. (2001) Alternative mating strategies in Atlantic salmon and brown trout. *Journal of Heredity*, **92**: 146–149.

García-Vázquez, E., Morán, P., Perez, J., Martinez, J.L., Izquierdo, J.I., de Gaudemar, B. and Beall, E. (2002) Interspecific barriers between salmonids when hybridisation is due to sneak mating. *Heredity*, **89**: 288–292.

García-Vázquez, E., Ayllon, F., Martinez, J.L., Perez, J. and Beall, E. (2003) Reproduction of interspecific hybrids of Atlantic salmon and brown trout in a stream environment. *Freshwater Biology*, **48**: 1100–1104.

Gephard, S., Morán, P. and García-Vázquez, E. (2000) Evidence of successful natural reproduction between brown trout and mature male Atlantic salmon parr. *Transactions of the American Fisheries Society*, **129**: 301–306.

Gibbs, H.L. and Weatherhead, P.J. (2001) Insights into population ecology and sexual selection in snakes through the application of DNA-based genetic markers. *Journal of Heredity*, **92**: 173–179.

Glaubitz, J.C., Rhodes, O.E. and DeWoody, A. (2003) Prospects for inferring pairwise relationships with single nucleotide polymorphisms. *Molecular Ecology*, **12**: 1039–1047.

Griffith, S.C., Owens, I.P.F. and Thuman, K.A. (2002) Extra pair paternity in birds: a review of interspecific variation and adaptive function. *Molecular Ecology*, **11**: 2195–2212.

Griffiths, S.W. and Armstrong, J.D. (2001) The benefits of genetic diversity outweigh those of kin association in a territorial animal. *Proceedings of the Royal Society of London, Series B*, **268**: 1293–1296.

Griffiths, S.W. and Armstrong, J.D. (2002) Kin-biased territory overlap and food sharing among Atlantic salmon juveniles. *Journal of Animal Ecology*, **71**: 480–486.

Griffiths, S.W., Armstrong, J.D. and Metcalfe, N.B. (2003) The cost of aggregation: juvenile salmon avoid sharing winter refuges with siblings. *Behavioral Ecology*, **14**: 602–606.

Hamilton, W.D. (1964a) The genetical evolution of social behaviour. I. *Journal of Theoretical Biology*, **7**: 1–16.

Hamilton, W.D. (1964b) The genetical evolution of social behaviour. II. *Journal of Theoretical Biology*, **7**: 17–52.

Hardy, O.J. (2003) Estimation of pairwise relatedness between individuals and characterization of isolation-by-distance processes using dominant genetic markers. *Molecular Ecology*, **12**: 1577–1588.

Hartley, S.E. (1996) High incidence of Atlantic salmon × brown trout hybrids in a Lake District stream. *Journal of Fish Biology*, **48**: 151–154.

Heggberget, T.G., Haukebø, T., Mork, J. and Ståhl, G. (1988) Temporal and spatial segregation of spawning in sympatric populations of Atlantic salmon, *Salmo salar* L., and brown trout, *Salmo trutta* L. *Journal of Fish Biology*, **33**: 347–356.

Hindar, K. and Balstad, T. (1994) Salmonid culture and interspecific hybridization. *Conservation Biology*, **8**: 881–882.

Hurrell, R.H. and Price, D.J. (1991) Natural hybrids between Atlantic salmon, *Salmo salar* L., and trout *Salmo trutta* L., in juvenile salmonid populations in south-west England. *Journal of Fish Biology*, **39A**: 335–341.

Hutchings, J.A. and Myers, R.A. (1988) Mating success of alternative maturation phenotypes in male Atlantic salmon, *Salmo salar*. *Oecologia*, **75**: 169–174.

Jansson, H. and Öst, T. (1997) Hybridization between Atlantic salmon (*Salmo salar*) and brown trout (*S. trutta*) in a restored section of the River Dalälven, Sweden. *Canadian Journal of Fisheries and Aquatic Sciences*, **54**: 2033–2039.

Jansson, H., Holmgren, I., Wedin, K. and Andersson, T. (1991) High frequency of natural hybrids between Atlantic salmon, *Salmo salar* L., and brown trout, *Salmo trutta* L., in a Swedish river. *Journal of Fish Biology*, **39A**: 343–348.

Jordan, W.C. and Bruford, M.W. (1998). New perspectives on mate choice and the MHC. *Heredity*, **81**: 127–133.

Jordan, W.C. and Verspoor, E. (1993) The incidence of natural hybrids between Atlantic salmon (*Salmo salar* L.) and brown trout (*Salmo trutta* L.) in Britain. *Aquaculture and Fisheries Management*, **24**: 441–445.

Jordan, W.C. and Youngson, A.F. (1992) The use of genetic marking to assess the reproductive success of mature male Atlantic salmon parr (*Salmo salar*, L.) under natural spawning conditions. *Journal of Fish Biology*, **41**: 613–618.

Jorde, P.E. and Ryman, N. (1995) Temporal allele frequency change and estimation of effective size in populations with overlapping generations. *Genetics*, **139**: 1077–1090.

Jones, J.W. (1959) *The Salmon*. Collins, London.

Jones, J.W. and King, G.M. (1950) Progeny of male salmon parr: a comparison with those from normal adults. *Salmon and Trout Magazine*, **128**: 24–26.

Jones, M.W. and Hutchings, J.A. (2001) The influence of male parr body size and mate competition on fertilization success and effective population size in Atlantic salmon. *Heredity*, **86**: 675–684.

Jones, M.W. and Hutchings, J.A. (2002) Individual variation in Atlantic salmon fertilization success: implications for effective population size. *Ecological Applications*, **12**: 184–193.

Kazakov, R.V. (1981) Peculiarities of sperm production by anadromous and parr Atlantic salmon (*Salmo salar* L.) and fish cultural characteristics of such sperm. *Journal of Fish Biology*, **18**: 1–8.

Landry, C., Garant, D., Duchesne, P. and Bernatchez, L. (2001) 'Good genes as heterozygosity': the major histocompatibility complex and mate choice in Atlantic salmon (*Salmo salar*). *Proceedings of the Royal Society of London, Series B*, **268**: 1279–1285.

Marshall, T.C., Slate, J., Kruuk, L.E.B. and Pemberton, J.M. (1998) Statistical confidence for likelihood-based paternity inference in natural populations. *Molecular Ecology*, **7**: 639–655.

Martinez, J.L., Morán, P., Perez, J., De Gaudemar, B., Beall, E. and García-Vázquez, E. (2000) Multiple paternity increases effective size of southern Atlantic salmon populations. *Molecular Ecology*, **9**: 293–298.

Matthews, M.A., Poole, W.R., Thompson, C.E., McKillen, J., Ferguson, A., Hindar, K. and Wheelan, K.F. (2000) Incidence of hybridisation between Atlantic salmon, *Salmo salar* L., and brown trout, *Salmo trutta* L., in Ireland. *Fisheries Management and Ecology*, **7**: 337–347.

McCarthy, I.D., Sanchez, J.A. and Blanco, G. (2003) Allozyme heterozygosity, date of first feeding and life history strategy in Atlantic salmon. *Journal of Fish Biology*, **62**: 341–357.

McGowan, C. and Davidson, W.S. (1992) Unidirectional natural hybridization between Atlantic salmon and brown trout in Newfoundland. *Canadian Journal of Fisheries and Aquatic Sciences*, **49**: 1953–1958.

Mjølnerød, I.B., Fleming, I.A., Refseth, U.H. and Hindar, K. (1998) Mate and sperm competition during multiple-male spawnings of Atlantic salmon. *Canadian Journal of Zoology*, **76**: 70–75.

Mjølnerød, I.B., Refseth, U.H. and Hindar, K. (1999) Spatial association of genetically similar Atlantic salmon juveniles and sex bias in spatial patterns in a river. *Journal of Fish Biology*, **55**: 1–8.

Moore, A., Ives, M.J. and Kell, L.T. (1994) The role of urine in sibling recognition in Atlantic salmon *Salmo salar* (L.) parr. *Proceedings of the Royal Society of London, Series B*, **225**: 173–180.

Morán, P. and García-Vázquez, E. (1998). Multiple paternity in Atlantic salmon: a way to maintain genetic variability in relicted populations. *Journal of Heredity*, **89**: 551–553.

Morán, P., Pendas, A.M., Beall, E. and García-Vázquez, E. (1996) Genetic assessment of the reproductive success of Atlantic salmon precocious parr by means of VNTR loci. *Heredity*, **77**: 655–660.

Morin, P.A., Wallis, J., Moore J.J. and Woodruff, D.S. (1994) Paternity exclusion in a community of wild chimpanzees using hypervariable simple sequence repeats. *Molecular Ecology*, **3**: 469–478.

Myers, R.A. and Hutchings, J.A. (1987) Mating of anadromous Atlantic salmon, *Salmo salar* L., with mature male parr. *Journal of Fish Biology*, **31**: 143–146.

Neff, B.D. and Pitcher, T.E. (2002) Assessing the statistical power of genetic analyses to detect multiple mating in fishes. *Journal of Fish Biology*, **61**: 739–750.

Neff, B.D., Repka, J. and Gross, M.R. (2001) A Bayesian framework for parentage analysis: the value of genetic and other biological data. *Theoretical Population Biology*, **59**: 315–331.

Nei, M. and Tajima, F. (1981) Genetic drift and estimation of effective population size. *Genetics*, **98**: 635–640.

Neigel, J.E. (1996) Estimation of effective population size and migration parameters from genetic data. In: T.B. Smith and R.K. Wayne (Ed.) *Molecular Genetic Approaches in Conservation*, pp. 329–346. Oxford University Press, Oxford.

Olsén, K.H., Grahn, M. and Langefors, A (1998) MHC and kin discrimination in juvenile arctic charr, *Salvelinus alpinus* (L.). *Animal Behaviour*, **56**: 319–327.

Olsén, K.H., Grahn, M. and Lohm, J. (2002) Influence of MHC on sibling discrimination in Arctic char, *Salvelinus alpinus* (L.). *Journal of Chemical Ecology*, **28**: 783–795.

Olsén, K.H., Petersson, E., Ragnarsson, B., Lundqvist, H. and Jarvi, T. (2004) Downstream migration in Atlantic salmon (*Salmo salar*) smolt sibling groups. *Canadian Journal of Fisheries and Aquatic Sciences*, **61**: 328–331.

Payne, R.H., Child, A.R. and Forrest, A. (1972) The existence of natural hybrids between the European brown trout and the Atlantic salmon. *Journal of Fish Biology*, **4**: 233–236.

Peuhkuri, N. and Seppa, P. (1998) Do three-spined sticklebacks group with kin? *Annales Zoologici Fennici*, **35**: 21–27.

Pineda, H., Borrell, Y.J., McCarthy, I., Vázquez, E., Sanchez, J.A. and Blanco, G. (2003) Timing of first feeding and life-history strategies in salmon: genetic data. *Hereditas*, **139**: 41–48.

Primmer, C.R., Landry, P-A., Ranta, E., Merilä, J., Piironen, J., Tiira, K., Peuhkuri, N., Pakkasmaa, S. and Eskelinen, P. (2003) Prediction of offspring fitness based on parental genetic diversity in endangered salmon populations. *Journal of Fish Biology*, **63**: 909–927.

Reed, D.H. and Frankham, R. (2003) Correlation between fitness and genetic diversity. *Conservation Biology*, **17**: 230–237.

Ritland, K. (2005) Multilocus estimation of pairwise relatedness with dominant markers. *Molecular Ecology*, **14**: 3157–3165.

Ross, K.G. (2001) Molecular ecology of social behaviour: analyses of breeding systems and genetic structure. *Molecular Ecology*, **10**: 265–284.

Saccheri, I., Kuussaari, M., Kankare, M., Vikman, P., Fortelius, W. and Hanski, I. (1998) Inbreeding and extinction in a butterfly metapopulation. *Nature*, **392**: 491–494.

San Cristobal, M. and Chevalet, C. (1997) Error tolerant parent identification from a finite set of parents. *Genetical Research*, **70**: 53–62.

Shuster, S.M. and Wade, M.J. (2003) *Mating Systems and Strategies*. Princeton University Press, Princeton, NJ.

Solomon, D.J. and Child, A.R. (1978) Identification of juvenile natural hybrids between Atlantic salmon (*Salmo salar* L.) and brown trout (*Salmo trutta* L.). *Journal of Fish Biology*, **12**: 499–501.

Taggart, J.B., Ferguson, A. and Mason, F.M. (1981) Genetic variation in Irish populations of brown trout (*Salmo trutta* L.): electrophoretic analysis of allozymes. *Comparative Biochemistry and Physiology*, **69B**: 393–412.

Taggart, J.B., McLaren, I.S., Hay, D.W., Webb, J.H. and Youngson, A.F. (2001) Spawning success in Atlantic salmon (*Salmo salar* L.): a long-term DNA profiling based study conducted in a natural stream. *Molecular Ecology*, **10**: 1047–1060.

Thomaz, D., Beall, E. and Burke, T. (1997) Alternative reproductive tactics in Atlantic salmon: factors affecting mature parr success. *Proceedings of the Royal Society of London, Series B*, **264**: 219–226.

Thompson, C.E., Poole, W.R., Matthews, M.A. and Ferguson, A. (1998) Comparison, using minisatellite DNA profiling, of secondary male contribution in the fertilisation of wild and ranched Atlantic salmon. *Canadian Journal of Fisheries and Aquatic Sciences*, **55**: 2011–2018.

Tiira, K., Laurila, A., Peuhkuri, N., Piironen, J., Ranta, E. and Primmer, C.R. (2003) Aggressiveness is associated with genetic diversity in landlocked salmon (*Salmo salar*). *Molecular Ecology*, **12**: 2399–2407.

Trivers, R. (1985) *Social Evolution*. Benjamin/Cummings Publishing, Menlo Park, CA.

Verspoor, E. (1988) Widespread hybridization between native Atlantic salmon, *Salmo salar*, and introduced brown trout, *Salmo trutta*, in eastern Newfoundland. *Journal of Fish Biology*, **32**: 327–334.

Verspoor, E. and Hammar, J. (1991) Introgressive hybridisation in fishes: the biochemical evidence. *Journal of Fish Biology*, **39A**: 309–334.

Villanueva, B., Verspoor, E. and Visscher, P.M. (2002) Parental assignment in fish using microsatellite genetic markers with finite numbers of parents and offspring. *Animal Genetics*, **33**: 33–41.

Vladic, T.V. and Järvi, T. (2001) Sperm quality in the alternative reproductive tactics of Atlantic salmon: the importance of the loaded raffle mechanism. *Proceedings of the Royal Society of London, Series B*, **268**: 2375–2381.

Vladic, T.V., Afzelius, B.A. and Bronnikov, G.E. (2002) Sperm quality as reflected through morphology in salmon alternative life histories. *Biology of Reproduction*, **66**: 98–105.

Vos, P., Hogers, R., Bleeker, M., Reijans, M., van de Lee, T., Hornes, M., Frijters, A., Pot, J., Peleman, J., Kuiper, M. and Zabeau, M. (1995) AFLP: a new technique for DNA fingerprinting. *Nucleic Acids Research*, **23**: 4407–4414.

Wang, J. (2002) An estimator for pairwise relatedness using molecular markers. *Genetics*, **160**: 1203–1215.

Wang, J. (2004a) Estimating pairwise relatedness from dominant genetic markers. *Molecular Ecology*, **13**: 3169–3178.

Wang, J. (2004b) Sibship reconstruction from genetic data with typing errors. *Genetics*, **166**: 1963–1979.

Wang, J. (2005) Estimation of effective population sizes from data on genetic markers. *Philosophical Transactions of the Royal Society, Series B*, **360**: 1395–1409.

Wang, J. and Whitlock, M.C. (2003) Estimating effective population size and migration rates from genetic samples over space and time. *Genetics*, **163**: 429–446.

Wang, S., Hard, J.J. and Utter, F. (2002a) Genetic variation and fitness in salmonids. *Conservation Genetics*, **3**: 321–333.

Wang, S., Hard, J.J. and Utter, F. (2002b) Salmonid inbreeding: a review. *Reviews in Fish Biology and Fisheries*, **11**: 301–319.

Waples, R.S. (1989) A generalized approach for estimating effective population size from temporal changes in allele frequency. *Genetics*, **121**: 379–391.

Waples, R.S. (2005) Genetic estimates of contemporary effective population size: to what time periods do the estimates apply? *Molecular Ecology*, **14**: 3335–3352.

Ward, A.J. and Hart, P.J.B. (2003) The effects of kin and familiarity on interactions between fish. *Fish and Fisheries*, **4**: 348–358.

Webb, J.H., Fryer, R.J., Taggart, J.B., Thompson, C.E. and Youngson, A.F. (2001) Dispersion of Atlantic salmon (*Salmo salar*) fry from competing families as revealed by DNA profiling. *Canadian Journal of Fisheries and Aquatic Sciences*, **58**: 2386–2395.

Weir, L.K., Hutchings, J.A., Fleming, I.A. and Einum, S. (2005) The influence of genetic origin on the spawning behaviour and fertilisation success of mature male Atlantic salmon parr. *Canadian Journal of Fisheries and Aquatic Sciences*, **62**: 1153–1160.

Youngson, A.F., Knox, D. and Johnstone, R. (1992) Wild adult hybrids of *Salmo salar* L. and *Salmo trutta* L. *Journal of Fish Biology*, **40**: 817–820.

Youngson, A.F., Webb, J.H., Thompson, C.E. and Knox, D. (1993) Spawning of escaped farmed Atlantic salmon (*Salmo salar*): hybridization of females with brown trout (*Salmo trutta*). *Canadian Journal of Fisheries and Aquatic Sciences*, **50**: 1986–1990.

7 Local Adaptation

*C. García de Leániz, I. A. Fleming, S. Einum, E. Verspoor,
S. Consuegra, W. C. Jordan, N. Aubin-Horth, D. L. Lajus,
B. Villanueva, A. Ferguson, A. F. Youngson and T. P. Quinn*

Upper: the River Asón at Marrón, in northern Spain, at the southern limit of the Atlantic salmon distribution. Lower: the Kapisidlit River in West Greenland, less than 10 km from the main ice sheet, Greenland's only river with a self-sustaining run of salmon. (Photos credit: E. Verspoor.)

The tendency of salmon, *Salmo salar*, to return from the sea to the river of hatching and form 'local populations' has apparently been known for a long time (Calderwood 1908). Writing in 1653, Izaak Walton described how juveniles marked with ribbons tied to their tails were later recaptured as adults in their home river. Since no two rivers are completely identical, salmon returning to spawn in different rivers will, with time, give rise to different strains or 'races' (Huntsman 1941). Those races, so the story goes, are presumably the ones best 'adapted' to the local river conditions. But, is this really the case? And if so, what are local salmon adapted to? And perhaps more importantly, why should we care about it?

In this chapter we will examine the evidence for (and against) the existence of local adaptations in Atlantic salmon, in particular those studies that have become available during the last decade, since Taylor (1991) reviewed this subject. We will also consider the implications of adaptive variation for the management and conservation of Atlantic salmon populations. But first we must define what we mean by 'adaptation'.

7.1 Introduction

Ever since Darwin (1859) first introduced the concept of 'adaptation', there has not been an entirely satisfactory definition of what adaptation really means (Reeve and Sherman 1993; De Jong 1994; Rose and Lauder 1996), which is perhaps surprising considering its central role in evolutionary theory and in the philosophy of biology (Hull 1974). Darwin described an adaptation as any feature of an organism that arose as a consequence of natural selection and hence enhances the fitness of the individual. He recognised that:

- one trait might be functionally linked to another, so that the response to selection of one trait could cause changes in others that were not necessarily adaptive;
- various forms of constraints could limit the ability to adapt or change in response to natural selection; and
- the traits of an organism are laden with its history and are not necessarily representative of optimal adaptation to current conditions.

However, Darwin himself was puzzled by how some adaptations came about in such apparently purposeful manner (the complex structure of the human eye being a case in point) and recognised the danger of a circular, tautological argument (Ruse 1982; Reid 1985). Some conceptual difficulties arise when adaptation is sometimes taken to mean a 'process' (birds may be said to have adapted to flying), while at other times it refers to an 'outcome' or 'feature' (feathers constitute an adaptation to flying). Problems are also encountered when, without the benefit of invoking teleological concepts of 'purpose' or 'design', it becomes necessary to determine which traits are adaptive and which traits are not (not everything is adaptive; Gould and Lewontin 1979). Finally, there is no consensus as to whether it is the genotype, the phenotype, the individual or the population that is 'adapted' (see Reeve and Sherman 1993), nor is it clear what are the relative roles of selection, mutation and chance in shaping the way organisms respond to local environmental conditions.

Perhaps the easiest way to view adaptation (the process) is to consider it as 'the good fit of organisms to their environment' (Gould and Lewontin 1979), and to regard adaptation (the outcome) as 'any feature that promotes fitness' (Gould and Vrba 1982; Mayr 2002).

7.1.1 Phenotypic diversity and fitness in a changing world

Environments are rarely constant or perfectly predictable, so there is never a single phenotype that can outperform the others under all environmental conditions. Frequency-dependence makes it possible for several phenotypes to coexist in an evolutionary stable state (Maynard Smith 1982) and phenotypic diversity is therefore the norm. Since natural selection can only act on existing designs, most phenotypes are also bound to be maladapted to some extent (see Box 7.1).

In his shifting-balance theory of evolution, Wright (1932) introduced the concept of 'adaptive landscape' to help visualise how the fitness of individuals would change under various conditions of selection intensity, mutation rates and environmental change. In Wright's

Box 7.1 Adaptation and fitness in changing environments.

Adaptation can be defined as the good fit of organisms to their environment (Gould and Lewontin 1979). At any given time how well adapted an organism is depends on both its phenotype (P) and the current environmental conditions (E). Fitness can be viewed as the degree of matching between the two, and natural selection can be thought of as a greyhound always attempting to track environmental change (Fig. B7.1). However, since the environment is not constant, and natural selection can only act on yesterday's designs, phenotypes are likely to be maladapted to some extent (i.e. natural selection is always 'late'). The better the phenotype matches the environment, the fitter the population (or organism) might be expected to be. For example, in the example illustrated below (Fig. B7.1) the population might be expected to perform 'better' (i.e. has a higher mean fitness) at time t_2 than at time t_1 since there is a better matching between the two (i.e. the vertical distance is smaller). Although both the environment (E) and the phenotype (P) can range widely for a given species, a population is subjected to only a small subset of possible environmental conditions and displays a relatively narrow range of possible phenotypes. Together these define an 'adaptive zone', contained between E_{max} (the upper environmental limit) and E_{min} (the lower environmental limit). Loss of fitness, and eventual extinction, may be expected to occur outside the adaptive zone (see Box 7.6).

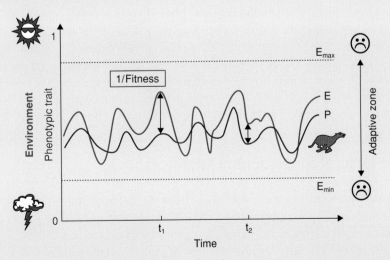

Fig. B7.1 Temporal changes in fitness in changing environments. Source: García de Leániz et al. (2007).

Box 7.2 Fisher's two opposing forces of evolution (Fisher 1958).

The fitness of an organism can be viewed as the degree of matching between the phenotype and the environment, being augmented in each generation by natural selection, and eroded by deleterious mutation and environmental change (Fig. B7.2a). As organisms are forever struggling to keep pace with environmental change, they are effectively 'fighting change with change' (Meyers and Bull 2002). Thus, when confronted with a change in environmental conditions of a given intensity and frequency, organisms may be expected to respond by adjusting their phenotype in the present or subsequent generations; the efficiency of such phenotypic adjustment will depend on the organism's adaptability (i.e. the magnitude by which organisms can adjust their phenotypes by evolutionary responses and phenotypic plasticity) and generation time (Fig. B7.2b).

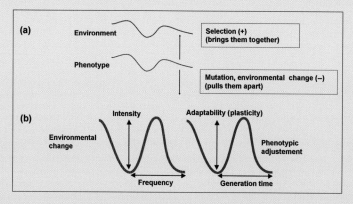

Fig. B7.2 (a) How the two opposing forces of evolution act together to determine fitness. (b) How changes in the environment (determined by their intensity and frequency) are matched by changes in the phenotype (determined by the degree of phenotypic plasticity and the organism's generation time).

three-dimensional model, allele frequencies are plotted against fitness and the resulting landscape resembles a topographical map, with adaptive peaks and valleys representing points of high and low fitness (see Hallerman 2003a). Similar, albeit simpler, diagrams showing how fitness may vary with temporal changes in the environment and in the phenotype are shown in Boxes 7.1 and 7.2. Here, fitness can be viewed as the degree of matching between phenotype and environment (Box 7.1, Fig. B7.1), being augmented in each generation by natural selection and eroded by mutations and changes in the environment (Box 7.2, Fig. B7.2).

7.2 Scope for local adaptations in Atlantic salmon

Local adaptations are presumed to exist when the average fitness of local individuals exceeds the average fitness of immigrants (Lenormand 2002). However, deciding where to draw the line between 'local' and 'foreign' individuals (i.e. setting geographical boundaries to populations) constitutes an unresolved problem for most salmonid populations (Riddell 1993; Waples 1995; Morán 2002, but see Chapter 5). Whatever their geographical limits, for local adaptations to develop, three essential conditions must first be met (Allendorf and Waples 1996):

- there must be genetic variation in fitness related traits, both within and among populations;
- populations must be subjected to different selective regimes with different genetic optima; and
- populations must be reproductively isolated to some extent (i.e. the strength of selection must exceed the level of gene flow).

The evidence for each of these in Atlantic salmon will be examined next.

7.2.1 Genetic variation in fitness-related traits

The existence of adaptive genetic variation is the first essential requirement for Atlantic salmon populations to evolve and for local adaptations to develop. For a given environment, some genotypes are more successful at surviving and reproducing than others and natural selection acts based on these differences. Relative fitness, the combined ability to survive and reproduce, is the measure of selection at the individual level.

Although the amount of genetic variation in Atlantic salmon for 'neutral' markers was once thought to be somewhat limited when compared to other salmonids (Altukhov *et al.* 2000), this is no longer considered to be the case. Significant genetic differences are found between major groupings separated by thousands of kilometres but also among populations inhabiting nearby tributaries of major river systems (e.g. Fontaine *et al.* 1997; Spidle *et al.* 2001; Verspoor *et al.* 2002, 2005; Chapter 5). Genetic variation in Atlantic salmon is distributed hierarchically among three major geographical areas (Western Atlantic, Eastern Atlantic and Baltic), among different lineages within each area (e.g. northern and southern lineages within the Baltic), among river systems and among tributaries within river systems (Chapters 5 and 6). Although not available specifically for Atlantic salmon, most estimates of mutation rates suggest that they play a negligible role in generating differences among populations on the postglacial time scale relevant for Atlantic salmon populations (Hartl and Clark 2000). Therefore, many studies have attempted to estimate only the relative strengths of genetic drift and gene flow by using molecular markers that are usually assumed to be selectively neutral, but in general results are consistent among studies and reveal limited levels of gene flow (Verspoor *et al.* 2005; Chapter 5).

While knowledge of levels and patterns of neutral genetic variation in Atlantic salmon is well developed (Chapters 5 and 6), there is relatively little comparable information on heritable variation in quantitative traits (acquired principally through rearing individuals from different populations in a common environment). Under most (but not all) conditions, heritable variation in quantitative traits may be expected to mimic genetic variation in 'neutral' markers (Reed and Frankham 2001; Hendry 2002). However, it appears that populations may be more differentiated at loci coding for quantitative traits than at neutral loci (Lynch *et al.* 1999; Merilä and Crnocrak 2001; McKay and Latta 2002). Estimates of population differentiation from molecular markers (which are available for Atlantic salmon) may, therefore, underestimate differentiation at quantitative trait loci (Latta and McKay 2002; Morán 2002), which are believed to be the major targets of selection. There is thus considerable scope for local adaptations to develop even in the absence of strong differentiation at neutral markers: it may only take a difference in one quantitative trait (controlled by a very small part of the genome).

A number of comparisons of populations and families of Atlantic salmon have identified genetic differences in several fitness-related traits, such as survival or energy content/acquisition,

Table 7.1 Evidence for genetic variation in fitness-related traits among and within populations of Atlantic salmon. Asterisks indicate studies suggesting G × E interactions.

Dependent trait	Reference
Among populations	
Body size	Jonasson (1993), Jonasson *et al.* (1997)*
Digestive rate	Nicieza *et al.* (1994b)
Growth efficiency	Jonsson *et al.* (2001)*
Growth rate	Gunnes & Gjedrem (1978)*, Holm & Fernö (1986), Torrissen *et al.* (1993), Nicieza *et al.* (1994a), Friedland *et al.* (1996), Jonsson *et al.* (2001)*
Survival	Gjedrem & Aulstad (1974), Hansen & Jonsson (1990), Jonasson (1993), Friedland *et al.* (1996), Jonasson (1996), Donaghy & Verspoor (1997), Jonasson *et al.* (1997), Rosseland *et al.* (2001)
Within populations	
Body size	Gjedrem (1979), Friars *et al.* (1990), Rye & Refstie (1995)
Feeding rate	Thodesen *et al.* (2001)
Timing of maturity	Gjerde (1984)
Stress	Fevolden *et al.* (1991)
Growth efficiency	Thodesen *et al.* (2001)
Growth rate	Thorpe & Morgan (1978), Gjerde (1986), Friars *et al.* (1990), Rye *et al.* (1990), Torrissen *et al.* (1993), Thodesen *et al.* (2001)
Sea louse infection	Mustafa & MacKinnon (1999)
Survival	Thorpe & Morgan (1978), Schom (1986)*, Standal & Gjerde (1987)*, Rye *et al.* (1990), Bailey *et al.* (1993)*, Fevolden *et al.* (1993), (1994)*, Gjedrem & Gjøen (1995)*, Lund *et al.* (1995)*, Gjøen *et al.* (1997)*, Langefors *et al.* (2001)*

in both freshwater and marine stages (Table 7.1). Many of these studies also indicate the existence of genotype-by-environment (G × E) interactions. For example, Jonsson *et al.* (2001) studied five Norwegian salmon populations under a range of temperatures, and found significant differences among populations in the optimal temperatures for both growth rate and growth efficiency. There did not seem to be any correlation between thermal optima and thermal conditions in the rivers that the populations originated from. However, maximum growth efficiencies were greatest in those populations with the lowest opportunities for feeding and growth, suggesting a possible adaptive advantage. G × E interactions in adult body size and survival have also been identified in studies where different populations have been released in different wild locations (e.g. Jonasson 1996; Jonasson *et al.* 1997). In these studies, the populations with the fastest growth and highest survival differed between locations, suggesting that different individual characteristics are required to maximise growth and survival in different environments. A similar conclusion was also indicated in a study of wild female Atlantic salmon; when controlling for differences in smolt age and size, individuals growing the fastest in fresh water grew more slowly at sea (Einum *et al.* 2002; see also Huntingford and García de Leániz 1997 for a discussion of this interaction hypothesis and Niva and Jokela 2000 for similar interactions in brown trout).

A special category of studies indicating G × E interactions include those showing differences in performance among populations or families under hostile abiotic conditions, or in susceptibility to diseases and parasites. These have identified distinct population and family responses to viral and bacterial infections and low pH levels (Table 7.1).

7.2.2 Environmental variation and differential selective pressures

Environmental conditions experienced by Atlantic salmon can differ considerably across the species range (Table 7.2), which is a necessary prerequisite for the existence of differential

Table 7.2 Range of environmental conditions, genetic variation, and phenotypic traits that are known to differ between Atlantic salmon populations inhabiting different locations. The list is not intended to be exhaustive.

Environmental variation	Genetic variation	Phenotypic variation
Physical	*Chromosome structure*	*Abundance and survival*
Latitude and longitude	Karyotype	Population size
Photoperiod	Banding patterns	Spatial distribution
Solar radiation	*Allozymes*	Density
Temperature regime	27 polymorphisms	Intraspecific competition
Rainfall	*Mitochondrial DNA*	Juvenile survival*
Discharge	*Microsatellite DNA*	Adult survival*
Stream order and size	*MHC*	Return rate*
Slope	Class I	Disease and parasite resistance*
Habitat type (lake, river, etc.)	Class II	Resistance to low pH*
Ice cover		*Morphology and physiology*
Microhabitat composition		Body size*
Geology		Body composition*
Water chemistry		Growth efficiency*
Water depth		Growth rate*
Water velocity		Digestive rate*
Substrate		Meristics*
Cover and shade		Morphometrics*
Droughts		Fluctuating asymmetry
Biological		Health condition*
Vegetation		Gamete quality*
Productivity		*Life history and behaviour*
Prey		Diet composition
Fish community structure		Habitat choice
Interspecific competition		Age at smolting*
Predators		Age at maturity*
Parasites		Longevity*
Pathogens		Movements and migrations*
Anthropogenic disturbances		Timing of spawning*
Human pressure		Timing of hatching/emergence*
Land use		Timing of smolting*
Water abstraction		Run timing*
Obstacles and accessible length		Egg size*
Introduction of foreign species		Fecundity*
Exploitation		Iteroparity*
Recreation		Sex ratio
Pollution and water quality		Reproductive effort
		Male parr maturation*

* Phenotypic traits with a known (or suspected) genetic component.

selective regimes and the development of local adaptations. There are at least 2321 salmon rivers (excluding those where the species is now extinct) in North America and Europe (WWF 2001) and the number of Atlantic salmon populations has conservatively been estimated at over 2000 (Saunders and Bailey 1980). In Europe, natural populations of Atlantic salmon are found from Iberia in the south (42°N) to Finnmark and arctic Russia (71°N) in the north (MacCrimmon and Gots 1979; Altukhov et al. 2000), thus covering over 3200 km. The natural distribution in North America is less extensive (43°N to 59°N or ~ 1800 km).

Salmon rivers can be grouped into five biogeographic regions according to geographic location, geology, flow regime, and climate (Elliott et al. 1998). Within each biogeographic region the freshwater environment varies somewhat predictably with latitude and altitude (Gibson 1993). Extreme cases refer, for example, to the Ungava Bay area of northern Quebec (Power 1981) or to some Norwegian rivers (Jensen and Johnsen 1986), where performance is constrained by low temperatures and long winters, and where overwinter habitat (Cunjak et al. 1998) might be of paramount importance for survival. In contrast, at the southern range of the distribution, Iberian rivers represent another form of extreme environment, where summer droughts and high water temperatures can impose severe constraints on survival (García de Leániz and Martinez 1988). Thus, it is clear that habitat differences among streams have the potential for creating local selection pressures (e.g. Riddell et al. 1981; Fleming and Gross 1989; Quinn et al. 2001a).

Environmental parameters also vary within a context-specific range defined by climatic patterns in continental or temperate zones. Within Europe, the stable seasonal profiles of the Baltic rivers, for example, contrast strongly with those of the British Isles. All these aspects of the physical environment, acting within the context set by predation and competition, can affect performance and have the potential for triggering specialised adaptations among juveniles and adults alike. For example, traits such as egg size, body morphology, run timing or breeding time are affected by water temperature and flow regime (e.g. Jonsson et al. 1991a); interpopulation variation, thus, is likely to reflect the outcome of natural selection acting against a background of climatic conditions of varying reliability. In addition to direct effects, temperature can also influence salmon performance indirectly because expression of life histories is tightly linked to temperature-dependent growth variation (Metcalfe 1998).

Following smolting, salmon spend an extended period at sea. Relatively little is known of the routes they take through the ocean (if indeed there are such routes) but the migrations may be extensive since, in the extreme case, fish of southern European origin are represented in fisheries on the western coast of Greenland. On these wide scales, the surface features of the ocean vary markedly, according to latitude and season and according to the dominant, oceanographic patterns of circulation. Both the outward and inward journeys therefore must involve passage through a sequence of marine environments in a manner that is determined in part by population-specific differences in migratory timing, marine routes and speeds of passage, variation which appears to be inherited and which is likely to affect fitness (Kallio-Nyberg and Koljonen 1999; Kallio-Nyberg et al. 1999, 2000).

In order to complete their life cycle, therefore, salmon must perform adequately in each of a sequence of disparate environments that fall into several or many main phases, e.g. freshwater growth, transitional migration, marine growth, transitional migration, spawning, embryo incubation and fry dispersal. Performance may be state-dependent and linked among phases (Budy et al. 2002; Einum et al. 2002). For each phase, the respective environments vary among years and they do so with some independence, tempered by the effect of large-scale climatic forcing processes. Against this background, it is evident that opportunities for selection, and for differential selection among population or regional groupings, are numerous. Constant, stabilising selection is expected to result from the more stable features (e.g. latitudinal effects operating over centuries) of the environmental sequence that fish must transit before reproduction. In addition, however, selection will also occur from environmental variation operating on shorter (decadal) time scales or from challenges that operate intermittently. Selective pressures, hence, may be expected to operate over a wide range of different spatial and temporal scales.

7.2.3 Reproductive isolation

Local adaptations can only develop if populations are isolated to some extent, since a substantial exchange of spawners would otherwise tend to homogenise any genetic differences resulting from local selective pressures (Holt and Gomulkiewicz 1997; Lenormand 2002). Reproductive isolation in salmonids is generally high and favoured by strong homing behaviour (reviewed by Stabell 1984; Quinn 1993). However, compared to resident fish, the scope for local adaptations among anadromous salmonids appears to be less extensive (Utter 2001), possibly because migration and anadromy tend to facilitate gene flow (Gyllensten 1985), while the common marine environment and a complex life cycle may impose a limit to excessive specialisation (Morán 1994; Nicieza 1995).

Nevertheless, homing accuracy in wild Atlantic salmon typically lies in the range 94–98% (Stabell 1984; Youngson et al. 1994; Altukhov et al. 2000; Jonsson et al. 2003), though it may vary between populations (Quinn 1993) and also between wild and hatchery-reared fish (Jóhannsson et al. 1998; Jonsson et al. 2003). Some degree of straying probably occurs in all salmon populations (Elo 1993; Jonsson et al. 2003), although it may not always result in gene flow, due to differences in survival (Borgstrøm et al. 2002), reproductive success (Tallman and Healey 1994) or in the timing of breeding (e.g. Quinn et al. 2000) of native and foreign fish. Gene flow, hence, may be an order of magnitude less than physical straying (Altukhov et al. 2000), although this is still a subject of considerable controversy (e.g. Hendry 2001; Howard et al. 2001) as it is difficult to obtain accurate estimates of asymmetric gene flow (but see Consuegra et al. 2005b).

The degree of iteroparity in salmon (i.e. the number of times an individual reproduces in a lifetime) is generally small (Jonsson et al. 1991a; Altukhov et al. 2000), and further restricts the possibilities for dispersing genes to other populations. Repeat spawners, moreover, tend to show very high homing accuracy (Foster and Schom 1989). Thus, homing behaviour, coupled with a tendency to reproduce only once or a few times, and the ability to become sexually mature without a marine phase (especially in males), facilitates the evolution and maintenance of population-specific traits (Hasler and Scholz 1983; Quinn and Dittman 1990).

Nevertheless, genetic evidence for reproductive isolation in Atlantic salmon is still equivocal, which is perhaps surprising considering the observed extent and scale of population structuring (Chapter 5). In general there is no consistent relationship between genetic differentiation at selectively neutral loci and geographical distance between Atlantic salmon populations (King et al. 2000, 2001), except when comparisons are made between major groups or lineages separated by hundreds or thousands of kilometres (Elo 1993; Fontaine et al. 1997; but see Spidle et al. 2003 and Chapter 5). Such lack of concordance between genetic and geographic distances at small spatial scales suggests that genetic drift is high relative to gene flow and/or that there is differential selection (or that the pattern of gene flow does not conform to a stepping-stone model which is counter to results from physical tagging studies: Stabell 1984). Therefore, the evidence from molecular markers also suggests that restricted gene flow among Atlantic salmon populations provides the necessary conditions for local adaptations to exist (Verspoor *et al.* 2002).

7.3 Evidence for the existence of local adaptations in Atlantic salmon

Atlantic salmon populations are, as we have seen, excellent candidates to show local adaptations as they seem to meet the three necessary conditions suggested by Allendorf and Waples (1996): (1) many of the life-history, morphological and behavioural traits that are important for fitness show significant genetic variation both within and among populations; (2) environmental variation among streams is extensive and has the potential for creating local selection pressures; and (3) populations are to a large extent reproductively isolated. Thus, ecological differences among streams, combined with restricted gene flow among populations, provide ample opportunities for population-specific differences to evolve. Yet the extent and significance of local adaptations in salmonids has been called into question by some (Bentsen 1994; Adkison 1995). When testing for local adaptations, it seems, the debate is not whether the *necessary* conditions are fulfilled, but whether the data are *sufficient* (Endler 1986; McPeek 1997), as we shall examine next.

7.3.1 Indirect, circumstantial evidence for local adaptations

Indirect evidence for the existence of local adaptations in Atlantic salmon is provided by studies of geographical variation in important, fitness-related traits (ecological correlates), by examining clines in genetic variation along environmental gradients, from the translocation of populations outside their native range, and by comparing the relative performance of wild and domesticated fish under relaxed or altered selective regimes (e.g. domestication). Additionally, it is becoming increasingly clear that the Atlantic salmon is not really that much different from other salmonids (Fleming 1998; Hendry and Stearns 2004; Quinn 2005). Thus, the substantial body of evidence that points to the existence of local adaptations in Pacific salmon (e.g. Taylor 1990; Blair *et al.* 1993; Hendry and Quinn 1997; Quinn 2005) and brown trout (e.g. Elliott 1994; Jonsson *et al.* 1994; Pakkasmaa and Piironen 2001a; Hansen *et al.* 2002) is probably also relevant to Atlantic salmon and should not be ignored.

Ecological correlates in fitness-related traits
Ecological correlates of phenotypic variation along environmental gradients provide one of the most common (albeit least powerful) methods for inferring the existence of local adaptations (Endler 1986). Thus, it is frequently suggested that phenotypic differences among Atlantic salmon populations are associated with variation in local environmental conditions (Youngson *et al.* 2003), often on a clinal basis, but few direct tests of those associations have been reported. Such direct tests generally require reciprocal transfers of individuals between sites or raising of fish from different populations in a common environment in order to identify heritable variation isolated from environmental variation (see below). In Atlantic salmon two types of phenotypic traits have been extensively studied using ecological correlates: variation in body morphology and variation in life-history traits (reviewed by Taylor 1991).

Differences in meristic and morphometric characters found in natural populations of Atlantic salmon (e.g. Thorpe and Mitchell 1981; MacCrimmon and Claytor 1985, 1986; Reddin *et al.* 1987, 1988) have been inferred to be adaptive in many cases. For example, a relationship appears to exist between water velocity and body shape in Atlantic salmon (Claytor *et al.* 1991) and also in pink (Beacham 1985) and coho salmon (Taylor and MacPhail 1985), and this may represent an adaptive response to water flow. Fish with longer heads and more streamlined bodies tend to predominate in high gradient rivers with higher water velocities (Riddell and Leggett, 1981; Riddell *et al.* 1981, Claytor *et al.* 1991). Such morphological variation was confirmed to be heritable by breeding experiments, for differences among Atlantic salmon populations persisted when fish were reared in the same environment (i.e. Riddell *et al.* 1981). In addition, the degree of phenotypic plasticity in shape appears to be high in juvenile salmonids experimentally reared in fast- or slow-flowing waters (Pakkasmaa and Piironen 2001b). Morphological variation in juvenile salmonids is thought to represent an adaptation to local environmental conditions (Riddell *et al.* 1981; Pakkasmaa and Piironen 2001a,b) as morphological differences between populations remain evident among returning adults (e.g. for coho salmon, Fleming and Gross 1989; Fleming *et al.* 1994; for sockeye, Blair *et al.* 1993), even if juvenile morphologies converge at smolting in preparation for the more homogeneous marine environment (Nicieza 1995).

Atlantic salmon can differ greatly with respect to important life-history traits such as age and size at maturity, reproductive investment (including egg size), age and size-specific survival and longevity (Thorpe and Stradmeyer 1995), not only among populations (Metcalfe and Thorpe 1990; Jonsson *et al.* 1991a; Hutchings and Jones 1998), but also within populations (Myers *et al.* 1986; Jonsson *et al.* 1996; Fleming 1998; Good *et al.* 2001). For example, the age at maturity may vary from a few months in mature parr at the southern end of the range to 10 or more years in large anadromous fish at the northern extreme (reviewed in Schaffer and Elson 1975; Gardner 1976; Hutchings and Jones 1998; see Chapter 2). Mature male parr that may be 1000 times smaller in weight than anadromous males also differ in the pattern of energy allocation, life-history traits and fertilisation success (Thomaz *et al.* 1997; Whalen and Parrish 1999; Arndt 2000; Garant *et al.* 2002; Letcher and Gries 2003).

Several studies indicate that age at maturity is partially inherited in Atlantic salmon (Nævdal 1983; Gjerde 1984). For example, Jonasson (2002) noted that differences in grilse rates between wild populations were maintained when fish were raised in a common environment and that in sea ranching the heritability for grilse rate could be as high as 0.65%. Estimates of heritability for age at maturity of different stocks of Atlantic salmon vary between 0.05 and 0.10 (Holm and Nævdal 1978; see Box 7.3), although heritability values are environment-specific and it is not clear to what extent heritabilities obtained in artificial conditions are applicable to the field (e.g. Hoffmann 2000), or what is the extent of phenotypic plasticity for age at maturity. For example, significant differences in grilse rates between artificial and natural conditions could be the result of the environmental differences and sea growth experienced by post-smolts (Saunders *et al.* 1983; Friedland *et al.* 1996) and sea age can be manipulated by altering ration levels in the preceding winter (Thorpe *et al.* 1990; Reimers *et al.* 1993).

The expression of early maturation in male parr appears to be inherited, but it also depends on attaining a certain body size threshold or growth during development (Prévost *et al.* 1992; Hutchings and Myers 1994; Gross 1996; Whalen and Parrish 1999; Aubin-Horth and Dodson 2004). Each male, hence, has the capability of becoming sexually mature as a parr,

Box 7.3 Estimating heritabilities: how much do like beget like?

The variance of the phenotype (i.e. the character we observe) can be partitioned into two components (Fig. B7.3a), one due to the effects of the environment (V_E) and one due to the effects of the genes (V_G); the genetic component of the phenotype is the only one that is passed from parents to offspring. Environmental effects can be subdivided into random (V_{ER}; those we cannot control) and systematic effects (V_{ES}; those we can recognise and control). Likewise, the inherited or genetic component of the phenotype (V_G) can be partitioned into additive and non-additive effects. Non-additive genetic effects represent the action of dominant (V_D) and epistatic (V_I) gene interactions: these change in each generation due to segregation and recombination and are of little predictive value. Additive genetic effects (V_A), on the other hand, are not disrupted in each generation, and represent the combined (additive) effect of multiple genes to the expression of a trait. The ratio of additive genetic variance to the total phenotypic variance is called the heritability (h^2) in the narrow sense (or simply the heritability); it measures how much of the phenotype is likely to resemble that of its parents. The higher the heritability, the more rapid the response to selection is likely to be (Mazer and Damuth 2001). The ratio of the total genetic variance (V_G) to the phenotypic variance is called the heritability in the broad sense (H^2). Although it measures the extent to which the phenotype is determined by the genotype, it is of little predictive value due to the unpredictable interactions of epistatic and dominant genes.

Most traits for which there are heritability estimates in Atlantic salmon are those that bear some economic significance for the salmon farming industry, in particular those related to growth, health condition and resistance to infectious diseases. Although heritability estimates are always population-specific and context-dependent, those traits related to life history and overall survival have relatively low heritabilities in Atlantic salmon, whereas those traits related to growth, body size and body composition generally yield higher heritabilities (Fig. B7.3b). Similar results are found in a variety of other organisms, both in laboratory and field conditions (Mousseau and Roff 1987; Hoffman 2000). In general those characters with the lowest heritability may be expected to be the ones most closely related to fitness (Falconer and McKay 1996; Merilä and Sheldon 1999).

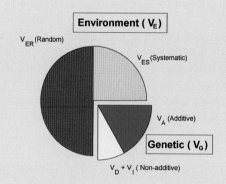

Fig. B7.3a Partition of the phenotypic (P) variance into its environmental (E) and genetic (G) components, and definition of the heritability value.

Box 7.3 (cont'd)

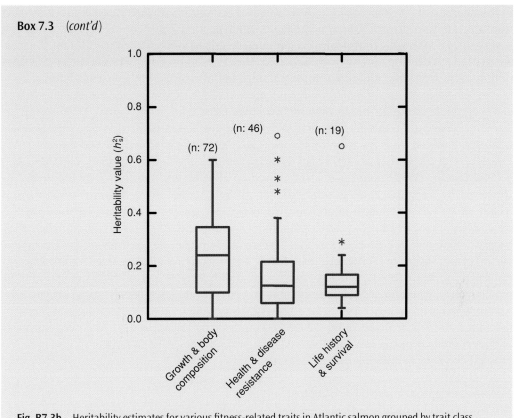

Fig. B7.3b Heritability estimates for various fitness-related traits in Atlantic salmon grouped by trait class.

but the size threshold for maturation (or some other measure of status, e.g. energy at a given time, Thorpe 1986; Thorpe *et al.* 1998) also appears to be genetically determined and varies among individuals and populations (Aubin-Horth and Dodson 2004). This suggests that there is a significant G × E interaction for age at maturity in Atlantic salmon, since different populations (genotypes) perform differently in different environments (see Hutchings 2004). Body size, growth efficiency, growth rate, survival and resistance to sea lice infections are also traits for which there is evidence of G × E interactions in salmon (Table 7.1).

Timing of hatching (Donaghy and Verspoor 1997), embryo muscle development (Johnston *et al.* 2000) and embryo mortality due to low pH (Donaghy and Verspoor 1997) are also heritable in Atlantic salmon, and trait values for different populations are generally consistent with hypotheses of stabilising selection under local environmental conditions. Analysis of post-hatch growth rates suggests that heritable rate differences are not found on small spatial scales (i.e. between tributaries within river systems) (Riddell *et al.* 1981; Johnston *et al.* 2000), but are apparent at larger spatial scales (Nicieza *et al.* 1994a; Jonsson *et al.* 2001). In summary, although associations between phenotypic and environmental variation have not always been easy to interpret, the combined evidence from ecological correlates of important, fitness-related traits suggests that natural selection (and therefore adaptation) probably plays an important, diversifying role in creating and maintaining phenotypic variation among salmonid populations (e.g. Koskinen *et al.* 2002). Furthermore, the detection of G × E interactions for many of these traits means that no single genotype can consistently perform best across all environments, thus providing the necessary (albeit insufficient) conditions for local adaptations to develop. Such results also suggest that Atlantic salmon populations differ not only in allelic frequencies at quantitative trait loci, but also in the way those loci interact.

Clines in genetic variation along environmental gradients
The clinal distribution of some allozyme loci or other non-neutral genetic variants along environmental gradients may be indicative of local adaptations (e.g. Crawford and Powers 1989; Powers 1990; Powers *et al.* 1991). In Atlantic salmon, the malic enzyme locus (*MEP-2**) provides the best circumstantial evidence based on allozymes in support of an adaptive cline (reviewed by Verspoor *et al.* 2005). *MEP-2** allelic distribution in Atlantic salmon forms a latitudinal cline in both Europe and North America (Verspoor and Jordan 1989) and populations inhabiting warm rivers tend to show higher frequencies of the *MEP-2* 100* allele than populations living in cold rivers, which tend to show higher frequencies of the alternative (**125*) allele. Significant differences in *MEP-2** frequencies are also observed among populations within river systems (Verspoor and Jordan 1989; Verspoor *et al.* 1991) and seem to be maintained by natural selection (Verspoor *et al.* 1991; Jordan *et al.* 1997), apparently in relation to juvenile growth (Jordan and Youngson 1991; Gilbey *et al.* 1999) and age at maturity (Jordan *et al.* 1990; Consuegra *et al.* 2005a). Thus, genetic variation at the malic enzyme locus seems probably adaptive and the observed differences among salmon populations may reflect local adaptations to different thermal regimes.

Performance of translocated populations
While successful transplantations of salmonids, including Atlantic salmon, have been reported all over the world (Kinnison *et al.* 1998; Quinn *et al.* 1998, 2001b; Cross 2000; Elliot and Reilly 2003), unsuccessful translocations have generally gone unnoticed making any comparative assessment difficult (Mills 1989; Youngson *et al.* 2003). In those instances where it has been possible to carry out a comparative assessment in natural conditions, foreign Atlantic salmon populations did not tend to survive as well as native fish (García de Leániz *et al.* 1989; García-Vázquez *et al.* 1991; Morán *et al.* 1994; Crozier *et al.* 1997; Verspoor and García de Leániz 1997).

Although the failure of transplanted stocks may indicate maladaptation (Altukhov *et al.* 2000), it is usually difficult to rule out alternative explanations. For example, failure of transplanted stocks may result from inferior performance due to hatchery rearing (Einum and Fleming 2001; Jutila *et al.* 2003a, see below). Moreover, such transfers are almost invariably carried out in only one direction (i.e. they are not *reciprocal*). Secondly, the transfer act itself may impair the performance of the transplanted population, especially if only one generation is examined, or if maternal effects carry over multiple generations (Mousseau 2000). Finally, for territorial salmonids such as Atlantic salmon, failure of the translocated stock (even wild fish) may simply reflect the superior competitive ability of residents, rather than the maladaptation of immigrants. For example, only one day of prior residency seems to be enough to provide Atlantic salmon half-sibs with a significant competitive advantage in territorial disputes (Huntingford and García de Leániz 1997). Thus failure of transplanted stocks, however compelling, should always be viewed with caution when used to support the existence of local adaptation.

Relative performance of wild and domesticated stocks
Since domestication involves an alteration or relaxation of natural selective pressures, the comparative analysis of wild and domesticated populations may provide some insights into the genetic basis of adaptation (e.g. Cruz and Wiley 1989; Shabalina *et al.* 1997).

Compared to wild conspecifics, hatchery-reared fish generally survive worse and stray more (Jonsson *et al.* 2003; Jutila *et al.* 2003a,b), but why? Domestication can greatly affect the body shape of Atlantic salmon (Kazakov and Semenova 1986; Salmanov 1986, 1989; Fleming 1994; Petersson *et al.* 1996). Compared to wild fish, cultured salmon are usually fed in excess and live in a low-exercise environment subject to artificial selection. Morphological responses to culture conditions include reductions in head size, fin length and caudal peduncle height (Fleming *et al.* 1994; Pelis and McCormick 2003), i.e. domestication affects those characters that show the highest level of inter-population variability in natural populations. Similar changes attributable to artificial rearing are also observed in other salmonid species (Taylor 1986; Swain *et al.* 1991), and are thought to reduce the survival of cultured fish in the wild (Johansson 1981; Jonsson *et al.* 1991b). The reasons for such changes are not well understood; domestication is known to cause physical damage (Pelis and McCormick 2003), but may also alter the normal development process (Fleming *et al.* 1994). Morphological changes due to domestication can occur within a single generation of culture and are also detected among sea-ranched salmon; however, they are particularly evident after several generations of farming. Responses to domestication are mediated by both genetic and environmental factors. Thus, although fin growth tends to be very similar in cultured and wild salmon when reared under similar conditions (probably due to fin regeneration), differences in fin length are still detectable between the two groups (Fleming *et al.* 1994), demonstrating that artificial rearing can result in important evolutionary changes (Petersson *et al.* 1996).

The higher levels of developmental instability (measured as fluctuating asymmetry – i.e. random deviations from perfect symmetry; Lajus *et al.* 2003) generally observed in cultured populations compared to wild conspecifics could be due both to environmental conditions such as high densities or low water quality, and to genetic effects induced by unintentional artificial selection. Maladaptation to the hatchery environment can also cause developmental instability (Vøllestad and Hindar 1997). Selection against asymmetric individuals in the wild, and thus decreased fluctuating symmetry in wild populations, has also been considered (Morán *et al.* 1997). Hatchery-reared Atlantic salmon tend to display a high degree of total phenotypic variance compared to wild fish (Salmanov 1986, 1989), possibly due to a higher degree of developmental instability. Cultured populations tend to be more asymmetrical than wild conspecifics (Kazakov *et al.* 1989; Morán *et al.* 1997), perhaps reflecting the more stressful conditions of the hatchery environment. However, other studies (Vøllestad and Hindar 1997; Lajus, unpubl.) found this to be true only in some cases. Another manifestation of developmental instability, the increased frequency of morphological abnormalities such as deformities in the jaw or the operculum, is also more common in cultured Atlantic salmon (Sadler *et al.* 2001 and references therein).

In brief, evidence from the culture of Atlantic salmon indicates that there are a wide number of morphological, behavioural and life-history changes associated with domestication. Changes in body morphology and in behaviour (notably an increase in aggression) are generally the traits that diverge the most following an alteration/relaxation of natural selective pressures. This suggests that local adaptations (or conversely, maladaptation) are probably related to phenotypic variation in morphology and behaviour, and thus that they may be widespread.

7.3.2 Direct evidence for local adaptations

Unlike the large body of circumstantial, indirect evidence that we examined before, direct evidence indicative of local adaptations in Atlantic salmon is still scant; it comes mostly from two sources: a few, carefully controlled experiments carried out in the field, and the pattern of inherited resistance to parasites and diseases observed in some populations.

Common garden experiments

In a 'common garden experiment', populations from different geographical locations are reared in a common environment, ideally for several generations (Reznick and Travis 2001), and are then tested for phenotypic differences. Since the environment is the same, any phenotypic differences among populations are presumed to reflect genetic differentiation (Mousseau 2000). Despite their apparent simplicity and potential explanatory power, common garden experiments have weaknesses as well, although these can be partially overcome by conducting the experiments in the wild rather than in the laboratory (Endler 1986; Mousseau 2000). Common garden studies, however, are notoriously difficult to carry out in the field and, not surprisingly, few have been undertaken with salmonids. One notable exception includes the large-scale study undertaken in the Burrishoole system (McGinnity et al. 1997, 2003), where the freshwater performance of wild, farmed and hybrid progeny of Atlantic salmon was compared in three cohorts, involving both local and foreign populations. No significant differences in survival were found in the hatchery, but there were pronounced differences in survival, growth, downstream movement, parr maturity and smolt age among groups in the field, indicating a genetic basis for these traits. Compared to the wild native population, all other groups (except the progeny of ranched fish) showed consistently poorer survival and produced fewer smolts. Adult return rates and overall lifetime survival were also higher for the native wild population than for most other groups, including wild fish from an adjacent river only a few kilometres away (see Chapter 12).

In another common garden study (García de Leániz, unpubl.), the freshwater performance of juvenile Atlantic salmon from southern (Spain; R. Ulla) and northern (Scotland, R. Shin and R. Oykel) populations was examined in stream tanks and under natural conditions in Spain over two consecutive years. Foreign Scottish ova were identified by a unique mtDNA marker and were temperature-accelerated to mimic developmental conditions in Spain and to ensure near-synchronous hatching. No difference in survival was observed under hatchery conditions. However, under natural conditions, Scottish alevins emerged earlier, had more yolk sac left, grew twice as fast, but had 1.5 times lower survival than native fish 3 months after hatching. At the end of their first year, smolt production was between 1.8 and 9.0 times higher for the native Spanish stock than for the two foreign Scottish stocks. Mean smolt age was also significantly lower for the Scottish stocks (which grew faster) than for the native Spanish stock. Thus, the results of this study indicate that northern salmon populations transplanted into southern rivers survived poorly, despite their superior growth performance.

Similarly, when eyed eggs of two Scottish populations (rivers Oykel and Shin) were planted together in a common garden experiment and their performance monitored both in the field and in the hatchery, embryo survival under low pH was found to be much greater in the native Oykel stock (subjected to recurrent problems of acidification) than in the foreign Shin population (Donaghy and Verspoor 1997).

Taken together, the results of a few carefully controlled field experiments strongly suggest that native Atlantic salmon performed better because they were locally adapted, though the nature of the local adaptations (i.e. what they were adapted to) is not always known. Reciprocal transfers (whereby native and foreign populations are reciprocally translocated) constitute an even stronger experimental way of testing for local adaptations (Endler 1986; Mousseau 2000). No such test seems to have been undertaken with Atlantic salmon, but experiments with Pacific salmon reveal the superior performance of local fish (e.g. Mayama *et al.* 1989) and the rapidity with which such evolution can occurr (Unwin *et al.* 2003).

Inherited resistance to parasites and diseases
The geographical pattern of inherited resistance to the external monogenean parasite, *Gyrodactylus salaris*, constitutes probably the most convincing example of local adaptation in Atlantic salmon (see Box 7.4). Comparative phylogenies of Atlantic salmon and *Gyrodactylus salaris* suggest that *G. salaris* was originally a parasite of grayling in the Baltic during the last Ice Age, and that Baltic salmon gradually acquired resistance through prolonged contact while salmon from the Atlantic basin did not. Currently, Baltic populations are generally resistant to infection by *Gyrodactylus salaris* whereas salmon populations migrating into the Atlantic are generally susceptible or partially susceptible to the parasite (Bakke and MacKenzie 1993; Rintamäki-Kinnunen and Valtonen 1996; Bakke *et al.* 2002; Dalgaard *et al.* 2003).

Other recent studies suggest that parasite-mediated balancing selection may drive genetic variation at the major histocompatibility complex (MHC; Consuegra *et al.* 2005c,d), and result in local adaptations with respect to pathogen resistance in Atlantic salmon and other fishes (Landry and Bernatchez 2001; Langefors *et al.* 2001; Lohm *et al.* 2002; Bernatchez and Landry 2003; Wegner *et al.* 2003).

7.3.3 Challenges to the local adaptation hypothesis

Although the prevailing view today, just as it was in 1991 when Taylor reviewed the subject, is that salmonid populations are locally adapted, the local adaptation hypothesis continues to be challenged (e.g. Bentsen 1994; Adkinson 1995). The following four questions are often raised.

(1) If salmon populations are indeed locally adapted, how is it that most of the observed genetic variation is within *populations, rather than* among *populations?*
In other words, most of the genetic differences are between individual fish, not between fish from different populations. Furthermore, why is most of the genetic variation observed among populations due to differences in allele frequencies, rather than to the occurrence of unique, private alleles in each population?

From the perspective of quantitative genetics theory, the above observations would imply that salmon populations should be largely overlapping in terms of adaptive traits; in other words they should be *universally* rather than *locally* adapted. If so, the mixing of populations might be expected to enhance population fitness. However, what evidence there is suggests the opposite is true: population mixing in Atlantic salmon and other salmonids tends to result in outbreeding depression and loss of fitness (e.g. Einum and Fleming 1997; Hallerman 2003b; McGinnity *et al.* 2003), thus implying the existence of locally adapted gene complexes.

Box 7.4 An example of local adaptation in Atlantic salmon: resistance to the parasite *Gyrodactylus*, by J. Lumme, J. Kinnsela and M. Meinilä – University of Oulu, Finland.

Gyrodactylosis is a fish disease caused by ectoparasitic monogenean flatworms of the genus *Gyrodactylus*. There are thousands of *Gyrodactylus* species in the world (Bakke *et al.* 2002), probably parasitising many fish species in freshwater, brackish and marine environments. These parasites attach to the host surface by means of hooks and feed on epithelial tissue, seldom causing death. Only in special circumstances, the propagation on host gills or skin can escalate and lead to heavy infestation, osmotic stress and secondary infections which may kill the fish, especially during the juvenile stages. Although most members of the genus appear to be host specific, some fish hosts may harbour more than one species of *Gyrodactylus* (Cone 1995), which in the case of *Gyrodactylus salaris* seems to have undergone host switch and rapid adaptive divergence (Meinilä *et al.* 2004). Gyrodactylosis is only one among the many infectious diseases of salmon but together with myxozoans, furunculosis and sea lice it is most likely to threaten wild and farmed salmon stocks in the future (Bakke and Harris 1998).

Resistance to *Gyrodactylus salaris* constitutes a classic example of local adaptation in salmonids and 'how a presumed harmless organism may become a pathogen if it is introduced to new areas where the host lacks effective responses against it' (Mo 1994). It is believed that *G. salaris* was probably introduced to Norwegian waters in the 1970s with infected salmon imported from a hatchery in Sweden (Johnsen and Jensen 1986, 1991; Bakke *et al.* 1990; Mo 1994). Within months the parasite spread quickly through the juvenile population and was soon detected in more than 40 salmon rivers, causing considerable damage in both wild and hatchery populations (Lund and Heggberget 1992). Two years after the introduction of *Gyrodactylus salaris*, salmon parr densities decreased by half, and Atlantic salmon became virtually extinct in most infected rivers 5–7 years later (Johnsen and Jensen 1986; Mo 1994). There appears to be no remedial treatment against infection by *Gyrodactylus salaris*, and the only control measure consists of applying rotenone (a poison which eliminates all the fish in the river), followed by stocking with parasite-free salmon stocks (Scholz 1999).

Atlantic salmon populations in the Baltic appear to be innately resistant to infection by *Gyrodactylus salaris*, whereas Norwegian, Scottish and other salmon populations migrating into the Atlantic are generally susceptible or partially susceptible (Bakke *et al.* 1990, 2002; Bakke 1991; Bakke and MacKenzie 1993; Rintamäki-Kinnunen and Valtonen 1996; Dalgaard *et al.* 2003). Under controlled laboratory conditions the parasite exhibits marked differences in fecundity, development and mortality when it infects different stocks (MacKenzie and Mo 1994; Cable *et al.* 2000), which may explain its variable virulence (Fig. B7.4a).

Host resistance to the parasite is known to be heritable (Jansen *et al.* 1991; Bakke *et al.* 1999) and probably under polygenic control (Bakke *et al.* 2002). Laboratory experiments have indicated possibly disastrous consequences of introducing *G. salaris* into new areas (MacKenzie and Mo 1994). However, why are salmon populations from the Baltic resistant to *G. salaris* while populations from the Atlantic are susceptible? Recent work on the phylogeography of *Salmo salar* in Europe (Verspoor *et al.* 1999) has shown that there are two distinct genetic lineages since the last ice age – Baltic and Atlantic. Most, if not all, of the Baltic salmon populations are derived from the Upper Volga ice lake (Koljonen *et al.* 1999; Nilsson *et al.* 2001), in contrast to the Atlantic lineage that probably originated from the Iberian Peninsula (Consuegra *et al.* 2002). The phylogeny of *Gyrodactylus* (Fig. B7.4b), on the other hand, suggests that *Gyrodactylus salaris* was originally a parasite of grayling, and 'jumped' to Baltic salmon in the large, Upper Volga ice lake. Although *G. salaris* cannot disperse easily in seawater, it can spread quickly in fresh water and it is likely that the refugial ice lake must have provided ample opportunities for spreading the parasite. During this episode, most salmon from the Baltic basin must have initially become infected, and then gradually acquired resistance due to the effect of natural selection. Salmon populations from the Atlantic lineage, in contrast, were never in contact with the parasite and did not develop resistance.

Phylogenetic analysis of *Gyrodactylus salaris* in rainbow trout, Atlantic salmon and grayling based on mtDNA sequences (Fig. B7.4b; Meinilä *et al.* 2002, 2004) indicates that there are four independent evolutionary lineages or strains of *Gyrodactylus salaris*. These are, however, not sufficiently differentiated to earn a species rank. One strain is commonly found in rainbow trout farms in Finland, Denmark and Sweden, and also in Norway and Russia. This 'rainbow trout strain' is genetically homogeneous and widely distributed among rainbow trout stocks, which are not harmed. Harm only comes when this parasite infects Atlantic salmon populations outside the Baltic area (Baltic salmon are resistant against the rainbow trout strain of *G. salaris*). The other phylogenetic lineage is the '*Gyrodactylus salaris* proper', the strain described by Malmberg in 1957. It is more variable and is found naturally not only in salmon stocks from the Baltic, but also in many rivers in Norway (e.g. R. Vefsna) and in

Box 7.4 (cont'd)

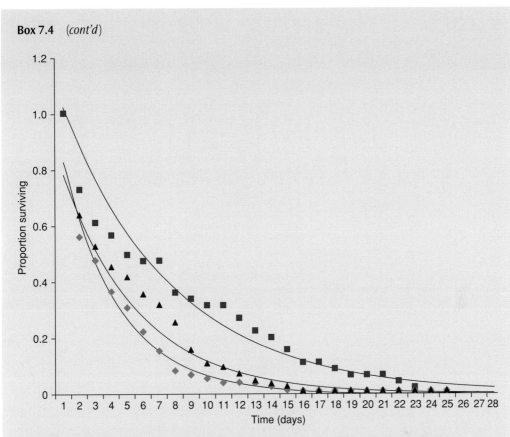

Fig. B7.4a Survivorship curves of *Gyrodactylus salaris* on two susceptible Atlantic salmon stocks from Norway (Alta ■, Lier ▲) and one resistant stock from the Baltic (Neva, ◆) infected experimentally. Parasite establishment success (proportion of worms surviving to give birth) was significantly lower in the resistant stock (45% with mean parasite survival of 3.5 days) than in the two susceptible stocks (60% with mean parasite survival of 7.9 and 5.2 days, respectively). Source: Cable *et al.* (2000).

the White Sea (R. Keret). The fact that Baltic salmon tolerate the Baltic strains of *G. salaris* and that the gyrodactylosis found in the Swedish west coast is 'balanced' and not fatal, suggests that there was some salmon gene flow out from the Baltic (Malmberg and Malmberg 1993). The malign *Gyrodactylus* infections on the Atlantic side and the benign ones in Baltic rivers are caused by parasites closely related to each other. Two geographically isolated evolutionary lineages are observed in grayling (*Thymallus thymallus*). One is found in the White Sea basin, and the other in the Baltic basin. The grayling strains are *sister* clades to those infecting salmon and rainbow trout, of similar evolutionary age (~ 150 000 years). The parasite strains detected in grayling appear to be strictly host specific, and do not infect Atlantic salmon in the same rivers (Tornionjoki, Finland/Sweden and Pistojoki, Russia). Norwegian graylings have their own strain, also found to be harmless for Atlantic salmon (Bakke *et al.* 2002; GenBank entries).

Box 7.4 (cont'd)

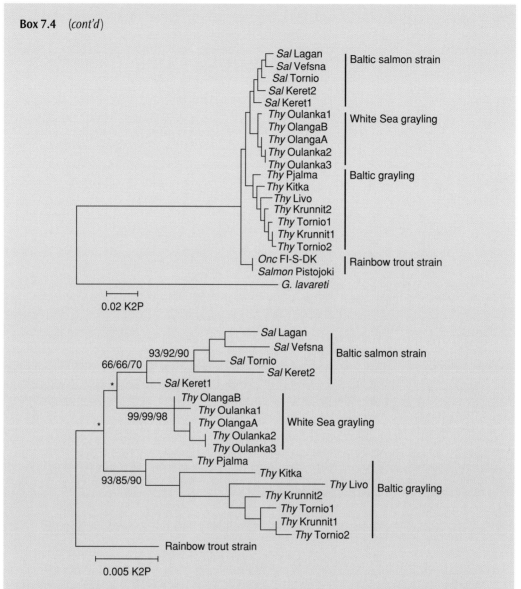

Fig. B7.4b Phylogenetic analysis of *Gyrodactylus salaris* in rainbow trout, Atlantic salmon and grayling based on mtDNA sequences. (upper) Tree 'rooted' with nearest relative *Gyrodactylus lavareti*. (lower) Only *G. salaris* clones. Based on sequencing about 800 bp of the mitochondrial gene *COI*; the scale is Kimura's two parameter distance (K2P) and the numbers along branches indicate the statistical significance (%) of each branch, achieved by different tree-building methods. Source: Meinilä *et al.* (2004).

Admittedly, little is known about the actual molecular basis of adaptive genetic variation in the wild (Mitchell-Olds 1995; Lynch and Walsh 1998), but adaptive differences among populations are certainly not contingent on populations having unique alleles or non-overlapping distributions of phenotypic traits, as shown in Box 7.5.

Box 7.5 Can allele frequency differences underlie phenotypic divergence?

The genetic requirements for local adaptation of populations are largely unknown and likely to be highly variable depending on the trait involved. However, adaptive genetic differentiation between populations for phenotypic traits which show both within- and between-population variation as commonly observed can, in principle as least, arise where genetic differences are solely generated by differences in allele frequencies, as the following hypothetical example shows. The two populations shown differ in mean gill raker number (Fig. B7.5a), a trait for which variation is known to be associated with different diets and whose variation can be both genetically and environmentally determined. These phenotypic distributions can be generated by genetic variation at three loci, each with two alleles, where trait heritability is 0.46, i.e. 46% of variation among individuals is genetic. In this model, the effect is additive. For each copy of allele p possessed by an individual at each locus it develops one gill raker while for each copy of allele q it generates two. Thus the genetically determined number of gill rakers can range from 6 (i.e. 6 copies of p – two at each locus) to 12 (i.e. 6 copies of q). The model also assumes that the actual distribution of phenotypes for a given genotype varies around the genotype value by two rakers due to the effect of the environment as shown below, with 50% of individuals having the type dictated by their genotype (Fig. B7.5b).

In this model, the phenotype distributions can be generated if the allele frequencies in population 1 are $p = 0.8$ and $q = 0.2$ at each of the three loci, and $p = 0.2$ and $q = 0.8$ at each locus in population 2. This gives the genotypic distributions shown below for the two populations (Fig. B7.5c).

All genotypes can occur in each population but expected frequencies in the populations are highly divergent: for example, the most common genotype in population 1, genotype 1, is expected to be possessed by over 25% of individuals in population 1 but by less than 0.01% of fish in population 2. The converse would be true for genotype 27. If each population has 1000 fish, 852 of the fish in each population would be expected to be of five genotypes which would be represented on average by less than two fish in the other population. This allele frequency divergence could be maintained, under certain conditions, by selection favouring individuals in each population which possess the mean gill raker number for that population.

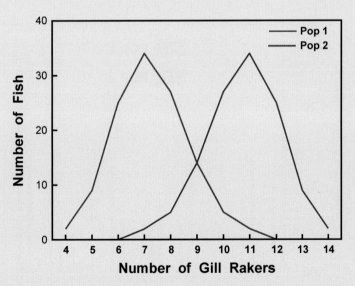

Fig. B7.5a Hypothetical distribution in the number of gill rakers in two salmon populations, a trait whose variation can be both genetically and environmentally determined.

Box 7.5 *(cont'd)*

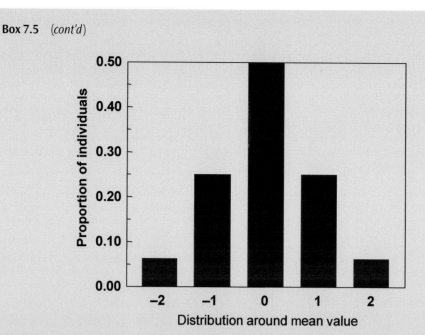

Fig. B7.5b Hypothetical degree of phenotypic variation in the number of gill rakers around the mean value for a given genotype due to environmental effects.

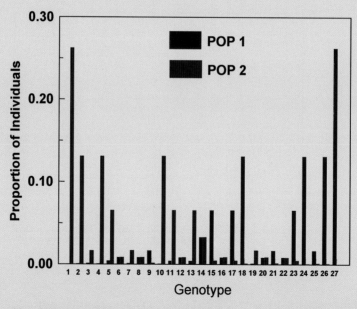

Fig. B7.5c Resulting distribution of genotypes when the phenotypic distributions in gill raker number are generated by genetic variation at three loci, each with two alleles, and a trait heritability of 0.46. The model assumes additive genetic variation. For each copy of allele p possessed by an individual at each locus it develops one gill raker while for each copy of allele q it generates two.

In only a few cases do the genetics of traits apparently linked to adaptation appear to be simple. For example, variation in coloration in the moth, *Biston betularia*, is determined by a single gene and thought to involve balancing selection, though the actual mechanisms of selection remain unclear (Hartl and Clark 2000; Grant and Grant 2002). A major assumption underpinning classical quantitative genetics theory (Lynch and Walsh 1998) is that most quantitative traits are controlled by many genes (i.e. they are polygenic). To what extent this is true is unknown. Certainly, modelling suggests that the continuous distributions observed for quantitative traits can arise from variation in as few as two or three genes when 50% of trait variation is due to the environment, and from as few as 12 genes when the influence of the environment is negligible (Tave 1993). Therefore, a high level of genetic variability within populations can neither prove nor disprove the existence of local adaptations.

(2) How can salmon populations be locally adapted and yet sometimes perform so well (and evolve so rapidly) outside their native habitats?
As pointed out before, some of the apparent success of translocated populations must undoubtedly be a result of what has been termed selective memory: only successful translocations tend to be reported, while unsuccessful attempts (and these must have been legion) go quietly unnoticed. Nevertheless, there are many examples of salmonids performing well outside their native habitats, even in such distant places as New Zealand or Australia. How is this possible? It seems that, for salmonids, one consequence of living in highly changing aquatic environments has been the development of considerable phenotypic plasticity (i.e. flexibility), which may itself have been the target of selection (Jørstad and Nævdal 1996; Pakkasmaa and Piironen 2001b). Thus, the same flexibility that may have allowed salmonids to adapt to local environmental conditions may have also allowed them to perform successfully in a variety of aquatic habitats (Klemetsen *et al.* 2003) and to evolve rapidly outside their native range (Hendry *et al.* 2000; Quinn *et al.* 2000, 2001b).

(3) How can migratory salmon adapt locally to an environment that changes so much spatially and temporally?
Although little is known about the patch dynamics of local adaptations in salmonids (what are salmon adapted to?, how long does it take to become adapted?; Rieman and Dunham 2000), genetic differentiation for adaptive behavioural traits can be surprisingly rapid in other organisms (e.g. Cousyn *et al.* 2001), providing a plausible mechanism for evolutionary change. Again, the ability shown by many salmonids to produce different phenotypes in different environments suggests that the costs of phenotypic plasticity are not too great (or that the benefits outweigh the costs; see DeWitt *et al.* 1998). Thus, when the environment is heterogeneous and unpredictable, selection may tend to favour a 'jack of all trades' strategy. It seems likely that the most reliable features of the salmon's niche, such as the temperature regime, the photoperiod, or the occurrence of predators, parasites and diseases, are perhaps the conditions most likely to trigger the development of local adaptations in salmon populations.

(4) If random genetic drift is the overriding diversifying factor in small, isolated populations (typical of many salmonids), then how can local adaptations be maintained under such conditions?
Undoubtedly, many salmon populations are fairly small and may be expected to have lost much genetic variation due to genetic drift (Adkison 1995; Bentsen 2000), unless there is

a high level of gene flow (which would then tend to erode any local selective advantage (Holt and Gomulkiewicz 1997; Lenormand 2002). However, the relationship between genetic variation and fitness in salmonids is a complex one (Wang *et al.* 2002a,b), and natural selection seems to be able to create and maintain adaptive phenotypic variation even among surprisingly small salmonid populations (Quinn 1999; Quinn *et al.* 1998, 2001b; Koskinen *et al.* 2002). Moreover, the possible existence of salmonid metapopulations (Fontaine *et al.* 1997; Rieman and Dunham 2000) means that local adaptations can probably persist in spite of some gene flow from adjacent, loosely related neighbours (Hanski 1999). Finally, there may be local adaptation for some traits but not for others.

7.4 Summary and conclusions

- The process of adaptation represents the end product of natural selection and can be viewed as the degree of matching between the phenotype and its environment.
- Local adaptations are revealed when the average performance of local fish exceeds that of immigrants; they evolve when populations are reproductively isolated, live in heterogeneous environments and are subjected to differential selective pressures. Local adaptations are likely to exist in Atlantic salmon, but their extent is probably very variable, depending on habitat heterogeneity and patterns of gene flow.
- Detailed knowledge of environmental variation across the Atlantic salmon range is scant, and this has no doubt hindered the identification of local selective pressures at biologically meaningful scales. Application of new techniques, such as those developed in geostatistics and landscape genetics (Manel *et al.* 2003), should prove useful in identifying the extent of local adaptations in Atlantic salmon.
- Water temperature and photoperiod (and variables related to them) are likely to be among the most important physical variables determining local selective pressures in Atlantic salmon populations. They are also relatively stable and predictable in a way that promotes the development of local adaptations.
- Heritability estimates for many fitness-related traits indicate that growth and body composition have the highest heritabilities among those examined, followed by health condition and resistance to diseases, and survival and life history variation, which have the lowest. $G \times E$ interactions are found for many of these traits, again suggesting that local adaptations might be important.
- Inferential evidence for the existence of local adaptations in Atlantic salmon comes from genetic correlates in fitness-related traits, the failure of many translocations, the poor performance of domesticated stocks, the results of common garden experiments, and the pattern of inherited resistance to some parasites in the wild, as well as from research in other salmonid species.
- The best example of a local adaptation in Atlantic salmon is perhaps the inherited resistance to the monogenean external parasite, *Gyrodactylus salaris*, shown by some Baltic populations but not by populations elsewhere. Many examples of this kind of adaptation are also found in other salmonid species.
- Since maladaptation often results from phenotype–environment mismatching, it becomes important to protect and maintain the native genotypes, as well as the original habitat conditions to which the populations have historically adapted.

7.5 Management recommendations

There is, we have seen, a substantial body of circumstantial evidence that suggests that populations of Atlantic salmon, like those of many other salmonids, are probably locally adapted. There are also some experimental results and certain patterns of inherited resistance to parasites and diseases that can be best viewed as local adaptations. However, what are the practical implications for conservation and management and, more precisely, how is adaptive genetic variation maintained and how is it lost?

Since the phenotype is the result of the interaction between the genotype and the environment, it follows that changes in either the genes or the habitat have the potential for altering the degree of adaptation and fitness of Atlantic salmon populations. Four general problems leading to the loss of adaptive variation can be envisaged, depending on whether the alteration is on the genes (Box 7.6, Figs B7.6a and B7.6b) or in the environment (Box 7.6, Figs 7.6c and B7.6d).

Collectively, the evidence suggests that local adaptation is an inevitable consequence of natural selection acting on genetic diversity and phenotypic plasticity in heterogenous habitats. If so, the management of Atlantic salmon should aim to minimise alterations in either the genotype or the environment of populations, while maintaining the conditions necessary for natural selection to operate efficiently and unhindered. This means that salmon populations should ideally be allowed to reach or extend beyond carrying capacity, and that competition and other sources of natural mortality (e.g. predation, diseases) may need to be allowed to develop. Furthermore, a number of specific management recommendations emerge (explored in more detail in chapters of Part III):

- View each breeding unit as largely but not entirely isolated, and manage each unit at the smallest (spatial and temporal) scales possible (i.e. tributaries, 1SW vs MSW, etc.).
- Maintain salmon populations at their largest possible size in order to retain genetic variation and adaptive potential, especially in the case of small populations inhabiting extreme, marginal habitats.
- Stop transplants of non-native fish (unless there is a clear genetic case for doing so), and limit or preferably eliminate aquaculture escapes.
- Critically examine 'supplementation' hatcheries and 'enhancement' programmes, especially the hidden costs of density-dependent competition and swamping of wild gene pools.
- Finally, minimise habitat change and where possible protect and maintain the habitat and natural conditions to which populations have historically adapted.

Acknowledgements

We are most grateful to Hans Bentsen for compiling most of the data on heritability values in Atlantic salmon and for pointing out some of the problems with the local adaptation hypothesis, and to J. Lumme, J. Kuusela and M. Meinilä for kindly providing the information in Box 7.4 and most of the information on adaptation to *Gyrodactylus*. Their help was invaluable and is gratefully acknowledged.

Box 7.6 Four ways to lose adaptive variation and fitness.

Problem 1

Loss of fitness, and eventual extinction, may occur if the genotype (and thus likely the phenotype) is allowed to shift 'out of bounds', i.e. outside the adaptive zone defined by the optimal environment for the population (Fig B7.6a). Such a situation could happen through random genetic drift (random loss of alleles) following a severe decline in population size, gene swamping or gene introgression due to stocking with maladapted individuals, or selective exploitation. In this example, the phenotype (P) first shifts beyond the adaptive zone at time t_c.

Problem 2

Loss of genetic variation may result in loss of fitness if it makes the population more vulnerable to environmental change. In this second example (Fig. B7.6b), changes in the genotype at time t_c result in a subsequent reduction in the population's tolerance limits, thereby reducing its ability to cope with future environmental changes and increasing the risk of extinction. Such a scenario is typical of small, bottlenecked populations following severe reductions in population size (see Chapter 9).

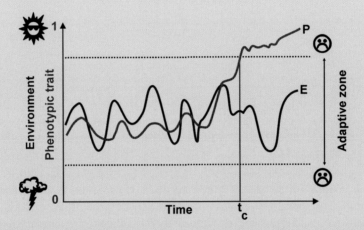

Fig. B7.6a Problem 1. Phenotype/genotype shifts beyond the adaptive zone.

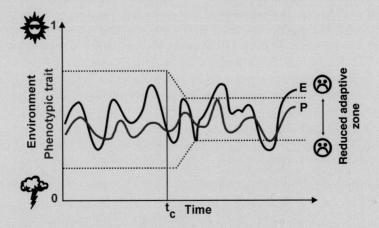

Fig. B7.6b Problem 2. Loss of genetic variation.

Box 7.6 *(cont'd)*

Problem 3

Just as the genotype can shift outside the adaptive zone in the preceding example (problem 1), loss of fitness can also occur if the environment (rather than the genotype) is the one that goes 'out of bounds', i.e. beyond the adaptive zone defined by the historical environment for the population (Fig. B7.6c). Such a condition could develop, for example, when key habitats are destroyed or the environment is altered beyond the population's tolerance limits.

Problem 4

The population may also fail to adapt if the environment (E) begins to change too rapidly in relation to the population's phenotypic plasticity (P) (Fig. B7.6d), which will always depend on generation time. Examples of rapid environmental changes include the sudden discharge of some power stations, and those brought about by deforestation, impoundment and stream regulation, siltation, point-source pollution, or blockage of migratory routes. Other, less rapid sources of enviromental change may include climate change (Carpenter *et al.* 1992).

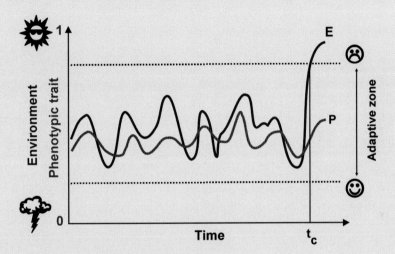

Fig. B7.6c   Problem 3. The environment changes too much.

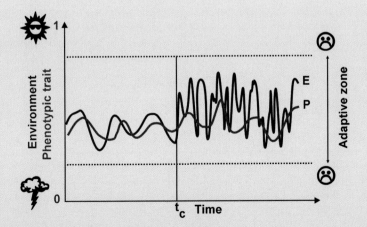

Fig. B7.6d   Problem 4. The environment changes too quickly.

Further reading

There are many specialised texts dealing with the subject of adaptation and natural selection, but few are easy to follow. Some of the most accessible are listed below:

Fox, C.W., Roff, D.A. and Fairbairn, D.J. (2001) *Evolutionary Ecology: Concepts and case studies*. Oxford University Press, Oxford.
Mayr, E. (2002) *What Evolution Is*. Phoenix, London.
Mousseau, T.A., Sinervo, B. and Endler, J. (Ed.) (2000) *Adaptive Genetic Variation in the Wild*. Oxford University Press, Oxford.
Rose, M.R. and Lauder, G.V. (Ed.) (1996) *Adaptation*. Academic Press, San Diego, CA.
Stearns, S.C. (1992) *The Evolution of Life Histories*. Oxford University Press, Oxford.

For a different (and provocative) look at adaptation and natural selection try:

Macbeth, N. (1971) *Darwin Retried: An appeal to reason*. Gambit, Boston, MA.

For local adaptations in salmonids see:

Altukhov, Y.P., Salmenkova, E.A. and Omelchenko, V.T. (2000) *Salmonid Fishes: Population biology, genetics, and management*. Blackwell Science, Oxford.
Hendry, A.P. and Stearns, S.C. (Ed.) (2004) *Evolution Illuminated: Salmon and their relatives*. Oxford University Press, Oxford.
National Research Council (1996) *Upstream: Salmon and society in the Pacific Northwest*. National Academy Press, Washington, DC.
Quinn, T.P. (2005) *The Behavior and Ecology of Pacific Salmon and Trout*. University of Washington Press, Seattle, WA.
Stouder, D.J., Bisson, P.A. and Naiman, R.J. (Ed.) (1997) *Pacific Salmon and their Ecosystems: Status and future options*. Chapman & Hall, Seattle, WA.
Taylor, E.B. (1991) A review of local adaptations in Salmonidae, with particular reference to Pacific and Atlantic salmon. *Aquaculture*, **98**: 185–207.

References

Adkison, M.D. (1995) Population differentiation in Pacific salmon: local adaptation, genetic drift, or the environment? *Canadian Journal of Fisheries and Aquatic Sciences*, **52**: 2762–2777.
Allendorf, F.W. and Waples, R.S. (1996) Conservation and genetics of salmonid fishes. In: J.C. Avise and J.L. Hamrick (Ed.) *Conservation Genetics: Case histories from nature*, pp. 238–280. Chapman & Hall, New York.
Altukhov, Y.P., Salmenkova, E.A. and Omelchenko, V.T. (2000) *Salmonid Fishes: Population biology, genetics, and management*. Blackwell Science, Oxford.
Arndt, S.K.A. (2000) Influence of sexual maturity on feeding, growth and energy stores of wild Atlantic salmon parr. *Journal of Fish Biology*, **57**: 589–596.
Aubin-Horth, N. and Dodson, J.J. (2004) Influence of individual body size and variable thresholds on the incidence of a sneaker male reproductive tactic in Atlantic salmon. *Evolution*, **58**: 136–144.
Bailey, J.K., Olivier, G. and Friars, G.W. (1993) Inheritance of furunculosis resistance in Atlantic salmon. In: K.T. Pirquet (Ed.) *Bulletin of the Aquaculture Association of Canada*, pp. 90–92. St Andrews, New Brunswick.
Bakke, T. (1991) A review of the inter- and intraspecific variability in salmonid hosts to laboratory infections with *Gyrodactylus salaris* Malmberg. *Aquaculture*, **98**: 303–310.
Bakke, T.A. and Harris, P.D. (1998) Diseases and parasites in wild Atlantic salmon (*Salmo salar*) populations. *Canadian Journal of Fisheries and Aquatic Sciences*, **55**: 247–266.

Bakke, T.A. and MacKenzie, K. (1993) Comparative susceptibility of native Scottish and Norwegian strains of Atlantic salmo *Salmo salar* L. to *Gyrodactylus salaris* Malmberg: laboratory experiments. *Fisheries Research*, **17**: 69–85.

Bakke, T.A., Jansen, P.A. and Hansen, L.P. (1990) Differences in the host resistance of Atlantic salmon, *Salmo salar* L., stocks to the monogenean *Gyrodactylus salaris* Malmberg, 1957. *Journal of Fish Biology*, **37**: 577–587.

Bakke, T.A., Soleng, A. and Harris, P.D. (1999) The susceptibility of Atlantic salmon (*Salmo salar* L.) × brown trout (*Salmo trutta* L.) hybrids to *Gyrodactylus salaris* Malmberg and *Gyrodactylus derjavini* Mikhailov. *Parasitology*, **119**: 467–481.

Bakke, T.A., Harris, P.D. and Cable, J. (2002) Host specificity dynamics: observations on gyrodactylid monogeneans. *International Journal for Parasitology*, **32**: 281–308.

Beacham, T.D. (1985) Meristic and morphometric variation in pink salmon (*Oncorhynchus gorbuscha*) in southern British Columbia and Puget Sound. *Canadian Journal of Zoology*, **63**: 366–372.

Bentsen, H.B. (1994) Genetic effects of selection on polygenic traits with examples from Atlantic salmon, *Salmo salar* L. *Aquaculture and Fisheries Management*, **25**: 89–102.

Bentsen, H.B. (2000) Bestandsgenetikk og effekter på fiskebestander av oppdrett, kultivering og beskatning. In: R. Borgstrøm and L.P. Hansen (Ed.) *Fisk i Ferskvann*, pp. 247–276. No-1349. Juul forlaget, Rykkinn, Norway.

Bernatchez, L. and Landry, C. (2003) MHC studies in nonmodel vertebrates: what have we learned about natural selection in 15 years? *Journal of Evolutionary Biology*, **16**: 363–377.

Blair, G.R., Rogers, D.E. and Quinn, T.P. (1993) Variation in life history characteristics and morphology of sockeye salmon in the Kvichak River system, Bristol Bay, Alaska. *Transactions of the American Fisheries Society*, **122**: 550–559.

Borgstrøm, R., Skaala, Ø. and Aastveit, A.H. (2002) High mortality in introduced brown trout depressed potential gene flow to a wild population. *Journal of Fish Biology*, **61**: 1085–1097.

Budy, P., Thiede, G.P., Bouwes, N., Petrosky, C.E. and Schaller, H. (2002) Evidence linking delayed mortality of Snake River salmon to their earlier hydrosystem experience. *North American Journal of Fisheries Management*, **22**: 35–51.

Cable, J., Harris, P.D. and Bakke, T.A. (2000) Population growth of *Gyrodactylus salaris* (Monogenea) on Norwegian and Baltic Atlantic salmon (*Salmo salar*) stocks. *Parasitology*, **121**: 621–629.

Calderwood, W.L. (1908) *The Life of the Salmon* (2nd edn). Edward Arnold, London.

Carpenter, S.R., Fisher, S.G., Grimm, N.B. and Kitchell, J.F. (1992) Global change and freshwater ecosystems. *Annual Review of Ecology and Systematics*, **23**: 119–139.

Claytor, R.R., MacCrimmon, H.R. and Gots, B.L. (1991) Continental and ecological variance components of European and North American Atlantic salmon (*Salmo salar*) phenotypes. *Biological Journal of the Linnean Society*, **44**: 203–229.

Cone, D.K. (1995) Monogenea (Phylum Platyhelminthes). In: P.T.K. Woo (Ed.) *Fish Diseases and Disorders. Volume 1. Protozoan and Metazoan Infections*, pp. 289–328. CAB International, Wallingford.

Consuegra, S., García de Leániz, C., Serdio, A., González Morales, M., Strauss, L.G., Knox, D. and Verspoor, E. (2002) Mitochondrial DNA variation in Pleistocene and modern Atlantic salmon from the Iberian glacial refugium. *Molecular Ecology*, **11**: 2037–2048.

Consuegra, S., García de Leániz, C., Serdio, A. and Verspoor, E. (2005a) Selective exploitation of early running fish may induce genetic and phenotypic changes in Atlantic salmon. *Journal of Fish Biology*, **67**: 130–146.

Consuegra, S., Verspoor, E., Knox, D. and García de Leániz, C. (2005b) Asymmetric gene flow and the evolutionary maintenance of genetic diversity in small, peripheral Atlantic salmon populations. *Conservation Genetics*, **6**: 823–842.

Consuegra, S., Megens, H.J., Leon, K., Stet, R.J.M. and Jordan, W.C. (2005c) Patterns of variability at the major histocompatibility class II alpha locus in Atlantic salmon contrast with those at the class I locus. *Immunogenetics*, **57**: 16–24.

Consuegra, S., Megens, H.J., Schaschl, H., Leon, K., Stet, R.J.M. and Jordan, W.C. (2005d) Rapid evolution of the MH class I locus results in different allelic compositions in recently diverged populations of Atlantic salmon. *Molecular Biology and Evolution*, **22**: 1095–1106.

Cousyn, C., De Meester, L., Colbourne, J.K., Brendonck, L., Verschuren, D. and Volckaert, F. (2001) Rapid, local adaptation of zooplankton behavior to changes in predation pressure in the absence of neutral genetic changes. *Proceedings of the National Academy of Sciences of the USA*, **98**: 6256–6260.

Crawford, D.L. and Powers, D.A. (1989) Molecular basis of evolutionary adaptation at the lactate dehydrogenase-B locus in the fish *Fundulus heteroclitus*. *Proceedings of the National Academy of Sciences of the USA*, **86**: 9365–9369.

Cross, T.F. (2000) Genetic implications of translocation and stocking of fish species, with particular reference to Western Australia. *Aquaculture Research*, **31**: 83–94.

Crozier, W.W., Moffett, I.J.J. and Kennedy, G.J.A. (1997) Comparative performance of native and non-native strains of Atlantic salmon (*Salmo salar* L.) ranched from the River Bush, Northern Ireland. *Fisheries Research*, **32**: 81–88.

Cruz, A. and Wiley, J.W. (1989) The decline of an adaptation in the absence of a presumed selection pressure. *Evolution*, **43**: 55–62.

Cunjak, R.A., Prowse, T.D. and Parrish, D.L. (1998) Atlantic salmon (*Salmo salar*) in winter: 'the season of parr discontent'? *Canadian Journal of Fisheries and Aquatic Sciences*, **55**: 161–180.

Dalgaard, M.B., Nielsen, C.V. and Buchmann, K. (2003) Comparative susceptibility of two races of *Salmo salar* (Baltic Lule river and Atlantic Conon river strains) to infection with *Gyrodactylus salaris*. *Diseases of Aquatic Organisms*, **53**: 173–176.

Darwin, C. (1859) *The Origin of Species*. John Murray, London.

De Jong, G. (1994) The fitness of fitness concepts and the description of natural selection. *Quarterly Review of Biology*, **69**: 3–29.

DeWitt, T.J., Sih, A. and Wilson, D.S. (1998) Costs and limits of phenotypic plasticity. *Trends in Ecology and Evolution*, **13**: 77–81.

Donaghy, M.J. and Verspoor, E. (1997) Egg survival and timing of hatch in two Scottish Atlantic salmon stocks. *Journal of Fish Biology*, **51**: 211–214.

Einum, S. and Fleming, I.A. (1997) Genetic divergence and interactions in the wild among native, farmed and hybrid Atlantic salmon. *Journal of Fish Biology*, **50**: 634–651.

Einum, S. and Fleming, I.A. (2001) Implications of stocking: ecological interactions between wild and released salmonids. *Nordic Journal of Freshwater Research*, **75**: 56–70.

Einum, S., Thorstad, E.B. and Næsje, T.F. (2002) Growth rate correlations across life-stages in female Atlantic salmon. *Journal of Fish Biology*, **60**: 780–784.

Elliott, J.M. (1994) *Quantitative Ecology and the Brown Trout*. Oxford University Press, Oxford.

Elliott, N.G. and Reilly, A. (2003) Likelihood of bottleneck event in the history of the Australian population of Atlantic salmon (*Salmo salar* L.). *Aquaculture*, **215**: 31–44.

Elliott, S.R., Coe, T.A., Helfield, J.M. and Naiman, R.J. (1998) Spatial variation in environmental characteristics of Atlantic salmon (*Salmo salar*) rivers. *Canadian Journal of Fisheries and Aquatic Sciences*, **55**: 267–280.

Elo, K. (1993) Gene flow and conservation of genetic variation in anadromous Atlantic salmon (*Salmo salar*). *Hereditas*, **119**: 149–159.

Endler, J.A. (1986) *Natural Selection in the Wild*. Princeton University Press, Princeton, NJ.

Falconer, D.S. and Mackay, T.F.C. (1996) *Introduction to Quantitative Genetics* (4th edn). Longman, Harlow, UK.

Fevolden, S.E., Refstie, T. and Røed, K.H. (1991) Selection for high and low stress response in Atlantic salmon (*Salmo salar*) and rainbow trout (*Oncorhynchus mykiss*). *Aquaculture*, **95**: 53–65.

Fevolden, S.E., Refstie, T. and Røed, K.H. (1993) Disease resistance in Atlantic salmon (*Salmo salar*) selected for high or low responses to stress. *Aquaculture*, **109**: 215–224.

Fevolden, S.E., Røed, K.H. and Gjerde, B. (1994) Genetic components of post-stress cortisol and lysozyme activity in Atlantic salmon; correlations to disease resistance. *Fish and Shellfish Immunology*, **4**: 507–519.

Fisher, R.A. (1958) *The Genetical Theory of Natural Selection* (2nd edn). Dover Press, New York.

Fleming, I.A. (1994) Captive breeding and the conservation of wild salmon populations. *Conservation Biology*, **8**: 886–888.

Fleming, I.A. (1998) Pattern and variability in the breeding system of Atlantic salmon (*Salmo salar*), with comparisons to other salmonids. *Canadian Journal of Fisheries and Aquatic Sciences*, **55** (Supplement 1): 59–76.

Fleming, I.A. and Gross, M.R. (1989) Evolution of adult female life history and morphology in a Pacific salmon (Coho: *Oncorhynchus kisutch*). *Evolution*, **43**: 141–157.

Fleming, I.A., Jonsson, B. and Gross, M.R. (1994) Phenotypic divergence of sea-ranched, farmed, and wild salmon. *Canadian Journal of Fisheries and Aquatic Sciences*, **51**: 2808–2824.

Fontaine, P.M., Dodson, J.J., Bernatchez, L. and Slettan, A. (1997) A genetic test of metapopulation structure in Atlantic salmon (*Salmo salar*) using microsatellites. *Canadian Journal of Fisheries and Aquatic Sciences*, **54**: 2434–2442.

Foster, J.R. and Schom, C.B. (1989) Imprinting and homing of Atlantic salmon (*Salmo salar*) kelts. *Canadian Journal of Fisheries and Aquatic Sciences*, **46**: 714–719.

Friars, G.W., Bailey, J.K. and Coombs, K.A. (1990) Correlated responses to selection for grilse length in Atlantic salmon. *Aquaculture*, **85**: 171–176.

Friedland, K.D., Haas, R.E. and Sheehan, T.F. (1996) Post-smolt growth, maturation, and survival of two stocks of Atlantic salmon. *Fishery Bulletin*, **94**: 654–663.

Garant, D., Fontaine, P-M., Good, S.P., Dodson, J.J. and Bernatchez, L. (2002) The influence of male parental identity on growth and survival of offspring in Atlantic salmon (*Salmo salar*). *Evolutionary Ecological Research*, **4**: 537–549.

García de Leániz, C. and Martínez, J.J. (1988) The Atlantic salmon in the rivers of Spain with particular reference to Cantabria. In: D. Mills and D. Piggins (Ed.) *Atlantic Salmon: Planning for the future*, pp. 179–209. Croom Helm, London.

García de Leániz, C., Verspoor, E. and Hawkins, A.D. (1989) Genetic determination of the contribution of stocked and wild Atlantic salmon, *Salmo salar* L., to the angling fisheries in two Spanish rivers. *Journal of Fish Biology*, **35**: 261–270.

García de Leániz, C., Fleming, I.A., Einum, S., Verspoor, E., Jordan, W.C., Consuegra, S., Aubin-Horth, N., Lajus, D.L., Letcher, B.H., Youngson, A.F., Webb, J.H., Vøllestad, L.A., Villanueva, B., Ferguson, A. and Quinn, T.P. (2007) A critical review of adaptive genetic variation in Atlantic salmon: implications for conservation. *Biological Review*, **82**, (in press).

García-Vázquez, E., Morán, P. and Pendás, A.M. (1991) Chromosome polymorphism patterns indicate failure of a Scottish stock of *Salmo salar* transplanted into a Spanish river. *Canadian Journal of Fisheries and Aquatic Sciences*, **48**: 170–172.

Gardner, M.L.G. (1976) A review of factors which may influence the sea-age of maturation of Atlantic salmon *Salmo salar* L. *Journal of Fish Biology*, **9**: 289–327.

Gibson, R.J. (1993) The Atlantic salmon in freshwater: spawning, rearing and production. *Reviews in Fish Biology and Fisheries*, **3**: 39–73.

Gilbey, J., Verspoor, E. and Summers, D. (1999) Size and *MEP-2** variation in juvenile Atlantic salmon (*Salmo salar*) in the River North Esk, Scotland. *Aquatic Living Resources*, **12**: 1–5.

Gjedrem, T. (1979) Selection for growth rate and domestication in Atlantic salmon. *Zeitshrift für Tierzücht und Züchtungsbiologie*, **96**: 56–59.

Gjedrem, T. and Aulstad, D. (1974) Selection experiments with salmon. I. Differences in resistance to vibrio disease of salmon parr (*Salmo salar*). *Aquaculture*, **3**: 51–59.

Gjedrem, T. and Gjøen, H.M. (1995) Genetic variation in susceptibility of Atlantic salmon, *Salmo salar* L., to furunculosis, BKD and cold water vibriosis. *Aquaculture Research*, **26**: 129–134.

Gjerde, B. (1984) Response to individual selection for age at sexual maturity in Atlantic salmon. *Aquaculture*, **38**: 229–240.

Gjerde, B. (1986) Growth and reproduction in fish and shellfish. *Aquaculture*, **57**: 37–55.

Gjøen, H.M., Refstie, T., Ulla, O. and Gjerde, B. (1997) Genetic correlations between survival of Atlantic salmon in challenge and field tests. *Aquaculture*, **158**: 277–288.

Good, S.P., Dodson, J.J., Meekan, M.G. and Ryan, D.A.J. (2001) Annual variation in size-selective mortality of Atlantic salmon (*Salmo salar*) fry. *Canadian Journal of Fisheries and Aquatic Sciences*, **58**: 1187–1195.

Gould, S.J. and Lewontin, R.C. (1979) The spandrels of San Marco and the Panglossian paradigm: a critique of the adaptationist programme. *Proceedings of the Royal Society of London, Series B*, **205**: 581–598.

Gould, S.J. and Vrba, E.S. (1982) Exaptation: a missing term in the science of form. *Paleobiology*, **8**: 4–15.

Grant, P.R. and Grant, B.R. (2002) Unpredictable evolution in a 30-year study of Darwin's finches. *Science*, **296**: 707–711.

Gross, M. (1996) Alternative reproductive strategies and tactics: diversity within sexes. *Trends in Ecology and Evolution*, **11**: 92–98.

Gunnes, K. and Gjedrem, T. (1978) Selection experiments with salmon. 4. Growth of Atlantic salmon during two years in the sea. *Aquaculture*, **15**: 19–33.

Gyllensten, U. (1985) The genetic structure of fish: differences in the intraspecific distribution of biochemical genetic variation between marine, anadromous, and freshwater species. *Journal of Fish Biology*, **26**: 691–699.

Hallerman, E. (2003a) Natural selection. In: E. Hallerman (Ed.) *Population Genetics: Principles and applications for fisheries scientists*, pp. 175–196. American Fisheries Society, Bethesda, MD.

Hallerman, E. (2003b) Coadaptation and outbreeding depression. In: E. Hallerman (Ed.) *Population Genetics: Principles and applications for fisheries scientists*, pp. 239–259. American Fisheries Society, Bethesda, MD.

Hansen, L.P. and Jonsson, B. (1990) Restocking the River Akerselv, Oslo with Atlantic salmon smolts *Salmo salar* L. of different stocks. *Fauna Norvegica, Series A*, **11**: 9–15.

Hansen, M.M., Ruzzante, D.E., Nielsen, E.E., Bekkevold, D. and Mensberg, K-L.D. (2002) Long-term effective population sizes, temporal stability of genetic composition and potential for local adaptation in anadromous brown trout (*Salmo trutta*) populations. *Molecular Ecology*, **11**: 2523–2535.

Hanski, I. (1999) *Metapopulation Ecology*. Oxford University Press, New York.

Hartl, D. and Clark, A.G. (2000) *Introduction to Population Genetics*. Sinauer Associates, Sunderland, MA.

Hasler, A.D. and Scholz, A.T. (1983) *Olfactory Imprinting and Homing in Salmon*. Springer-Verlag, Berlin.

Hendry, A.P. (2001) Adaptive divergence and the evolution of reproductive isolation in the wild: an empirical demonstration using introduced sockeye salmon. *Genetica*, **112–113**: 515–534.

Hendry, A.P. (2002) $Q_{ST} > = \neq < F_{ST}$? *Trends in Ecology and Evolution*, **17**: 502.

Hendry, A.P. and Quinn, T.P. (1997) Variation in adult life history and morphology among Lake Washington sockeye salmon (*Oncorhynchus nerka*) populations in relation to habitat features and ancestral affinities. *Canadian Journal of Fisheries and Aquatic Sciences*, **54**: 75–84.

Hendry, A.P. and Stearns, S.C. (Ed.) (2004) *Evolution Illuminated: Salmon and their relatives*. Oxford University Press, Oxford.

Hendry, A.P., Wenburg, J.K., Bentzen, P., Volk, E.C. and Quinn, T.P. (2000) Rapid evolution of reproductive isolation in the wild: evidence from introduced salmon. *Science*, **290**: 516–518.

Hoffmann, A.A. (2000) Laboratory and field heritabilities: Some lessons from *Drosophila*. In: T.A. Mousseau, B. Sinervo and J. Endler (Ed.) *Adaptive Genetic Variation in the Wild*, pp. 200–218. Oxford University Press, Oxford.

Holm, M. and Fernö, A. (1986) Aggression and growth of Atlantic salmon parr. II. Different populations in pure and mixed groups. *Fiskeridirektoratets skrifter, Serie Havundersøkelser*, **18**: 123–129.

Holm, M. and Nævdal, G. (1978) Quantitative genetic variation in fish: its significance for salmonid culture. In: B. Battaglia and J.A. Beardmore (Ed.) *Marine Organisms: Genetics, ecology and evolution*, pp. 678–698. Plenum Press, New York.

Holt, R.D. and Gomulkiewicz, R. (1997) How does immigration influence local adaptation? A reexamination of a familiar paradigm. *American Naturalist*, **149**: 563–572.

Howard, D.J., Marshall, J.L., Braswell, W.E. and Coyne, J.A. (2001) Examining evidence of reproductive isolation in sockeye salmon. *Science*, **291**: 1853–1855.

Hull, D. (1974) *Philosophy of Biological Science*. Prentice-Hall, Englewood Cliffs, NJ.

Huntingford, F.A. and García de Leániz, C. (1997) Social dominance, prior residence and the acquisition of profitable feeding sites in juvenile Atlantic salmon. *Journal of Fish Biology*, **51**: 1009–1014.

Huntsman, A.G. (1941) Races and homing instinct. *Salmon and Trout Magazine (London)*, 3–7.

Hutchings, J.A. (2004) Norms of reaction and phenotypic plasticity in salmonid life histories. In: A.P. Hendry and S.C. Stearns (Ed.) *Evolution Illuminated: Salmon and their relatives*, pp. 154–174. Oxford University Press, Oxford.

Hutchings, J.A. and Jones, M.E.B. (1998) Life history variation and growth rate thresholds for maturity in Atlantic salmon, *Salmo salar*. *Canadian Journal of Fisheries and Aquatic Sciences*, **55** (Supplement 1): 22–47.

Hutchings, J.A. and Myers, R.A. (1994) The evolution of alternative mating strategies in variable environments. *Evolutionary Ecology*, **8**: 256–268.

Jansen, P.A., Bakke, T.A. and Hansen, L.P. (1991) Resistance to *Gyrodactylus salaris* Malmberg 1957 (Monogenea) in *Salmo salar*: a genetic component. *Bulletin of the Scandinavian Society for Parasitology*, **1**: 1–50.

Jensen, A.J. and Johnsen, B.O. (1986) Different adaptation strategies of Atlantic salmon (*Salmo salar*) populations to extreme climates with special reference to some cold Norwegian rivers. *Canadian Journal of Fisheries and Aquatic Sciences*, **43**: 980–984.

Jóhannsson, V., Jónasson, J., Ósakarsson, S. and Ísaksson, Á. (1998) The straying of Icelandic ranched Atlantic salmon, *Salmo salar* L.: release and recapture techniques. *Aquaculture Research*, **29**: 679–686.

Johansson, N. (1981) General problems in Atlantic salmon rearing in Sweden. *Ecological Bulletin*, **34**: 75–83.

Johnsen, B.O. and Jensen, A. (1986) Infestations of Atlantic salmon, *Salmo salar*, by *Gyrodactylus salaris* in Norwegian rivers. *Journal of Fish Biology*, **29**: 233–241.

Johnsen, B.O. and Jensen. A. (1991) The *Gyrodactylus* story in Norway. *Aquaculture*, **98**: 289–302.

Johnston, I.A., McLay, H.A., Abercromby, M. and Robins, D. (2000) Phenotypic plasticity of early myogenesis and satellite cell numbers in Atlantic salmon spawning in upland and lowland tributaries of a river system. *Journal of Experimental Biology*, **203**: 2539–2552.

Jonasson, J. (1993) Selection experiments in salmon ranching. I. Genetic and environmental sources of variation in survival and growth in freshwater. *Aquaculture*, **109**: 225–236.

Jonasson, J. (1996) Selection experiments on Atlantic salmon ranching. 2. Variation among release sites and strains for return rate, body weight and ratio of grilse to total return. *Aquaculture*, **144**: 277–294.

Jonasson, J. (2002) Case study of Atlantic salmon in sea ranching (in manuscript).

Jonasson, J., Gjerde, B. and Gjedrem, T. (1997) Genetic parameters for return rate and body weight of sea-ranched Atlantic salmon. *Aquaculture*, **154**: 219–231.

Jonsson, N., Hansen, L.P. and Jonsson, B. (1991a) Variation in age, size and repeat spawning of adult Atlantic salmon in relation to river discharge. *Journal of Animal Ecology*, **60**: 937–947.

Jonsson, B., Jonsson, N. and Hansen, L.P. (1991b) Differences in life history and migratory behaviour between wild and hatchery reared Atlantic salmon in nature. *Aquaculture*, **98**: 69–78.

Jonsson, N., Jonsson, B., Skurdal, J. and Hansen, L.P. (1994) Differential response to water current in offspring of inlet- and outlet-spawning brown trout *Salmo trutta*. *Journal of Fish Biology*, **45**: 356–359.

Jonsson, N., Jonsson, B. and Fleming, I.A. (1996) Does early growth cause a phenotypically plastic response in egg production of Atlantic salmon? *Functional Ecology*, **10**: 89–96.

Jonsson, B., Forseth, T., Jensen, A.J. and Næsje, T.F. (2001) Thermal performance of juvenile Atlantic salmon, *Salmo salar* L. *Functional Ecology*, **15**: 701–711.

Jonsson, B., Jonsson, N. and Hansen, L.P. (2003) Atlantic salmon straying from the River Imsa. *Journal of Fish Biology*, **62**: 641–657.

Jordan, W.C. and Youngson, A.F. (1991) Genetic protein variation and natural selection in Atlantic salmon (*Salmo salar* L.) parr. *Journal of Fish Biology*, **39**: 185–192.

Jordan, W.C., Youngson, A.F. and Webb, J.H. (1990) Genetic variation in the malic enzyme-2 locus and age at maturity in sea-run Atlantic salmon (*Salmo salar*). *Canadian Journal of Fisheries and Aquatic Sciences*, **47**: 1672–1677.

Jordan, W.C., Verspoor, E. and Youngson, A.F. (1997) The effect of natural selection on estimates of genetic divergence among populations of the Atlantic salmon. *Journal of Fish Biology*, **51**: 546–560.

Jørstad, K.E. and Nævdal, G. (1996) Breeding and genetics. In: W. Pennell and B.A. Barton (Ed.) *Principles of Salmonid Culture*, 655–726. Elsevier, Amsterdam.

Jutila, E., Jokikokko, E., Kallio-Nyberg, I., Saloniemi, I. and Pasanen, P. (2003a) Differences in sea migration between wild and reared Atlantic salmon (*Salmo salar* L.) in the Baltic Sea. *Fisheries Research*, **60**: 333–343.

Jutila, E., Jokikokko, E. and Julkunen, M. (2003b) Management of Atlantic salmon in the Simojoki river, northern Gulf of Bothnia: effects of stocking and fishing regulation. *Fisheries Research*, **64**: 5–17.

Kallio-Nyberg, I., and Koljonen, M-L. (1999) Sea migration patterns in the Atlantic salmon: a comparative study of two stocks and their hybrids. *Boreal Environment Research*, **4**: 163–174.

Kallio-Nyberg, I., Peltonen, H., and Rita, H. (1999) Effects of stock-specific and environmental factors on the feeding migration of Atlantic salmon (*Salmo salar*) in the Baltic Sea. *Canadian Journal of Fisheries and Aquatic Sciences*, **56**: 853–861.

Kallio-Nyberg, I., Koljonen, M-L. and Saloniemi, I. (2000) Effect of maternal and paternal line on spatial and temporal marine distribution in Atlantic salmon. *Animal Behaviour*, **60**: 377–384.

Kazakov, R.V. and Semenova, O.V. (1986) Morphological characteristics of cultured and wild Atlantic salmon *Salmo salar* juveniles. *Trudy Zoologicheskogo Instituta AN SSSR*, **154**: 75–85. (in Russian)

Kazakov, R.V., Liashenko, A.N. and Titov, S.F. (1989) Use of fluctuating asymmetry to control ecological and genetic status of populations of Atlantic salmon *Salmo salar* and brown trout *Salmo trutta*. Vsesoyuznoe soveshchanie po genetike, selektsii i gibridizatsii ryb, 3th. Tartu, 1986. Proceedings, Leningrad. (in Russian)

King, T.L., Spidle, A.P., Eackles, M., Lubinski, B.A. and Schill, W.B. (2000) Mitochondrial DNA diversity in North American and European Atlantic salmon with emphasis on the Downeast rivers of Maine. *Journal of Fish Biology*, **57**: 614–630.

King, T.L., Kalinowski, S., Schill, W.B., Spidle, A.P. and Lubinski, B.A. (2001) Population structure of Atlantic salmon (*Salmo salar* L.): a range-wide perspective from microsatellite DNA variation. *Molecular Ecology*, **10**: 807–821.

Kinnison, M.T., Unwin, M.J., Boustead, N. and Quinn, T.P. (1998) Population-specific variation in body dimensions of adult chinook salmon (*Oncorhynchus tshawytscha*) from New Zealand and their source population, 90 years after introduction. *Canadian Journal of Fisheries and Aquatic Sciences*, **55**: 554–563.

Klemetsen, A., Amundsen, P-A., Dempson, J.B., Jonsson, B., Jonsson, N., O'Connell, M.F. and Mortensen, E. (2003) Atlantic salmon *Salmo salar* L., brown trout *Salmo trutta* L. and Arctic charr *Salvelinus alpinus* (L.): a review of aspects of their life histories. *Ecology of Freshwater Fish*, **12**: 1–59.

Koljonen, M.L., Jansson, H., Paaver, T., Vasin, O. and Koskiniemi, J. (1999) Phylogeographic lineages and differentiation pattern of Atlantic salmon (*Salmo salar*) in the Baltic Sea with management implications. *Canadian Journal of Fisheries and Aquatic Sciences*, **56**: 1766–1780.

Koskinen, M.T., Haugen, T.O. and Primmer, C.R. (2002) Contemporary fisherian life-history evolution in small salmonid populations. *Nature*, **419**: 826–830.

Lajus, D.L., Graham, J.H. and Kozhara, A.V. (2003) Developmental instability and the stochastic component of total phenotypic variance. In: M. Polak (Ed.) *Developmental Instability: Causes and consequences*, pp. 343–363. Oxford University Press, Oxford.

Landry, C. and Bernatchez, L. (2001) Comparative analysis of population structure across environments and geographical scales at major histocompatibility complex and microsatellite loci in Atlantic salmon (*Salmo salar*). *Molecular Ecology*, **10**: 2525–2539.

Langefors, Å., Lohm, J., Grahn, M., Andersen, Ø. and von Schantz, T. (2001) Association between major histocompatibility complex class IIB alleles and resistance to *Aeromonas salmonicida* in Atlantic salmon. *Proceedings of the Royal Society of London, Series B*, **268**: 479–485.

Latta, R.G. and McKay, J.K. (2002) Genetic population divergence: markers and traits. A response from Latta and McKay. *Trends in Ecology and Evolution*, **17**: 501–502.

Lenormand, T. (2002) Gene flow and the limits to natural selection. *Trends in Ecology and Evolution*, **17**: 183–189.

Letcher, B.H. and Gries, G. (2003) Effects of life history variation on size and growth in stream-dwelling Atlantic salmon. *Journal of Fish Biology*, **62**: 97–114.

Lohm, J., Grahn, M., Langefors, Å., Andersen, Ø., Storset, A. and von Schantz, T. (2002) Experimental evidence for major histocompatibility complex-allele-specific resistance to a bacterial infection. *Proceedings of the Royal Society of London, Series B*, **269**: 2029–2033.

Lund, R.A. and Heggberget, T.G. (1992) Migration of Atlantic salmon, *Salmo salar* L, parr through a Norwegian fjord: potential infection path of *Gyrodactylus salaris*. *Aquaculture and Fisheries Management*, **23**: 367–372.

Lund, T., Gjedrem, T., Bentsen, H.B., Eide, D.M., Larsen, H.J.S. and Røed, K.H. (1995) Genetic variation in immune parameters and associations to survival in Atlantic salmon. *Journal of Fish Biology*, **46**: 748–758.

Lynch, M. and Walsh, B. (1998) *Genetics and Analysis of Quantitative Traits*. Sinuaer Associates, Sunderland, MA.

Lynch, M., Pfrender, M., Spitze, K., Lehman, N., Hicks, J., Allen, D., Latta, L., Ottene, M., Bogue, F. and Colbourne, J. (1999) The quantitative and molecular genetic architecture of a sub-divided species. *Evolution*, **53**: 100–110.

MacCrimmon, H.R. and Claytor, R.R. (1985) Meristic and morphometric identity of Baltic stocks of Atlantic salmon (*Salmo salar*). *Canadian Journal of Zoology*, **63**: 2032–2037.

MacCrimmon, H.R. and Claytor, R.R. (1986) Possible use of taxonomic characters to identify Newfoundland and Scottish stocks of Atlantic salmon, *Salmo salar* L. *Aquaculture and Fisheries Management*, **17**: 1–17.

MacCrimmon, H.R. and Gots, B.L. (1979) World distribution of Atlantic salmon, *Salmo salar*. *Journal of the Fisheries Research Board of Canada*, **36**: 422–457.

MacKenzie, K. and Mo, T.A. (1994) Comparative susceptibility of native Scottish and Norwegian stocks of Atlantic salmon, *Salmo salar* L., to *Gyrodactylus salaris* Malmberg: laboratory experiments. In: A.W. Pike and J.W. Lewis (Ed.) *Parasitic Diseases of Fish*, pp. 57–58. Samara Publishers, Tresaith, Dyfed, UK.

Malmberg, G. and Malmberg, M. (1993) Species of *Gyrodactylus* (Platyhelminthes, Monogenea) on salmonids in Sweden. *Fisheries Research*, **17**: 59–68.

Manel, S., Schwartz, M.K., Luikart, G. and Taberlet, P. (2003) Landscape genetics: combining landscape ecology and population genetics. *Trends in Ecology and Evolution*, **18**: 189–197.

Mayama, H., Nomura, T. and Ohkuma, K. (1989) Reciprocal transplantation experiment of masu salmon (*Oncorhynchus masou*) population. 2. Comparison of seaward migrations and adult returns of local stock and transplanted stock of masu salmon. *Scientific Reports of the Hokkaido Salmon Hatchery*, **43**: 99–113.

Maynard Smith, J. (1982) *Evolution and the Theory of Games*. Cambridge University Press, Cambridge.

Mayr, E. (2002) *What Evolution Is*. Phoenix, London.

Mazer, S.J. and Damuth, J. (2001) Evolutionary significance of variation. In: C.W. Fox, D.A. Roff and D.J. Fairburn (Ed.) *Evolutionary Ecology*, pp. 16–28. Oxford University Press, Oxford.

McGinnity, P., Stone, C., Taggart, J.B., Cooke, D., Cotter, D., Hynes, R., McCamley, C., Cross, T. and Ferguson, A. (1997) Genetic impact of escaped farmed Atlantic salmon (*Salmo salar* L.) on native populations: use of DNA profiling to assess freshwater performance of wild, farmed, and hybrid progeny in a natural river environment. *ICES Journal of Marine Science*, **54**: 998–1008.

McGinnity, P., Prodöhl, P., Ferguson, A., Hynes, R., Ó Maoiléidigh, N., Baker, N., Cotter, D., O'Hea, B., Cooke, D., Rogan, G., Taggart, J. and Cross, T. (2003) Fitness reduction and potential extinction of wild populations of Atlantic salmon, *Salmo salar*, as a result of interactions with escaped farm salmon. *Proceedings of the Royal Society of London, Series B*, **270**: 2443–2450. (Published online, DOI: 10.1098/rspb.2003.2520.)

McKay, J.K. and Latta, R.G. (2002) Adaptive population divergence: markers, QTL and traits. *Trends in Ecology and Evolution*, **17**: 285–291.

McPeek, M.A. (1997) Measuring phenotypic selection on an adaptation: lamellae of damselflies experiencing dragonfly predation. *Evolution*, **51**: 459–466.

Meinilä, M., Kuusela, J., Ziętara, M.S. and Lumme. J. (2002) Primers for amplifying ~ 820 bp of highly polymorphic mitochondrial COI gene of *Gyrodactylus salaris*. *Hereditas*, **137**: 72–74.

Meinilä, M., Kuusela, J., Zietara, M.S. and Lumme, J. (2004) Initial steps of speciation by geographic isolation and host switch in salmonid pathogen *Gyrodactylus salaris* (Monogenea: Gyrodactylidae) *International Journal for Parasitology*, **34**: 515–526.

Merilä, J. and Crnokrak, P. (2001) Comparison of genetic differentiation at marker loci and quantitative traits. *Journal of Evolutionary Biology*, **14**: 892–903.

Merilä, J. and Sheldon, B. (1999) Genetic architecture of fitness and nonfitness traits: empirical patterns and development of ideas. *Heredity*, **83**: 103–109.

Metcalfe, N.B. (1998) The interaction between behavior and physiology in determining life history patterns in Atlantic salmon (*Salmo salar*). *Canadian Journal of Fisheries and Aquatic Sciences*, **55** (Supplement 1): 93–103.

Metcalfe, N.B. and Thorpe, J.E. (1990) Determinants of geographical variation in the age of seaward-migrating salmon, *Salmo salar*. *Journal of Animal Ecology*, **59**: 135–145.

Meyers, L.A. and Bull, J.J. (2002) Fighting change with change: adaptive variation in an uncertain world. *Trends in Ecology and Evolution*, **17**: 551–557.

Mills, D.H. (1989) *Ecology and Management of Atlantic Salmon*. Chapman & Hall, London.

Mitchell-Olds, T. (1995) The molecular basis of quantitative genetic variation in natural populations. *Trends in Ecology and Evolution*, **10**: 324–328.

Mo, T.A. (1994) Status of *Gyrodactylus salaris* problems and research in Norway. In: A.W. Pike and J.W. Lewis (Ed.) *Parasitic Diseases of Fish*, pp. 43–56. Samara Publishers, Tresaith, Dyfed, UK.

Morán, N.A. (1994) Adaptation and constraint in the complex life cycles of animals. *Annual Review of Ecology and Systematics*, **25**: 573–600.

Morán, P. (2002) Current conservation genetics: building an ecological approach to the synthesis of molecular and quantitative genetic methods. *Ecology of Freshwater Fish*, **11**: 30–55.

Morán, P., Pendás, A.M., García-Vázquez, E., Izquierdo, J.T. and Rutherford, D.T. (1994) Electrophoretic assessment of the contribution of transplanted Scottish Atlantic salmon (*Salmo salar*) to the Esva river (northern Spain). *Canadian Journal of Fisheries and Aquatic Sciences*, **51**: 248–252.

Morán, P., Izquierdo J.I., Pendas, A.M. and García-Vazquez, E. (1997) Fluctuating asymmetry and isozyme variation in Atlantic salmon: relation to age of wild and hatchery fish. *Transactions of the American Fisheries Society*, **126**: 194–199.

Mousseau, T.A. (2000) Intra- and interpopulation genetic variation: explaining the past and predicting the future. In: T.A. Mousseau, B. Sinervo and J. Endler (Ed.) *Adaptive Genetic Variation in the Wild*, pp. 219–250. Oxford University Press, Oxford.

Mousseau, T.A. and Roff, D.A. (1987) Natural selection and the heritability of fitness components. *Heredity*, **59**: 181–197.

Mustafa, A. and MacKinnon, B.M. (1999) Genetic variation in susceptibility of Atlantic salmon to the sea louse *Caligus elongatus* Nordmann, 1832. *Canadian Journal of Zoology*, **77**: 1332–1335.

Myers, R.A., Hutchings, J.A. and Gibson, R.J. (1986) Variation in male parr maturation within and among populations of Atlantic salmon, *Salmo salar*. *Canadian Journal of Fisheries and Aquatic Sciences*, **43**: 1242–1248.

Nævdal, G. (1983) Genetic factors in connection with age at maturation. *Aquaculture*, **33**: 97–106.

Nicieza, A.G. (1995) Morphological variation between geographically disjunct populations of Atlantic salmon: the effects of ontogeny and habitat shift. *Functional Ecology*, **9**: 448–456.

Nicieza, A.G., Reyes-Gavilán, F. and Braña, F. (1994a) Differentiation in juvenile growth and bimodality patterns between northern and southern populations of Atlantic salmon (*Salmo salar* L). *Canadian Journal of Zoology*, **72**: 1603–1610.

Nicieza, A.G., Reiriz, L. and Braña, F. (1994b) Variation in digestive performance between geographically disjunct populations of Atlantic salmon: countergradient in passage time and digestion rate. *Oecologia*, **99**: 243–251.

Nilsson, J., Gross, R., Asplund, T., Dove, O., Jansson, H., Kelloniemi, J., Kohlmann, K., Löytynoja, A., Nielsen, E.E., Paaver, T., Primmer, C.R., Titov, S., Vasemägi, A., Veselov, A., Öst, T. and Lumme, J. (2001) Matrilinear phylogeography of Atlantic salmon (*Salmo salar* L.) in Europe and postglacial colonization of the Baltic Sea area. *Molecular Ecology*, **10**: 89–102.

Niva, T., and Jokela, J. (2000) Phenotypic correlation of juvenile growth rate between different consecutive foraging environments in a salmonid fish: a field experiment. *Evolutionary Ecology*, **14**: 111–126.

Pakkasmaa, S. and Piironen, J. (2001a) Morphological differentiation among local trout (*Salmo trutta*) populations. *Biological Journal of the Linnean Society*, **72**: 231–239.

Pakkasmaa, S. and Piironen, J. (2001b) Water velocity shapes juvenile salmonids. *Evolutionary Ecology*, **14**: 721–730.

Pelis, R.M. and McCormick, S.D. (2003) Fin development in stream- and hatchery-reared Atlantic salmon. *Aquaculture*, **220**: 525–536.

Petersson, E., Järvi, T., Steffner, N.G. and Ragnarsson, B. (1996) The effect of domestication on some life history traits of sea trout and Atlantic salmon. *Journal of Fish Biology*, **48**: 776–791.

Power, G. (1981) Stock characteristics and catches of Atlantic salmon (*Salmo salar*) in Quebec and Newfoundland and Labrador in relation to environmental variables. *Canadian Journal of Fisheries and Aquatic Sciences*, **38**: 1601–1611.

Powers, D.A. (1990) The adaptive significance of allelic isozyme variation in natural populations. In: D.H. Whitmore (Ed.) *Electrophoretic and Isoelectric Focusing Techniques in Fisheries Management*, pp. 323–340. CRC Press, Boca Raton, FL.

Powers, D.A., Lauerman, T., Crawford, D. and DiMichele, L. (1991) Genetic mechanisms for adapting to a changing environment. *Annual Review of Genetics*, **25**: 629–659.

Prévost, E., Chadwick, E.M.P. and Claytor, R.R. (1992) Influence of size, winter duration, and density on sexual maturation of Atlantic salmon (*Salmo salar*) juveniles in Little Codroy River (southwest Newfoundland). *Journal of Fish Biology*, **41**: 1013–1019.

Quinn, T.P. (1993) A review of homing and straying of wild and hatchery-produced salmon. *Fisheries Research*, **18**: 29–44.

Quinn, T.P. (1999) Revisiting the stock concept in Pacific salmon: insights from Alaska and New Zealand. *Northwest Science*, **73**: 312–324.

Quinn, T.P. (2005) *Pacific Salmon Ecology and Evolution*. University of Washington Press, Seattle, WA.

Quinn, T.P. and Dittman, A.H. (1990) Pacific salmon migrations and homing: mechanisms and adaptive significance. *Trends in Ecology and Evolution*, **5**: 174–177.

Quinn, T.P., Graynoth, E., Wood, C.C. and Foote, C.J. (1998) Genotypic and phenotypic divergence of sockeye salmon in New Zealand and their ancestral British Columbia populations. *Transactions of the American Fisheries Society*, **127**: 517–534.

Quinn, T.P., Unwin, M.J. and Kinnison, M.T. (2000) Evolution of temporal isolation in the wild: genetic divergence in timing of migration and breeding by introduced chinook salmon populations. *Evolution*, 54: 1372–1385.

Quinn, T.P., Hendry, A.P. and Buck, G.B. (2001a) Balancing natural and sexual selection in sockeye salmon: interactions between body size, reproductive opportunity and vulnerability to predation by bears. *Evolutionary Ecology Research*, 3: 917–937.

Quinn, T.P., Kinnison, M.T. and Unwin, M.J. (2001b) Evolution of chinook salmon (*Oncorhynchus tshawytscha*) populations in New Zealand: pattern, rate and process. *Genetica*, 112–113: 493–513.

Reddin, D.G., Verspoor, E. and Downton, P.R. (1987) An integrated phenotypic and genotypic approach to discriminating Atlantic salmon. *International Council for the Exploration of the Sea*, **1987/M: 16**.

Reddin, D.G., Stansbury, D.E. and Short, P.B. (1988) Continent of origin of Atlantic salmon (*Salmo salar* L.) at West Greenland. *Journal du Conseil*, 44: 180–188.

Reed, D.H. and Frankham, R. (2001) How closely correlated are molecular and quantitative measures of genetic variation? A meta-analysis. *Evolution*, 55: 1095–1103.

Reeve, H.K. and Sherman, P.W. (1993) Adaptation and the goals of evolutionary research. *Quarterly Review of Biology*, 68: 1–32.

Reid, R.G.B. (1985) *Evolutionary Theory: The unfinished synthesis*. Croom Helm, London.

Reimers, E., Kjørrefjord, A.G. and Stavøstrand, S. (1993) Compensatory growth and reduced maturation in second sea winter farmed Atlantic salmon following starvation in February and March. *Journal of Fish Biology*, 43: 805–810.

Reznick, D. and Travis, J. (2001) Adaptation. In: C.W. Fox, D.A. Roff and D.J. Fairburn (Ed.) *Evolutionary Ecology*, pp. 44–57. Oxford University Press, Oxford.

Riddell, B.E. (1993) Spatial organization of Pacific salmon: what to conserve? In: J.G. Cloud and G.H. Thorgaard (Ed.) *Genetic Conservation of Salmonid Fishes*, pp. 23–41. Plenum Press, London.

Riddell, B.E. and Leggett, W.C. (1981) Evidence of an adaptive basis for geographic variation in body morphology and time of downstream of juvenile Atlantic salmon (*Salmo salar*). *Canadian Journal of Fisheries and Aquatic Sciences*, 38: 308–320.

Riddell, B.E., Leggett, W.C. and Saunders, R.L. (1981) Evidence of adaptive polygenic variation between two populations of Atlantic salmon (*Salmo salar*) native to tributaries of the S.W. Miramichi River, N.B. *Canadian Journal of Fisheries and Aquatic Sciences*, 38: 321–333.

Rieman, B.E. and Dunham, J.B. (2000) Metapopulations and salmonids: a synthesis of life history patterns and empirical observations. *Ecology of Freshwater Fish*, 9: 51–64.

Rintamäki-Kinnunen, P. and Valtonen, E.T. (1996) Finnish salmon resistant to *Gyrodactylus salaris*: a long-term study at fish farms. *International Journal for Parasitology*, 26: 723–732.

Rose, M.R. and Lauder, G.V. (1996) *Adaptation*. Academic Press, London.

Rosseland, B., Kroglund, F., Staurnes, M., Hindar, K. and Kvellestad, A. (2001) Tolerance to acid water among strains and life stages of Atlantic salmon (*Salmo salar* L.). *Water, Air, and Soil Pollution*, 130: 899–904.

Ruse, M. (1982) *Darwinism Defended: A guide to the evolution controversies*. Addison-Wesley Publishing Company, London.

Rye, M. and Refstie, T. (1995) Phenotypic and genetic parameters of body size traits in Atlantic salmon *Salmo salar* L. *Aquaculture Research*, 26: 875–885.

Rye, M., Lillevik, K.M. and Gjerde, B. (1990) Survival in early life of Atlantic salmon and rainbow trout: estimates of heritabilities and genetic correlations. *Aquaculture*, 89: 209–216.

Sadler, J., Pankhurst, P.M. and King, H.R. (2001) High prevalence of skeletal deformity and reduced gill surface area in triploid Atlantic salmon (*Salmo salar* L.). *Aquaculture*, 198: 369–386.

Salmanov, A.V. (1986) Osteological peculiarities of cultured and wild juveniles of Atlantic salmon (*Salmo salar* L.) from Luvenga River. *Trudy Zoologicheskogo Instituta AN SSSR*, 154: 87–98. (in Russian)

Salmanov, A.V. (1989) Analysis of variability of morphometric parameters in cultured and wild juveniles of Atlantic salmon (*Salmo salar* L.). *Trudy Zoologichaskogo Instituta AN SSSR*, **192**: 126–144. (in Russian)

Saunders, R.L. and Bailey, J.K. (1980) The role of genetics in Atlantic salmon management. In: A.E.J. Went (Ed.) *Atlantic Salmon: Its future*, pp. 182–200. Fishing News Books, Farnham, UK.

Saunders, R.L., Henderson, E.B., Glebe, B.D. and Loudenslager, E.J. (1983) Evidence of a major environmental component in determination of the grilse : larger salmon ratio in Atlantic salmon (*Salmo salar*). *Aquaculture*, **33**: 107–118.

Schaffer, W.M. and Elson, P.F. (1975) The adaptive significance of variations in life history among local populations of Atlantic salmon in North America. *Ecology*, **56**: 577–590.

Scholz, T. (1999) Parasites in cultured and feral fish. *Veterinary Parasitology*, **84**: 317–335.

Schom, C.B. (1986) Genetic, environmental, and maturational effects on Atlantic salmon (*Salmo salar*) survival in acute low pH trials. *Canadian Journal of Fisheries and Aquatic Sciences*, **43**: 1547–1555.

Shabalina, S., Yampolsky, L.Y. and Kondrashov, A.S. (1997) Rapid decline of fitness in panmictic populations of *Drosophila melanogaster* maintained under relaxed natural selection. *Proceedings of the National Academy of Sciences of the USA*, **94**: 13034–13039.

Spidle, A.P., Schill, W.B., Lubinski, B.A. and King, T.L. (2001) Fine-scale population structure in Atlantic salmon from Maine's Penobscot River drainage. *Conservation Genetics*, **2**: 11–24.

Spidle, A.P., Kalinowski, S.T., Lubinski, B.A., Perkins, D.L., Beland, K.F., Kocik, J.F. and King, T.L. (2003) Population structure of Atlantic salmon in Maine with reference to populations from Atlantic Canada. *Transactions of the American Fisheries Society*, **132**: 196–209.

Stabell, O.B. (1984) Homing and olfaction in salmonids: a critical review with special reference to the Atlantic salmon. *Biological Reviews*, **59**: 333–388.

Standal, M. and Gjerde, B. (1987) Genetic variation in survival of Atlantic salmon during the sea-rearing period. *Aquaculture*, **66**: 197–207.

Swain, D.P., Riddell, B.E. and Murray, C.B. (1991) Morphological differences between hatchery and wild populations of coho salmon (*Oncorhynchus kisutch*): environmental versus genetic origin. *Canadian Journal of Fisheries and Aquatic Sciences*, **48**: 1783–1791.

Tallman, R.F. and Healey, M.C. (1994) Homing, straying, and gene flow among seasonally separated populations of chum salmon (*Oncorhynchus keta*). *Canadian Journal of Fisheries and Aquatic Sciences*, **51**: 577–588.

Tave, D. (1993) *Genetics for Fish Hatchery Managers* (2nd edn). Van Nostrand Reinhold, New York.

Taylor, E.B. (1986) Differences in morphology between wild and hatchery populations of juvenile coho salmon. *The Progressive Fish-Culturist*, **48**: 171–176.

Taylor, E.B. (1990) Phenotypic correlates of life-history variation in juvenile chinook salmon, *Oncorhynchus tshawytscha*. *Journal of Animal Ecology*, **59**: 455–468.

Taylor, E.B. (1991) A review of local adaptations in Salmonidae, with particular reference to Pacific and Atlantic salmon. *Aquaculture*, **98**: 185–207.

Taylor, E.B. and MacPhail, J.D. (1985) Variation in body morphology among British Columbia populations of coho salmon (*Oncorhynchus kisutch*). *Canadian Journal of Fisheries and Aquatic Sciences*, **42**: 2020–2028.

Thodesen, J., Gjerde, B., Grisdale-Helland, B. and Storebakken, T. (2001) Genetic variation in feed intake, growth and feed utilization in Atlantic salmon (*Salmo salar*). *Aquaculture*, **194**: 273–281.

Thomaz, D., Beall, E. and Burke, T. (1997) Alternative reproductive tactics in Atlantic salmon: factors affecting mature parr success. *Proceedings of the Royal Society of London, Series B*, **264**: 219–226.

Thorpe, J.E. (1986) Age at first maturity in Atlantic salmon, *Salmo salar* L.: freshwater period influences and conflicts with smolting. *Canadian Special Publication in Fisheries and Aquatic Sciences*, **89**: 7–14.

Thorpe, J.E. and Mitchell, K.A. (1981) Stocks of Atlantic salmon (*Salmo salar*) in Britain and Ireland: discreteness, and current management. *Canadian Journal of Fisheries and Aquatic Sciences*, **38**: 1576–1590.

Thorpe, J.E. and Morgan, R.I.G. (1978) Parental influence on growth rate, smolting rate and survival in hatchery reared juvenile Atlantic salmon, *Salmo salar*. *Journal of Fish Biology*, **13**: 549–556.

Thorpe, J.E. and Stradmeyer, L. (1995) The Atlantic salmon. In: J.E. Thorpe, G. Gall, J. Lannan and C. Nash (Ed.) *Conservation of Fish and Shellfish Resources: Managing diversity*, pp. 79–114. Academic Press, London.

Thorpe, J.E., Talbot, C., Miles, M.S. and Keay, D.S. (1990) Control of maturation in cultured Atlantic salmon, *Salmo salar*, in pumped seawater tanks, by restricting food intake. *Aquaculture*, **86**: 315–326.

Thorpe, J.E., Mangel, M.S., Metcalfe, N.B. and Huntingford, F.A. (1998) Modeling the proximate basis of salmonid life-history variation, with application to Atlantic salmon, *Salmo salar* L.. *Evolutionary Ecology*, **12**: 581–599.

Torrissen, K.R., Male, R. and Nævdal, G. (1993) Trypsin isozymes in Atlantic salmon, *Salmo salar* L.: studies of heredity, egg quality and effect on growth of three different populations. *Aquaculture and Fisheries Management*, **24**: 407–415.

Unwin, M.J., Kinnison, M.T., Boustead, N.C. and Quinn, T.P. (2003) Genetic control over survival in Pacific salmon (*Oncorhynchus* spp.): experimental evidence between and within populations of New Zealand chinook salmon (*O. tshawytscha*). *Canadian Journal of Fisheries and Aquatic Sciences*, **60**: 1–11.

Utter, F. (2001) Patterns of subspecific anthropogenic introgression in two salmonid genera. *Reviews in Fish Biology and Fisheries*, **10**: 265–279.

Verspoor, E. and García de Leániz, C. (1997) Stocking success of Scottish Atlantic salmon in two Spanish rivers. *Journal of Fish Biology*, **51**: 1265–1269.

Verspoor, E. and Jordan, W.C. (1989) Genetic variation at the Me-2 locus in the Atlantic salmon within and between rivers: evidence for its selective maintenance. *Journal of Fish Biology*, **35** (Supplement A): 205–213.

Verspoor, E., Fraser, N.H.C. and Youngson, A.F. (1991) Protein polymorphisms in the Atlantic salmon within a Scottish river: evidence for selection and estimates of gene flow between tributaries. *Aquaculture*, **98**: 217–230.

Verspoor, E., McCarthy, E.M., Knox, D., Bourke, E. and Cross, T.F. (1999) The phylogeography of European Atlantic salmon (*Salmo salar* L.) based on RFLP analysis of the ND1/16sRNA region of the mtDNA. *Biological Journal of the Linnean Society*, **68**: 129–146.

Verspoor, E., O'Sullivan, M., Arnold, A.L., Knox, D. and Amiro, P.G. (2002) Restricted matrilineal gene flow and regional differentiation among Atlantic salmon (*Salmo salar* L.) populations within the Bay of Fundy, eastern Canada. *Heredity*, **89**: 465–472.

Verspoor, E., Beardmore, J.A., Consuegra, S., García de Leániz, C., Hindar, K., Jordan, W.C., Koljonen, M-L., Mahkrov, A.A., Paaver, T., Sánchez, J.A., Skaala, Ø., Titov, S. and Cross, T.F. (2005) Genetic protein variation in the Atlantic salmon: population insights gained from 40 years of research. *Journal of Fish Biology*, **67**: 3–54.

Vøllestad, L.A. and Hindar, K. (1997) Developmental stability and environmental stress in Atlantic salmon *Salmo salar*. *Heredity*, **78**: 215–222.

Wang, S., Hard, J.J. and Utter, F. (2002a) Salmonid inbreeding: a review. *Reviews in Fish Biology and Fisheries*, **11**, 301–319.

Wang, S., Hard, J.J. and Utter, F. (2002b) Genetic variation and fitness in salmonids. *Conservation Genetics*, **3**: 321–333.

Waples, R.S. (1995) Evolutionary significant units and the conservation of biological diversity under the Endangered Species Act. *American Fisheries Society Symposium*, **17**: 8–27.

Wegner, K.M., Reusch, T.B.H. and Kalbe, M. (2003) Multiple parasites are driving major histocompatibility complex polymorphism in the wild. *Journal of Evolutionary Biology*, **16**: 224–232.

Whalen, K.G. and Parrish, D.L. (1999) Effect of maturation on parr growth and smolt recruitment of Atlantic salmon. *Canadian Journal of Fisheries and Aquatic Sciences*, **56**: 79–86.

Wright, S. (1932) The role of mutation, inbreeding, crossbreeding, and selection on evolution. *Proceedings of the Sixth International Congress of Genetics*, **1**: 356–366.

WWF (2001) *The Status of Wild Atlantic Salmon: A river by river assessment.* (Available at: http://www.wwf.org.uk/filelibrary/pdf/atlanticsalmon.pdf)

Youngson, A.F., Jordan, W.C. and Hay, D.W. (1994) Homing of adult Atlantic salmon (*Salmo salar* L.) to a tributary stream in a major river catchment. *Aquaculture*, **121**: 259–267.

Youngson, A.F., Jordan, W.C., Verspoor, E., McGinnity, P., Cross, T.F. and Ferguson, A. (2003) Management of salmonid fisheries in the British Isles: towards a practical approach based on population genetics. *Fisheries Research*, **62**: 193–209.

Part III
Management Issues

8 Population Size Reductions

S. Consuegra and E. E. Nielsen

Upper: Atlantic salmon in a river pool. (Photo credit: G. van Ryckevorsel.) Lower: the Economy River, one of the many small rivers of the Minas Basin in the Inner Bay of Fundy, Canada, whose populations are depleted and on the verge of extinction. (Photo credit: E. Verspoor.)

Atlantic salmon management and conservation is concerned with maintaining local species abundance and diversity. Where the abundance of the populations contributing to diversity declines, they become more vulnerable in the short term to extinction from other impacts. However, reductions may also have an impact on genetic diversity, with longer-term implications for a population's character, survival and reproductive success. Population reductions, among other things, are one of the main impacts of various human activities, such as fishing and stocking, and it is worth considering in detail, on its own, just what the genetic implications of population reductions may be.

8.1 Introduction

Many wild salmonid populations have been severely reduced in numbers during the last decades across their range (Nielsen *et al.* 1997; Consuegra *et al.* 2002; Koljonen *et al.* 2002), often as a result of human activities such as habitat destruction, pollution, overexploitation and/or introduction of exotic species. Such reductions in population size, on the basis of population genetics theory, are expected to have a direct impact on the genetic variation present: smaller populations have less variation, and are potentially more prone to extinction for demographic and genetic reasons than are larger ones (Taylor *et al.* 1994). Whether this is actually true remains to be fully explored and may be difficult to establish given the diverse and entangled web of genetic and demographic change occurring in most wild populations. However, what is known suggests it is likely to be important, though processes such as metapopulation structuring may help to counter such changes. This chapter focuses on the potential genetic consequences of population reductions, how such reductions could affect the survival of natural populations of Atlantic salmon, and what can be done from a management perspective to avoid or reduce losses of genetic variation and the evolutionary potential of the population.

8.2 Loss of genetic variation in small populations

Genetic theory predicts that small populations are more prone to extinction than large populations. This is because of the loss of genetic variation, which reduces their ability to respond to changes in the environment, and the cost of mating between close relatives, or inbreeding.

8.2.1 Importance of genetic diversity in natural populations

Within a species, genetic diversity is important to its capacity to exploit a variety of habitats throughout its distribution, protect it against environmental changes in the short term, and provide the mechanisms necessary for its long-term survival (Lande and Barrowclough 1987; Carvalho 1993). In pink salmon, *Oncorhynchus gorbuscha*, it has been directly linked to fisheries productivity (Geiger *et al.* 1997).

When the population size is reduced, genetic variation is lost at a higher rate mainly by random changes in allele frequencies. The reduction in the population size can be due to sustained depressions in numbers of individuals or to short-term extreme reductions in numbers (population bottlenecks) that cause changes in frequencies and loss of rare genetic variants (Daniels *et al.* 2000; Sherwin and Moritz 2000). This is likely to reduce the evolutionary potential of the populations as has been shown experimentally in *Drosophila melanogaster* (Frankham *et al.* 1999), where populations that had suffered a bottleneck showed reduced survival to adverse experimental conditions.

The importance of maintaining genetic diversity was underlined by a detailed collective analysis (i.e. meta-analysis) of 34 data sets from different species (Reed and Frankham 2003). The results of this study indicated that genetic variation and population size are correlated with fitness and that the future adaptability of small populations may be compromised by the negative effects of the two main consequences of the reduction in numbers of the populations: loss of genetic diversity (loss of mean heterozygosity, alleles and multilocus genotype combinations) and inbreeding depression.

The foundation of a new population from a few individuals can be considered as a special case of bottleneck and is referred to as a founder effect. This can occur, for example, in the case of colonisation of rivers from adjacent ones (as suggested by Vasemägi *et al.* 2001) establishing a new stable population or, in the case of the foundation of hatchery stocks, from a few individuals derived from one population (Elliott and Reilly 2003). In these cases the variability of the new population is reduced as a consequence of the small number of founders and is lower than the variability of the original population.

In small natural populations rare alleles will be easily lost. This is due to the distribution of the genetic variants (alleles) in most populations, where the majority of them occur at either very high or very low frequencies and many rare alleles are readily susceptible to being lost during a bottleneck, even if of short duration (Waples 1990a; Fig. 8.1). Although the genetic variation could be regained by mutation (or through immigration from adjacent populations), the number of new variants regained by mutation is so low in small populations that it has little importance on a short time scale for restoring genetic variation.

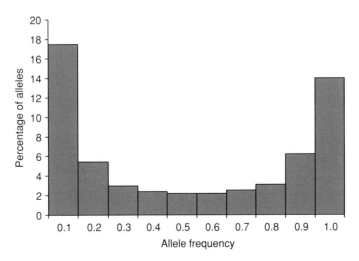

Fig. 8.1 The distribution of the alleles according to their frequencies based on the analysis of 177 chinook salmon populations; most of the alleles occur at very low or very high frequencies, meaning that many rare alleles are susceptible to be lost by genetic drift. Adapted from Waples (1990a).

The loss of variation is dependent on both the size and duration of the reduction in population size, meaning that less variation is lost in a population that has undergone a sudden reduction in size than in populations that have been chronically depressed over time. This is because in one generation usually only the rarest variants are lost, while in the case of several generations what is lost is most of the original variation that has been accumulated over years and is irreplaceable (Lande and Barrowclough 1987). Thus, some species may seem to have recovered even after severe and long-lasting bottlenecks; for example, northern elephant seals recovered after a severe decline from a small population of 20–30 individuals (Bonnell and Selander 1974); the Mauritius kestrel suffered six generations of severe bottleneck before a single breeding pair contributed to the recovery of the population (Groombridge *et al.* 2000). However, their genetic diversity continued to be reduced and the fitness of the populations was negatively affected by inbreeding, probably compromising their future evolutionary potential (Waples 1990b; Beaumont 2001), thereby increasing the probability of population extirpations.

8.2.2 Measuring loss of genetic variation in small populations: heterozygosity and allelic diversity

Heterozygosity and allelic diversity are common measures of genetic variation within a population. The heterozygosity of a population can be defined as the proportion of individuals carrying different alleles (genetic variants) at one or several loci, while the allelic diversity or richness is the number of different alleles in the population adjusted by sample size. Both measures are commonly estimated by neutral genetic markers. Allozymes were used for many years to make inferences about the genetic diversity of the populations (Allendorf and Phelps 1981) but recently DNA markers have been more widely used (mitochondrial DNA, minisatellites, see Chapter 4). Of these markers, DNA microsatellites are probably at present the most commonly used method to assess genetic variation in natural populations, not only in salmonids but also in many endangered species (Nielsen *et al.* 1997; Groombridge *et al.* 2000; Koljonen *et al.* 2002). Heterozygosity has traditionally been used as a measure of genetic variation relevant to the evolutionary potential (Franklin 1980), as it is correlated with fitness (Reed and Frankham 2003). In general, the correlation appears to be positive but in some studies, such as those on pink salmon *Oncorhynchus gorbuscha*, a parabolic relationship with fitness was found, with survival of juvenile salmon maximal at intermediate levels (Altukhov 2006).

Allelic richness or allelic diversity is a more sensitive indicator of population bottleneck and loss of genetic diversity than mean heterozygosity (Cornuet and Luikart 1997). This is because mean heterozygosity is related to the alleles present at higher frequencies (Ryman *et al.* 1995) while rare alleles are more prone to be lost by genetic drift (Waples 1990b). These rare alleles that do not contribute very much to the contemporary mean heterozygosity may prove valuable under future environmental conditions. After a bottleneck there is a temporary excess of heterozygosity due to the loss of rare alleles, which make a small contribution to the overall heterozygosity. Testing for excess of heterozygotes in populations is a way to identify populations that have suffered recent bottlenecks (Cornuet and Luikart 1997; Luikart *et al.* 1998). Ideally, loss of genetic variation could be tested using samples spanning the period of the supposed genetic bottleneck. This can be achieved by using historical collections of scales or otoliths as a source of DNA for genetic studies (Nielsen *et al.* 1997, 1999; Consuegra *et al.* 2002).

Heterozygosity is lost at a greater rate in smaller than in larger populations, and the shorter the generation length is the faster the loss of genetic variation tends to be. Genetic diversity, as measured by heterozygosity or allelic richness, may also serve as an indicator of the evolutionary potential of the populations. For example, endangered populations tend to show lower levels of heterozygosity at microsatellite loci, and fewer alleles, than related, non-endangered populations (Frankham et al. 2002).

However, the use of neutral genetic markers for assessing the loss of variation in relation to fitness has to be made with caution. Although heterozygosity can be a good indicator of population fitness and adaptability (Soulé and Wilcox 1980; Allendorf and Leary 1986) estimates are based only on a part of the entire genome (Reed and Frankham 2001), and if the loci examined are not directly or indirectly related with fitness traits, only the effects of genetic drift can be detected.

Few studies have documented a correlation between heterozygosity and fitness in fish. Vrijenhoek (1994) showed that a reduction in the abundance and heterozygosity of desert topminnows was accompanied by a reduction in fitness, as evidenced by an increase in the incidence of deformities and greater susceptibility to parasites (see Box 8.1). In salmonids a number of studies, some of them on Atlantic salmon, have compared allozyme heterozygosity and fitness-related traits with different results (Blanco et al. 1990, 1998; Crozier 1997). A negative correlation between fluctuating asymmetry (as a measure of development instability) and heterozygosity was found by Blanco et al. (1990) for juvenile Atlantic salmon, while no correlation was found by Crozier (1997). Positive correlations between heterozygosity and fitness traits have been found in some salmonid populations, although these can be affected by environmental and demographic factors and even by the markers employed (Wang et al. 2002a). In contrast, as mentioned, the work by Altukhov (2006) on pink salmon found a strong non-linear association, with low and high heterozygosity juveniles having higher mortality than those with intermediate levels.

The main reason for the low number of studies of correlation between genetic variation and fitness in the wild is due to general difficulties in measuring fitness of individuals in the wild (Endler 1986). This is particularly true for species with a complex life cycle, such as anadromous salmonids. Atlantic salmon spend different stages of their life in different environments (see Chapter 2) and the length of the life cycle varies among individuals, depending on the environmental conditions, which makes it especially difficult to estimate the lifetime reproductive success (fitness) of the individuals.

8.3 Effective population size

Central to the understanding of how genetic variation is lost within populations is the concept of genetically effective size of a population (N_e). Lage and Kornfield (2006) provide an example of historical reduction in genetic diversity parallel to a reduction in the effective population size of an endangered Maine Atlantic salmon population that had undergone a long-term decline in the number of spawners. As suggested by Lage and Kornfield (2006) historical reductions in genetic diversity and effective population size could negatively influence current restoration efforts. Informally, the concept of effective population size embodies a measure of the ability of a population to pass on genetic variation from one generation to the next. This ability is influenced by a large number of factors. N_e is formally defined as *the number of*

Box 8.1 Loss of heterozygosity and fitness in the Sonoran topminnow *Poeciliopsis monacha*.

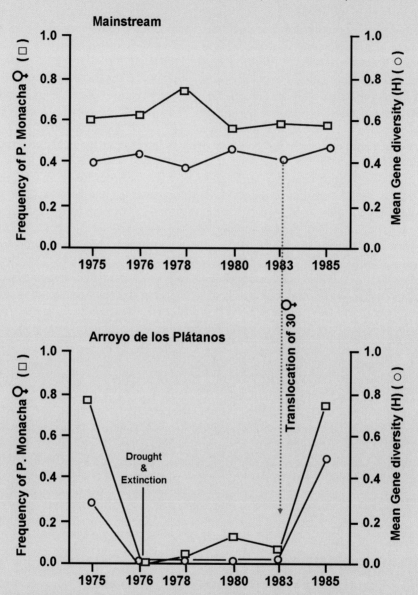

Fig. B8.1 Evolution of the genetic variability in the *P. monacha* females after the desiccation of the pool and the posterior translocation of 30 females from the mainstream. The blue line shows the percentage of *P. monacha* females in the pools in relation with the clonal fish and the red line shows the mean heterozygosity averaged across four polymorphic allozyme loci. The figure shows how, after the desiccations, the two forms (sexual and clonal) of topminnow are extinct and how after the recolonisation by a low number of individuals the *P. monacha* females have very low genetic diversity and abundance. The genetic variability and abundance were restored after transferring 30 variable females from the mainstream. Source: Vrijenhoek (1996).

Box 8.1 *(cont'd)*

The long-term study of the topminnow fish in Sonora, Mexico, is one of the best examples of the correlation between fitness and heterozygosity in fish and one of the clearest examples of inbreeding depression in the wild (Vrijenhoek 1994, 1996). The population of *Poeciliopsis monacha* in the Arroyo de los Plátanos has been followed by Vrijenhoek since 1975. In 1976 some pools of the river were desiccated during a severe drought. The pools were recolonised two years later by a small number of *P. monacha* gravid females and probably the same number of a clonal fish of the same genus with parthenogenetic reproduction. The new sexual colony showed extremely reduced heterozygosity compared to other permanent colonies in the downstream pools and with the clonal fish (Fig. B8.1). The *P. monacha* fish showed deformities and poor survival to hypoxic stress compared with both the clonal form and with conspecifics from the pools that did not suffer population reductions.

The loss of genetic variation also compromised the competitive ability of *P. monacha* with respect to the asexual form. *P. monacha* was more abundant than the clonal form before the drought and after the recolonisation the situation reversed. That this shift in abundance was related to genetic variability was suggested by two facts: (1) in the pools downstream that did not suffer desiccation the abundances of the two forms did not change; (2) with the introduction of 30 variable females from one of the downstream pools the population regained variability as well as ecological dominance, going back to the original abundance (80% of the fish in the pool). The increase in genetic variation affected the parasite load of the species as well; after the restoration of the variation, *P. monacha* showed a lower loading of a common parasite (the trematode *Uvulifer* sp.) than before and lower than the asexual form (now less variable).

In summary, the loss of genetic variation after a founder effect reduced the fitness of the *P. monacha* population which showed classical signs of inbreeding depression. The recovery of genetic variation through the translocation of females from another more variable population resulted in the recovery of fitness and competitive ability of the population.

individuals that would give rise to the same amount of inbreeding or genetic drift if they behaved in the manner of an idealised population (Frankham *et al.* 2002). In an idealised population there is no genetic migration, selection or mutation. There are also non-overlapping generations, each with the same number of individuals and where all individuals would be potential breeders and hermaphrodites. Furthermore, the number of offspring per adult averages 1 with variance of 1. Deviations from these conditions can variously affect the success with which variation is passed on to future generations.

Idealised populations do not occur in nature but N_e can usefully be viewed as the common genetic currency to which all other populations can be related and subsequently standardised. So if two populations have the same estimated genetically effective population size we could infer that genetic drift and inbreeding takes place at the same rates.

8.3.1 Minimum effective population size

One of the central questions in conservation of small populations is determining the minimum number of individuals needed to maintain existing levels of genetic diversity. In the conservation genetic literature, the '50-500 rule' (Franklin 1980) is often mentioned as a general guide for how large genetically effective population sizes should be to ensure the genetic health of a population.

The '50' refers to the minimum short-term effective population size to avoid effects of inbreeding. An N_e of 50 corresponds to an increase in the inbreeding coefficient of 1% per generation. It is assumed that in the short term natural selection can counteract the harmful effects of inbreeding by only 1% per generation. Therefore, over a short period of perhaps

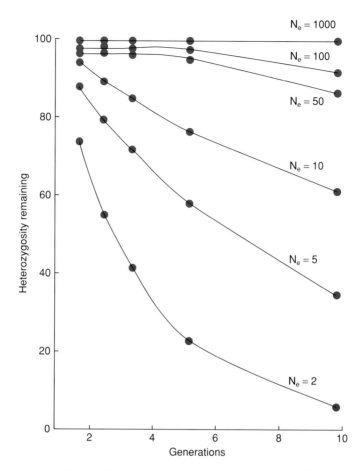

Fig. 8.2 Loss of genetic variability after 10 generations in simulated populations with different effective population size (N_e). Populations with large N_e maintain the main part of their genetic diversity after 10 generations while small populations lose a larger proportion (e.g. up to 65% in a population with N_e = 5). Source: Meffe and Carroll (1997).

5–10 generations, it is not expected at this 1% level that the effects of inbreeding (inbreeding depression) should reduce a population's viability. However, over longer time periods inbreeding will continue, resulting in increased inbreeding depression. Based on long-term experiments with houseflies, the number of generations to population extinction, due to genetic effects, has been estimated to be approximately the same as the effective population size. So, in the case of an N_e of 50 the average time to extinction was 54 generations (Reed and Bryant 2000). Such estimates cannot be directly transferred to other species such as Atlantic salmon with different evolutionary histories. However, they illustrate that populations managed at an effective size of 50 individuals will face problems in the not too distant future if nothing is done to increase the N_e (Fig. 8.2).

The '500' refers to the minimum effective size required to avoid loss of genetic variation, i.e. the evolutionary potential. It was originally derived by Franklin (1980) based on the equilibrium between genetic drift and mutation for additive genetic variation and represents the effective population size at which new genetic variation brought by mutation is expected to replenish the variation lost by random genetic drift. Since then there has been a lot of debate about this number in the conservation genetics literature (Lande 1995). In brief, 500 must be considered as an absolute minimum that might be too small under some conditions. For Atlantic salmon with an average generation time of around 3–5 years (depending on geographical region), this corresponds to an effective number of breeders (N_b) of between 100 and 167 per year. These numbers are similar to the number of breeders per year estimated by Waples (1990b) for Pacific salmon to reduce the loss of rare alleles. However, the '50-500 rule' must be taken with caution in fish due to the large effect of variance in progeny size.

8.3.2 Relationship between census and effective population sizes

How does the effective population size relate to the actual number of individuals present in the population, i.e. the census size (N_c)? In general the effective number of individuals in a population is much lower than the census number. An overarching, meta-analysis of more than 45 estimates of the N_e/N_c ratio for species from various taxa, by Frankham (1995), estimated an average ratio of 0.11. Thus if a population consists of 100 individuals, its effective size can be as low as 10.

Estimates of N_e/N_c for salmonids are rare due to the inherent difficulties in estimating effective population size and census size in natural populations. However, some estimates of N_e and N_e/N_c ratios from both wild and captive populations are available (Bartley *et al.* 1992; Hedrick *et al.* 1995; Hansen *et al.* 2002; Koljonen *et al.* 2002; Ardren and Kapuscinski 2003; Säisä *et al.* 2003; Shrimpton and Heath 2003). N_e/N_c ratios vary widely, i.e. from > 0.8 in captive Atlantic salmon populations (Koljonen *et al.* 2002) to 0.1 in wild Pacific salmon (Bartley *et al.* 1992) depending on the species, on the population type (wild or captive) and on the method used to calculate N_e. Assuming a ratio for N_e/N_c of 0.1, in order to ensure an effective number of breeders of between 100 and 167, approximately 1000–2000 individuals available at spawning time may be needed each year.

8.3.3 Factors influencing genetically effective population size in Atlantic salmon

As we have seen, the census size is often much larger than the effective population size and salmonid populations are likely to be far from the idealised population. A number of characteristics of the biology of salmonids in general, and for Atlantic salmon in particular, are likely to account for the large difference.

One obvious contributing factor is how the population size is estimated, which often varies. Estimates may be based only on the number of returning spawners or include the mature male parr that can contribute to reproduction. Depending on the method employed the discrepancy between census and effective size will change. For Atlantic salmon the most relevant number is the number of spawners, including mature parr, that is, the number of all potentially breeding individuals.

Probably the most important factor which lowers N_e in natural populations of Atlantic salmon is *fluctuations in population size* over time. The effective size of a population over time is generally determined by the years with the smallest numbers and it is well known that population size varies from generation to generation mainly in response to environmental instability. The long-term N_e of a population is the harmonic mean over generations (Box 8.2)

> **Box 8.2** Mathematical equations quantifying effective population size N_e.
>
> The long-term N_e of a population over a given time period is the harmonic mean of the individual values for the intervening generations,
>
> $$N_e = t / \sum (1/N_{ei}) \qquad (8.1)$$
>
> rather than the simple arithmetic average value, where t is the time in generations and N_{ei} is the effective size in the i'th generation.
>
> The effect of variance in family size on N_e is generally estimated as follows:
>
> $$N_e = (4N - 2)/(V_k + 2) \qquad (8.2)$$
>
> where V_k is the variance in family size and N is the number of individuals born in each generation.
>
> The effect of sex ratio on the effective population size can be estimated by:
>
> $$N_e = 4N_f N_m / (N_f + N_m) \qquad (8.3)$$
>
> where N_f is the number of females in the population, N_m the number of males.

rather than the simple average value. This means that genetic variation lost in one generation due to low effective size cannot be regained quickly, even if the population rebounds in numbers in the next generation.

Another very important contributor to the low N_e/N_c ratio is the large *variance in family size* commonly observed in Atlantic salmon and most other salmonid fishes. This is caused by the high fecundity of both sexes, a high degree of sexual selection and potential for high survival difference of eggs and fry caused either by environmental stochasticity or natural selection. An example of the latter, associated with differential marine survival among families of pink salmon, can be found in Geiger *et al.* (1997).

In an idealised population each family replaces itself every generation so the average family size is 2 with a variance of 2 (Poisson distribution). From the equation in Box 8.2, it can be seen that if the variance increases above 2, then this leads to a lowering of the effective population size. Alternatively, if family sizes are equalised so that the variance becomes close to zero, N_e can be doubled. This latter feature offers a means to increase effective population sizes in supportive breeding programmes by appropriate mating designs.

Sex ratios also have a major impact on N_e. The further ratios deviate from 1 : 1 the more N_e will be reduced, the extent to which is determined by the formulation given in Box 8.2. Skewed sex ratios can be expected to be important as, in many species populations, the number of breeding males and females are not equal. Some of the best-known examples are provided by animals that have harems, such as red deer or pheasants. Such *unequal sex ratios* can originate from stochastic events caused by low population size, but also by the biology of the species, and so the life history, the mating system or the sex determining system can influence the sex ratio.

For a species such as Atlantic salmon with many small populations, stochastic events can be important in creating unequal sex ratios; the special life history of Atlantic salmon is also important. For instance, the proportion of males and females among grilse and multi-sea-winter fish has been shown to be different, with a higher proportion of males among grilse. More important, perhaps, is the occurrence of mature male parr. For some populations a very high proportion of the male parr mature and participate in spawning (Taggart *et al.* 2001), leading to a highly skewed sex ratio. The contribution of mature male parr to increasing the effective size can be very important, in particular when N_e is low (Martínez *et al.* 2000), but it will not be proportional to their often large numbers.

8.3.4 Calculating effective population size

As mentioned earlier, precise estimation of N_e is very difficult for natural populations. In principle there are two types of method: demographic and genetic methods.

(1) Demographic methods
These are based on the equations in Box 8.2 (Nunney and Elam 1994). These methods require good quality and detailed data, which can be difficult to collect for natural populations of Atlantic salmon. For instance, it is practically impossible to determine the average and variance for family size under natural conditions. However, as rough guidelines, demographic estimates are highly relevant, in particular to the potential consequences of various management initiatives.

(2) Genetic methods
These are based on the theoretical expected relationship between a number of genetic parameters and N_e (reviewed in Caballero 1994 and Beaumont 2001). Conceptually there are two different types of method. 'Two-sample methods' are where the genetic parameters of interest such as changes in allele frequencies, loss of alleles, loss of heterozygosity, rate of decay of linkage disequilibrium and increase in inbreeding coefficients are measured at intervals and used to infer N_e. At present the most widely applied method is the so-called 'temporal method' (Waples 1990a,b; Beaumont 2001; Wang 2001; Wang and Whitlock 2003). This method estimates the genetic variance induced by genetic drift between temporal samples, taking the average generation time into account, and uses this to infer N_e. The precision of the method critically depends on the number of generations between samples and the number of samples collected, i.e. the best results can be obtained with several temporally spaced samples covering a relatively long time interval.

The combination of these methods with DNA analysis from archived scales, bones and otoliths collected over longer time spans give powerful tools for gaining insight about N_e in salmonid fishes (see Hansen *et al.* 2002 for an example from brown trout *Salmo trutta*). However, often the period of sampling is restricted to a single year, leaving no possibility of gaining a good estimate using the temporal method. Nevertheless, estimates of N_e can be obtained using 'one-sample methods' (Beaumont 2001). The principles behind these methods are based on detecting deviations from Hardy–Weinberg, and interlocus genotype linkage, equilibria within a population sample. In other words, as the effective population size decreases, increased deviations are expected.

Genetic methods have now become faster and more easily applicable than demographic methods, but the outcomes should also be treated with caution. First, the described methods are very sensitive to sampling variance. Additionally, phenomena in anadromous salmon

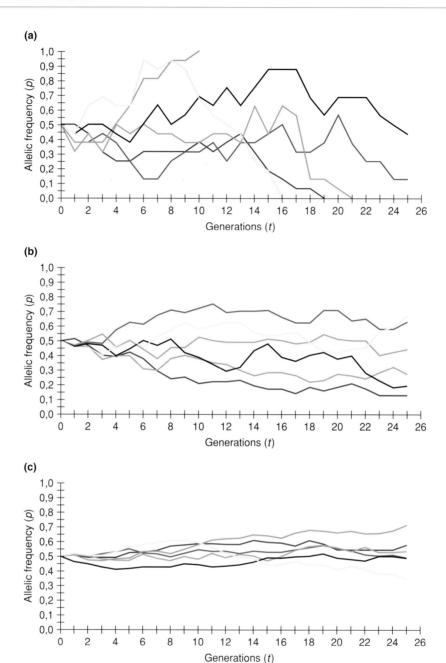

Fig. 8.3 The effect of the genetic drift in the allelic frequencies at a single locus with two alleles in six simulated populations with sizes $n = 10$ (a), $n = 50$ (b) and $n = 250$ (c) after 25 generations. In the model the effect of random genetic drift in two alleles with initial frequency of 0.5 is analysed with no migration, mutation or selective advantage of any of the alleles. In the smallest population, each allele has the same probability of being fixed after 25 generations and, in the $n = 50$ population, both alleles have a similar probability of increasing in frequency after the same number of generations. However, in larger populations ($n = 250$), genetic drift has a very small effect in changing the allelic frequencies under the conditions considered. Simulations from Populus 5.3 3.1 by Don Alstad (http://ecology.umn.edu/populus).

populations, such as genetic migration, can bias the estimates of N_e and non-random sampling of individuals with respect to life stage, location (subpopulation structure within drainages) and time (different 'runs') are also potential sources of error, which will lead to an underestimation of the effective population size.

8.4 The effects of genetic drift and selection in small populations

Genetic drift refers to the random sampling of gametes from generation to generation, that can cause random changes in allelic frequencies (see Chapter 4). The magnitude of genetic drift depends inversely on the effective population size. Thus, in large populations, genetic drift causes only small changes in allelic frequencies between generations while in small populations the allele frequencies fluctuate more between generations, resulting in unpredictable fixation and loss of different alleles (Fig. 8.3). The expected proportion of selectively neutral genetic variation lost from a population per generation is related to the effective population size by $1/(2N_e)$. For this reason, the probability of losing rare alleles due to genetic drift will tend to be higher in small populations than in large ones.

Genetic drift plays an important role in the genetic differentiation of small populations, even in cases where natural selection pressures tend to favour the development of local adaptations. This is because when the populations are small the effects of genetic drift on weakly selected genes are expected to be similar to neutral genes. This is particularly relevant in salmonids whose homing behaviour tends to produce locally adapted populations.

One of the consequences of the weaker effect of selection in small populations is that deleterious alleles are less likely to be removed by natural selection and are more prone to fixation, thereby lowering the reproductive fitness of the individuals within the populations. However, the higher frequency of homozygotes can also increase the selection against some recessive deleterious alleles with strong effect, reducing its frequency (so-called 'purging').

In natural populations the effects of genetic drift may be counterbalanced by migration and mutation, both of which tend to increase genetic variation. However, while mutation rates in nature are probably too small to have a significant effect to counteract the effects of genetic drift in small populations, gene flow between neighbouring populations may prevent the loss of intrapopulation genetic variation. This may account for the fact that in some cases small populations seem to maintain relatively high genetic diversity despite having very low effective population size (Nielsen 1999; Säisä *et al.* 2003; Consuegra *et al.* 2005; see Box 8.3). Nevertheless, in many cases Atlantic salmon populations that have experienced severe bottlenecks tend to show a net loss of alleles, probably due to random drift (Nielsen *et al.* 1997).

The tetraploid nature of the genome could provide some buffering against the effects of genetic drift (Allendorf and Waples 1996), even though in Atlantic salmon approximately 50% of the duplicated genes have lost their expression (Allendorf and Thorgaard 1984). A phenomenon called 'founder-flush' (Templeton 1980) is also seen as a possible reason some small salmonid populations maintain high levels of genetic variation (Nielsen 1999). Controversial 'founder-flush' models (Templeton 1980; Slatkin 1996; Charlesworth 1997) propose that a population suffering a severe bottleneck, followed by a rapid population growth, has a greater opportunity to respond to selection and potential for reproductive isolation than a population of constant size (Slatkin 1996). The newly founded population would contain a small sample of the alleles from the ancestral population from which genetic drift

Box 8.3 Effective population size and genetic diversity in small Iberian Atlantic salmon populations.

Consuegra *et al.* (2005) compared microsatellite variation in four small Atlantic salmon populations from the Iberian Peninsula (Fig. B8.3.1) and three larger populations from Scotland. They also examined the evolution of genetic diversity over a 50-year period in one Iberian population that suffered a severe bottleneck (River Asón). The relation between census size, effective population size and genetic diversity was estimated in order to test (a) whether the endangered Iberian salmon populations displayed reduced genetic diversity in relation to the other larger European populations, and (b) whether a historical population bottleneck in one of the populations was accompanied by a corresponding decrease in genetic diversity.

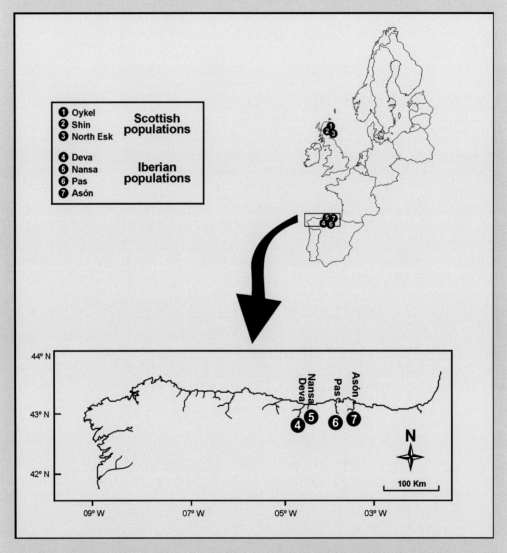

Fig. B8.3.1 Location of the four small, peripheral Atlantic salmon populations from the Iberian Peninsula, and the three relatively large populations from Scotland, analysed for microsatellite diversity. Source: Consuegra *et al.* (2005).

Box 8.3 *(cont'd)*

Despite low effective population sizes of the four Iberian salmon populations (with N_e estimates ranging from 12 to 175 individuals using temporal methods; Table B8.3) they seemed to maintain relatively high levels of genetic diversity (as measured by allelic richness and heterozygosity), comparable to those found in three larger, more stable Scottish populations (Fig. B8.3.2). N_e was in all cases below the minimum of 500 estimated by Franklin (1980) to be necessary for avoiding the loss of genetic variation, and well under the 5000 suggested by Lande (1995). Historical estimates of N_e in the River Asón seemed to track well the observed demographic decline in census size (Table B8.3).

However, despite the evidence of genetic bottlenecks in all four Iberian populations, little or no relationship between estimates of effective population size (or census size) and measures of genetic diversity was found. Furthermore, the temporal analysis with archival scales from one of the rivers (R. Asón) failed to detect any

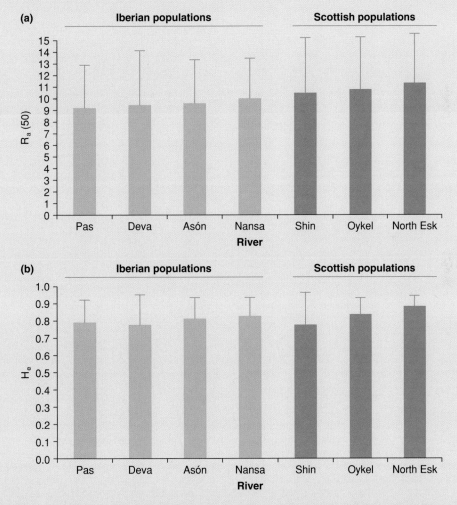

Fig. B8.3.2 Estimates of (a) allelic richness approximated to a minimum sample size of 50 genes (Ra50; SD indicated) and (b) heterozygosity (H_e; SE indicated) in four small populations from the Iberian Peninsula (Asón, Pas, Nansa and Deva) and three larger populations from Scotland (Shin, Oykel, North Esk). Source: Consuegra *et al.* (2005).

Box 8.3 (cont'd)

Table B8.3 Demographic and genetic estimates of census size (N_c; average no. of breeders, per generation) and effective population size per generation (95% CI in parentheses), based on Wright (1938) expression for unequal sex ratio of spawners (N_{eH}, harmonic mean), and the temporal method as implemented by Waples (1990) ($N_{e(1)}$) and by Wang and Whitlock (2003) ($N_{e(2)}$, $N_{e(3)}$). For the historical analysis, the River Asón has been considered only as an isolated population, while for the analysis of present populations migration has also been taken into account. Source: Consuegra et al. (2005).

River	Period	Demographic estimates			Closed populations without immigration		Closed populations without immigration		Open populations with immigration	
		N_c	N_{eH}	N_{eH}/N_c	$N_{e(1)}$	$N_{e(1)}/N_c$	$N_{e(2)}$	$N_{e(2)}/N_c$	$N_{e(3)}$	$N_{e(3)}/N_c$
Asón	1960–86	9686	445	0.04	109 (60–168)	0.01	90 (59–149)	0.009	–	–
Asón	1986–96	2375	386	0.16	95 (46–157)	0.04	55 (34–109)	0.02	–	–
Asón	1996–2000	477	121	0.25	42 (21–67)	0.09	38 (27–87)	0.08	12 (11–16)	0.02
Pas	1996–2000	961	217	0.22	175 (67–284)	0.18	98 (81–645)	0.10	29 (23–37)	0.03
Nansa	1996–2000	1541	499	0.32	68 (33–110)	0.04	74 (38–313)	0.05	31 (24–42)	0.02
Deva	1996–2000	655	189	0.29	113 (49–197)	0.17	84 (42–506)	0.13	28 (22–37)	0.04

historical loss of allelic richness or heterozygosity, despite the apparent intensity and duration of the population crash in this river (Table B8.3).

Such apparent lack of correspondence between population size and genetic diversity is a common phenomenon in salmonids (Heath et al. 2002; Østergaard et al. 2003), where levels of genetic variation may only be weakly related to population size or to fitness (Wang et al. 2002a), possibly due to genetic compensation (e.g. Ardren and Kapuscinski 2003). While Palm et al. (2003) found a positive relation between effective population size and genetic diversity in brown trout, Jorde and Ryman (1996) reported the same level of heterozygosity across a wide range of N_e values and Säisa et al. (2003) found high heterozygosity in an Atlantic salmon population with N_e below 100 for several generations, even with N_e as low as 13 fish.

Several reasons were considered for the maintenance of relatively high levels of genetic diversity in Iberian salmon populations. First, Iberian populations pre-date the last glacial maximum (Consuegra et al. 2002) and present levels of genetic diversity in Iberian populations may reflect the eroding effect of bottlenecks acting on formerly higher levels of genetic variation, although no such erosion was detected within the time frame of the study. Second, the complex life cycle of Atlantic salmon with overlapping generations and multiple paternities might help to minimise or delay the loss of genetic diversity in the face of demographic catastrophes (Waples 1991). Third, the contribution of mature male parr to reproduction may help to increase N_e and genetic diversity (Jones and Hutchings 2001), especially in southern rivers, where maturation rates are high. However, N_e appeared to be below 100 individuals even considering the presence of mature male parr. Finally, asymmetric dispersal and gene flow may have facilitated the maintenance of genetic diversity in these small, peripheral populations, especially during periods of low abundance, as suggested by physical tagging and microsatellite data. Results indicated that the exchange of migrants between Iberian populations was high and not symmetrical (Fig. B8.3.3), but appeared instead to conform to a 'source-sink' metapopulation scenario (e.g. Hanski 1999; Rieman and Dunham 2000).

In summary, the results of the study indicated that the effective population size of the four Iberian salmon populations analysed was well under the size suggested necessary to maintain genetic diversity and long-term population viability. Yet, the four Iberian populations seemed to maintain relatively high levels of genetic

Box 8.3 (cont'd)

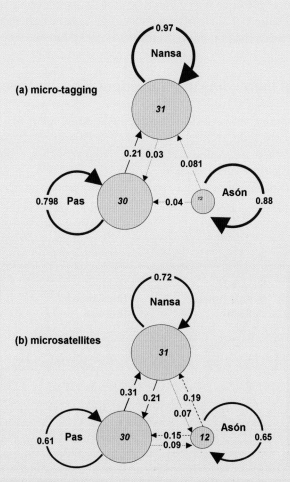

Fig. B8.3.3 Patterns of dispersal and asymmetric gene flow between the three small Iberian rivers based on (a) recaptures of micro-tagged adults stocked as juveniles in each river, and (b) microsatellite variation and Bayesian assignment tests. Size of circles is proportional to N_e estimated by the temporal method (Wang and Whitlock 2003), and size of arrows is proportional to the inferred proportion of fish migrating between rivers. Continuous lines denote strong directional flows (> 0.20) while dotted lines denote weaker relationships (< 0.20).

diversity. Furthermore, in the R. Asón, historical loss of genetic diversity could not be detected despite a dramatic decline in abundance. Although several mechanisms may have helped to minimise or delay the loss of genetic variation, the results suggested that asymmetric gene flow resulting from source-sink metapopulation dynamics (Hanski 1999; Fraser et al. 2004) had probably been the dominant evolutionary strategy for maintaining genetic diversity in Iberian salmon populations living in marginal, peripheral habitats.

and recombination can originate new co-adapted gene complexes. Rapid population growth will make selection more effective than it could be in the constant population size by acting on those alleles or combination of alleles that were in low frequency after the bottleneck but can be in multiple copies when the population stabilises in size (Slatkin 1996).

8.5 The effects of inbreeding in small populations: inbreeding depression

Inbreeding refers to mating between related individuals. Although in many species individuals avoid mating with relatives, the average relatedness among individuals increases as the population size decreases, thereby increasing inbreeding and the probability that one individual gets the same copy of a gene from each one of its parents (Fig. 8.4).

The main consequence of inbreeding in wild populations is an increased risk of extinction due to inbreeding depression (Saccheri *et al.* 1998; see Keller and Waller 2002 for an extensive

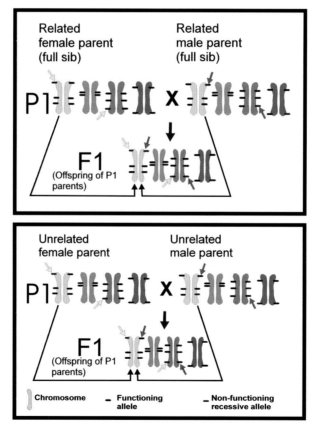

Fig. 8.4 The offspring from matings between relatives have a higher probability of receiving the same copy of the gene from each one of the parents compared with offspring from outbreeding matings as shown in the figure. The increase in the proportion of homozygotes, and the exposure of recessive deleterious alleles, can reduce the fitness of the populations, although in some cases it can serve to purge harmful alleles.

review on inbreeding in wild populations). Inbreeding depression is the decline in the value of a trait related to the fitness of the individuals as a direct consequence of inbreeding (Fig. 8.4). The reduction of fitness after close inbreeding can result from an increase in the frequency of recessive deleterious alleles or from an increase in homozygosity in cases where heterozygotes are superior. Although an accelerated rate of inbreeding can potentially drive the population towards extinction, it has also been suggested that high levels of inbreeding can purge the genome of deleterious recessive alleles, giving a higher resistance to inbreeding depression in future generations (Hedrick 1994).

In summary, when populations suffer a reduction in size the expected consequence is increased inbreeding (see Box 8.4) and subsequent inbreeding depression (Ralls *et al.* 1988). However, if population size is recovered rapidly, it can result in a fitness rebound as a consequence of the loss of deleterious alleles during inbreeding, but usually the recovery of reproductive fitness is not to the same level (Frankham *et al.* 2002). To avoid compromising the future survival of the population, and to maintain enough evolutionary potential, the effective population size has to be large enough not only to counteract the effects of inbreeding, but also genetic drift (Caballero 1994; Franklin and Frankham 1998).

In spite of its relevance to conservation, there are few studies of inbreeding depression in wild salmonids (reviewed by Wang *et al.* 2002b), and very few studies on Atlantic salmon. Most studies refer to other commercial species such as rainbow trout *Oncorhynchus mykiss* in hatchery/farmed populations (Su *et al.* 1996; Rye and Mao 1998; Pante *et al.* 2001). There are only two studies on Atlantic salmon: the study of Rye and Mao (1998) looking at the importance of non-additive effects and inbreeding depression on body weight, and the work of Ryman (1970), which indicated a reduction in return rates among inbred compared to non-inbred lines. Despite such scarcity of studies, the available data do suggest that inbreeding depression can be influential in both captive and natural populations of salmonids.

Most studies of inbreeding depression in salmonids have used experimental designs to detect inbreeding in the short term. However, the most probable scenario in the wild is one of slow rates of inbreeding that would not be detected by the methods employed to date. Although empirical studies suggest that lower rates of inbreeding, like those that can occur in nature, produce lower inbreeding depression (Su *et al.* 1996; Rye and Mao 1998; Pante *et al.* 2001), it has been suggested that natural conditions increase the cost of inbreeding compared to captivity (Frankham and Ralls 1998; Crnokrak and Roff 1999).

Estimating the effects of inbreeding depression in fitness-related traits can be difficult in species with a complex life cycle such as Atlantic salmon. Basing management decisions solely on measures of only a few fitness components may be inappropriate for endangered salmonid populations (Wang *et al.* 2002a,b). The differential advantage of genotypes with the highest fitness in one stage of the life cycle (e.g. the survival from alevin to smolt) might not be advantageous in another life stage (see discussion of differences in performance of wild and farmed fish at various life-cycle stages in Chapter 12). Thus doing so could lead to inappropriate management decisions.

While, as suggested before, the tetraploid nature and overlapping generations of salmonids might in theory protect them from loss of genetic variation to a certain degree (Allendorf and Thorgaard 1984; Waples 1991), estimates of inbreeding depression in salmonids seem to be similar to those for other species (Wang *et al.* 2002b; see Box 8.5 for estimating inbreeding coefficient) and the evidence suggests that the growth and survival of salmonids can be affected by inbreeding both as juveniles (Wang *et al.* 2002b) and adults (Ryman 1970).

Box 8.4 Inbreeding accumulation over generations.

Inbreeding accumulates in populations. It does so at a rate that depends on the population size, increasing more quickly in small populations. Assume we start with N individuals from an idealised population who are initially totally unrelated (i.e. $F_0 = 0$) and that produce N offspring for the next generation (generation t). We can define F (the inbreeding coefficient) as the average probability of one individual getting two alleles that are identical by descent.

From the original population a total of $2N$ gametes are drawn at random. Given that the initial generation was not inbred, the only way to get two identical alleles this generation is to have both gametes coming from the same parental individual (Fig. B8.4). Once one gamete has been sampled, the probability of obtaining a second gamete from the same parent is $1/N$. However, due to Mendelian segregation, only $1/2$ of the sampled gametes from the same parent will be identical by descent. Thus the total probability of the second gamete matching the first is simply $1/2N$ and the chance of drawing two alleles that are identical by descent (IBD) is $F_t = 1/2N$; the probability that they are not identical is $1 - 1/2N$. If we consider now the next generation ($t + 1$) the probability that two gametes will be identical by random sampling is again $1/2N$.

Even if gametes are not copies of the same allele, due to random sampling in the current generation, they can still be identical due to inbreeding in previous generations. Therefore the probability of one individual getting two identical alleles in generation t is the sum of (1) the probability that the alleles are identical in the current generation ($1/2N$) and (2) the probability that they are not identical in this generation ($1 - 1/2N$) but from a previous generation (multiplied by the inbreeding coefficient in the previous generation F_{t-1}):

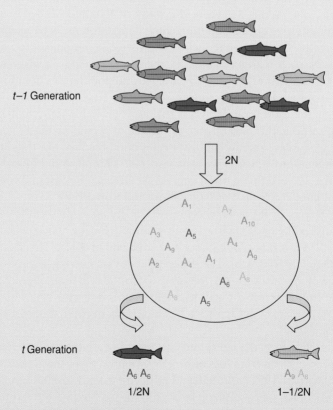

Fig. B8.4 Inbreeding accumulation over generations.

Box 8.4 (cont'd)

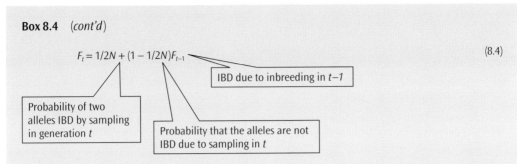

(8.4)

If the initial population is not inbred ($F_0 = 0$) the inbreeding coefficient in the next generations can be estimated as follows:

$$F_t = 1 - (1 - 1/(2N))^t \qquad (8.5)$$

The accumulation of inbreeding depends on the population size (N) and that is why it is accumulated more rapidly in smaller populations.

To illustrate this point we can imagine a hatchery population founded by 10 females and 10 males, all of them unrelated, and we can estimate what would happen if we use a broodstock of 20 individuals during several generations.

Generation 0 $F = 0$
Generation 1 $F = 1 - (1 - 1/20)^1 = 0.050$
Generation 2 $F = 1 - (1 - 1/20)^2 = 0.098$
Generation 3 $F = 1 - (1 - 1/20)^3 = 0.143$
.... to
Generation 10 $F = 1 - (1 - 1/20)^{10} = 0.401$

In 10 generations the inbreeding coefficient would increase to 40%. As effective population size (N_e) is usually lower than the census population size (N_c), this means that in small natural populations, without gene flow, the rate of increase of inbreeding will be even more rapid as it will depend on N_e rather than on N_c.

8.6 Population reductions, gene flow and local adaptation

Gene flow is an important factor to counteract the effects of inbreeding and also to reduce the genetic divergence among populations; since populations interconnected by even small rates of gene flow are less likely to go extinct than isolated populations of similar size.

Natural straying in salmonids provides a possible source for augmenting genetic variation. However, to increase the level of genetic variation effectively, strayers have to contribute to the next generation. Several studies, however, indicate that not all strayers reproduce successfully (Tallman and Healey 1994), possibly due to maladaptation (Quinn *et al.* 2000).

The homing behaviour of salmonids and the different timing of migration and spawning between populations can promote population subdivision (Quinn *et al.* 2000) and local adaptations (Taylor 1991; see Chapter 7) but also inbreeding (Wang *et al.* 2002b). However small

Box 8.5 Calculating the inbreeding coefficient.

The inbreeding coefficient (F) is the probability that two alleles in the same individual are identical by descent (IBD), i.e. that they are copies of the same allele coming from a common ancestor.

Estimates of individual F from pedigrees

Pedigrees can be used to estimate the inbreeding coefficient (F) of an individual as it is shown in Figs B8.5a and B8.5b. In this case the parents of the individual X are half-sibs (in red) and their parents are half-sibs as well and

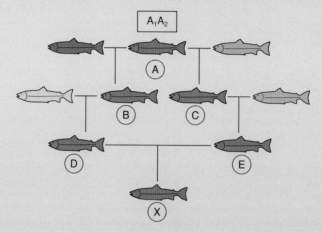

Fig. B8.5a Pedigree showing two matings between half-sibs; red represents individuals related by a common ancestor. The inbreeding coefficient of the individual X (F_X) is the probability that the two alleles carried are identical by descent.

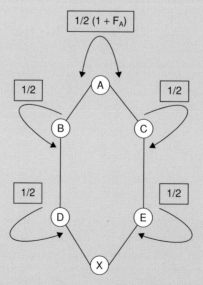

Fig. B8.5b Diagram showing the probabilities of X receiving one allele (A_1 or A_2) from A, including the probability that the two alleles of the common ancestor (A) are identical by descent from previous inbreeding (F_A, inbreeding coefficient of A).

Box 8.5 *(cont'd)*

the genotype of the common ancestor A (for one gene with two alleles) is A_1A_2. The inbreeding coefficient of individual X (F_X) is then the probability that the two alleles carried by X are copies identical by descent (both come from A), this is that they are A_1A_1 or A_2A_2.

The probability that A_1 is transmitted from A (the common ancestor) to B is 1/2 (following a normal Mendelian segregation, see Chapter 4) and it is the same for the transmission of A_1 from A to C. The probability that A_1 is now transmitted from the parental B to D is again 1/2 and the same from C to E. Finally, the probability that X receives the allele A_1 from D is 1/2 and from E is 1/2 as well (Fig. B8.5b).

Thus the probability that X is A_1A_1 is the product of the probabilities of receiving the alleles from all the paths:

$$P(X = A_1A_1) = (1/2)^6 \tag{8.6}$$

The probability of X being A_2A_2 is the same and then the probability that the two alleles carried by X are identical by descent is the sum of the probabilities of X being A_1A_1 or A_2A_2:

$$P(X = A_1A_1 \text{ or } A_2A_2) = (1/2)^6 + (1/2)^6 = (1/2)^5 \tag{8.7}$$

However, this would be the inbreeding in X if there was no previous inbreeding accumulated in A. To estimate the actual inbreeding in X we have to take into account the possibility that the alleles carried by A are identical by descent as a product of previous inbreeding, so we have to take into account the possibility that A's parents were related as well. If this was the case, the probability that A_1 and A_2 are identical is the inbreeding coefficient of A (F_A). The probability that X was A_1A_2 the two alleles being identical is then the sum of the probabilities of X receiving A_1 from the father and A_2 from the mother and vice versa and the probability that the two alleles of X are identical from previous inbreeding is the product of the probability that A's alleles were identical and that they are transmitted to X:

$$P(X = A_1A_2 \text{ identical from previous inbreeding}) = ((1/2)^6 + (1/2)^6) \, F_A = (1/2)^5 \, F_A \tag{8.8}$$

> Probability that X receives A_1 from the father A_2 from the mother or A_1 from the mother A_2 from the father

> Probability that A_1 and A_2 alleles were identical

And finally, the probability of X carrying two alleles identical by descent from previous or actual inbreeding, the inbreeding coefficient of X, is the sum of both:

$$F_X = (1/2)^5 + F_A(1/2)^5 = (1/2)^5 (1 + F_A) \tag{8.9}$$

> Probability that X receives two copies of the same allele from the common ancestor (IBD)

> Probability that X receives two copies of the same allele IBD due to previous inbreeding

In general, the inbreeding coefficient of an individual can be calculated from the number of individuals (N) in each path that connect with the common ancestor (A) and the inbreeding coefficient of this common ancestor (F_A):

$$F_X = \sum (1/2)^N (1 + F_A) \tag{8.10}$$

Box 8.5 *(cont'd)*

Indirect estimates of F in populations

Although inbreeding by itself does not cause a change in allele frequencies, it causes a reorganisation in the genotype, and the result is a change in the genotype frequencies. The result of inbreeding is in general a deficiency of heterozygotes compared to Hardy–Weinberg equilibrium (see Chapter 4) that can be used to estimate the inbreeding coefficient of the population.

The changes in the genotype frequencies caused by inbreeding (as shown in Table B8.5) can be used in the calculations of the inbreeding coefficient in the population. By calculating the ratio between the proportion of heterozygotes in an inbred population (H_I) in relation with a non-inbred (H_o) population we can estimate the inbreeding coefficient of the population (F):

$$H_I/H_o = 2pq(1-F)/2pq = 1 - F \tag{8.11}$$

An estimate of the average inbreeding coefficient of a population can be calculated as well from the loss of genetic diversity over time as related with the effective population size, smaller populations have usually more probability of inbreeding as shown in the following relation:

$$H_t/H_o = (1 - 1/2N_e) = 1 - F \tag{8.12}$$

Table B8.5 Comparison between the Hardy–Weinberg equilibrium (Chapter 4) genotype frequencies of populations with random mating and with inbreeding. F is the inbreeding coefficient of the population, in the case of $F=0$ (outbred population) the frequencies correspond to those in Hardy–Weinberg equilibrium and extreme case of $F=1$ (maximum value of inbreeding) the proportion of heterozygotes is 0.

Mating type	Genotypes		
	AA	Aa	aa
Random mating	p^2	$2pq$	q^2
Inbreeding	$p^2 + Fpq$	$2pq(1-F)$	$q^2 + Fpq$

rates of gene flow between adjacent populations can reduce the negative effects of inbreeding and maintain population variability (Elo 1993) without suppressing the local adaptation advantage (Hendry 2001). This natural situation where salmon populations are connected with small but significant levels of gene flow can be interrupted by human disturbance by reducing the number of populations (stepping stones for gene flow) and population sizes (number of potential migrants). This can lead to increased risk of inbreeding, as seems to have happened in some salmonid populations following severe human disturbance and habitat fragmentation (Allendorf and Waples 1996).

The fact that gene flow is fundamental for maintaining the variability in many small populations could lead to the idea that artificially introducing individuals could produce a beneficial increase in the variability of small, declining populations. Although the introduction of individuals from another population could result in positive effects in fitness-related traits (counteracting the effects of inbreeding depression and genetic drift) it could also have negative consequences as a result of the phenomenon called outbreeding depression. Outbreeding depression refers to a potential fitness reduction (i.e. reduced survival or fertility) in hybrid individuals between two different populations (or species, e.g. Atlantic salmon and brown trout).

Three different mechanisms can be responsible for outbreeding depression in the offspring: (1) the maladaptation of the genome of the hybrids to the environmental conditions of the original population; (2) the breakdown of co-adapted gene complexes; or (3) differences in number or structure of the chromosomes of different species, or very distant populations that can affect the gamete formation and can result in a decrease in fertility. These gene complexes are combinations of alleles at different genes that have an overall beneficial effect on the fitness of the individuals. They can be broken by subsequent generations of recombination in the hybrids.

In this sense the existence of local adaptations suggests that a certain level of outbreeding depression could be the consequence of gene flow between populations and this has to be considered in the case of translocation between populations. Another consideration in relation to translocation is that a low rate of gene flow is sufficient to prevent genetic differentiation at neutral or weakly selected loci, but adaptive differences between populations can still be maintained in the population. This means that populations that do not differ at neutral marker loci should not necessarily be treated as one population (leading to extensive translocations) since they could still have adaptive variation in morphological or behavioural traits with a genetic background.

8.6.1 Small populations of Atlantic salmon and the metapopulation models

When considering the conservation of the species and the extinction of small populations in salmonids, the implications of the metapopulation theory need to be taken into account. Metapopulations are defined as groups of local populations with similar but independent risks of extinction, where the local extinctions are balanced by migration and recolonisation from remaining extant populations within the group (Hanski 1999). The metapopulation structure can influence the effective population size, the amount of genetic variation and the distribution of the genetic variation of the local populations, because the extinction–recolonisation dynamics can act as a population bottleneck, reducing the effective population size. Due to their life history salmonids seem to be ideal candidates for structuring into metapopulations, and the evidence that exists (Chapter 5) supports this, at least in some local and regional contexts.

In Atlantic salmon short-term extinctions and recolonisations, as found in other species where metapopulations have been defined (Saccheri *et al.* 1998), have not been reported. Homing behaviour in Atlantic salmon can promote local adaptations and population structuring, and the natural straying can be seen as a way of dispersal between local populations (Cooper and Mangel 1999). However, the main question remains whether the persistence of many small populations of salmonids can be seen in the context of metapopulation theory, with large populations acting as 'sources' that provide fish to the neighbouring small populations that would act as 'sinks'. It has been suggested that the generalisation of the model to all salmonid populations is not appropriate (Rieman and Dunham 2000) and in some cases the extinction and recolonisation of local salmonid populations cannot fit clearly into the extinction–recolonisation pattern that defines a metapopulation.

It is likely that some population structuring in the Atlantic salmon is at the level of metapopulations (Chapter 5; Garant *et al.* 2000), based on evidence for variation in gene flow and population structuring through the distribution range of Atlantic salmon. Therefore, meta-structure should be a consideration in the management and conservation of small salmonid populations. Failing to take it into account could potentially lead to their extinction (Cooper and Mangel 1999).

8.7 Summary and conclusions

- Many Atlantic salmon populations have suffered reductions in population size during the last decade due mainly to overexploitation and habitat destruction. These small populations are more prone to extinction due to the loss of genetic variation that can reduce the ability of the populations to respond to the changes in the environment, and to the negative costs of mating between close relatives or inbreeding (inbreeding depression).
- In small populations, genetic variation is lost in the main from random changes in allele frequencies (genetic drift) and inbreeding. Effective population size, an estimate of the number of individuals contributing to the next generation, quantifies the rate at which genetic variation is lost by genetic drift in the population.
- For Atlantic salmon the recommended minimum effective number of breeders (N_b) per year is not less than 150. The effective population size is often lower than the population size because it is affected by several factors such as fluctuations in the population size over time, unequal sex ratio or variation among individuals in the number of offspring produced.
- The main consequence of mating between relatives (inbreeding) in wild populations is an increased risk of extinction due to inbreeding depression caused by an increase in the frequency of recessive deleterious alleles or in the homozygosity when heterozygote individuals are superior. Inbreeding depression can be influential in both captive and natural populations of Atlantic salmon and the cost of inbreeding can be higher in the latter, increasing the risk of extinction of natural populations.
- Gene flow is an important factor in counteracting the effects of inbreeding and also in reducing genetic divergence among populations. In Atlantic salmon natural straying provides a possible source of increasing genetic variation although not all the strayers contribute effectively to the next generation, probably due to the existence of local adaptations. Although it is still not clear whether the metapopulation theory can be applied to salmon populations, ignoring the possible structure of salmonids in metapopulations could result in inappropriate management of local populations.

8.8 Management recommendations

- Maintain genetic variation within small native salmon populations, by keeping effective population sizes as large as possible (see 50-500 rule) and by avoiding artificial selection and unnatural migration rates.
- Maintain large wild census population size by:
  - ensuring sufficient suitable natural spawning and nursery habitat in rivers;
  - providing free upstream and downstream passage for migrating salmon adults and smolts;
  - regulating harvest by commercial and sports fishing both with regard to numbers of fish (escapement) and selective fishing for specific individuals (for instance large multi-sea-winter fish).

- Avoid stocking unless the above-mentioned habitat options have been exhausted (see also Cowx 1994).
- Use supportive breeding when possible and avoid stocking with non-native fish of either hatchery or wild origin.
- Only carry out supportive breeding as a conservation measure when the population is in danger of going extinct for demographic reasons and then ensure a *high effective population size* by:
  – using a high number of individuals (50–500 per generation) since the total effective population size of the whole population (wild and hatchery reproduction) will depend heavily on the proportion in the hatchery due to their relatively high reproductive output (see Ryman and Laikre 1991);
  – ensuring a high number of breeders in each generation/year to avoid fluctuations in population size over time;
  – avoiding unequal sex ratios by focusing on collecting individuals of the rarer sex;
  – minimising variance in family size by using eggs from all families and equalising contribution from breeders.
- When using supportive breeding *avoid domestication selection* by:
  – using fish representing natural variation (e.g. life history and morphological variation); avoid selecting broodstock with specific traits of human interest;
  – avoiding selection in the hatchery by ensuring low mortality;
  – simulating natural environmental conditions in the hatchery as much as possible (light, temperature, current and so on);
  – releasing eggs/fry/juveniles as early as possible to allow natural selection.
- When carrying out supportive breeding also:
  – evaluate connection with other populations. Have natural levels of migration been disturbed by reduction or elimination of many populations in the area?
  – consider induced gene flow in cases where N_e is extremely small or the population objectively suffers from inbreeding depression.

Further reading

Altukhov, Y.P. (2006) *Intraspecific Genetic Diversity: Monitoring, conservation and management.* Springer Verlag, Berlin.

Frankham, R., Ballou, J.D. and Briscoe, D.A. (2002) *Introduction to Conservation Genetics.* Cambridge University Press, Cambridge.

Hendry, A. and Stearns, S.C. (2004) *Evolution Illuminated: Salmon and their relatives.* Oxford University Press, New York.

References

Allendorf, F.W. and Leary, R.F. (1986) Heterozygosity and fitness in natural populations of animals. In: M.E. Soulé (Ed.) *Conservation Biology: The science of scarcity and diversity*, pp. 57–76. Sinauer Associates, Sunderland, MA.

Allendorf, F.W. and Phelps, S.R. (1981) Use of allelic frequencies to describe population structure. *Canadian Journal of Fisheries and Aquatic Sciences*, **38**: 1507–1514.

Allendorf, F.W. and Thorgaard, G.H. (1984) Tetraploidy and the evolution of salmonid fishes. In: B.J. Turner (Ed.) *Evolutionary Genetics of Fishes*, pp.1–53. Plenum, New York.

Allendorf, F.W. and Waples, R.S. (1996) Conservation and genetics of salmonid fishes. In: J.C. Avise and J.L. Hamrick (Ed.) *Conservation Genetics: Case histories from nature*, pp. 238–501. Chapman & Hall, New York.

Altukhov, Y.P. (2006) *Intraspecific Genetic Diversity: Monitoring, conservation and management*. Springer Verlag, Berlin.

Ardren, W.R. and Kapuscinski, A.R. (2003) Demographic and genetic estimates of effective population size (N_e) reveals genetic compensation in steelhead trout. *Molecular Ecology*, **12**: 35–49.

Bartley, D., Bagley, M., Gall, G. and Bentley, B. (1992) Use of linkage disequilibrium data to estimate effective size of hatchery and natural fish populations. *Conservation Biology*, **6**: 365–375.

Beaumont, M.A. (2001) Conservation genetics. In: D.J. Balding, M. Bishop and C. Cannings (Ed.) *Handbook of Statistical Genetics*, pp. 779–812. John Wiley & Sons, Chichester.

Blanco, G., Sanchez, J.A., Vazquez, E., García, E. and Rubio, J. (1990) Superior developmental stability of heterozygotes at enzyme loci in *Salmo salar* L. *Aquaculture*, **84**: 199–209.

Blanco, G., Presa, P. and Sanchez, J.A. (1998) Allozyme heterozygosity and development in Atlantic salmon, *Salmo salar*. *Fish Physiology and Biochemistry*, **19**: 163–169.

Bonnell, M.L. and Selander, R.K. (1974) Elephant seals: genetic variation and near extinction. *Science*, **184**: 908–909.

Caballero, A. (1994) Developments in the prediction of effective population size. *Heredity*, **73**: 657–679.

Carvalho, G.R. (1993) Evolutionary aspects of fish distribution: genetic variation and adaptation. *Journal of Fish Biology*, **43**: 53–73.

Charlesworth, B. (1997) Is founder-flush speciation defensible? *American Naturalist*, **149**: 600–603.

Consuegra, S., García de Leániz, C., Serdio, A., Gonzalez Morales, M., Straus, L.G., Knox, D. and Verspoor, E. (2002) Mitochondrial DNA variation in Pleistocene and modern Atlantic salmon from the Iberian glacial refugium. *Molecular Ecology*, **11**: 2037–2048.

Consuegra, S., Verspoor, E., Knox, D. and García de Leániz, C. (2005) Asymmetric gene flow and the evolutionary maintenance of genetic diversity in small, peripheral Atlantic salmon populations. *Conservation Genetics*, **6**: 823–842.

Cooper, A.B. and Mangel, M. (1999) The dangers of ignoring metapopulation structure for the conservation of salmonids. *Fisheries Bulletin*, **97**: 213–226.

Cornuet, J.M. and Luikart, G. (1997) Description and power analysis of two tests for detecting recent population bottlenecks from allele frequency data. *Genetics*, **144**: 2001–2014.

Cowx, I.G. (1994) Stocking strategies. *Fisheries Management and Ecology*, **1**: 15–30.

Crnokrak, P. and Roff, D.A. (1999) Inbreeding depression in the wild. *Heredity*, **83**: 260–270.

Crozier, W.W. (1997) Genetic heterozygosity and meristic character variance in a wild Atlantic salmon population and a hatchery strain derived from it. *Aquaculture International*, **5**: 407–414.

Daniels, S.J., Triddy, J.A. and Walters, J.R. (2000) Inbreeding in small populations of red-cockaded woodpeckers: insights from a spatially explicit individual-based model. In: A.G. Young and G.M. Clarke (Ed.) *Genetics, Demography and Viability of Fragmented Populations*, pp. 129–148. Cambridge University Press, Cambridge.

Elliott, N.G. and Reilly, A. (2003) Likelihood of bottleneck event in the history of the Australian population of Atlantic salmon (*Salmo salar* L.). *Aquaculture*, **215**: 31–44.

Elo, K. (1993) Gene flow and conservation of genetic variation in anadromous Atlantic salmon (*Salmo salar*). *Hereditas*, **119**: 149–159.

Endler, J.A. (1986) *Natural Selection in the Wild*. Princeton University Press, Princeton, NJ.

Frankham, R. (1995) Effective population size/adult population size ratios in wildlife: a review. *Genetical Research*, **66**: 95–107.

Frankham, R. and Ralls, K. (1998) Inbreeding leads to extinction. *Nature*, **392**: 441–442.

Frankham, R., Lees, K., Montgomery, M.E., England, P.R., Lowe, E.H. and Briscoe, D.A. (1999) Do population bottlenecks reduce evolutionary potential? *Animal Conservation*, 2: 255–260.

Frankham, R., Ballou, J.D. and Briscoe, D.A. (2002) *Introduction to Conservation Genetics*. Cambridge University Press, Cambridge.

Franklin, I.R. (1980) Evolutionary change in small populations. In: M.E. Soulé and B.A. Wilcox (Ed.) *Conservation Biology: An evolutionary-ecological perspective*, pp. 135–149. SinauerAssociates, Sunderland, MA.

Franklin, I.R. and Frankham, R. (1998) How large must populations be to retain evolutionary potential. *Animal Conservation*, 1: 69–71.

Fraser, D.J., Lippé, C. and Bernatchez, L. (2004) Consequences of unequal population size, asymmetric gene flow and sex-biased dispersal on population structure in brook charr (*Salvelinus fontinalis*). *Molecular Ecology*, 13: 67–80.

Garant, D., Dodson, J.J. and Bernatchez, L. (2000) Ecological determinants and temporal stability of the within-river population structure in Atlantic salmon (*Salmo salar* L.). *Molecular Ecology*, 9: 615–628.

Geiger, H.J., Smoker, W.W., Zhivotovsky, L.A. and Gharrett, A.J. 1997. Variability of family size and marine survival in pink salmon (*Oncorhynchus gorbuscha*) has implications for conservation biology and human use. *Canadian Journal of Fisheries and Aquatic Sciences*, 54: 2684–2690.

Groombridge, J.J., Jones, C.G., Bruford, M.W. and Nichols, R.A. (2000) Ghost alleles of the Mauritius kestrel. *Nature*, **403**: 616.

Hansen, M.M., Ruzzante, D.E., Nielsen, E.E., Bekkevold, D. and Mensberg, K-L.D. (2002) Long-term effective population sizes, temporal stability of genetic composition and potential for local adaptation in anadromous brown trout (*Salmo trutta*) populations. *Molecular Ecology*, 11: 2523–2535.

Hanski, I. (1999) *Metapopulation Ecology*. Oxford University Press, New York.

Heath, D.D., Busch, C., Kelly, J. and Atagi, D.Y. (2002) Temporal change in genetic structure and effective population size in steelhead trout (*Oncorhynchus mykiss*). *Molecular Ecology*, 11: 197–214.

Hedrick, P.W. (1994) Purging inbreeding depression and the probability of extinction: full-sib mating. *Heredity*, 73: 363–372.

Hedrick, P.W., Hedgecock, D. and Hamelberg, S. (1995) Effective population size in winter-run chinook salmon. *Conservation Biology*, 9: 615–624.

Hendry, A.P. (2001) Adaptive divergence and the evolution of reproductive isolation in the wild: an empirical demonstration using introduced sockeye salmon. *Genetica*, 112–113: 515–534.

Jones, M.W. and Hutchings, J.A. (2001) The influence of male parr body size and mate competition on fertilization success and effective population size in Atlantic salmon. *Heredity*, 86: 675–684.

Jorde, P.E. and Ryman, N. (1996) Demographic genetics of brown trout (*Salmo trutta*) and estimation of effective population size from temporal change of allele frequencies. *Genetics*, 143: 1369–1381.

Keller, L. and Waller, D.M. (2002) Inbreeding effects in wild populations. *Trends in Ecology and Evolution*, 17: 230–241.

Koljonen, M-L., Tähtinen, J., Säisä, M. and Koskiniemi, J. (2002) Maintenance of genetic diversity of Atlantic salmon (*Salmo salar*) by captive breeding programmes and the geographic distribution of microsatellite variation. *Aquaculture*, 212: 69–92.

Lage, C. and Kornfield, I. (2006) Reduced genetic diversity and effective population size in an endangered Atlantic salmon (*Salmo salar*) population from Maine, USA. *Conservation Genetics*, 7: 91–104.

Lande, R. (1995) Mutation and conservation. *Conservation Biology*, 9: 782–791.

Lande, R. and Barrowclough, G.F. (1987) Effective population size, genetic variation, and their use in population management. In: M.E. Soulé (Ed.) *Viable Populations for Conservation*, pp. 87–123. Cambridge University Press, New York.

Luikart, G., Allendorf, F.W., Cornuet, J.M. and Sherwin, W.B. (1998) Distortion of allele frequency distributions provides a test for recent population bottlenecks. *Journal of Heredity*, 89: 238–247.

Martínez, J.L., Morán, P., Perez, J., de Gaudemar, B., Beall, E. and García-Vázquez, E. (2000) Multiple paternity increases effective size of southern Atlantic salmon populations. *Molecular Ecology*, **9**: 293–298.

Meffe, G. and Carroll, C.R. (1997) *Principles of Conservation Biology*. Sinauer Associates, Sunderland, MA.

Nielsen, E.E., Hansen, M.M. and Loeschcke, V. (1997) Analysis of microsatellite DNA from old scale samples of Atlantic salmon *Salmo salar*: a comparison of genetic composition over 60 years. *Molecular Ecology*, **6**: 487–492.

Nielsen, E.E., Hansen, M.M. and Loeschcke, V. (1999) Genetic variation in time and space: microsatellite analysis of extinct and extant populations of Atlantic salmon. *Evolution*, **53**: 261–268.

Nielsen, J.L. (1999) The evolutionary history of steelhead (*Oncorynchus mykiss*) along the US Pacific Coast: developing a conservation strategy using genetic diversity. *ICES Journal of Marine Science*, **56**: 449–458.

Nunney, L. and Elam, D.R. (1994) Estimating the effective population size of conserved populations. *Conservation Biology*, **8**: 175–184.

Østergaard, S., Hansen, M.M., Loeschcke, V. and Nielsen, E.E. (2003) Long-term temporal changes of genetic composition in brown trout (*Salmo trutta* L.) populations inhabiting an unstable environment. *Molecular Ecology*, **12**: 3123–3135.

Palm, S., Laikre, L., Jorde, P.E. and Ryman, N. (2003) Effective population size and temporal genetic change in stream resident brown trout (*Salmo trutta*, L.). *Conservation Genetics*, **4**: 249–264.

Pante, M.J.R., Gjerde, B. and McMillan, I. (2001) Effect of inbreeding on body weight at harvest in rainbow trout, *Oncorhynchus mykiss*. *Aquaculture*, **192**: 201–211.

Quinn, T.P., Unwin, M.J. and Kinninson, M.T. (2000) Evolution of temporal isolation in the wild: genetic divergence timing of migration and breeding by introduced chinook salmon populations. *Evolution*, **54**: 1372–1385.

Ralls, K., Ballou, J.D. and Templeton, A.R. (1988) Estimates of lethal equivalents and the cost of inbreeding in mammals. *Conservation Biology*, **2**: 185–193.

Reed, D.H. and Bryant, E.H. (2000) Experimental tests of minimum viable population size. *Animal Conservation*, **3**: 7–13.

Reed, D.H. and Frankham, R. (2001) How closely correlated are molecular and quantitative measures of genetic variation? A meta-analysis. *Evolution*, **55**: 1095–1103.

Reed, D.H. and Frankham, R. (2003) Correlation between fitness and genetic diversity. *Conservation Biology*, **17**: 230–237.

Rieman, B.E. and Dunham, J.B. (2000) Metapopulations and salmonids: a synthesis of life history patterns and empirical observations. *Ecology of Freshwater Fish*, **9**: 51–64.

Rye, M. and Mao, I.L. (1998) Nonadditive genetic effects of inbreeding depression for body weight in Atlantic salmon (*Salmo salar* L.). *Livestock Production Science*, **57**: 15–22.

Ryman, N. (1970) A genetic analysis of recapture frequencies of released young of salmon (*Salmo salar* L.). *Hereditas*, **65**: 159–160.

Ryman, N. and Laikre, L. (1991). Effects of supportive breeding on the genetically effective population size. *Conservation Biology*, **5**: 325–329.

Ryman, N., Jorde, P.E. and Laikre, L. (1995) Supportive breeding and variance effective population size. *Conservation Biology*, **9**: 1619–1628.

Saccheri, I., Kuussaari, M., Kankare, M., Vikman, P., Fortelius, W. and Hanski, I. (1998) Inbreeding and extinction in a butterfly metapopulation. *Nature*, **392**: 491–494.

Säisä, M., Koljonen, M-L. and Tähtinen, J. (2003) Genetic changes in Atlantic salmon stocks since historical times and the effective population size of a long-term captive breeding programme. *Conservation Genetics*, **4**: 613–627.

Sherwin, W.B. and Moritz, C. (2000) Managing and monitoring genetic erosion. In: A.G. Young and G.M. Clarke (Ed.) *Genetics, Demography and Viability of Fragmented Populations*, pp. 9–34. Cambridge University Press, Cambridge.

Shrimpton, J.M. and Heath, D.D. (2003) Census vs. effective population size in chinook salmon: large and small-scale environmental perturbation effects. *Molecular Ecology*, **12**: 2571–2583.

Slatkin, M. (1996) In defense of founder-flush theories of speciation. *American Naturalist*, **147**: 493–505.

Soulé, M.E. and Wilcox, B.A. (Ed.) (1980) *Conservation Biology: An evolutionary-ecological perspective*. Sinauer Associates, Sunderland, MA.

Su, G.S., Liljedahl, L.E. and Gall, G.A.E. (1996) Effects of inbreeding on growth and reproductive traits in rainbow trout (*Oncorhynchus mykiss*). *Aquaculture*, **142**: 139–148.

Taggart, J.B., McLaren, I.S., Hay, D.W., Webb, J.H. and Youngson, A.F. (2001) Spawning success in Atlantic salmon (*Salmo salar* L.): a long-term DNA profiling-based study conducted in a natural stream. *Molecular Ecology*, **10**: 1047–1060

Tallman, R.F. and Healey, M.C. (1994) Homing, straying, and gene flow among seasonally separated populations of chum salmon (*Oncorhynchus keta*). *Canadian Journal of Fisheries and Aquatic Sciences*, **51**: 577–588.

Taylor, A.C., Sherwin, W.B. and Wayne, R.K. (1994) The use of simple sequence loci to measure genetic variation in bottlenecked species: the decline of the hairy-nosed wombat (*Lasiorhinus krefftii*). *Molecular Ecology*, **3**: 277–290.

Taylor, E.B. (1991) A review of local adaptations in Salmonidae, with particular reference to Pacific and Atlantic salmon. *Aquaculture*, **98**: 185–207.

Templeton, A.R. (1980) The theory of speciation *via* the founder principle. *Genetics*, **94**: 1011–1038.

Vasemägi, A., Gross, R., Paaver, T., Kangur, M., Nilsson, J. and Eriksson, L.-O. (2001) Identification of the origin of an Atlantic salmon (*Salmo salar* L.) population in a recently recolonized river in the Baltic Sea. *Molecular Ecology*, **10**: 2877–2882.

Vrijenhoek, R.C. (1994) Genetic diversity and fitness in small populations. In: V. Loeschcke, J. Tomuik and S.K. Jain (Ed.) *Conservation Genetics*, pp. 37–53. Birkhauser Verlag, Basel.

Vrijenhoek, R.C. (1996) Conservation genetics of North American desert fishes. In: J.C. Avise and J.L. Hamrick (Ed.) *Conservation Genetics: Case histories from nature*, pp. 367–397. Chapman & Hall, New York.

Wang, J. (2001) Optimal marker-assisted selection to increase the effective size of small populations. *Genetics*, **157**: 867–874.

Wang, J. and Whitlock, M.C. (2003) Estimating effective population size and migration rates from genetic samples over space and time. *Genetics*, **163**: 429–446.

Wang, S., Hard, J.J. and Utter, F. (2002a) Genetic variation and fitness in salmonids. *Conservation Genetics*, **3**: 321–333.

Wang, S., Hard, J.J. and Utter, F. (2002b) Salmonid inbreeding: a review. *Reviews in Fish Biology and Fisheries*, **11**: 301–319.

Waples, R.S. (1990a) Conservation genetics of Pacific salmon. II. Effective population size and rate of loss of genetic variability. *Journal of Heredity*, **81**: 267–276.

Waples, R.S. (1990b) Conservation genetics of Pacific salmon. III. Estimating effective population size. *Journal of Heredity*, **81**: 277–289.

Waples, R.S. (1991) Pacific salmon, *Oncorhynchus* spp., and the definition of 'species' under the Endangered Species Act. *Marine Fisheries Reviews*, **53**: 11–21.

Wright, S. (1938) Size of population and breeding structure in relation to evolution. *Science*, **87**: 430–431.

9 Genetic Identification of Individuals and Populations

M-L. Koljonen, T. L. King and E. E. Nielsen

Upper: fishing for Atlantic salmon using drift nets off the shore of Northumberland in the north-east of England. Lower: a salmon caught by one of the drift nets of the boat shown above. (Photos credit: Crown copyright FRS, reproduced with the permission of FRS Freshwater Laboratory, Pitlochry.)

There are many circumstances when populations of Atlantic salmon, *Salmo salar*, mix naturally, such as during the migratory phases of its life cycle; on other occasions mixing occurs as a result of human activities such as stocking or farming. In such situations, understanding the extent to which different populations or regional stock groups mix is often essential for management and conservation. Understanding can only be gained if the population of origin of individual fish, or the proportional contribution of different populations to a mixture, can be determined. Such determinations can sometimes be achieved using visual or external characteristics, or applied physical tags. However, over the last 15 years, this issue has been increasingly successfully addressed by exploiting molecular genetic diversity among populations.

9.1 Introduction

Genetic studies have demonstrated extensive population structuring in many fish species and the Atlantic salmon is no exception (Chapter 5). The genetic differences among populations associated with structuring can be exploited to identify the origin of a fish, or sample, in situations where the source population is unknown. This is often the case for marine fishes where populations are frequently found to mix. In such contexts, where exploitation rates are high and population productivity is unequal, harvesting risks the survival of the smallest populations (Ricker 1958; Hilborn 1985; see Chapter 10). Yet for sustainable exploitation it is essential to maintain the viability of all populations contributing to the fishery, and fisheries should be managed to allow adequate escapement to ensure the continuing reproductive success of all contributing populations. What is required is to achieve a balance between sufficient protection of weak populations and effective harvesting of strong populations. Information on spatial and temporal variations in stock composition in mixed-population fisheries is therefore essential for effective fisheries management and conservation (Begg *et al.* 1999; Shaklee *et al.* 1999).

Hatchery rearing and releases also create a need for population-specific information on the origin of fisheries catches. Wild and reared fish have to be managed with different objectives and with different harvest strategies. Managers need to know when, where, in what amounts and by whom fish populations and stocks are exploited. Population- or stock-specific harvest strategies are needed for resource management and for dividing fish resources and fishing rights among nations and other potential user groups. Resource allocation may require regulation of mixed-stock fisheries. Post-season information on the proportions of populations in catches is useful in efforts to direct and regulate future fisheries. Population or stock proportion estimates can sometimes even be used as a direct tool to regulate fishing, for example, when deciding on the closure of a fishery or other regulatory measures.

Traditionally, extensive external tagging and recapture programmes have been employed to acquire information on the origin of fish caught. Those programmes encompass commonly used mechanical techniques, such as Carlin tagging, coded wire tagging and scale analyses. However, stock proportions can also be estimated and the origin of individual fishes determined by using genetic differences among fish stocks. Usually naturally occurring genetic differences have been used, but controlled, selective matings can also be used to increase the differences in allele frequencies and to create induced or artificial genetic marks.

Genetic methods have several advantages over those using traditional tagging. First and foremost, all individuals, from eggs to adults, carry the genetic signature of their population of origin and so all individuals are marked – for life. Thus it is possible to study fishes that cannot be tagged by other methods, for example, wild fish in remote areas or newly hatched fish in release programmes (Mathias *et al.* 1992). With genetic methods there is no loss of tags and no bias due to viability or catchability effects as with external tags. In addition, the time and place of sampling can be chosen more freely and precisely, as preceding tag-and-release programmes are not required. Moreover, genetic identification is not dependent on fishermen in returning tags or on the detection of internal tags. Genetic stock composition analysis acts as a link between genetic stock structure information and the practical fisheries management decision-making process when the definition of management units is based on the same genetic information as that used in population or stock identification (Koljonen *et al.* 1999; Koljonen 2001).

Genetic methods also have limitations. The most obvious constraint on the use of genetics is that temporally stable genetic differences will not arise without a sufficient level of reproductive isolation among populations or stocks. The extent of differentiation will affect the resolution which can be achieved. Statistically significant differences in allele frequencies often occur, but quantitatively they may be too small to meet the needs of the managers for assignment accuracy and precision.

In recent years, studies of DNA variation have greatly increased the amount of detectable genetic variation available for stock identification work. In many cases, DNA methods have a greater resolution than the traditional methods based on allozyme data. The large numbers of alleles, detected by direct analysis of DNA, and multilocus genotypes they define, have sometimes made it possible to assign the population of origin for individual fish. However, even with DNA methods, populations or stocks may be too alike to be discriminated with a high level of reliability. The high number of alleles in DNA microsatellite loci has also created new problems. The larger the number of alleles at one gene locus, the more likely it is that not all the alleles will be observed in the samplings, even if they occur in the populations studied. Non-detection of rare alleles in the sampling of potentially contributing stocks may thus cause bias in the estimation and present a challenge to statistical estimation methods.

In general, salmonids such as the Atlantic salmon show a high degree of genetic differentiation at both the population and regional stock level (Chapter 5). Thus, for the Atlantic salmon, genetic analysis can generally be expected to provide a useful tool, at minimum, for estimating the contribution made by regional groups which will often be composed of genetically similar populations. Even so, genetic tags in most cases will only give statistical or probabilistic information about the origin of individuals or populations. Diagnostic differences enabling the origin of each individual fish to be identified with effectively 100% probability is only likely to be possible with respect to Eastern and Western Atlantic salmon populations and, possibly, in a few local circumstances where mixing populations show extreme phylogenetic divergence.

Mixed-stock or mixed-populations fisheries for Atlantic salmon still occur though there is a growing trend towards stock (or population)-specific, terminal fisheries close to or within rivers. The marine fisheries target salmon on their feeding migration in offshore areas, or in coastal areas during their return spawning migration (Fig. 9.1). In the Baltic Sea, only about 7% of the total salmon catch was taken in rivers in 2002 (ICES 2003). In such mixed-stock fisheries, genetic methods can be particularly useful in providing information on the origin of fish caught to help with effective regulation of exploitation to protect naturally reproducing stocks.

Fig. 9.1 Catch of a large adult salmon from a trap net at the Gulf of Finland. (Photo credit: E. Lehtonen.)

Active research into Atlantic salmon stock identification in fisheries has been under way since the mid-1960s, when effective commercial exploitation of salmon started at West Greenland (Reddin and Friedland 1999). The proportional harvesting of salmon from Europe and North America in this fishery remains an important management issue.

Diagnostic markers are unlikely to exist in most cases for populations within a species. However, it has increasingly proven possible to use the information present in 'multilocus genotype' distributions, for non-diagnostic genetic loci, to carry out 'individual assignment (IA)' or 'mixed-stock analysis (MSA)' and estimate the proportional contribution of different populations or stocks to a fishery. This chapter looks at how these methods can be, and have been, exploited in applications to fish in general, and to Atlantic salmon in particular. It also describes the different mathematical approaches and the principles which underlie them, and examines their reliability and resolving power.

9.2 Assignment of individuals

Individual assignment (IA) is based on estimates of the probability of encountering the genotype of an individual of unknown origin in a number of potential source populations (baseline populations) whose genotype composition has been determined by sampling. The individual is then assigned to the source population in which its genotype is expected to be most frequent, based on baseline allele frequency information (e.g. Paetkau *et al.* 1995). The main application of assignment tests is for establishing the origin of one or a few fish when all baseline stocks are equally likely to be the source. Individual assignment can be used, for instance, in the selection of individuals for broodstock when they are collected from a mixture of wild and hatchery fish.

A classical example of how IA can be applied is the study of migrating birds by Haig *et al.* (1997), who used assignment tests to determine the population of origin of migrating shorebirds that breed in genetically differentiated breeding colonies. The advantage of this approach over banding techniques is that all migrating birds have multilocus genotypes but not all have bands. The scenario of collecting migrating birds from a number of breeding populations is similar to that of collecting fish from a mixed-stock fishery with many species of anadromous and freshwater fish. Individual assignment has been used to identify individuals in samples from mixed-stock fisheries and other mixed samples, too (Banks *et al.* 2000; Olsen *et al.* 2000). It has also been used to assign parentage to individual offspring in hatchery populations (Estoup *et al.* 1998a).

Assignment methods have even been used with high accuracy for marine fishes, in which genetic differentiation is generally much lower than in freshwater and anadromous species. For instance, Nielsen *et al.* (2001a) were able to assign Atlantic cod *Gadus morhua* individuals to a North Sea, Baltic Sea or north-east Arctic Ocean population with almost 100% certainty using STRUCTURE, a Bayesian method developed by Pritchard *et al.* (2000). The differentiation level, F_{ST}, between North Sea and Baltic Sea populations was 0.045 and between North Sea and Arctic Ocean populations 0.035. In addition to this conventional application of assignment tests to the study of population differentiation and migration, it is possible to apply assignment methods to a number of problems in fish and fisheries biology:

- discrimination among species (and hybridisation);
- hybridisation among fish from different populations (Nielsen *et al.* 2004);
- sex-biased dispersal, i.e. whether dispersal is higher for males or females;
- identification of indigenous populations with the aid of archived DNA material;
- fish forensics for legal cases such as poaching.

For examples of these applications with regard to fish, see Hansen *et al.* (2001).

The application of IA to the Atlantic salmon is reviewed below. This is followed by a more detailed description of the mathematical methodology. This includes a description of the basic principles on which IA is founded, the different approaches which can be used, and the power of the methods.

9.2.1 Application to Atlantic salmon

Monitoring commercial marine harvests of individual Atlantic salmon stocks is problematic due to the relatively high degree of morphological similarity among stocks. Identification of diagnostic genetic characters could facilitate the management of important fisheries by fine-tuning management regulation to allow a larger number of spawners to return to their natal stream.

Previous genetic surveys of the Atlantic salmon have shown relatively little genetic divergence across the range of the species, due in all likelihood to the recent (8000–10 000 years BP) colonisation of its present habitats (Crossman and McAllister 1986; Ståhl 1987; Davidson *et al.* 1989; Verspoor 1997). Frequency variation in transferrin proteins (Payne *et al.* 1971; Thorpe and Mitchell 1981), allozymes (Ståhl 1987; Verspoor 1988; Bourke *et al.* 1997;

Koljonen *et al.* 1999), restriction fragment length polymorphisms of mitochondrial DNA (Bermingham *et al.* 1991; King *et al.* 2000) and chromosome numbers (Hartley and Horne 1984) have suggested only a moderate degree of genetic differentiation between Atlantic salmon inhabiting North America and those in Europe. In DNA microsatellite loci, however, the differences have been clear (McConnell *et al.* 1995).

In a recent study, King *et al.* (2001) surveyed the microsatellite DNA variation in Atlantic salmon collected from a large part of their range in North America and Europe. Individual assignment tests conducted on the data set permitted 100% correct assignment to continent of origin and, on average, almost 83% correct classification to province of origin across continents. In addition, a diagnostic difference between the continents was found at only one microsatellite locus (SSOSL311) (Koljonen *et al.* 2002). The frequency of a single allele was as high as 0.96–0.93 in samples from Maine and Labrador, however this allele did not occur in European populations at all. Identification of individual fish is therefore already possible with very high precision on the basis of information from this one locus alone. The F_{ST} value for populations of different continents in that study was as high as 0.22 for nine microsatellite DNA loci, which is about twice as high as that between European and Baltic Sea salmon stocks (0.09). These findings illustrate the potential use of assignment methods in mixed-stock fisheries such as the Atlantic salmon subsistence fishery off the west coast of Greenland. This fishery is composed of one-sea-winter age fish of both North American and European origin. Fish harvested in this fishery would otherwise attempt a return to their natal waters as two-sea-winter old fish. Therefore, proper management of this valuable resource on both continents requires assessment of the relative contributions of these stocks to the fishery.

A microsatellite database of 4749 Atlantic salmon genotypes of known origin has been used as a baseline to assign the salmon in the harvest of West Greenland to their continent of origin. In 2002, a total of 501 caught salmon were sampled from Maniitsoq, Nuuk and Qaqortoq, Greenland, and were genotyped at 11 microsatellite DNA loci for assignment to continent of origin. In total, 338 (67.5%) of these salmon were of North American origin and 163 (32.5%) of European origin (Table 9.1, Fig. 9.2). The southernmost samples, those taken from Qaqortoq, yielded the highest proportion of salmon of European origin (96 fish or 63.6%).

The variation in the proportion of continental representation across these collections underscores the need to sample a number of regions to achieve an accurate estimate of the contribution of fish from each continent to the entire mixed fishery. The samples should also be representative of the fishery. It also raises the question of whether European fish tend to congregate in the more southerly feeding grounds.

Table 9.1 Numbers and proportions of North American and European fish in samples from West Greenland Atlantic salmon fishery in 2002.

Sample site	N	North American		European	
		N	%	N	%
Maniitsoq	146	102	69.9	44	30.1
Nuuk	204	181	88.7	23	11.3
Qaqortoq	151	55	36.4	96	63.9
Total	501	338	67.5	163	32.5

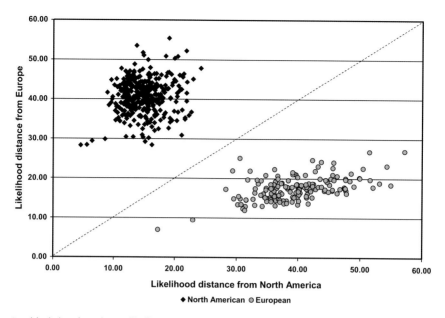

Fig. 9.2 Graphical plot of maximum likelihood assignment scores generated by a suite of 11 microsatellite DNA markers in 500 fish sampled from the West Greenland mixed fishery in 2002. The baseline data set consisted of more than 4700 salmon collected throughout the range.

Individual assignment has further been used to assist the hatchery breeding of North American salmon stocks. In an effort to eliminate European-origin Atlantic salmon and to eradicate their potential negative impacts on the restoration and supplementation programme under way in Maine rivers, resource management agencies are requiring the local industry to demonstrate that their brood fish are of North American heritage. This is being done by assignment tests similar to those employed in continent-of-origin determinations.

Commercial aquaculture of Atlantic salmon was established in Maine in 1987. The salmon used by the aquaculture industry include North American strains from the Penobscot and St John Rivers, as well as the Landcatch strain derived from European Atlantic salmon. These aquaculture strains have been hybridised, increasing the genetic distance between hatchery and native strains even more than by mere domestication selection. Despite the best efforts of producers to prevent escapes, fish of all life stages are lost into the adjacent aquatic environment.

By means of IA, fish genotyped at seven microsatellite DNA loci (from King *et al.* 2001) are assigned to their continent of origin on the basis of a database of known-origin genotypes from about 2100 Atlantic salmon. The North American baseline database includes fish from Maine, all Canadian provinces, and the Penobscot and St John River aquaculture strains. The European database contains fish from Iceland, Norway, Finland, Scotland, Ireland, Spain, one of the main European aquaculture strains, and the hybrid North American and European strain. Management agencies are currently establishing a threshold likelihood ratio (e.g. European/North American score) that will minimise the chances of escaped second-generation (F_2) hybrids and various backcrosses from being incorporated into restoration programmes.

One more example of the use of assignment methods in a mixed-fishery situation was the assessment of the spatio-temporal distribution of four landlocked Atlantic salmon populations during their sympatric feeding phase in Lake St-Jean (Quebec, Canada) (Potvin and Bernatchez 2001). It was demonstrated that individuals from the four populations were not randomly distributed in the lake and that there was a weak negative correlation between genetic distance and spatial overlap in the lake. Another example illustrating the power of assignment methods on a small geographical scale is the study of Spidle *et al.* (2001). This showed that salmon from the unimpounded lower reaches of the Penobscot River in Maine are so distinct from those in two of the river's tributaries that the source of 90.4–96.1% of the individuals from the Penobscot drainage could be identified by IA methods.

Assignment tests can also be used to identify individuals on a temporal scale, i.e. to identify indigenous populations and individuals. In fact, this was the first application of assignment methods in studies of Atlantic salmon. With the aid of historical samples, Nielsen *et al.* (1997) used genetic information from four microsatellite loci to assign contemporary salmon from the highly threatened Skjern River population in Denmark to baseline samples consisting of historical scale samples from the same river collected in the 1930s and samples from two other populations. A very high proportion of the contemporary individuals were assigned to the historical samples, revealing that the indigenous population had persisted in the river. Similarly, Nielsen *et al.* (2001b) investigated whether remains of indigenous salmon could be found in other Danish rivers where exogenous salmon of a known source had been stocked in high numbers.

Assignment of adults and fry from the contemporary populations to a baseline of historical and exogenous source populations showed that, in two of the five populations studied, a significantly higher number than would be expected from misclassification alone was assigned to the indigenous populations. In particular, the proportion of fry was higher, indicating the better fitness of fish from the local population. As a result, indigenous salmon are now selected for broodstock, and exogenous fish are culled in both rivers.

Likewise, Vasemägi *et al.* (2001) used assignment tests to identify the genetic origin of salmon in the Selja, a river in Estonia that had been subject to such massive releases of hatchery fish that these now outnumbered wild fish. Four wild populations and two hatchery stocks served as the baseline for the assignment, showing that the most likely recolonisers of the river were from a nearby native river, the Kunda. These results suggest that native populations may still play an important role in the colonisation of former salmon rivers draining into the Baltic Sea.

9.2.2 Background to methodology

Principle of assignment tests

Assignment tests estimate the probability of encountering the genotype of an individual in a number of potential source populations. The individual is then assigned to the population in which its genotype is expected to be most frequent or from which the genetic distance of the individual is the shortest (Box 9.1). However, genetic distances are not usually intended for measuring distances between individuals; for that purpose, likelihood-based methods generally perform better than genetic distance methods. The highest resolution power has been attained with a Bayesian method estimating baseline allele frequencies to accommodate potentially missing rare alleles (Cornuet *et al.* 1999).

Box 9.1 Classical individual assignment.

In the classical individual assignment the multilocus genotype of an individual fish (X1, X2, X3, . . .) is compared with the multilocus genotype frequencies (MLGT fr.) of contributing baseline stocks (S1, S2 and S3), and the fish is assigned to the baseline stock in which it has the highest probability (W) of occurring (Fig. B9.1). The proportion of each individual baseline stock in the sample (p) is then calculated from the number of individuals assigned to each of them. The occurrence of missing rare alleles can be corrected by assuming even distribution for all alleles before baseline sampling.

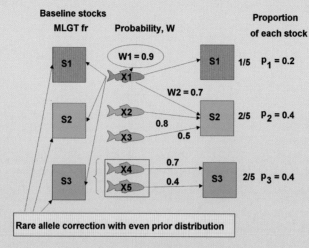

Fig. B9.1 A schematic picture of classical individual assignment.

An assignment test has three steps. First, a number of individuals, the baseline samples, are collected from the potential source populations. The allele frequencies for a number of (microsatellite) loci are estimated for all baseline stocks. Each individual in the baseline samples is then assigned to the population where its multilocus genotype is most frequent. Known as 'self assignment', this is done to assess the statistical power of the assignment. If all individuals in the baseline samples are assigned to a population of known origin, we have 100% assignment power. This is, however, rarely the case. The next step is to assign a number of individuals of unknown origin to the baseline populations. Again the unknown individuals are assigned to the population with the highest frequency of their multilocus genotypes. With the above estimated statistical power of assignment in mind, it can then be concluded, with varying degrees of confidence, from which of the populations the unknown individuals originate.

Different assignment methods

The original assignment test was frequency based. A number of computer programs are available for conducting 'classical assignment tests' (e.g. Banks and Eichert 2000; see also Hansen *et al.* 2001; Table 9.1), but recently developed probabilistic Bayesian assignment methods have higher power and have therefore gained wide application. The partly Bayesian exclusion test of Cornuet *et al.* (1999) included in the GeneClass software uses a Bayesian approach to estimate baseline allele frequencies by assuming an equal prior probability of occurrence for the allelic frequencies at each locus in each population. This assumption corrects for the occurrence of missing rare alleles in the baseline sampling. The method of Pritchard et al. (2000) is a fully Bayesian model based on a clustering method that groups individuals to minimise Hardy–Weinberg and linkage disequilibria. Individuals are assigned probabilistically to a set of populations or jointly to two or more populations if their genotypes indicate that they are admixed, i.e. they are hybrids of individuals from different populations. To assist the grouping of individuals, the program allows for the incorporation of additional prior, existing information on baseline populations, such as the level of genetic differentiation among identifiable groups, possible migration routes and the location of samples.

Power of assignment tests
The power of the assignment test depends on: (1) the level of genetic differentiation among populations, (2) the level of polymorphism at microsatellite loci, (3) the number of loci studied, (4) the baseline sample sizes, (5) the number of populations and (6) the statistical method used. It is difficult to give a single optimal design for assignment analyses, since the design also depends on the question addressed, and it is impossible to assess the power of assignment before at least some of the baseline samples have been collected. Unlike criminal cases, in which high exclusion power is of paramount importance, many biological questions do not need 100% assignment power. On the basis of both simulated and real population data, Manel et al. (2002) found that nearly all individuals can be assigned with high statistical certainty (99.9%) to two highly differentiated populations (differentiation level $F_{ST} \approx 0.15-0.20$) using 10 loci with heterozygosity > 0.6. The most thorough treatment of the subject is probably that by Cornuet et al. (1999) who found that, with 10 simulated populations, sample sizes of 30 individuals, 10 loci and F_{ST} of 0.1, assignment success was close to 100%. Nevertheless, certain factors can bias assignment results even when the assignment power is apparently high.

First, the baseline samples may not be representative of the individuals to be assigned. For instance, if the baseline samples were collected several years previously, they do not necessarily represent the present population. Especially in small and hatchery-reared populations, allele frequencies may vary considerably over time due to genetic drift caused by a low number of breeders.

Second, it is not always possible to collect baseline samples from all potential source populations. Although some programmes incorporate methods for assessing whether or not an individual belongs to any of the sampled baseline populations, unsampled populations are still a 'black box' that needs to be considered when interpreting the results.

Individual assignment seems to work relatively well if the F_{ST} values exceed 0.1, but bias may easily occur if the separation between stocks is less than that. The F_{ST} for Atlantic salmon stocks of the same phylogeographic lineage and from the same geographic area is often less than 0.1. As F_{ST} estimation does not, however, take into account the identity of alleles, but only differences in frequencies, the estimates are the same, irrespective of whether the sets of alleles overlap or not. Thus the separation power might actually be higher than that indicated by the F_{ST} value. However, the proportion of correct assignment according to a self-assignment test for nine microsatellite loci of 26 Atlantic salmon stocks in the Baltic Sea area was only 63% when the F_{ST} among the stocks was 0.07.

The drawback of traditional IA is that it does not take into account the multilocus genotype distribution of the mixture sample. Thus individuals may be assigned to baseline stocks with zero probabilities of occurring in the mixture on the basis of the genotype distribution in the mixture as a whole. Individuals are always assigned to the stocks in which they have the highest frequency, even if they have a relatively high frequency in some other baseline stocks as well. The classical IA approach is also unable to improve the initial assignment of each fish as the analysis is done only once. It is, however, possible to define the probabilities of individual fish belonging to any of the baseline stocks with a Bayesian mixed-stock analysis program (Pella and Masuda 2001). The program based on Bayesian mixture modelling does not suffer from the above shortcomings.

9.3 Identification of population contributions

Mixed-stock analysis (MSA) is used to estimate the proportion of each baseline population in a mixed sample of fish of unknown origin, and to determine the probability of each individual belonging to each of the baseline stocks. In contrast to individual assignment, MSA applies mixture modelling. It takes into account the genotypes of individual fish across multiple loci, the multilocus genotype distributions of the baseline samples and, in addition, the multilocus genotype distribution in the mixture sample.

The probability that an individual fish is from a given population or stock depends on (1) its particular genotype, (2) the relative frequency of that genotype in each of the baseline stocks, and (3) the proportions of the baseline stocks in the mixture; it equals the proportion of fish in the mixture sample with a particular genotype whose source is the stock in question (the posterior probability in equation 5 of Pella and Milner 1987). The probability of each stock occurring in the mixture is thus also reflected in the probability of each individual fish originating from any of the baseline stocks, which increases the separation power compared with classical individual assignment, IA. Mixed-stock analysis is commonly applied when fisheries are targeted at mixed aggregations of fish from several populations. It was initially based on allozyme data and maximum likelihood estimation (MLE). However, today, allozyme data are often replaced by DNA microsatellite variation information and the estimation method by Markov chain Monte Carlo (MCMC) and Bayesian methods, both of which increase the resolving power.

9.3.1 Application to Pacific salmon fisheries

MSA methods were first and most widely applied to Pacific salmon. The general recognition of a large number of more-or-less discrete breeding populations within each Pacific salmon species together with the existence of a substantial ocean fishery harvest in mixed-stock areas has underscored the need to determine the numbers, geographic distributions and fishery contributions of individual salmon stocks with considerable accuracy and precision. MSA is recognised as an important source of information for salmon management programmes in the USA and Canada. Some salmon mixed-stock fisheries, those of chinook *Oncorhynchus tshawytscha*, chum *Oncorhynchus keta*, pink *Oncorhynchus gorbuscha* and sockeye *Oncorhynchus nerka*, at least, have been analysed by this method. In general, MSA results provide information on (1) major differences in stock contributions at the level of country-of-origin from fishery to fishery, (2) temporal differences in stock-group contributions throughout the duration of an individual fishery, and (3) substantial differences in important component stock groups among fisheries (Shaklee *et al.* 1990).

For example, the MSA of the chum salmon mixed fishery has long been used in southern British Columbia and northern Washington to identify country-of-origin in order to (1) fulfil the accounting obligations of the Pacific Salmon Treaty and (2) identify geographic locations and time periods when fisheries could be open so as to minimise the catch of stocks of Fraser River origin. The occurrence of Fraser River chum salmon in catches is a criterion of quota limitations for many chum fisheries. Traditionally, allozyme data of 22 loci have been analysed by the maximum likelihood estimation (MLE) method.

The Columbia River chinook salmon gill-net fishery has been regulated on the basis of allozyme information since 1990. Commercial gill-net fisheries are specifically managed to target hatchery stocks originating from the lower Columbia River. Mixed-stock analysis is based on 27-stock, 22-loci allozyme baseline data, and applies the MLE method (Shaklee 1991; Shaklee et al. 1999). The chinook fishery in the lower river was reduced to get adequate numbers of upper river salmon stocks to return to their spawning grounds. The harvest rate at the lower river was set at 4.1% of the total upriver salmon run. Since 1990, the primary data used to manage the fishery have been in-season allozyme analysis data together with daily counts of the total fish harvest. Estimates of the cumulative impacts on upper river stocks are determined and when the harvest rate approaches the maximum acceptable impact level (4.1%), based on pre-season run size predictions, the fishery is closed.

Allozyme-based MSA has been used for decades to estimate stock proportions or determine region-of-origin in many mixed-stock fisheries of Pacific salmonids. For pink salmon, baseline data of 25 baseline stocks and 11 allozyme loci were sufficient to estimate the proportions of three genetically different stock groups: (1) Fraser River; (2) Canadian, non-Fraser River; and (3) Puget Sound stocks (Beacham et al. 1985). In this case, the estimation was checked with a comparison test in which the true stock proportions were known. For the estimation of sockeye salmon stock proportions, several biological markers (e.g. freshwater age, scale patterns and prevalence of a brain parasite) were used in addition to allozyme variation (Wood et al. 1989). When all markers were used in combination, the reliabilities of the proportion estimates for all test mixtures were acceptable with the criteria ± 10% at 95% confidence, when the mixture sample sizes were at least 300.

The pink salmon fishery in the Fraser River is an example of the systematic use of allozyme analysis in fisheries management (Shaklee et al. 1999). In 1987, the Pacific Salmon Commission began to use genetic MSA to identify Fraser River pink salmon in catches to the south of Alaska. In-season tissue samples were usually analysed within 3 days, thus providing fishery managers with up-to-date information on the contributions of Fraser River pink salmon to important fisheries and permitting the implementation of management measures to meet in-season catch and escapement requirements. Post-season estimates of actual Fraser River pink salmon catches by country and within US user groups were compared with the allocated catches. Shortfalls or excesses in allocation by country or user group were to be remedied in future catch paybacks. The information of MSA was also used for in-season run size estimation and migration behaviour studies of Fraser River pink salmon.

In many cases, separation power is better with DNA methods than with allozyme data. Mitochondrial DNA analysis has been used in combination with allozyme data to distinguish between Asian and North American stocks of chum salmon along the South Alaskan peninsula (Seeb and Crane 1999), and minisatellite analysis alone has been applied to three Pacific salmonids: sockeye salmon (Beacham et al. 1995), chum salmon (Beacham 1996) and coho salmon, *Oncorhynchus kisutch* (Miller et al. 1996).

Microsatellites have been tested for estimating the stock composition of coho salmon, sockeye salmon and steelhead trout O. *mykiss* in British Columbia. For coho salmon, three microsatellite loci were sufficient to estimate stock group proportions at the level of 'evolutionary significant units' (Small *et al.* 1998). But, for this species, the major histocompatibility complex variation has proved to be more powerful for stock identification than any of the eight microsatellite loci examined in the tests, and almost as powerful as all microsatellite loci together (Beacham *et al.* 2001). However, the number of microsatellite loci studied can be increased from the eight used in this study.

When four methods (i.e. parasite infection rate, allozyme loci, minisatellite loci and microsatellite loci) were compared for estimating the stock composition of three sockeye salmon stocks, the highest accuracy and greatest precision in the long term were obtained by analysis of only four microsatellite loci. However, parasite infection rate would have been more powerful for some earlier period. Minisatellites gave less reliable estimates, and allozyme variation gave the least reliable estimates of all four methods evaluated (Beacham *et al.* 1998).

For sockeye salmon, six microsatellite loci were sufficient to identify stocks of three lakes in the same area. The differences between lakes were, on average, 12 times greater than the variation within populations, confirming the relative stability of microsatellite loci (Beacham and Wood 1999). For steelhead, the eight DNA loci surveyed could also be used to provide relatively accurate and precise estimates of stock composition for fishery management purposes in one case (Beacham *et al.* 1999), but not between the Nass and Skeena rivers in British Columbia (Beacham *et al.* 2000). In general, the annual variation in allele frequencies relative to population differentiation was higher among steelhead than sockeye salmon stocks.

9.3.2 Application to Atlantic salmon fisheries

In principle, the methods of genetic mixed-stock analysis available for Atlantic salmon are the same as those for Pacific salmonids. However, the amount of allozyme variation and the number of observed variable loci in the former have been relatively low compared with those in some Pacific salmonids. Still, allozyme variation has been used successfully in some cases. In combination with scale characteristics, it has been used in the West Greenland fishery to estimate the contribution of stocks from different continents (Reddin *et al.* 1987) and, in combination with freshwater age, to define the proportion of wild stocks in the Baltic Sea fishery (Koljonen and McKinnell 1996; Koljonen and Pella 1997). The variation at four minisatellite loci has also been tested for MSA of Atlantic salmon in the River Shannon system in Ireland (Galvin *et al.* 1995).

In the Baltic Sea, MSA has been used to estimate the timing of spawning migration of wild Atlantic salmon stocks. The information can be used to ease the fishing pressure on this threatened catch component of Finland's coastal fisheries. The timing of wild and hatchery fish differ somewhat, wild fish tending to return earlier than hatchery fish. Opening the fishery after the majority of the wild fish have passed would enable the escapement of wild fish to be increased and thus natural smolt production in the rivers to be enhanced.

By combining the data of allele frequencies at seven allozyme loci with the freshwater age distribution, it is possible to identify three stock groups in Finnish coastal Atlantic salmon fisheries that are interesting from the management point of view (Fig. 9.3). Different harvest strategies are required in spring for: (1) wild stocks, (2) spawning migrating hatchery stocks and (3) locally sea-ranched stocks (Neva). Exploitation of wild stocks needs to be minimised

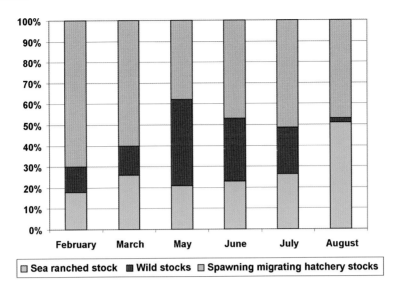

Fig. 9.3 Occurrence of three stock groups, important from the management point of view, in Atlantic salmon catches along the Finnish coast in the Baltic Sea (Koljonen and Pella 1997). Stock group proportions are based on data at seven allozyme loci and freshwater-age distributions, and were estimated by a maximum likelihood estimation method.

in offshore areas; some escapement is needed from the second group – the migrating hatchery stocks – to secure a sufficient number of spawners for hatchery production, but the third group, the sea-ranched Neva stock, could be exploited without restrictions.

Genetic mixed-stock analysis showed that the proportion of wild fish was at its highest in May and that almost all the wild fish had disappeared from catches by August. In May, more than 40% of the total coastal catch comprised wild fish (Fig. 9.3). The proportion of migrating hatchery fish was high at the beginning of the year, but in August the sea-ranched Neva stock was the major contributor, at over 50% of the catch. Its proportion of the catches increased as autumn approached.

The number of wild fish caught can be estimated from the total number of fish caught. Catch numbers were highest in the coastal fishery during the three summer months, when the proportion of wild fish caught was also high. Catches were at their highest in July, but the proportion of wild fish caught was highest in May (Table 9.2). In all, 28% of the total catch in the area studied was of wild origin, and 43% of the total wild catch was taken in May. So, opening the coastal fishery after May would have saved 55% of the passing wild salmon, and catch loss would have amounted to 30.3% of the total catch. Valuable information about the amounts of wild fish and the timing of their spawning migration in spring was thus available from genetic MSA.

Bayesian methods have been used to estimate stock proportions in Atlantic salmon catches in the Baltic Sea since 2000 (Koljonen 2004, 2006; Koljonen *et al.* 2005). With a baseline data set of 26 stocks and eight microsatellite loci, the proportions of five stock groups important from the management point of view could be estimated with high precision when the mixture sample sizes were near 200 (Table 9.3). The catch samples were representative of Finnish catches in three areas. In June, the bulk of the catch in the Åland Sea – the southernmost area – was composed of wild fish originating from rivers draining into the Gulf of Bothnia (69%),

Table 9.2 Trap net catches in one statistical square at the Finnish coast. Total number of fish caught, estimated percentage of wild fish and estimated number of wild fish caught during the year (Koljonen and Pella 1997).

Month	Catch numbers	Wild %	Wild catch numbers
April	71	14	10
May	5 706	41	2 340
June	5 116	30	1 535
July	6 461	23	1 486
August	1 325	2	27
September	279	2	6
October	56	2	1
Total	19 014		5 404

Table 9.3 Stock group proportions (%) in a Finnish Atlantic salmon catch sample in 2002 on the basis of genetic analysis. The Bayesian procedure BAYES (Pella and Masuda 2001) and variation at eight microsatellite loci were used for the estimation (ICES 2003; Koljonen 2004).

Origin of Atlantic salmon stock group	Fishery samples					
	Åland Sea 4 May to 27 June 60°10′N, 19°20′E $n = 218$		Bothnian Sea 29 May to 7 August 62°15′N, 21°15′E $n = 179$		Bothnian Bay 23 May to 23 August 65°00′N, 24°30′E $n = 180$	
	Mean	SD	Mean	SD	Mean	SD
Gulf of Bothnia, wild	68.8	5.1	38.9	6.0	43.1	5.2
Gulf of Bothnia, Finnish hatchery	16.5	4.4	56.8	6.1	47.6	5.2
Gulf of Bothnia, Swedish hatchery	13.2	3.2	1.6	1.8	9.1	3.3
Gulf of Finland, wild	0.0	0.1	0.6	0.6	0.0	0.2
Baltic Main Basin	1.5	0.8	2.1	1.1	0.1	0.3

and the contributions of Finnish and Swedish hatchery fish were more or less equal. In the more northerly samples – from the Bothnian Sea and Bothnian Bay – the contribution of wild fish was lower (39–43%), as was the proportion of Swedish fish in these Finnish coastal catches (Fig. 9.4).

9.3.3 Background to mixed-stock analysis

Maximum likelihood estimation

The identification of stocks and populations in catches, i.e. genetic stock proportion estimation, has traditionally been based on the multilocus genotype distributions in the mixture sample and in the known baseline stocks contributing, and potentially contributing, to the sample, using a conditional maximum likelihood estimation (MLE) method (Fournier *et al.*

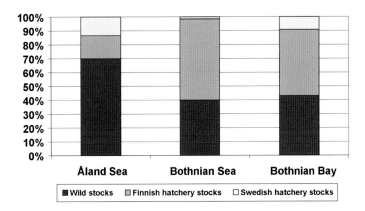

Fig. 9.4 Proportion estimates of Atlantic salmon stock groups in Finnish catches sampled in 2002 from three different sea areas on the basis of variation at eight DNA microsatellite loci and the Bayesian estimation.

1984; Pella and Milner 1987). The method determines the relative contributions of baseline stocks with the highest likelihood of providing the observed multilocus genotypic frequencies in the catch sample (see Box 9.2).

A statistical program for analysing mixtures (SPAM) is available for the Windows environment (Debevec *et al.* 2000). SPAM searches for maximum likelihood estimates of population proportions using three numerical algorithms: conjugate gradient (CG), iteratively reweighed least squares (IRLS) and expectation-maximization (EM). The algorithms are described in detail in Masuda *et al.* (1991) and Pella *et al.* (1996). Maximum likelihood-based methods are not usually suitable for identifying the stock of origin of individual fish and SPAM does not explicitly assign individuals to stocks. However, the EM algorithm used in SPAM (see equations 4 and 5 in Pella and Milner 1987) assigns individuals fractionally to baseline stocks with the fractions being equal to the estimated probabilities resulting from the analysis (i.e. posterior probabilities) for the source stocks. It also includes the 'conditional population probabilities' option, and one can assign individuals to the stocks for which this posterior probability is the greatest.

The most likely proportions can always be solved with likelihood-based methods, so the problem is not getting the estimates but rather assessing their reliability. If genetic differentiation among baseline stocks is inadequate, the baseline sampling is biased or the sample sizes are too small, the information obtained may be misleading or heavily biased. Reliability simulations with actual baseline data and test samplings are therefore necessary to assess the expected reliability of the estimates before the genetic estimation (Koljonen *et al.* 2005). The sizes of the baseline and mixture samples, as well as the number of variable loci needed, can be assessed with the simulation, depending on the accuracy and precision of the proportion estimates required for management purposes. Genetic estimation can also be validated by comparing genetic estimates with other types of information such as that obtained from tagging or blind tests (Brodziak *et al.* 1992).

Bayesian method

A new probabilistic method based on Bayesian statistics has been developed for stock composition estimation by Pella and Masuda (2001). Prior to the analysis, information on multilocus

Box 9.2 Maximum likelihood estimation of stock proportions.

Maximum likelihood estimation of stock proportions is based on the assumption that the sum of baseline stock multilocus genotype distributions (S1, S2 and S3) times their proportions ($p1$, $p2$ and $p3$) has to be the same as the observed mixture multilocus genotype frequency distribution (Mixture, MLGT fr.) (Fig. B9.2).

Baseline genotype frequencies are obtained from the observed allele frequencies of the baseline stocks by assuming Hardy–Weinberg equilibrium and independence of loci within the stocks.

Given the multilocus genotypic distributions in c baseline stocks and the distribution observed in the catch mixture sample, the proportions of baseline stocks (p_i, $i = 1,2, \ldots, c$, $\sum_{i=1}^{c} p_i = 1$) comprising the stock mixture are best estimated as those for which the observed mixture genotype distribution is most probable. The probability of sampling a particular multilocus genotype h from any mixture is $\lambda_h = \sum_{i=1}^{c} p_i g_{hi}$, where g_{hi} is the frequency of that genotype in the baseline stock i. The probability of the sample, or the likelihood function that is maximised with respect to the p_i's, is obtained as the product of the individual probabilities of the N observed genotypes:

$L = \prod_h \lambda_h^{m_h}$, where m_h is the observed number of fish of multilocus genotype h in the catch sample. The resulting estimate represents the proportion of baseline stocks that provides the statistically most likely fit to the distribution of multilocus genotypes observed in the fishery sample.

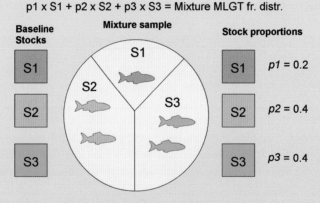

Fig. B9.2 The principle of conditional maximum likelihood estimation of stock proportions.

genotype distributions is obtained from baseline samples (i.e. prior for the baseline stocks), while no prior knowledge is assumed for the stock proportions in the mixture (i.e. prior for stock proportions is even distribution). This knowledge is updated by the multilocus genotype distribution in the catch mixture sample, and makes it possible to estimate both the stock composition of the mixture sample and the genotypic composition of the baseline stocks. The program also provides probabilities for each individual of belonging to each of the baseline stocks (Box 9.3). These probabilities can be used to assign the individuals to their stock of origin, just as can the posterior probabilities computed by SPAM.

Bayesian mixture modelling has several advantages over plain individual assignment and traditional MLE. First of all, mixture modelling takes into account the multilocus genotype distribution in the mixture, as does the maximum likelihood, but not the individual assignment,

Box 9.3 A schematic picture of Bayesian estimation.

The multilocus genotype of each individual fish (X1, X2, . . .) is assigned (with probability w) repeatedly to a stock of origin by using the probability distribution of stock proportions (p) and multilocus genotype distributions (MLGT fr.) (Fig. B9.3). The first round is done by using prior distributions for both stock proportions (Prior p1), which is uninformative, and baseline stocks (Prior MLGT fr. Q1), which is informative and is derived from baseline sampling. They are then converted into posterior distributions (Posterior Q2 and Posterior p2) according to assignment results.

As a result of the iterative process, the Bayes program outputs MCMC chains of stock proportions for each baseline stock in the mixture and also the posterior distribution of the baseline allele frequencies. The iterative process alternately assigns the mixture individuals to the possible stocks and, from the assignments, updates and draws from the revised distributions for baseline allele frequency distributions and stock proportions. The assignments to stocks are random, with probabilities equal to source posterior probabilities (see equation 5 in Pella and Milner 1987) computed from current estimates of the unknowns. The chains are run until the estimates have stabilised and the chains converged. The number of iterations needed depends on the difficulty of the problem, but it may be as many as several thousand. To monitor the convergence of the chains to the posterior density, a univariate statistic called the shrink factor (Gelman and Rubin 1992) is computed for each of the stock proportions. The shrink factor compares the variation within a single chain with the total variation among the chains.

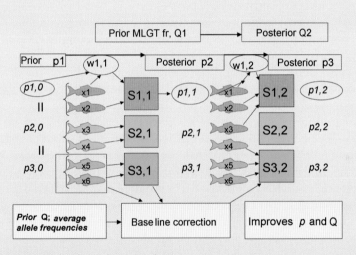

Fig.B9.3 A schematic picture of Bayesian estimation.

method. In addition, the methodologically inherent bias typical of MLE, which underestimates major stocks and overestimates minor stocks, is smaller in the Bayesian estimation, especially when the mode of the posterior distribution is used as a point estimate. Further, it can improve the estimation from the first initial assignment by repeating the assignment run after run when fitting the multilocus genotype distribution of a mixture sample to that of individual assignments of individual fishes. Another important advantage of the Bayesian method is that the potentially missing rare alleles in microsatellite data in baseline sampling can be corrected for, as all are assumed to occur in all baseline samples, albeit only at a very low frequency. Moreover, direct probability statements can be made about stock proportions. The Bayesian application can also include information about the genetic similarity of the baseline stocks or their geographical distances.

Reliability of stock proportion estimates

Variation in the composition estimates may derive from the mixture sample, the baseline sample, or both. When the samples are large enough, these two errors are additive in theory (Pella and Robertson 1979). However, the bias of the proportion estimates is mainly due to the baseline data, and is at its greatest when genetically similar stocks differ markedly in abundance (Millar 1987). In general, the maximum likelihood method tends to underestimate the proportions of major stocks and overestimate those of minor stocks.

The variance resulting from mixture sampling depends on the size and stock composition of the sample. Wood *et al.* (1987) observed that the number of stocks contributing to a mixture has little or no effect on the model performance, although the number of fish per stock sample is important. If, however, the number of contributing stocks is high, larger mixture samples are needed for reliable estimates. According to Wood *et al.* (1987), in the maximum likelihood estimation the critical sample size for several levels of separation is about 40 fish per stock; below that the reliability of the estimates is greatly reduced. In the mixed-stock analysis of Atlantic salmon catches in the Baltic Sea, the proportions of six stock groups could be estimated with less than 10% error in absolute value using a mixture sample size of 300 fish and seven allozyme loci (Koljonen and Pella 1997).

Baseline sampling is generally more complicated than mixture sampling. Several factors may affect the amount of variation and bias in the estimates: completeness of baseline data, conformity with Hardy–Weinberg expectations, sample sizes, temporal variation in allele frequencies, number of loci, genetic differentiation among stocks and pooling strategies. In the event that not all baseline stocks are sampled, fish cannot be assigned to unknown stocks. This might be a marked source of error if not all of the contributing stocks are known. The multilocus genotype distributions of the baseline stocks are calculated from allele frequency data assuming Hardy–Weinberg equilibrium; if this assumption is not met it might cause bias. The recommended minimum sample size is 50 fish per baseline stock (Wood *et al.* 1987). The precision and accuracy of the estimates are often easily improved by increasing the baseline sample sizes, say, to 100 fish for each baseline stock. The required number of loci depends on the level of differentiation among stocks and is case-specific. As the number of variable loci available might cause limitations in allozyme analysis of Atlantic salmon, it is worthwhile using all potentially variable loci. In microsatellite analysis, the number of variable loci is not a limiting factor.

Temporal changes in baseline allele frequencies should also be taken into account as a potential source of error. In general, however, any effects of changes caused by genetic drift can be compensated for to a marked extent by collecting baseline data over several years (Waples 1990). The importance of repeated sampling depends on the life history of the species concerned and on the degree of overlapping in the year classes. The Atlantic salmon has overlapping generations, and partly for that reason the temporal variation of allele frequencies in large natural stocks may well be insignificant, whereas in small natural stocks or in hatchery stocks genetic drift can cause pronounced changes.

When the identification of individual stocks has been unsuccessful, groups of genetically similar stocks can still often be estimated accurately enough for management purposes (Seeb and Crane 1999). Intentional changes in allele frequencies by controlled matings in hatcheries can also be applied to create a genetically identifiable population. These changes do not need to be dramatic. Even a relatively small increase in the frequency of a rare allele – a quantitative change – may improve separation of the stock of interest (Seeb *et al.* 1986, 1990; Gharrett and Seeb 1990). A qualitative change resulting in identification of all the individuals of a certain group – the offspring of planned matings – is also possible, but it frequently requires the use of a very small number of parents and thus causes a bottleneck in the population history.

If genetic information alone is insufficient for accurate stock identification, additional information such as parasite infection rates and/or scale pattern characteristics (Rutherford *et al.* 1994; Wood *et al.* 1988, 1989; Pella *et al.* 1998; Wilmot *et al.* 1999) or freshwater age distributions (Koljonen and Pella 1997) may be useful in increasing the separation power. The traditional MLE is incapable of correcting for potentially missing rare alleles in the samples and hence might cause bias when microsatellite data are used. However, SPAM (Debevec *et al.* 2000) now includes an option in which baseline allele frequencies are estimated by Bayesian methods.

9.4 Resolving power of different markers

The most important factor for successful application of MSA and IA is sufficient genetic differentiation among baseline populations. Various statistical methodologies have different powers in separation. However, the genetic markers used are also important for the successful outcome of these analyses. Allozymes have been the markers of choice for genetic studies of fishes since the 1960s and have also been used for MSA. Unfortunately, the levels of genetic variation at allozyme loci are generally low, usually with only two or three alleles at each polymorphic locus. Further, for Atlantic salmon the number of polymorphic loci is also low. As a result, allozyme diversity rarely allows individual river stocks of Atlantic salmon to be identified. For example, in the northern Baltic Sea, two out of 16 baseline stocks could be identified as individual river stocks on the basis of plain allozyme data, whereas the contribution of the other stocks could be assessed only as part of a stock group composed of genetically similar stocks (Koljonen and Pella 1997).

With the recent revolution in molecular biology, the use of DNA-based genetic markers has gained wide application in population genetic studies of fish in general and also in MSA and IA. The currently preferred markers, microsatellites, have a number of attractive features. First of all, they are widely distributed in the genomes of most organisms, which means that a large number of genetic markers can be isolated relatively easily, thus increasing precision. Further, microsatellite loci are, in general, much more variable than allozyme loci, and the larger number of alleles enhances the genetic resolution power. For brown trout *Salmo trutta* it has been estimated that, on average, one variable microsatellite locus provides the same discriminatory power as eight allozyme loci (Estoup *et al.* 1998b).

The greater variability of microsatellites than of allozymes is illustrated in the study of Baltic salmon of Koljonen *et al.* (2002). They found that the average mean heterozygosity was more than ten times as high at microsatellite (0.704) as at allozyme loci (0.058). Further, the number of different alleles was generally only two or three at allozyme loci whereas the number of alleles ranges from 8 to 32, with a mean of 17.8 alleles per microsatellite locus within Baltic salmon populations. This has not changed much since historical times in the large wild stocks (Säisä *et al.* 2003). Allele numbers of up to 50 have been reported in Atlantic salmon from the east coast of Canada (McConnell *et al.* 1997) and numbers from 2 to 29 in European Atlantic salmon populations (Nielsen *et al.* 1999; Norris *et al.* 1999). This high allelic diversity increases the number of multilocus genotypes in populations to a level larger than the actual sample sizes. For mixed-stock analysis, the loci used can be chosen to have only a reasonable number of alleles. For microsatellite studies, the number of loci available is not a limiting factor.

Table 9.4 Simulation test for the separation of Atlantic salmon baseline stocks in the Baltic Sea using maximum likelihood estimation, data of eight DNA microsatellite loci and a sample size of 400. In the simulations, each of the baseline stocks contributed 50% to the catch sample, the contributions of the remaining stocks being distributed evenly. Bootstrap resampling was done 300 times for both mixture and baseline samples. The table shows the means of 300 simulations, observed estimate, bias, standard deviation of proportion estimates, 96% confidence interval and mean number of fish (over eight loci) sampled for each baseline stock.

Stock		Expected %	Observed %	Bias	Std	0.95 CI	Mean N
1	Tornionjoki, hatchery	50	42.2	−7.8	4.2	8.2	125.8
2	Tornionjoki, wild	50	38.8	−11.2	4.5	8.8	131.5
3	Simojoki	50	46.4	−3.6	3.2	6.3	63.9
4	Iijoki	50	42.0	−8.0	3.8	7.5	67.3
5	Oulujoki	50	46.0	−4.0	3.5	6.8	63.5
6	Kalixälven	50	45.1	−5.0	3.9	7.6	189.1
7	Luleälven	50	42.0	−8.0	4.1	8.1	67.1
8	Byske	50	43.8	−6.2	3.4	6.7	81.6
9	Skellefteälven	50	45.6	−4.4	3.7	7.2	51.1
10	Vindelälven	50	40.0	−10.0	4.2	8.3	56.3
11	Umeälven	50	39.9	−10.2	4.0	7.9	54.4
12	Lögde	50	47.2	−2.8	3.3	6.5	50.3
13	Ångermanälven	50	42.2	−7.8	3.9	7.7	65.4
14	Indalsälven	50	44.6	−5.5	3.8	7.4	71.4
15	Ljusnan	50	40.5	−9.5	4.0	7.8	58.1
16	Dalälven	50	44.5	−5.5	3.9	7.7	66.6
17	Neva	50	48.7	−1.3	2.9	5.7	114.3
	Mean		43.5	−6.5	3.8	7.4	

The increased resolving power of microsatellite data can be demonstrated with a simulation comparison in which individual stock proportions of Baltic salmon stocks were estimated by MLE. Variance estimates of stock proportions for both allozyme and DNA microsatellite data were assessed for 17 Atlantic salmon stocks. The allozyme baseline data set, which had seven loci, was the same as that used by Koljonen and Pella (1997); the microsatellite data set had eight loci (Table 9.4). The mixture sample size was set at 400 in all cases. Each catch sample was made up of 50% of one particular baseline stock whereas the rest of the catch sample was evenly distributed among the remaining stocks (50%/16 = 3.1%). For variance estimation, bootstrap resampling was conducted 300 times on both the baseline stock and the mixture samples. The results showed that, in all cases, the proportion of the major stock was underestimated, causing a downward bias. In the allozyme data, the 50% proportion estimates ranged from 31% to 49% and the mean bias was − 10% in absolute value. In the microsatellite data, the estimates ranged from 39% to 48%, with a mean bias of − 6.5% (Tables 9.4 and 9.5).

The precision of the 50% estimates was three times as high for the microsatellite data as for the allozyme data when assessed as the range of the confidence interval. The means of the estimated standard errors of the proportion estimates were 13.7% for the allozyme and 3.8% for the microsatellite data. F_{ST} values for the tested data sets were clearly below 0.1, being 0.02 for the allozyme loci and 0.04 for the microsatellite loci.

Mitochondrial DNA is also a highly variable marker, but it has the disadvantage of being effectively only a single locus, and so application is limited to highly separated populations

Table 9.5 Simulated comparison of maximum likelihood estimation errors using allozyme and microsatellite data from 17 Atlantic salmon baseline stocks in the Baltic Sea. In the simulations, the proportion of each of the 17 baseline stocks was set at 50% in turn, the others having even distribution. Means of variance estimates derive from 17 simulations.

Variance estimates	7 allozyme loci	8 microsatellite loci
Mean standard error of estimates	13.7	3.8
Range of standard errors	3.5–19.9	2.9–4.5
Mean range of 95% confidence interval	26.8	7.4
Range of confidence interval of individual stock proportion estimates	5 stocks < 10% 10 stocks < 15%	5% > all stocks < 10%
Mean bias for 50% estimate	−10.5	−6.5

or population groups such as those of North American and European Atlantic salmon (see review by Reddin and Friedland (1999) and references therein). Mitochondrial variation would most likely give the best results when combined with either allozyme or microsatellite loci. Minisatellites are highly variable, but they are not so numerous or easy to isolate from the genome. They have nevertheless been used for discriminating between European and North American Atlantic salmon (Taggart *et al.* 1995). Mitochondrial DNA is not inherited according to a Mendelian pattern and thus genotypes do not follow Hardy–Weinberg distribution, making them less powerful than microsatellites for MSA. At present, then, the markers of choice for both MSA and IA are microsatellites.

As the development of molecular markers continues, the preferred marker for MSA and IA could change. One potential marker type is single nucleotide polymorphism (SNP). In general SNPs are much less variable than microsatellites, but their high numbers in the genome and the ease with which automated scoring can be carried out may outweigh their lack of resolution power. A recent study demonstrated that, for reliable estimates of relatedness, 100 SNP loci had to be employed to achieve the same power as that given by 16 microsatellite loci (Brumfield *et al.* 2003). Another very promising method is amplified fragment length polymorphism (AFLP). Campbell *et al.* (2003) found that, with a given comparable analytical effort in the laboratory, AFLPs were much more effective than microsatellites at discriminating among potential source populations for a species (whitefish, *Coregonus clupeaformis*) with low levels of inter-population differentiation.

9.5 Summary and conclusions

- Information on the spatial and temporal variation in stock composition in mixed-stock fisheries is essential for effective fisheries management. Genetic differences among fish stocks can be, and have been, used to estimate stock and stock group proportions in mixtures and to identify the stock of origin of individual fishes.
- Genetic identification of fish stocks has several advantages over traditional external tagging provided that the genetic differentiation among stocks is sufficient for the estimation. Genetic identification programs give statistical and probabilistic information about the stock composition of catches and about the origin of individual fishes.

- Atlantic salmon are often the target of mixed-stock fisheries either offshore or in coastal areas, notably in West Greenland waters and the Baltic Sea.
- In principle, genetic identification can be based on individual assignment (IA) or mixture modelling, which is applied in genetic mixed-stock analysis (MSA). In IA, each fish is assigned to the known stock in which its multilocus genotype is most likely to occur. MSA, in contrast, also takes into account the multilocus genotype distribution in the mixture sample. The probability of each stock occurring in the mixture is thus reflected in the probability of each individual fish originating from any of the baseline stocks, which increases the separation power over that of IA.
- For individual assignment it has been demonstrated that with 10 simulated populations assignment success has been close to 100%, with baseline sample sizes of 30 individuals, 10 loci and an F_{ST} of 0.1 among baseline stocks. Individual assignment is a simple and useful genetic identification method when differentiation among contributing stocks is considerable ($F_{ST} \geq 0.1$) and when only very few individuals are assigned.
- Mixed-stock analysis has been sufficiently reliable when using mixture sample sizes of 300, baseline sample sizes of 50 and 7 allozyme loci to estimate six stock groups in Baltic salmon fisheries by maximum likelihood estimation. As well as genetic variables, a combination of freshwater age, scale characteristics and parasite infection rates has been used to increase the separation power of allozyme variation.
- A probabilistic method based on Bayesian statistics and developed for stock composition estimation gives more precise and accurate estimates than does maximum likelihood estimation. It gives probability distributions for baseline stock proportions in the mixture sample, for baseline allele frequencies and also for the origin of individual fishes.
- Exploitation of DNA variation has greatly increased the amount of genetic variation available for stock identification. However, the high number of alleles in DNA microsatellite loci has created problems, as the estimate may be biased if rare alleles are missed in the sampling of potentially contributing stocks. For the present, the markers of choice are microsatellites.
- By applying mixture modelling of the microsatellite variation of nine loci with a Bayesian approach, it has been possible to obtain stock group proportion estimates for five Baltic salmon stock groups with errors ≤ 6% in absolute value, when the F_{ST} for the baseline stocks was 0.04.
- Genetic results have been used for Pacific salmonids to determine fishery openings and closures, to provide harvest benefits or meet conservation needs, to address catch allocation and equity issues among user groups and between countries, and also to provide data for in-season run size updates, and investigate migration patterns and timing.
- Genetic identification has been used to assess the continents of origin of Atlantic salmon in the West Greenland fishery and to estimate wild stock proportions and national contributions to Baltic salmon catches.

9.6 Management recommendations

Genetic methods can be used to determine either the proportional contribution of populations to mixtures, or the origin of individual fish. The genetic differentiation observed among Atlantic salmon stocks on average seems to offer sufficient resolution power for the use of genetic identification. In many contexts, genetic analysis is likely to be more practical and cost-effective than traditional methods such as physical tagging or scale analysis and it also can be used as an independent source of information in addition to that available from more traditional methods. However, the use of genetic identification has been underutilised. Wider exploitation of genetic methods would help to provide increased management insights into patterns and levels of exploitation, as well as to advance basic understanding of the differences in the spatial and temporal distribution and movements of Atlantic salmon populations.

Acknowledgements

The authors are grateful to Jerry Pella, who read a preliminary version of the manuscript and made a number of valuable proposals for improvements.

Further reading

Cadrin, S.X., Friedland, K.D. and Waldman, J.R. (2005) *Stock Identification Methods: Applications in fishery science*. Elsevier Academic Press, Amsterdam and London.

References

Banks, M.A. and Eichert, W. (2000) WHICHRUN (Version 3.2) a computer program for population assignment of individuals based on multilocus genotype data. *Journal of Heredity*, **91**: 87–89.

Banks, M.A., Rashbrook, V.K., Calavetta, M.J., Dean, C.A. and Hedgecock, D. (2000) Analysis of microsatellite DNA resolves genetic structure and diversity of chinook salmon (*Oncorhynchus tshawytscha*) in California's Central Valley. *Canadian Journal of Fisheries and Aquatic Sciences*, **57**: 915–927.

Beacham, T.D. (1996) The use of minisatellite DNA variation for stock identification of chum salmon, *Oncorhynchus keta*. *Fishery Bulletin*, **94**: 611–627.

Beacham, T.D. and Wood, C.C. (1999) Application of microsatellite DNA variation to estimation of stock composition and escapement of Nass River sockeye salmon (*Oncorhynchus nerka*). *Canadian Journal of Fisheries and Aquatic Sciences*, **56**: 297–310.

Beacham, T.D., Withler, R.E. and Gould, A.P. (1985) Biochemical genetic stock identification of pink salmon (*Oncorhynchus gorbusha*) in South British Columbia and Puget Sound. *Canadian Journal of Fisheries and Aquatic Sciences*, **42**: 1474–1483.

Beacham, T.D., Withler, R.E. and Wood, C.C. (1995) Stock identification of sockeye salmon by means of minisatellite DNA variation. *North American Journal of Fisheries Management*, **15**: 249–265.

Beacham, T.D., Margolis, L. and Nelson, R.J. (1998) A comparison of methods of stock identification for sockeye salmon (*Oncorhynchus nerka*) in Barkley Sound, British Columbia. *NPAFC Bulletin. Assessment and status of Pacific Rim salmon stocks*. **No 1**, 227–240.

Beacham, T.D., Pollard, S. and Le, K.D. (1999) Population structure and stock identification of steelhead in Southern British Columbia, Washington, and the Columbia River based on microsatellite DNA variation. *Transactions of the American Fisheries Society*, **128**: 1068–1084.

Beacham, T.D., Pollard, S. and Le, K.D. (2000) Microsatellite DNA population structure and stock identification of steelhead trout (*Oncorhynchus mykiss*) in the Nass and Skeena rivers in Northern British Columbia. *Marine Biotechnology*, **2**: 587–600.

Beacham, T.D., Candy, J.R., Supernault, K.J., Ming, T., Deadle, B., Shulze, A., Tuck, D., Kaukinen, K.H., Irvine, J.R., Miller, K.M. and Withler, R.E. (2001) Evaluation and application of microsatellite and major histocompatibility complex variation for stock identification. *Transactions of the American Fisheries Society*, **130**: 1116–1149.

Begg, G.A., Friedland, K.D. and Pearce, J.B. (1999) Stock identification – its role in stock assessment and fisheries management: a selection of papers presented at a symposium of the 28th annual meeting of the American Fisheries Society in Hartford, Connecticut, USA, 23–27 August 1998. *Fisheries Research*, **43/1–3**: 249 pp.

Bermingham, E., Forbes, S.H., Friedland, K. and Pla, C. (1991) Discrimination between Atlantic salmon (*Salmo salar*) of North American and European origin using restriction analyses of mitochondrial DNA. *Canadian Journal of Fisheries and Aquatic Sciences*, **48**: 884–893.

Bourke, E.A., Coughlan, J., Jansson, H., Galvin, P. and Cross, T.F. (1997) Allozyme variation in populations of Atlantic salmon located throughout Europe: diversity that could be compromised by introduction of reared fish. *ICES Journal of Marine Science*, **54**: 974–985.

Brodziak, J., Bentley, B., Bartley, D., Gall, G.A.E., Gomulkiewicz, R. and Mangel, M. (1992) Test of genetic stock identification using coded wire tagged fish. *Canadian Journal of Fisheries and Aquatic Sciences*, **49**: 1507–1517.

Brumfield, R.T., Beerli, P., Nickerson, D.A. and Edwards, S.V. (2003) The utility of single nucleotide polymorphisms in inferences of population history. *Trends in Ecology and Evolution*, **5**: 249–256.

Campbell, D., Duchesne, P. and Bernatchez, L. (2003) AFLP utility for population assignment studies: analytical investigation and empirical comparison with microsatellites. *Molecular Ecology*, **12**: 1979–1991.

Cornuet, J-M., Piry, S., Luikart, G., Estoup, A. and Solignac, M. (1999) New methods employing multi-locus genotypes to select or exclude populations as origins of individuals. *Genetics*, **153**: 1989–2000.

Crossman, E.J. and McAllister, D.E. (1986) Zoogeography of freshwater fishes of the Hudson Bay drainage, Ungava Bay and the Arctic Archipelago. In: C.H. Hocutt and E.O. Wiley (Ed.) *Zoogeography of North American Freshwater Fishes*, pp. 53–104. John Wiley & Sons, New York.

Davidson, W.S., Birt, T.P. and Green, J.M. (1989) A review of genetic variation in Atlantic salmon, *Salmo salar* L. and its importance for stock identification, enhancement programmes and aquaculture. *Journal of Fish Biology*, **34**: 547–560.

Debevec, E.M., Gates, R.B., Masuda, M., Pella, J., Reynolds, J. and Seeb, L.W. (2000) SPAM (Version 3.2): statistics program for analyzing mixtures. *Journal of Heredity*, **91**: 509–510.

Estoup, A., Gharbi, K., San Cristobal, M., Chevalet, C., Haffray, P. and Guyomard, R. (1998a) Parentage assignment using microsatellites in turbot (*Scophthalmus maximus*) and rainbow trout (*Oncorhynchus mykiss*) hatchery populations. *Canadian Journal of Fisheries and Aquatic Sciences*, **55**: 715–725.

Estoup, A., Rousset, F., Michalakis, Y., Cornuet, J-M., Adriamanga, M. and Guyomard, R. (1998b) Comparative analysis of microsatellite and allozyme markers: a case study investigating microgeographic differentiation in brown trout (*Salmo trutta*). *Molecular Ecology*, **7**: 339–353.

Fournier, D.A., Beacham, T.D., Riddell, B.E. and Busack, C.A. (1984) Estimating stock composition in mixed stock fisheries using morphometric, meristic, and electrophoretic characteristics. *Canadian Journal of Fisheries and Aquatic Sciences*, **41**: 400–408.

Galvin, P., McKinnell, S., Taggart, J.B., Ferguson, A., O'Farrell, M. and Cross, T.F. (1995) Genetic stock identification of Atlantic salmon (*Salmo salar* L.) using single locus minisatellite DNA profiles. *Journal of Fish Biology*, **47** (Supplement A): 186–199.

Gelman, A. and Rubin, D.B. (1992) Inference from iterative simulation using multiple sequences. *Statistical Science*, **7**: 457–511.

Gharrett, A.J. and Seeb, J.E. (1990) Practical and theoretical guidelines for genetically marking fish populations. *American Fisheries Society Symposium*, **7**: 407–417.

Haig, S.M., Gratto-Trevor, C.L., Mullins, T.D. and Colwell, M.A. (1997) Population identification of western hemisphere shorebirds throughout the annual cycle. *Molecular Ecology*, **6**: 413–427.

Hansen, M.M., Kenchington, E. and Nielsen, E.E. (2001) Assigning individual fish to populations using microsatellite DNA markers: methods and applications. *Fish and Fisheries*, **2**: 93–112.

Hartley, S.E. and Horne, M.T. (1984) Chromosome polymorphism and constitutive heterochromatin in the Atlantic salmon, *Salmo salar*. *Chromosoma*, **89**: 377–380.

Hilborn, R. (1985) Apparent stock recruitment relationship in mixed stock fisheries. *Canadian Journal of Fisheries and Aquatic Sciences*, **42**: 718–723.

ICES (2003) *Report of the Baltic Salmon and Trout Assessment Working Group*. International Council for the Exploration of the Sea, Advisory Group on Fishery Management, *ICES CM*, **2003/ACFM:20**, Karlskrona, Sweden, 2–11 April 2003.

King, T.L., Spidle, A.P., Eackles, M.S., Lubinski, B.A. and Schill, W.B. (2000) Mitochondrial DNA diversity in North American and European Atlantic salmon with emphasis on the Downeast rivers of Maine. *Journal of Fish Biology*, **57**: 614–630.

King, T.L., Kalinowski, S.T., Schill, W.B., Spidle, A.P. and Lubinski, B.A. (2001) Population structure of Atlantic salmon (*Salmo salar* L.): a range wide perspective from microsatellite DNA variation. *Molecular Ecology*, **10**: 807–821.

Koljonen, M-L. (2001) Conservation goals and fisheries management units for Atlantic salmon in the Baltic Sea area. *Journal of Fish Biology*, **59** (Supplement A): 269–288. (Published online, DOI: 10.1006/jfbi.2001.1757)

Koljonen, M-L. (2004) Changes in stock composition of annual Atlantic salmon catches in the Baltic Sea on basis of DNA-microsatellite data and Bayesian estimation. *ICES CM* 2004/Stock Identification Methods/**EE:08**: 1–21.

Koljonen, M-L. (2006) Annual changes in the proportions of wild and hatchery Atlantic salmon (*Salmo salar*) caught in the Baltic Sea. *ICES Journal of Marine Science*, **63**: 1274–1285.

Koljonen, M-L. and McKinnell, S. (1996) Assessing seasonal changes in stock composition of Atlantic salmon catches in the Baltic Sea with genetic stock identification. *Journal of Fish Biology*, **49**: 998–1018.

Koljonen, M-L. and Pella, J.J. (1997) The advantage of using smolt age with allozymes for assessing wild stock contributions to Atlantic salmon catches in the Baltic Sea. *ICES Journal of Marine Science*, **54**: 1015–1030.

Koljonen, M-L., Jansson, H., Paaver, T., Vasin, O. and Koskiniemi, J. (1999) Phylogeographic lineages and differentiation pattern of Atlantic salmon in the Baltic Sea with management implications. *Canadian Journal of Fisheries and Aquatic Sciences*, **56**: 1766–1780.

Koljonen, M-L., Tähtinen, J., Säisä, M. and Koskiniemi, J. (2002) Maintenance of genetic diversity of Atlantic salmon (*Salmo salar*) by captive breeding programmes and the geographic distribution of microsatellite variation. *Aquaculture*, **212**: 69–92.

Koljonen, M-L., Pella, J.J. and Masuda, M. (2005) Classical individual assignments versus mixture modelling to estimate stock proportions in Atlantic salmon (*Salmo salar*) catches from DNA microsatellite data. *Canadian Journal of Fisheries and Aquatic Sciences*, **62**: 2143–2158.

Manel, A., Berthier, P. and Luikart, G. (2002) Detecting wildlife poaching: identifying the origin of individuals with Bayesian assignment tests and multilocus genotypes. *Conservation Biology*, **16**: 650–659.

Masuda, M., Nelson, S. and Pella, J. (1991) *User's manual for GIRLSEM, GIRLSYM, and CONSQRT*. The Computer Programs for Computing Conditional Maximum Likelihood Estimates of Stock Composition from Discrete Characters. Personal Computer Version. USA-DOC-NOAA-NMFS Auke Bay Laboratory, Auke Bay. Release September 1991.

Mathias, J.A., Franzin, W.G., Craig, J.F., Babaluk, J.A. and Flannagan, J.F. (1992) Evaluation of stocking walleye fry to enhance a commercial fishery in a large, Canadian prairie lake. *North American Journal of Fisheries Management*, **12**: 299–306.

McConnell, S.K., O'Reilly, P., Hamilton, L., Wright, J.N. and Bentzen, P. (1995) Polymorphic microsatellite loci from Atlantic salmon (*Salmo salar*): genetic differentiation of North American and European populations. *Canadian Journal of Fisheries and Aquatic Sciences*, **52**: 1863–1872.

McConnell, S.K.J., Ruzzante, D.E., O'Reilly, T.O. and Hamilton, L. (1997) Microsatellite loci reveal highly significant genetic differentiation among Atlantic salmon (*Salmo salar* L.) stocks from east coast of Canada. *Molecular Ecology*, **6**: 1075–1089.

Millar, R.B. (1987) Maximum likelihood estimation of mixed stock fishery composition. *Canadian Journal of Fisheries and Aquatic Sciences*, **44**: 583–590.

Miller, K.M., Withler, R.E. and Beacham, T.D. (1996) Stock identification of coho salmon (*Oncorhynchus kisutch*) using minisatellite DNA variation. *Canadian Journal of Fisheries and Aquatic Sciences*, **53**: 181–195.

Nielsen, E.E., Hansen, M.M. and Loeschke, V. (1997) Analysis of microsatellite DNA from old scale samples of Atlantic salmon: a comparison of genetic composition over sixty years. *Molecular Ecology*, **6**: 487–492.

Nielsen, E.E., Hansen, M.M. and Loeschke, V. (1999) Genetic variation in time and space: microsatellite analyses of extinct and extant populations of Atlantic salmon. *Evolution*, **53**: 261–268.

Nielsen, E.E., Hansen, M.M., Schmidt, C., Meldrup, D. and Grønkjær, P. (2001a) Population of origin of Atlantic cod. *Nature*, **413**: 272.

Nielsen, E.E., Hansen, M.M. and Bach, L. (2001b) Looking for a needle in a haystack: discovery of indigenous salmon in heavily stocked populations. *Conservation Genetics*, **2**: 219–232.

Nielsen, E.E., Nielsen, P.H., Meldrup, D. and Hansen, M.M. (2004) Genetic population structure of turbot (*Scophtalmus maximus* L.) supports the presence of multiple hybrid zones for marine fishes in the transition zone between the Baltic Sea and North Sea. *Molecular Ecology*, **13**: 585–595.

Norris, A.T., Bradley, D.G. and Cunningham, E.P. (1999) Microsatellite genetic variation between and within farmed and wild Atlantic salmon (*Salmo salar*) populations. *Aquaculture*, **180**: 247–264.

Olsen, J.B., Bentzen, P., Banks, M.A., Shaklee, J.B. and Young, S. (2000) Microsatellites reveal population identity of individual pink salmon to allow supportive breeding of a population at risk of extinction. *Transactions of the American Fisheries Society*, **129**: 232–242.

Paetkau, D., Calvert, W., Stirling, I. and Strobeck, C. (1995) Microsatellite analysis of population structure in Canadian polar bears. *Molecular Ecology*, **4**: 347–354.

Payne, R.H., Child, A.R. and Forrest, A. (1971) Geographical variation in the Atlantic salmon. *Nature*, **231**: 250–252.

Pella, J. and Masuda, M. (2001) Bayesian methods for analysis of stock mixtures from genetic characters. *Fisheries Bulletin*, **99**: 151–167.

Pella, J.J. and Milner, G.B. (1987) Use of genetic marks in stock composition analysis. In: N. Ryman and F. Utter (Ed.) *Population Genetics and Fishery Management*, pp. 247–276. Washington Sea Grant Program. University of Washington Press, Seattle, WA.

Pella, J.J. and Robertson, T. (1979) Assessment of composition of stock mixtures. *National Marine Fisheries Service Fishery Bulletin*, **77**: 387–398.

Pella, J.J., Masuda, M. and Nelson, S. (1996) *Search algorithms for computing stock composition of a mixture from traits of individuals by maximum likelihood*. NOAA technical memo NMFS-AFSC-61. US Department of Commerce, Washington, DC.

Pella, J., Masuda, M., Guthrie, C., Kondzela, C., Gharrett, A., Moles, A. and Winans, G. (1998) *Stock composition of some sockeye salmon, Oncorhynchus nerka, catches in southeast Alaska, based on incidence of allozyme variants, freshwater ages, and a brain-tissue parasite*. NOAA Technical Report NMFS 132. Auke Bay Laboratory, Alaska Fisheries Science Center, Juneau, AK.

Potvin, C. and Bernatchez, L. (2001) Lacustrine spatial distribution of landlocked Atlantic salmon populations assessed across generations by multi-locus individual assignment and mixed-stock analyses. *Molecular Ecology*, **10**: 2375–2388.

Pritchard, J.K., Stephens, M. and Donnelly, P. (2000) Inference of population structure using multilocus genotype data. *Genetics*, 155: 945–959.

Reddin, D.G. and Friedland, K.D. (1999) A history of identification to continent of origin of Atlantic salmon (*Salmo salar* L.) at west Greenland, 1969–1997. *Fisheries Research*, 43: 221–235.

Reddin, D.G., Verspoor, E. and Downton, P.R. (1987) An integrated phenotypic and genotypic approach to discriminating Atlantic salmon. *ICES CM* **1987/M:16**.

Ricker, W.E. (1958) Maximum sustainable yields from fluctuating environments and mixed stocks. *Journal of Fisheries Research Board of Canada*, 15: 991–1006.

Rutherford, D.T., Wood, C.C., Jantz, A.L. and Southgate, D.R. (1994) Biological characteristics of Nass River sockeye salmon (*Oncorhynchus nerka*) and their utility for stock composition analysis of test fishery samples. *Canadian Technical Report of Fisheries and Aquatic Sciences*, no. **1988**, 1–72.

Säisä, M., Koljonen, M-L. and Tähtinen, J. (2003) Genetic changes in Atlantic salmon stocks since historical times and the effective population sizes of the long-term captive breeding programmes. *Conservation Genetics*, 4: 613–627.

Seeb, L.W. and Crane, P.A. (1999) Allozymes and mtDNA discriminate Asian and North American populations of chum salmon in mixed-stock fisheries along the South Alaska peninsula. *Transactions of the American Fisheries Society*, 128: 88–103.

Seeb, J.E., Seeb, L.W. and Utter, F.M. (1986) Use of genetic marks to assess stock dynamics and management programs for chum salmon. *Transactions of the American Fisheries Society*, 115: 448–454.

Seeb, L.W., Seeb, J.E., Allen, R.L. and Hershberger, W.K. (1990) Evaluation of adult returns of genetically marked chum salmon, with suggested future applications. *American Fisheries Society Symposium*, 7: 418–425.

Shaklee, J.B. (1991) *Simulations and other analysis of the 1991 Columbia River spring chinook GSI baseline.* WDF Technical Report 115. Washington Department of Fisheries, Olympia, WA.

Shaklee, J.B., Busack, C., Marshall, A., Miller, M. and Phelps, S.R. (1990) The electrophoretic analysis of mixed-stock fisheries of Pacific salmon. In: Z-I. Ogita and C.L. Markert (Ed.) *Isozymes: Structure, function, and use in biology and medicine*, pp. 235–265. *Progress in Clinical and Biological Research*, **344**. Wiley-Liss Inc., New York.

Shaklee, J.B., Beacham, T.D., Seeb, L. and White, B.A. (1999) Managing fisheries using genetic data: case studies from four species of Pacific salmon. *Fisheries Research*, 43: 45–78.

Small, M.P., Withler, R.E. and Beacham, T.D. (1998) Population structure and stock identification of British Columbia coho salmon, *Oncorhynchus kisutch*, based on microsatellite DNA variation. *Fishery Bulletin*, 96: 843–858.

Spidle, A.P., Schill, W.B., Lubinski, B.A. and King, T.L. (2001). Fine-scale population structure in Atlantic salmon from Maine's Penobscot River drainage. *Conservation Genetics*, 2: 11–24.

Ståhl, G. (1987) Genetic population structure of Atlantic salmon. In: N. Ryman and F. Utter (Ed.) *Population Genetics and Fishery Management*, pp. 121–141. University of Washington Press, Seattle, WA.

Taggart, J.B., Verspoor, E., Galvin, P.T., Moran, P. and Ferguson, A. (1995) A minisatellite DNA marker for discriminating between European and North American Atlantic salmon (*Salmo salar*). *Canadian Journal of Fisheries and Aquatic Sciences*, 52: 2305–2311.

Thorpe, J.E. and Mitchell, K.A. (1981) Stocks of Atlantic salmon (*Salmo salar*) in Britain and Ireland: discreteness, and current management. *Canadian Journal of Fisheries and Aquatic Sciences*, 38: 1576–1590.

Vasemägi, A., Gross, R., Paaver, T., Kangur, M., Nilsson, J. and Eriksson L-O. (2001) Identification of the origin of an Atlantic salmon (*Salmo salar* L.) population in a recently recolonized river in the Baltic Sea. *Molecular Ecology*, 10: 777–782.

Verspoor, E. (1988) Reduced genetic variability in first-generation hatchery populations of Atlantic salmon. *Canadian Journal of Fisheries and Aquatic Sciences*, 45: 1686–1690.

Verspoor, E. (1997) Genetic diversity among Atlantic salmon (*Salmo salar* L.) populations. *ICES Journal of Marine Sciences*, **54**: 965–973.

Waples, R.S. (1990) Temporal changes of allele frequency in Pacific salmon: implications for mixed-stock fishery analysis. *Canadian Journal of Fisheries and Aquatic Sciences*, **47**: 968–976.

Wilmot, R., Kondzela, C., Guthrie, C., Moles, A., Martinson, E. and. Helle, J. (1999) Origins of sockeye and chum salmon seized from the Chinese vessel *Ying Fa* (NPAFC document). Auke Bay Fisheries Laboratory, Alaska Fisheries Science Center, Juneau, AK.

Wood, C.C., McKinnell, S., Mulligan, T.J. and Fournier, D.A. (1987) Stock identification with the maximum-likelihood mixture model: sensitivity analysis and application to complex problems. *Canadian Journal of Fisheries and Aquatic Sciences*, **44**: 866–881.

Wood, C.C., Oliver, G.T. and Rutherford, D.T. (1988) Comparison of several biological markers used in stock identification of sockeye salmon (*Oncorhynchus nerka*) in northern British Columbia and southeast Alaska. *Canadian Technical Report of Fisheries and Aquatic Sciences*, **1624**, 1–49.

Wood, C.C., Rutherford, D.T. and McKinnell, S. (1989) Identification of sockeye salmon (*Oncorhynchus nerka*) stocks in mixed-stock fisheries in British Columbia and southeast Alaska using biological markers. *Canadian Journal of Fisheries and Aquatic Sciences*, **46**: 2108–2120.

10 Fisheries Exploitation

K. Hindar, C. García de Leániz, M-L. Koljonen, J. Tufto and A. F. Youngson

Upper: capture of salmon from a boat in the Faroes long line fishery in the early 1980s (photo credit: A. Youngson.)
Lower: net and coble fishery at Bonar Bridge on the Kyle of Sutherland in Northeast Scotland (photo credit: D. Hay.)

The exploitation of Atlantic salmon, *Salmo salar*, in food and recreational fisheries has a long history in the rivers and coastal waters of the species range. Prior to the last few centuries, when river stocks were largely healthy, exploitation was not a serious management concern. However, increasingly with many populations in decline, fisheries exploitation has come to be a major concern for managers of salmon fisheries, both in rivers and in the sea. Until recently, most of this concern focused on the direct threat posed by the reduction in numbers. However, recent research suggests that the impacts are likely to be more complex and insidious through effects on the genetic structure of the affected populations.

10.1 Introduction

The Atlantic salmon is a highly sought-after species, and has probably been so since humans colonised the Atlantic coasts of Europe and North America many thousands of years ago. Atlantic salmon can be found as 40 000-year-old bone remains in Spanish caves (Consuegra *et al.* 2002), and as 6000 to 10 000-year-old rock paintings and rock carvings in Norway and Sweden (Fig. 10.1). Moreover, the species is one among very few that are mentioned in the earliest national legislation (e.g. from the thirteenth century in Norway), and its exploitation is continually a matter of strong national and international regulation (NASCO 2005). The species is exploited for food and recreation, both with strong economic implications. This generates potential conflicts not only among nations, but also between commercial and recreational interests within nations. The high value of Atlantic salmon as a resource invites overexploitation, which has led to concerns about population viability and evolutionary change in exploited populations (Schaffer and Elson 1975).

Exploitation can have evolutionary consequences. One is the selective removal of highly valued species, often large-sized and late-maturing top predators (Atlantic salmon being one of them), followed by successive concentration on smaller species, often at lower trophic levels (Pauly *et al.* 1998). Other effects of exploitation include selection of particular populations within species and selection of particular phenotypes within populations, both of which may have genetic consequences. Selective removal of populations can occur even if the fishery itself is nonselective. When fishing mortality is high and several populations contribute to the fishery, differences in life history and productivity may lead to the less productive populations being selectively removed (Larkin 1977). Finally, undirected (random) loss of genetic variation within populations occurs when exploitation reduces spawning escapement and subsequently

Fig. 10.1 Rock paintings of salmonids from Tingvoll, Norway, 6000 years BP. Actual size is ~ 1 m body length. (Image credit: K. Hindar.)

the effective population size of the population (Wright 1969). All types of genetic effects (selection among and within populations, and increased genetic drift) can be viewed as loss of genetic variation. The topic has previously been reviewed by Miller (1957) who did not find evidence of genetic change caused by selective fishing, and more recently by Nelson and Soulé (1987) and Thorpe (1993) who found some evidence of genetic effects caused by fishery exploitation. Recently, genetic changes in life-history traits caused by selective fishing have been implicated in the collapse of the northern cod, *Gadus morhua*, off Labrador and Newfoundland (Olsen *et al.* 2004).

The number of wild Atlantic salmon may now be at an all-time low (Kellogg 1999; Crozier *et al.* 2003). Small population sizes may lead to increased rates of inbreeding and loss of genetic variation, which may in turn have direct negative consequences for a number of fitness-related traits (Nelson and Soulé 1987; Chapter 8). Moreover, loss of genetic variation can reduce the potential for a population to adapt to changing environments (Lande and Shannon 1996; Chapter 7). Indeed, the genetically effective size of a population has been used as one criterion for determining the extinction risk, e.g. in the World Conservation Union (IUCN) guidelines for categorising threatened species (Mace and Lande 1991).

In fisheries management, consideration of the effective population size translates into finding conservation limits that set constraints on the maximum yield that can be sustainably harvested from the populations. A problem, however, is that conservation guidelines have been designed for isolated populations, whereas anadromous salmon populations – although genetically different – are only partially isolated from one another (Ståhl 1987; Chapter 5). This means that the level of genetic variation is a function not only of the population size and demography of the local population, but also of the neighbouring populations and of the level of migration (gene flow) among them. A few migrants between populations can provide input to reduce the risk of inbreeding and loss of genetic variation even though they do not contribute much to reduce demographic risks in a local population (Waples 1998).

In this chapter, we present fisheries exploitation in a biological and management context, review harvest rates of Atlantic salmon populations, outline possible genetic effects of fishing, and examine the evidence for genetic changes caused by fishing of Atlantic salmon.

10.2 A historical perspective on fisheries exploitation

The exploitation of Atlantic salmon in fresh water constitutes probably the oldest kind of fishery (Cleyet-Merle 1990). Spears and fixed engines have been used in a number of rivers, especially where salmon had to pass through rapids and other narrow parts. In wider and more slow-flowing rivers, seines and other nets have been more common. Today, most of the freshwater catch of Atlantic salmon is by rod and line (Fig. 10.2), a tradition that was brought by Britons to most of the world's salmon-producing rivers in the nineteenth century and which has reached high popularity throughout the range of the species.

Along the coasts, bag nets (Fig. 10.3), bend nets and other fixed engine methods have century-old traditions. These methods developed in areas where the migratory routes of salmon from many rivers coincide, and can be highly efficient in narrow fjords and in other locations where migrating salmon seek near shore. In the open ocean, exploitation of Atlantic salmon took place from the late 1950s. The methods used were drift nets and long-line fishing operated from ocean-going vessels.

Fig. 10.2 Rod-caught Atlantic salmon from the River Verdalselva, Norway. (Photo credit: D.H. Karlsen.)

Fig. 10.3 Bag net in the Drammensfjorden, Norway. (Photo credit: L.P. Hansen, NINA.)

Concerns about the decline of salmon populations from the 1980s, and the mixed-stock nature of fishing in the ocean, led to strong regulations of both coastal and oceanic fisheries. Quotas for the fisheries around the Faroes and off West Greenland have since 1984 been set by an international body (North Atlantic Salmon Conservation Organization, NASCO) based on scientific advice from the International Council for the Exploration of the Sea (ICES). Although Atlantic salmon use vast areas in the North Atlantic Ocean during their feeding migration, populations from different regions are concentrated in different oceanic areas. In the West Greenland fishery, salmon from North America (Canada, USA) are represented along with European populations from Scotland, Ireland, the United Kingdom (UK) (excluding Scotland), France and Spain, whereas in the Faroes fishery, populations from Norway, Scotland, Russia, Ireland and the UK are represented (Crozier *et al.* 2003). Genetic methods (allozymes and DNA) can be used to distinguish between continental origins of salmon in the West Greenland fishery (Verspoor 1988; Chapter 5). Such a mixed-fishery analysis has been taken a step further in the Baltic Sea, where the relative contributions of all stock groups that are considered important from a fisheries management point of view are estimated on the basis of DNA microsatellite variation (Chapter 9; Koljonen *et al.* 2005).

Following the regulation of coastal and oceanic fisheries, in-river fisheries account for an increasing proportion of the salmon catches in the North Atlantic. In 2004, 66% of North American catches took place in rivers compared to 42% in north-east Atlantic countries (ICES 2005). Another trend is an increase in the use of catch and release by anglers. While Russia reported that 76% of the total rod catch was released during the 2004 angling season, this practice is viewed with scepticism by animal health authorities in Norway. They consider catch and release in numerically strong populations as being unethical, whereas selective release of wild salmon can be acceptable where a vulnerable population coexists with sea trout and/or escaped farm salmon that are targeted for capture (Statens Dyrehelsetilsyn 2002). Countries also differ widely in other types of fisheries regulation. However, a common pattern is that regulation is based on concepts of catch rate and/or spawning escapement.

10.2.1 Catch statistics

The reported catch of Atlantic salmon in the North Atlantic peaked at about 12 000 tonnes annually in 1973–75, and thereafter showed a steady decline to less than 2500 tonnes during the last few years (Fig. 10.4). In the peak years during the 1960s and 1970s, catches reached 2000 tonnes or more in several countries, including Canada, Norway, Ireland and the UK (Scotland, in particular). Recently, only the Norwegian catches have reached 1000 tonnes. The decline in total catch from the 1980s is partly explained by the closure or regulation of

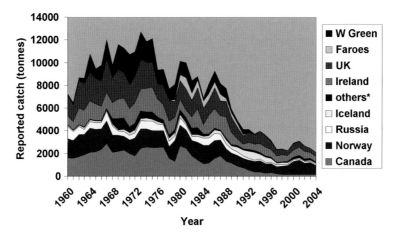

Fig. 10.4 Total reported catches 1960–2004 of Atlantic salmon in the North Atlantic, including river, coastal and oceanic fisheries. * = USA, Sweden, Denmark, Finland, France and Spain (1960–2004), and catches in the Norwegian Sea (1968–1984). Based on data from ICES (2005).

some fisheries, especially at sea. For example, the West Greenland fishery accounted for 2000 tonnes or more in the early 1970s (Fig. 10.4), but now accounts for less than 20 tonnes (ICES 2005). Likewise, the significance of the Faroes fishery, which reached 1025 tonnes in 1981, is now negligible. In the Norwegian Sea, two other fisheries peaked at around 1000 tonnes within this period: long-lining in 1970 (closed from 1984 onwards) and drift nets in 1981 (closed from 1989). However, the decline in catches also reflects lower survival rates of Atlantic salmon in the ocean (Friedland *et al.* 1998), and possibly reduced smolt production caused by habitat degradation (WWF 2001).

10.2.2 Exploitation rates

Estimates of exploitation rates on Atlantic salmon populations suggest that variable and sometimes high proportions of returning adults are caught in sea and river fisheries (Table 10.1a,b). The highest exploitation rates for rod and line are found in Norway, Iceland and Spain, where in-river exploitation rates often exceed 0.30. Salmon in rivers of the United Kingdom and Ireland typically experience lower rates of exploitation, but may in addition be exploited by net and coble used in the estuaries (e.g. Shearer 1992).

Exploitation rates can also be high in the marine environment (Table 10.1a,b). During the early 1980s, marine exploitation rates were estimated at 0.70 or above for many salmon populations on either side of the North Atlantic Ocean (Hansen 1988; Crozier and Kennedy 1993; Dempson *et al.* 2001), Iceland being a possible exception (Scarnecchia *et al.* 1989). The rise and fall of the Faroes and West Greenland fisheries (Fig. 10.4) suggest temporally variable exploitation rates of salmon in the sea, with a significant decline during recent years.

10.2.3 Potential for selection

River fishing is targeted on sexually maturing anadromous fish. Although the methods used may not directly select for body size, indirect selection may occur through selection on run timing which varies both within and between populations. A common pattern is that early-run, large fish are more heavily exploited than late-run, smaller fish (Gee and Milner 1980; Consuegra *et al.* 2005a). Consistent with this is that subpopulations spawning in the upper tributaries may be subject to higher exploitation rates than those spawning downstream (Youngson *et al.* 2003). An opposite pattern is seen in some Norwegian rivers, where 1-sea-winter (1SW) fish seem to be more heavily exploited than multi-sea-winter fish (Sættem 1995; Sandhaugen and Hansen 2001). An apparently nonselective river fishery is the net fishing on some Russian rivers, where the entire river is closed every second or third day. Another example may be catch-and-release fishing. Catch and release may impose little mortality on the fish, provided water temperatures are low, but may affect the behaviour after release (Dempson *et al.* 2002; Thorstad *et al.* 2003). At any rate, all fishing is likely to act more strongly on anadromous individuals than on individuals maturing in fresh water (Caswell *et al.* 1984).

Most fishing methods used in the marine environment are size-selective, especially as a particular mesh size of the nets catches fish with a certain girth size with higher probability than either larger or smaller fish. Also, the potential for selection is stronger in marine than in river fisheries, as both immature and maturing salmon may be targeted for capture. By being size-selective, these methods can also select among populations, for example by not catching populations that return as 1-sea-winter salmon to spawn.

Table 10.1a Exploitation rates sustained by different Atlantic salmon populations.

Population (Country)	Period	Environment	Type of fishery	Stock origin	Age class	Exploitation rate Mean	Exploitation rate Range	Reference
North America								
Newfoundland								
Little Codroy River	—	Marine	Nets	W	1SW	0.47	—	Murray (1968)
	—	Marine	Nets	W	MSW	0.75	—	"
Western Arm Brook	—	Marine	Nets	W	1SW	0.62	—	Reddin (1981)
	—	Marine	Nets	W	1SW	0.65	—	Chadwick et al. (1985)
	1984–91	Marine	Nets	W	1SW	0.57	0.40–0.64[1]	Dempson et al. (2001)
	1984–91	Marine	Nets	W	MSW	0.96	0.94–0.99[1]	"
Exploits River	1984–91	Marine	Nets	W	1SW	0.46	0.24–0.62[1]	"
	1984–91	Marine	Nets	W	MSW	0.76	0.55–0.86[1]	"
Gander River	1984–91	Marine	Nets	W	1SW	0.66	0.66–0.70[1]	"
	1984–91	Marine	Nets	W	MSW	0.72	0.55–0.82[1]	"
Middle Brook	1984–91	Marine	Nets	W	1SW	0.37	0.22–0.49[1]	"
	1984–91	Marine	Nets	W	MSW	0.80	0.66–0.88[1]	"
Terra Nova River	1984–91	Marine	Nets	W	1SW	0.35	0.22–0.45[1]	"
	1984–91	Marine	Nets	W	MSW	0.69	0.60–0.78[1]	"
NE Placentia River	1984–91	Marine	Nets	W	1SW	0.39	0.20–0.53[1]	"
	1984–91	Marine	Nets	W	MSW	0.74	0.58–0.85[1]	"
	—	River	Rods	W	1SW+MSW	0.38	—	Chadwick (1982)
Humber River	1984–91	Marine	Nets	W	1SW	0.47	0.23–0.63[1]	Dempson et al. (2001)
	1984–91	Marine	Nets	W	MSW	0.60	0.34–0.72[1]	"
Lomond River	1984–91	Marine	Nets	W	1SW	0.28	0.12–0.41[1]	"
	1984–91	Marine	Nets	W	MSW	0.63	0.42–0.77[1]	"
Torrent River	1984–91	Marine	Nets	W	1SW	0.56	0.39–0.66[1]	"
	1984–91	Marine	Nets	W	MSW	0.75	0.57–0.86[1]	"
Middle Brook	1984–99	River	Rods	W?	1SW	0.26	0.06–0.53	O'Connell (2003)
Indian Bay Brook	1997–99	River	Rods	W?	1SW	0.14	0.12–0.16	"
Labrador								
Sand Hill River	—	Marine	Nets	W	1SW	0.33	—	Peet & Pratt (1972)
	—	Marine	Nets	W	1SW	0.36	—	Reddin (1981)
	—	Marine	Nets	W	MSW	0.90	—	Peet & Pratt (1972)
	—	Marine	Nets	W	MSW	0.92	—	Reddin (1981)

Table 10.1a (cont'd)

Population (Country)	Period	Environment	Type of fishery	Stock origin	Age class	Exploitation rate Mean	Exploitation rate Range	Reference
New Brunswick								
NW Miramichi River	—	Marine	Nets	W	1SW	0.32	—	Saunders (1969)
	—	Marine	Nets	W	1SW	0.34	—	Kerswill (1971)
	—	Marine	Nets	W	MSW	0.87	—	Saunders (1969)
	—	Marine	Nets	W	MSW	0.78	—	Kerswill (1971)
SW Miramichi River	—	Marine	Nets	W	1SW	0.36	—	"
	—	Marine	Nets	W	MSW	0.92	—	"
Salmon River	—	River	Rods	W	1SW+MSW	0.16	—	Chadwick (1982)
Nova Scotia								
Liscomb River	—	Marine	Nets	H	1SW	0.36	—	Semple & Cameron (1990)
	—	Marine	Nets	H	MSW	0.79	—	"
Maine								
Lakes		Lakes (landlocked)	Rods	H/W	—	—	0.18–0.41	Warner & Havey (1985)
Machias River	1962–72	River	Rods	H/W	1SW+MSW	0.20	0.12–0.22	Baum (1997)
Narraguagus River	1962–74	River	Rods	H/W	1SW+MSW	0.26	0.11–0.41	"
Penobscot River	1969–94	River	Rods	H/W	1SW+MSW	0.09	0.01–0.28	"
Europe								
Iceland								
R. Haukadalsá		Marine	Nets	W	MSW	0.16	—	Scarnecchia et al. (1989)
Laxá í Leirársveit		Marine	Nets	W	MSW	0.48	—	"
Laxá í Kjós		Marine	Nets	W	MSW	0.29	—	"
Thverá		Marine	Nets	W	MSW	0.21	—	"
Nordurá	—	Marine	Nets	W	MSW	0.28	—	"
Laxá í Dölum	1972–85	River	Rods	W	1SW+MSW	0.25	0.11–0.82	Gudjónsson (1988)
Fáskrúd		Marine	Nets	W	MSW	0.27	—	Scarnecchia et al. (1989)
		Marine	Nets	W	MSW	0.28	—	"
R. Ellidaár	1965–76	River	Rods	W	1SW+MSW	0.35	0.18–0.58	Gudjónsson (1988)
R. Úlfarsá	1955–63	River	Rods	W	1SW+MSW	0.29	0.14–0.46	"
R. Blandá	1982–85	River	Rods	W	1SW+MSW	0.65	0.55–0.82	"
avg of 3 rivers	—	River	Rods	W	1SW	—	0.39–0.73	Gudjónsson et al. (1996)
	—	River	Rods	W	MSW	—	0.75–0.84	"

Region/River	Years	Location	Gear	H/W	Sea age	Value	Range	Reference
Northern Ireland								
R. Bush	1983–90	Marine+coastal	Nets	H	1SW	0.77	0.46–0.94	Crozier & Kennedy (1993)
	1983–90	Marine+coastal	Nets	W	1SW	0.70	0.62–0.89	"
	1983–90	Marine+coastal	Nets	H/W	MSW	0.45	0.36–0.60	"
	1973–88	River	Rods	—	1SW+MSW	0.11	0.05–0.17	Crozier & Kennedy (2001)
R. Burrishoole	—	Marine+coastal	Nets	H	1SW	—	0.52–0.90	Crozier & Kennedy (1994)
	1970–80	River	Rods	H	1SW	0.11	0.03–0.24	Mills & Piggins (1983)
	1970–80	River	Rods	W	1SW	0.14	0.07–0.23	"
	1970–81	River	Rods	W?	1SW+MSW	0.12	0.06–0.20	Mills *et al.* (1986)
R. Erne		Marine+coastal	Nets	H	1SW		0.54–0.64	Crozier & Kennedy (1994)
Scotland								
West coast	1981–83	Coastal	Nets	W	1SW	0.04	0.02–0.04[2]	Shearer (1992)
		Estuary	Nets	W	1SW	0.03	0.04–0.05[2]	"
		River	Rods	W	1SW	0.05	0.03–0.14[2]	"
North-west coast	1979–81	Coastal	Nets	W	1SW	0.06	0.06–0.08[2]	"
		Estuary	Nets	W	1SW	0.03	0.02–0.04[2]	"
		River	Rods	W	1SW	0.02	0.02–0.04[2]	"
North coast	1977–79	Coastal	Nets	W	1SW	0.06	0.05–0.07[2]	"
		Estuary	Nets	W	1SW	0.10	—	"
		River	Rods	W	1SW	0.03	0.03–0.04[2]	"
North-east coast	1985–88	Coastal	Nets	W	1SW	0.17	0.05–0.23[2]	"
		Estuary	Nets	W	1SW	0.02	0.00–0.03[2]	"
		River	Rods	W	1SW	0.05	0.03–0.08[2]	"
Moray Firth	1978–83	Coastal	Nets	W	1SW	0.11	0.07–0.15[2]	"
	1978–83	Estuary	Nets	W	1SW	0.09	0.04–0.15[2]	"
	1978–83	River	Rods	W	1SW	0.05	0.02–0.06[2]	"
	1983	Coastal	Nets	W	MSW	0.08	0.04–0.12[2,3]	"
	1983	Estuary	Nets	W	MSW	0.06	0.03–0.09[2,3]	"
	1983	River	Rods	W	MSW	0.04	—	"
East coast	1954–78	Coastal	Nets	W	1SW	0.29	0.23–0.37[2]	"
	1954–78	Estuary	Nets	W	1SW	0.26	0.19–0.35[2]	"
	1954–78	River	Rods	W	1SW	0.01	0.01–0.02[2]	"
	1952–78	Coastal	Nets	W	MSW	0.27	0.24–0.32[2]	"
	1952–78	Estuary	Nets	W	MSW	0.28	0.16–0.36[2]	"
	1952–78	River	Rods	W	MSW	0.04	0.04–0.08[2]	"
R. North Esk	1981–87	Estuary	Nets	W	1SW	0.27	0.15–0.40	"
	1981–87	Estuary	Nets	W	MSW	0.40	0.29–0.59	"
	1981–91	Sea (Faroes)	L.Line	W	1SW	—	0.00–0.04	MAFF/SOAEFD/WOAD (1999)
	1981–91	Sea (Faroes)	L.Line	W	MSW	—	0.00–0.18	"

Table 10.1a (cont'd)

Population (Country)	Period	Environment	Type of fishery	Stock origin	Age class	Exploitation rate Mean	Exploitation rate Range	Reference
R. Spey	1983–85	Estuary	Nets	W	1SW+MSW	0.09	0.04–0.11[2]	Shearer (1988)
	1983–85	River	Rods	W	1SW+MSW	0.06	0.05–0.08	"
Wales								
R. Wye	1925–34	River	Rods	W	1SW+MSW	0.25	—	Gee & Milner (1980)
	1965–74	River	Rods	W	1SW+MSW	0.47	—	"
Norway								
R. Imsa	1981–84	Marine+coastal	Nets	H	1SW	0.88	0.82–0.96	Hansen (1988)
	1981–84	Marine+coastal	Nets	W	1SW	0.77	0.66–0.84	"
	1981–84	Marine+coastal	Nets	H	MSW	0.97	0.94–1.00	"
	1981–84	Marine+coastal	Nets	W	MSW	0.94	0.94–0.95	"
	—	Marine+coastal	Nets	H	1SW	—	0.37–0.81	"
	—	Marine+coastal	Nets	H	MSW	—	0.22–0.70	"
R. Drammenselv	—	River (downstream)	Rods	H	1SW+MSW	0.33	—	Hansen et al. (1986)
	—	River (upstream)	Rods	H	1SW+MSW	0.04	—	"
	1985–90	River	Rods	H	1SW+MSW	—	0.33–0.53	Hansen (1990)
R. Lærdalselv	1960–74	River	Rods	W	1SW+MSW	0.54	0.44–0.61	Jensen (1981)
R. Eira	1966–74	River	Rods	W	1SW+MSW	—	0.46–0.83	"
Finland								
R. Tana	1995	River	Nets	W	1SW	0.51	—	Karppinen et al. (2004)
	1995	River	Rods	W	1SW	0.21	—	"
	1992–93	River	Nets	W	MSW	0.27	0.27–0.28	Erkinaro et al. (1999)
	1992–93	River	Rods	W	MSW	0.27	0.13–0.41	"
Spain								
R. Bidasoa	1980–2000	River	Rods	H/W	1SW+MSW	0.25	0.08–0.53	Alvarez & Lamuela (2001)
	1992–2000	River	Rods	H/W	1SW	0.10	0.06–0.14	"
	1992–2000	River	Rods	H/W	MSW	0.29	0.07–0.45	"
R. Nansa	1986–95	River	Rods	W	1SW+MSW	0.21	0.12–0.36	García de Leániz et al. (1992, unpbl)
R. Asón	1982–96	River	Rods	W	1SW+MSW	0.40	0.17–0.64	"

[1] 5th–95th percentiles.
[2] Exploitation of vulnerable population available to the fishery (rather than total population).
[3] 95% CL.

Table 10.1b Summary of differential exploitation ratios calculated from paired comparisons in Table 10.1a.

Character examined in paired comparisons of exploitation rates	Mean ratio of differential exploitation	Range	No. of studies (n)
(a) By fishery (nets/rods)	3.05	1.00–5.33	9
(b) By age class (MSW/1SW)			
in rod fisheries	2.25	0.80–4.00	4
in net fisheries	1.65	0.53–2.73	24
(c) By stock origin (wild/hatchery)	1.04	0.88–1.27	3

10.3 Fisheries exploitation as an ecological and evolutionary force

Fishing affects the biomass of most of the world's major fish resources, with large ecosystem effects through fishing down marine food webs (Pauly et al. 1998). During recent decades, several of the world's major fisheries have shown significant collapses, and in 1997, the Food and Agriculture Organization (FAO) estimated that 60% of the major marine fisheries were either fully exploited or overexploited, some of them even to the point where they would be designated 'vulnerable' by the threat categories of IUCN (1999). The recovery of overexploited populations is not necessarily as rapid as was generally believed for high-fecundity species (Hutchings 2001), and it is possible to exploit salmon populations to extinction. In common stock–recruitment (SR) models, this occurs when the exploitation rate exceeds the maximum reproductive rate of the population, which is determined by the slope of the stock–recruitment curve near the origin. Estimates of this slope (R/S near the origin) in salmonid fishes suggest that maximum sustained exploitation rates are between about 65 and 80% (Potter et al. 2003). With small numbers of spawners, however, other factors such as local inbreeding depression (Lynch 1991) and local demographic and environmental stochasticity (Lande et al. 1999) need to be taken into consideration, as they may reduce the reproductive capacity below what is inferred from deterministic SR models.

Here we are concerned with two types of loss of genetic variation: undirected genetic erosion (genetic drift) caused by reduced spawning escapement and directed genetic change (selection) caused by fishing on particular components of the population. Both of them are easy to demonstrate theoretically and experimentally (e.g. Law 2000; Conover and Munch 2002; Tufto and Hindar 2003). However, there are few clear examples of loss of genetic variation in marine fisheries (Hauser et al. 2002; Hutchinson et al. 2003; Kenchington 2003; Olsen et al. 2004) and even fewer that we know of in salmon fisheries (Hard 2004; Consuegra et al. 2005a).

10.3.1 Undirected genetic erosion

The genetic diversity of most marine and anadromous fishes has generally been thought to be unaffected by exploitation because, even at 'collapsed' total population sizes, they are so numerous that changes in diversity are unlikely to occur (Crow and Kimura 1970). However, when a population is reduced from a very large to moderate size, which would have negligible effect on heterozygosity or inbreeding (Chapter 8), genetic variation can still be lost, as the population will harbour a lower expected number of alleles per locus (Ryman *et al.* 1995). This may be significant, for example in some major loci such as the immune response genes (termed MHC, major histocompatibility complex) where adaptability seems to depend on a high number of alleles at a small number of loci (Chapter 7).

Loss of genetic variation through reduction in population sizes must be considered both at the level of individual populations (Wang *et al.* 2002a,b) and for several populations viewed together. For anadromous fishes being harvested together in the sea, the management problem is to optimise harvest while maintaining effective population size in several populations interconnected by migration. To study this, Tufto and Hindar (2003) developed a model using numerical methods to compute the total effective population size for a set of local populations with known population sizes and migration patterns. This was combined with a population dynamic model that includes the harvest rates of each population. The population dynamic/genetic model was used to investigate:

- how the biological reference points (e.g. spawning escapement) for a group of populations relates to river-specific reference points;
- how harvesting can be strategically used to minimise genetic loss; and
- how the population genetic structure affects the answers to these questions.

It is first necessary to understand the dynamics of effective population size in a group of interconnected populations (Box 10.1). Such a group of populations is what population geneticists refer to as a 'subdivided population' (Wright 1969), and what ecologists have termed a 'metapopulation' (Levins 1969). It is shown that low, symmetric migration rates between component populations increase the total effective size (relative to the component population sizes, Box 10.1). In some idealised situations, it is possible to relate the effective size of the total population to the sum of the effective component population sizes and the

---

Box 10.1   Effective population size in a subdivided population (Tufto and Hindar 2003).

Consider a two-population system, each having a local effective population size of 10, and symmetric migration so that subpopulation 1 receives 1% migration from subpopulation 2 and vice versa. For this population system, the total effective population size becomes $N_e = 38.7$, that is, almost twice as high as the sum of the local effective sizes. For the same population system but with free interbreeding between the two subpopulations (panmixia, which corresponds to migration rates equal to 50%), the total effective population size becomes $N_e = 20.5$, approximately equal to the sum of the local effective sizes.

In situations with asymmetric migration, the total effective population size can become greatly reduced. Consider the same two-population system with local effective sizes of 10, but where subpopulation 1 receives 10% migration from subpopulation 2 and subpopulation 2 receives 1% from subpopulation 1. For this system, we get $N_e = 13.2$. Here, the total effective population size is not much higher than the size of one of the subpopulations. In the extreme case of one-way migration, the fate of the two-population system will be completely governed by the subpopulation emitting migrants and the total effective size equal to the effective size of this donor population.

The effective population size of any pattern of migration can be computed. As long as subpopulations are of constant size over time, the only limitation of this approach is the amount of computer memory needed (proportional to n^4) and central processing unit time (proportional to n^6). For example, for a system of $n = 40$ subpopulations, the numerical computations involve 820×820 matrices (for details, see Tufto *et al.* 1996).

migration rate among the component populations. In the so-called finite island model of Wright (1969), where component populations of fixed effective size N_e exchange a fraction m through a common pool of migrants, the ratio of effective size to total component population size can be approximated by:

$$N_e/(\Sigma N) \cong 1 + 1/(4Nm) \quad \text{(Waples 2002)} \tag{10.1}$$

From this it can be seen that the smaller the migration rate between subpopulations, the higher becomes the ratio between the total effective size and the sum of the local sizes.

Asymmetric migration, on the other hand, decreases the total effective size (Box 10.1). In the extreme case, that is, one-way migration, the total effective size eventually becomes equal to the effective size of the subpopulation emitting migrants (Tufto and Hindar 2003).

The second step to consider is the relationship between maximum sustainable yield (MSY) in a set of interconnected populations and MSY of the total population (Box 10.2). With deterministic population dynamics and full knowledge of each subpopulation, it is shown that the effect of migration can be ignored and the optimal harvesting strategy is to harvest each subpopulation to half of its carrying capacity (Tufto and Hindar 2003). This is the classical MSY solution applied to each subpopulation, and can be used as a starting point for setting 'conservation limits'.

Box 10.2 Maximum sustainable yield in a subdivided population (Tufto and Hindar 2003).

We consider a simple population dynamic model where the change in population size in subpopulation i is given by:

$$\Delta N_i = r_i N_i \left(1 - \frac{N_i}{K_i}\right) - Y_i(N_i) - N_i + \sum_{j=1}^{n} \tilde{m}_{ij} N_j \tag{10.2.1}$$

The first term on the right-hand side is the standard Lotka–Volterra logistic growth model, where r_i is the intrinsic growth rate and K_i is the carrying capacity of subpopulation i. The second term represents the annual reduction in population size as a result of the chosen harvesting strategy. The two last terms represent emigration and immigration. Note that $\tilde{m}_{ij}$ is the probability that an individual migrates to subpopulation j given that it originates from subpopulation i, which gives the entries of the forward migration matrix (as opposed to the backward migration matrix above). It follows (see Tufto and Hindar 2003 for details) that the total yield from the population can be written:

$$Y(N_1, N_2, \ldots, N_n) = \sum_{i=1}^{n} r_i N_i \left(1 - \frac{N_i}{K_i}\right) \tag{10.2.2}$$

that is, the effect of migration can be ignored and the optimal harvesting strategy with no constraints on the effective size is to harvest each sub-population to half of its carrying capacity such that $N_i = K_i/2$.

Our primary interest here is in the optimal solution for harvesting with constraints on the total effective size. We want to maximise (10.2.2) subject to the constraint:

$$N_e(N_1, N_2, \ldots, N_n) - N_e^* = 0 \tag{10.2.3}$$

where N_e^* is the chosen required total effective size. This problem must be solved numerically. Details are given in Tufto and Hindar (2003).

By developing a model that maximises harvesting yield of a group of populations, subject to constraints set by maintaining the total effective size, Tufto and Hindar (2003) showed that:

- considerable gain can be made in total effective size in a group of populations when exploitation is based on knowledge about population structure;
- in source-sink population systems, the total effective size can be increased without reducing total harvesting yield by first reducing the harvest in the smallest population(s), while keeping the harvest in the largest population;
- when populations differ in their degree of isolation, it pays to harvest relatively less in isolated populations because these contribute more to the total effective size; and
- in cases with moderate or strong directionality in the migration pattern, the total effective size can become less than the sum of the subpopulation sizes.

10.3.2 Directed genetic change

Selective harvesting of fish populations potentially affects a number of ecological characters. Among these are body size, growth rate, age at maturity, reproductive effort, repeat spawning and run timing. These characters are known to vary both among and within Atlantic salmon populations (Nordqvist 1924; Taylor 1991), and part of this phenotypic variation has been shown to be heritable (Chapter 7). Moreover, when different components of a run consist of fish with different genotypic proportions, selection can act directly or indirectly on single-locus genotypes (Nelson and Soulé 1987).

The selective effects of fishing can be argued as follows (Law 2000): if a particular phenotype is selected in the fisheries and part of the phenotypic variation is heritable (e.g. body size, Gjerde 1993), then fishing causes evolutionary change. In quantitative genetic terms, the response to selection (R) is a function of the selection differential (S) and the heritability (h^2) of the trait:

$$R = h^2 \times S \qquad (10.2)$$

where S is the difference between the mean phenotypic value after and before selection, and R is the difference between the mean value (before selection) in the current generation and the next generation (Falconer 1989).

Studies of selective harvesting have been motivated by the substantial changes in growth and maturation that have been observed in heavily exploited fish populations during the twentieth century (Law 2000). Such changes are important for yield and, thereby, fisheries management. If they cannot be satisfactorily explained by changes in the environment, selection due to exploitation may be a contributing factor. In that case, the possibility exists that fishing generates selection, causing evolution that changes the sustainable yield (Law 2000).

Experimental studies of both Atlantic salmon (Gjerde 1993) and Pacific salmon, e.g. chinook, *Oncorhynchus tshawytscha* (Hard 2004), show responses to selection for change in adult size and age at maturity. This clearly shows the potential for selective changes caused by a fishery that consistently targets a particular phenotype. It is, however, quite another task to demonstrate that the phenotypic changes observed in salmon populations reflect genetic changes caused by selection. Riddell (1986) discusses five reasons why realised responses to selection would be less than predicted from models of single-trait response. Among these are:

(1) limitations to data quality and/or the duration of monitoring change; (2) no additive variance for age at maturity in the wild; (3) the inability of single-trait models to account for genetic covariances among traits, or the inadequacy of the harvested portion of a population as a measure of selection intensity; (4) demographic optimisation models (Schaffer 2004) are inadequate to account for the realised response to selection; and (5) the tetraploid ancestry of salmonids is not accounted for and the association between genotype and phenotype is poorly understood. Some of these discussion points find support in long-term studies of growth rate (Friedland *et al.* 2000) and/or sea age distribution (Summers 1995) that indicate a strong environmental component to phenotypic change in salmon. Others, however, seem to be counteracted by careful studies that estimate genetic parameters from individually tagged fish released into nature (Hard 2004), and by documented changes in studies of species with a simple and invariant life history (Ricker 1981).

10.4 Fishing and effective population size: the evidence

In some species, overexploitation has led to local or global extinction (Hutchings 2001) which represents irreplaceable losses of genetic variation. The next point to consider is whether exploitation leads to a reduction of genetic diversity. Obviously, any harvesting that reduces the number of spawners will lead to a reduction in the effective population size, unless there are compensating mechanisms, e.g. in the spawning behaviour. In isolated, numerically small populations, this relates to the same phenomenon as small founder populations used in hatchery propagation (Chapter 11). A recent genetic study of a marine fish species, the New Zealand snapper, *Pagrus auratus*, even suggests that genetic variation (heterozygosity) declines significantly as a result of exploitation of an abundant species (Hauser *et al.* 2002). One reason why this could occur seems to be related to the effective population size in this species being as much as five orders of magnitude smaller than the census population size from fishery data. We are not aware, however, that loss of heterozygosity has been linked to harvesting in anadromous Atlantic salmon populations.

In ten rivers located along the Sognefjorden, western Norway, Hindar *et al.* (2004) attempted to model the total effective population size as a function of the effective population size in the most numerous population, the River Lærdalselva. This river used to harbour more than 60% of the spawners in this system (Sættem 1995), but has recently been infected by the parasite, *Gyrodactylus salaris*, which is likely to reduce the number of spawners in the Lærdalselva by 85% or more. Using the model developed by Tufto and Hindar (2003) together with estimates of local population sizes based on river-bank counts (Sættem 1995), and estimates of migration patterns from limited tagging and allozyme studies, it can be shown that the total effective size in this system is linearly related to the effective size of the R. Lærdalselva (Hindar *et al.* 2004). Moreover, as long as the system is dominated by one river population (which probably acts as a source emitting more migrants than it receives from the small neighbouring populations), the total effective size is not greatly dependent on whether the fishery takes place in the fjord or in the rivers. The latter finding should be used with caution, as this metapopulation approach does not take into account that subpopulations may show inbreeding effects (Lynch 1991) or genetic adaptations to local environments, such as timing of return to a particular river (Hansen and Jonsson 1991) or tributary (Stewart *et al.* 2002). The need to protect the smaller populations in the Sognefjorden is also supported by the finding that one fish from each of these 'sink' populations, contributes more to the total effective size than one fish from the R. Lærdalselva (Hindar *et al.* 2004). Similarly, Consuegra *et al.* (2005b) found no evidence for a historical reduction in genetic diversity in an Iberian salmon population despite a drastic reduction in abundance. This finding was attributed to source-sink metapopulation dynamics and high levels of asymmetric gene flow.

10.5 Phenotypic and evolutionary changes in exploited populations

Pink salmon, *Oncorhynchus gorbuscha*, provide a classical example of the effects of fishing (Ricker 1981). Because of their uniform life history (all returning to spawn after 2 years, and all dying after spawning), changes in body size reflect growth changes only and are not complicated by changes in the age composition of the catch. Ricker (1981) found that the size (weight) of pink salmon maturing in even-numbered years decreased from 2.1 to 1.4 kg between 1951 and 1975 (32 g/year) and that those maturing in odd-numbered years decreased from 2.5 to 2.0 kg (19 g/year). Pink salmon are caught by gill nets, seine nets and by trolling, and it appears that the decline in body size started when gill-netters shifted to a larger mesh size after new market policies from 1945 onwards made large-sized fish more valuable. Environmentally induced changes are unlikely, as a more intensive fishery should result in less dense populations and increased, rather than decreased, individual growth rate. The size difference between the pink salmon harvested and those available to the fishery (the selection differential) was compared with the observed decrease in size (the response to selection, if all of the decrease resulted from selection), giving a ratio of 0.22 for odd years and 0.30 for even years (Ricker 1981). These figures represent estimates of heritability (eqn 10.2) that lie within the range of heritabilities of body size in rearing experiments (Gjerde 1993; Hard 2004).

Similar evidence of size and/or age at maturity being affected by selective harvesting exists for other species (e.g. whitefish, cod, whiting) although perhaps less convincing than for pink salmon (Law 2000). In sockeye salmon, *Oncorhynchus nerka*, a massive numerical decline in anadromous sockeye in Lake Dal'neye between the 1930s and 1970s coincided with an increase in the proportion of fish maturing in fresh water without going to sea (see Thorpe 1993). Moreover, Altukhov and Salmenkova (1991) noted a higher enzyme heterozygosity in early maturing males, suggesting that oceanic fishing affected life history as well as molecular genetic characteristics of the population. In chinook salmon, a species with a life history not very different from Atlantic salmon, Hard (2004) showed through a quantitative genetic study of a large-scale crossing and release experiment that strong directional selection on body size was likely to produce modest short-term reductions in size. The magnitude of this effect depended (among other factors) on harvest rate, harvest size threshold (i.e. minimum size captured by the nets), and the strength of stabilising natural selection on size. Another important result was that disruptive selection, which would occur if the fisheries captured an intermediate window of the size distribution, could substantially reduce the strength of selection on size (Hard 2004).

Considerable evidence exists from Atlantic salmon populations that fishing captures a non-random portion of the population. For example, some Spanish populations, that have been harvested by rod-and-line only for the last 50 years, show significant differences between the size of fish caught and the spawning population (Consuegra *et al.* 2005a; Table 10.2). Also, these rivers show a long-term decline in the body size of salmon (Fig. 10.5). Moreover,

Table 10.2 Phenotypic and genetic traits of Atlantic salmon that are actually or potentially affected by selective harvesting.

Trait	Observed change attributed to selective harvesting	Reference
1. Timing of entry into the river	Delayed entry	Consuegra et al. (2005a)
2. Adult body size	Reduction in average size	Schaffer & Elson (1975) Gee & Milner (1980) Bielak & Power (1986) Consuegra et al. (2005a)
3. Age structure	Increased smolt age	Ritter & Newbould (1977) Consuegra et al. (2005a)
	Increased incidence of grilse	Schaffer & Elson (1975) Gee & Milner (1980) Consuegra et al. (2005a)
4. Life span	Reduced life span and longevity	Consuegra et al. (2005a)
5. Sexual maturation	Increased incidence of mature male parr Maturation at earlier ages	Caswell et al. (1984) Porter et al. (1986) Wohlfarth (1986) Consuegra et al. (2005a)
6. Allozyme variation	Change in MEP-2* frequencies	Consuegra et al. (2005a)
7. mtDNA variation	Change in ND1/16sRNA haplotypes	Consuegra et al. (2005a)

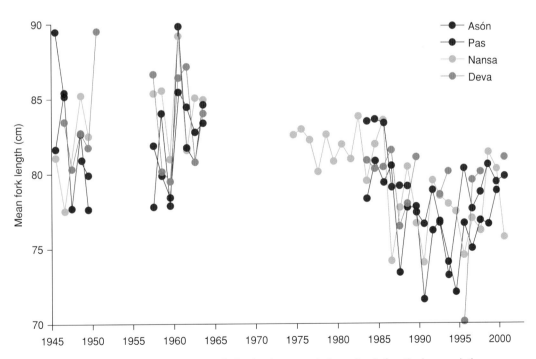

Fig. 10.5 Historical trends in the average size of Atlantic salmon caught by anglers in four Iberian populations exploited during a fairly constant fishing season from February/March to July. Based on García de Leániz et al. (2001).

Table 10.3 Reared smolts released as 2+ in Imsa 1981 and recaptured as 1SW fish in coastal fishery and in river trap (data based on Hansen 1984).

Stock	Sea fishery		Caught in trap		Proportion caught in fishery	Est. wt. before fishery	Selection differential	Predicted response
	weight	N	weight	N				
Lone	1.76	40	1.55	68	0.37	1.63	−0.08	−0.03
Imsa	2.54	213	2.35	114	0.65	2.47	−0.12	−0.04
Figgjo	2.60	201	2.26	79	0.72	2.50	−0.24	−0.09
Suldal*	2.94	46	2.49	15	0.75	2.83	−0.34	−0.12
Sandvika*	2.96	118	2.69	33	0.78	2.90	−0.21	−0.08
Alta*	3.15	155	2.72	11	0.93	3.12	−0.40	−0.14
Eira*	3.28	49	2.68	7	0.88	3.21	−0.52	−0.19
Årøy*	3.42	33	3.19	7	0.83	3.38	−0.19	−0.07

* Multi-sea-winter stock.

calculation of the selection differential on body size of released salmon returning as 1SW fish to the Imsa in Norway (Hansen 1984; Table 10.3) suggests that strong directional selection could act on this trait (ignoring that some of these populations mature largely as MSW salmon). For example, three 1SW populations were predicted to decline by 30–90 g per generation by the coastal fishery alone (Table 10.3).

Changes in numbers and phenotype of Norwegian salmon populations were recorded following the cessation of the Norwegian drift net fishery in 1989 (Jensen et al. 1999). Drift nets are known to select large-sized 1SW fish and small-sized MSW salmon. In three out of four rivers, the catches of 1SW salmon increased and their mean weight increased in all four rivers. Higher sea ages were less affected. In Russian rivers draining to the Barents Sea, a similar but less obvious trend was found, whereas White Sea populations showed no change. These results showed good correspondence to actual drift-net selectivity curves and the mean weights of drift-net catches.

Nevertheless, it should be noted that one of the best-documented declines in adult body size of Atlantic salmon (Friedland et al. 2000) is more likely to be a result of changing environmental conditions in the sea. Another impressive data set, showing parallel, long-term trends in the proportion of 1SW fish in several eastern Scottish rivers (Summers 1995), is also likely to show an environmentally induced effect, the causative agent being unknown. It remains a challenge, therefore, to demonstrate phenotypic changes in Atlantic salmon populations that are best interpreted as a response to selection. An inability to demonstrate a response to selection does not imply that fisheries selection is unimportant, and is likely to be caused mainly by inadequate data (see Riddell 1986) because major changes in the fishery have taken place within a few salmon generations.

On the River Spey in Scotland, tagging of rod-caught fish and monitoring of subsequent recaptures indicates re-exploitation rates of ~ 30% for fish caught first, for example, in February and March, but re-exploitation rates of only < 2% for fish caught in July and August. Differences in re-exploitation rate appear to depend partly on seasonal differences in intrinsic catchability and partly on the duration of the period over which post-entrant fish remain catchable (J. Thorley, pers.comm.). Radio-tracking studies on a number of rivers (Webb and Campbell 2000) demonstrate that run timing and spawning location are spatially correlated, with earlier-running fish spawning at higher elevations that are generally more distant from the sea. Since sub-catchment populations spawning in the upper parts of catchments tend to comprise early-running fish, they are therefore susceptible to higher exploitation rates than those spawning in down-river locations.

In addition, early-running fish have suffered particularly marked recent declines in abundance, probably due to differential rates of marine mortality (Youngson *et al.* 2003). Reduced abundance, coupled with high intrinsic catchability, means that, under current conditions, spawning populations of early-running fish are particularly susceptible to overexploitation by rod fisheries. It follows that corresponding latitude exists, under conditions such as those that apply at present, for anthropogenic effects on the relative strength of different spawning populations, even when they are exposed to constant rates of rod fishing effort throughout the fishing season. These considerations have a well-defined genetic aspect since run timing (and therefore exploitation rate, escapement rate and local spawning adequacy) is strongly affected by the genetic characteristics of the populations (Hansen and Jonsson 1991; Stewart *et al.* 2002).

10.6 Future management of salmon fisheries

Genetic considerations in fisheries resemble in many ways the considerations that would arise from hunting. Harris *et al.* (2002) reviewed the potential effects of hunting on terrestrial vertebrates, and saw no urgency as they found only a few documented cases of undesirable genetic consequences. Against this background, harvesting of salmon seems to represent a smaller threat to the genetic constitution of Atlantic salmon populations today than gene flow from escaped farm salmon invading wild populations to spawn (Chapter 12), and habitat degradation (WWF 2001).

Nonetheless, the recent collapse of some of the world's major fish stocks and the 30-year-long decline of wild Atlantic salmon populations suggests that a precautionary approach be taken to salmon management (NASCO 1998; Crozier *et al.* 2003). To this end, a better ability to judge whether spawning escapement is above a threshold necessary to secure long-term population viability, is a first step. A good example is the recent catch advice provided for Irish fishing districts, based on estimates of pre-fishery abundance and stock–recruitment relationships for Irish rivers (Ó Maoiléidigh *et al.* 2004). Secondly, the potential for selective changes should be considered by salmon managers (Table 10.4). Selection should be considered when

Table 10.4 Some possible consequences of fishery regulations and how they can lead to selective harvesting of Atlantic salmon populations.

Regulation	Possible consequences
1. Catch quotas	Fish in excess of catch quotas are not exploited by the fishery
2. Regulation of fishing gear (lines, lures, nets)	Individuals can vary in their vulnerability to the fishery, depending on body size, sex, activity, etc.
3. Size limits	Small fish can be underexploited while large, trophy fish can be overexploited (or vice versa)
4. Fishing areas and fishing effort	Some fish are overexploited while others are underexploited due to unequal spatial distribution of fishing effort
5. Fishing season and fishing times	Early fish are overexploited while late fish are underexploited due to unequal temporal distribution of fishing effort

designing new regulations for river and coastal fisheries: for example, size limits and fishing season in relation to run timing of different population components (Stewart *et al.* 2002; Consuegra *et al.* 2005a). Selective changes towards sexual maturation at a smaller body size should be a prime concern, as they have implications for population fecundity (see Olsen *et al.* 2004). Finally, strong spawning populations of wild Atlantic salmon should always be considered as a buffer against natural and human-made environmental factors that threaten the viability and genetic integrity of Atlantic salmon populations.

10.7 Summary and conclusions

- Fishing can result in loss of genetic diversity by selectively removing species and populations, and by causing undirected and directed changes within populations.
- Microevolution caused by fishing may be important even when masked by other factors.
- Run timing and body size are prime candidates for studies of genetic response to selective fishing.
- Genetic responses in life-history characters (e.g. age at maturity) are notoriously difficult to demonstrate.
- Isolation increases total effective population size whereas asymmetry in gene flow (and population size) decreases it.
- If population sizes and patterns of gene flow (or genetic structure) are known, harvesting can be used strategically to increase total effective size without much loss of yield.
- Harvesting of salmon populations today seems to represent a smaller threat to the genetic constitution of many populations than gene flow from escaped farm salmon, and habitat degradation.

10.8 Management recommendations

- The management of salmon fisheries needs to take into account the genetic effects caused by reduced population size, as well as those caused by selectively removing populations or population components.
- Undirected (random) loss of genetic diversity should be considered at the single-population level as well as at the level of a set of interconnected populations.
- Knowledge about local population sizes and migration patterns, or alternatively, studies of the genetic structure of the species, should be used to assess the relationship between local effective population sizes and the effective size of the total population.
- Care should be taken to avoid selective fishing on population components that have a hereditary basis.
- Reduced harvesting can be used strategically to avoid (or reduce) genetic effects of other factors affecting salmon populations.

Acknowledgements

We thank A. Serdio for help in compiling the exploitation data, and three referees for helpful comments.

Further reading

Crozier, W.W., Potter, E.C.E., Prévost, E., Schön, P-J. and Ó Maoiléidigh, N. (Ed.) (2003) *A coordinated approach to the development of a scientific basis for management of wild Atlantic salmon in the North-East Atlantic (SALMODEL)*. Queen's University, Belfast.

Hendry, A.P. and Stearns, S.C. (Ed.) (2004) *Evolution Illuminated: Salmon and their relatives*. Oxford University Press, Oxford.

Reynolds, J.D., Mace, G.M., Redford, K.H. and Robinson, J.G. (Ed.) (2001) *Conservation of Exploited Species*. Cambridge University Press, Cambridge.

References

Altukhov, Yu.P. and Salmenkova, E.A. (1991) The genetic structure of salmon populations. *Aquaculture*, **98**: 11–40.

Alvarez, J. and Lamuela, M. (2001) Situación actual del salmón atlántico en Navarra. In: C. García de Leániz, A. Serdio and S. Consuegra (Ed.) *El Salmón, Joya de Nuestros Ríos*, pp. 97–110. Consejería de Ganadería, Agricultura y Pesca, Santander. (in Spanish with English summary)

Baum, E. (1997) *Maine Atlantic Salmon: A national treasure*. Atlantic Salmon Unlimited, Hermon, ME.

Bielak, A.T. and Power, G. (1986) Changes in mean weight, sea-age composition and catch per unit effort of Atlantic salmon (*Salmo salar*) angled in the Godbout River, Quebec. *Canadian Journal of Fisheries and Aquatic Sciences*, **43**: 281–287.

Caswell, H., Naiman, R.J. and Morin, R. (1984) Evaluating the consequences of reproduction in complex salmonid life cycles. *Aquaculture*, **43**: 123–134.

Chadwick, E.M.P. (1982) Stock recruitment relationship for Atlantic salmon (*Salmo salar*) in Newfoundland. *Canadian Journal of Fisheries and Aquatic Sciences*, **39**: 1496–1501.

Chadwick, E.M.P., Reddin, D.G. and Burfitt, R.F. (1985) Fishing and natural mortality rates for 1SW Atlantic salmon *(Salmo salar* L.). *ICES CM* **1985/M:18**.

Cleyet-Merle, J.-J. (1990) *La Préhistoire de la Pêche*. Editions Errance, Paris.

Conover, D.O. and Munch, S.B. (2002) Sustaining fisheries yields over evolutionary time scales. *Science*, **297**: 94–96.

Consuegra, S., García de Leániz, C., Serdio, A., Gonzalez Morales, M., Straus, L.G., Knox, D. and Verspoor, E. (2002) Mitochondrial DNA variation in Pleistocene and modern Atlantic salmon from the Iberian glacial refugium. *Molecular Ecology*, **11**: 2037–2048.

Consuegra, S., García de Leániz, C., Serdio, A. and Verspoor, E. (2005a) Selective exploitation of early running fish may induce genetic and phenotypic changes in Atlantic salmon. *Journal of Fish Biology*, **67** (Supplement A): 129–145.

Consuegra, S., Verspoor, E., Knox, D. and García de Leániz, C. (2005b) Asymmetric gene flow and the evolutionary maintenance of genetic diversity in small, peripheral Atlantic salmon populations. *Conservation Genetics*, **6(5)**: 823–842. (Published online 25 August 2005, DOI: 10.1007/s10592-005-9042-4)

Crow, J.F. and Kimura, M. (1970) *An Introduction to Population Genetics Theory*. Burgess Publishing, Minneapolis, MN.

Crozier, W.W. and Kennedy, G.J.A. (1993) Marine survival of wild and hatchery-reared Atlantic salmon (*Salmo salar* L.) from the River Bush, Northern Ireland. In D. Mills (Ed.) *Salmon in the Sea and New Enhancement Strategies*, pp. 139–162. Fishing News Books, Oxford.

Crozier, W.W. and Kennedy, G.J.A. (1994) Marine exploitation of Atlantic salmon (*Salmo salar* L.) from the River Bush, Northern Ireland. *Fisheries Research*, **19**: 141–155.

Crozier, W.W. and Kennedy, G.J.A. (2001) Relationship between freshwater angling catch of Atlantic salmon and stock size in the River Bush, Northern Ireland. *Journal of Fish Biology* **58**: 240–247.

Crozier, W.W., Potter, E.C.E., Prévost, E., Schön, P-J. and Ó Maoiléidigh, N. (Ed.) (2003) *A coordinated approach to the development of a scientific basis for management of wild Atlantic salmon in the North-East Atlantic (SALMODEL)*. Queen's University, Belfast.

Dempson, J.B., Schwarz, C.J., Reddin, D.G., O'Connell, M.F., Mullins, C.C. and Bourgeois, C.E. (2001) Estimation of marine exploitation rates on Atlantic salmon (*Salmo salar* L.) stocks in Newfoundland, Canada. *ICES Journal of Marine Science*, **58**: 331–341.

Dempson, J.B., Furey, G. and Bloom, M. (2002) Effects of catch and release angling on Atlantic salmon, *Salmo salar* L., of the Conne River, Newfoundland. *Fisheries Management and Ecology*, **9**: 139–147.

Erkinaro, J., Økland, F., Moen, K. and Niemelä, E. (1999) Return migration of the Atlantic salmon in the Tana River: distribution and exploitation of radiotagged multi-sea-winter salmon. *Boreal Environmental Research*, **4**: 115–124.

Falconer, D.S. (1989) *Introduction to Quantitative Genetics* (3rd edn). Longman Scientific and Technical, Harlow, UK.

Friedland, K.D., Hansen, L.P. and Dunkley, D.A. (1998) Marine temperatures experienced by postsmolts and the survival of Atlantic salmon, *Salmo salar* L., in the North Sea area. *Fisheries Oceanography*, **7**: 22–34.

Friedland, K.D., Hansen, L.P., Dunkley, D.A. and MacLean, J.C. (2000) Linkage between ocean climate, post-smolt growth, and survival of Atlantic salmon (*Salmo salar* L.) in the North Sea area. *ICES Journal of Marine Science*, **57**: 419–429.

García de Leániz, C., Caballero, P., Valero, E., Martínez, J.J. and Hawkins, A.D. (1992) Historical changes in Spanish Atlantic salmon (*Salmo salar* L.) rod and line fisheries: why are large multi-seawinter fish becoming scarcer? *Journal of Fish Biology*, **41**: 179.

García de Leániz, C., Serdio, A. and Consuegra, S. (2001) Present status of Atlantic salmon in Cantabria. In: C. García de Leániz, A. Serdio and S. Consuegra (Ed.) *El Salmón, Joya de Nuestros Ríos*, pp. 55–82. Consejería de Ganadería, Agricultura y Pesca, Santander. (in Spanish with English summary)

Gee, A.S. and Milner, N.J. (1980) Analysis of 70-year catch statistics for Atlantic salmon (*Salmo salar*) in the River Wye and implications for management of stocks. *Journal of Applied Ecology*, **17**: 41–57.

Gjerde, B. (1993) Breeding and selection. In: K. Heen, R.L. Monahan and F. Utter (Ed.) *Salmon Aquaculture*, pp. 187–208. Fishing News Books, Oxford.

Gudjónsson, T. (1988) Exploitation of salmon in Iceland. In: D. Mills and D. Piggins (Ed.) *Atlantic Salmon: Planning for the future*, pp. 162–178. Croom Helm, London.

Gudjónsson, S., Antonsson, Th. and Tomasson, T. (1996) Exploitation ratio of salmon in relation to salmon run in three Icelandic rivers. *ICES CM* **1996/M:8**.

Hansen, L.P. (1984) *A preliminary analysis of the exploitation pattern of Atlantic salmon tagged and released as smolts in River Imsa, Norway, 1981*. Working paper, ICES North Atlantic Salmon Working Group.

Hansen, L.P. (1988) Status of exploitation of Atlantic salmon in Norway. In: D. Mills and D. Piggins (Ed.) *Atlantic Salmon: Planning for the future*, pp. 143–161. Croom Helm, London.

Hansen, L.P. (1990) Exploitation of Atlantic salmon *(Salmo salar* L.) from the River Drammenselv, SE Norway. *Fisheries Research*, **10**: 125–135.

Hansen, L.P. and Jonsson, B. (1991) Evidence of a genetic component in the seasonal return pattern of Atlantic salmon, *Salmo salar* L. *Journal of Fish Biology*, **38**: 251–258.

Hansen, L.P., Næsje, T.F. and Garnås, E. (1986) Stock assessment and exploitation of Atlantic salmon *Salmo salar* L. in the river Drammenselv. *Fauna Norvegica, Series A*, 7: 23–26.

Hard, J.J. (2004) Evolution of Chinook salmon life histories under size-selective harvest. In: A.P. Hendry and S.C. Stearns (Ed.) *Evolution Illuminated: Salmon and their relatives*, pp. 315–337. Oxford University Press, Oxford.

Harris, R.B., Wall, W.A. and Allendorf, F.W. (2002) Genetic consequences of hunting: what do we know and what should we do? *Wildlife Society Bulletin*, 30: 634–643.

Hauser, L., Adcock, G.J., Smith, P.J., Ramirez, J.H.B. and Carvalho, G.R. (2002) Loss of microsatellite diversity and low effective population size in an overexploited population of New Zealand snapper (*Pagrus auratus*). *Proceedings of the National Academy of Sciences of the United States of America*, 99: 11742–11747.

Hindar, K., Tufto, J., Sættem, L.M. and Balstad, T. (2004) Conservation of genetic variation in harvested salmon populations. *ICES Journal of Marine Science*, 61: 1389–1397.

Hutchings, J.A. (2001) Conservation biology of marine fishes: perceptions and caveats regarding assignment of extinction risk. *Canadian Journal of Fisheries and Aquatic Sciences*, 58: 108–121.

Hutchinson, W.F., van Oosterhout, C., Rogers, S.I. and Carvalho, G. R. (2003) Temporal analysis of archived samples indicates marked genetic changes in declining North Sea cod (*Gadus morhua*). *Proceedings of the Royal Society of London, Series B*, 270: 2125–2132.

ICES (2005) *Report of the ICES Advisory Committee on Fishery Management*. Annex 9 to NASCO (2005), pp. 77–97.

IUCN (1999) *IUCN red list categories*. International Union for the Conservation of Nature, London.

Jensen, A.J., Zubchenko, A.V., Heggberget, T.G., Hvidsten, N.A., Johnsen, B.O., Kuzmin, O., Loenko, A.A., Lund, R.A., Martynov, V.G., Næsje, T.F., Sharov, A.F. and Økland, F. (1999) Cessation of the Norwegian drift net fishery: changes observed in Norwegian and Russian populations of Atlantic salmon. *ICES Journal of Marine Science*, 56: 84–95.

Jensen, K.W. (1981) On the rate of exploitation of salmon from two Norwegian rivers. *ICES CM* **1981/M:11**.

Karppinen, P., Erkinaro, J., Niemelä, E., Moen, K. and Økland, F. (2004) Return migration of one-sea-winter Atlantic salmon in the River Tana. *Journal of Fish Biology*, 64: 1179–1192.

Kellogg, K.A. (1999) Salmon on the edge. *Trends in Ecology and Evolution*, 14: 45–46.

Kenchington, E.L. (2003). The effects of fishing on species and genetic diversity. In: M. Sinclair and G. Valdimarsson (Ed.) *Responsible Fisheries in the Marine Ecosystem*, pp. 235–253. CABI Publishing, Wallingford, UK.

Kerswill, C.J. (1971) Relative rates of utilization by commercial and sport fisheries of Atlantic salmon (*Salmo salar*) from the Miramichi River, New Brunswick. *Journal of the Fisheries Research Board of Canada*, 28: 351–363.

Koljonen, M.-L., Pella, J.J. and Masuda, M. (2005) Classical individual assignments versus mixture modeling to estimate stock proportions in Atlantic salmon (*Salmo salar*) catches from DNA microsatellite data. *Canadian Journal of Fisheries and Aquatic Sciences*, 62: 2143–2158.

Lande, R. and Shannon, S. (1996) The role of genetic variation in adaptation and population persistence in a changing environment. *Evolution*, 50: 434–437.

Lande, R., Engen, S. and Sæther, B.-E. (1999) Spatial scale of population synchrony: correlation versus dispersal and density regulation. *American Naturalist*, 154: 271–281.

Larkin, P.A. (1977) An epitaph for the concept of maximum sustainable yield. *Transactions of the American Fisheries Society*, 106: 1–11.

Law, R. (2000) Fishing, selection, and phenotypic evolution. *ICES Journal of Marine Science*, 57: 659–668.

Levins, R. (1969) Some demographic and genetic consequences of environmental heterogeneity for biological control. *Bulletin of the Entomological Society of America*, 15: 237–240.

Lynch, M. (1991) The genetic interpretation of inbreeding depression and outbreeding depression. *Evolution*, 45: 622–629.

Mace, G.M. and Lande, R. (1991) Assessing extinction threats: towards a reassessment of IUCN endangered species categories. *Conservation Biology*, **5**: 148–157.

MAFF, SOAEFD and WOAD (1999) *Factors Affecting Salmon in the Sea: Report of the Salmon Advisory Committee*, pp. 1–60. Ministry of Agriculture, Fisheries and Food, London; Scottish Office Agriculture, Environment and Fisheries Department, Edinburgh; Welsh Office Agriculture Department, Cardiff.

Miller, R.B. (1957) Have the genetic patterns of fish been altered by introductions or by selective fishing. *Journal of the Fisheries Research Board of Canada*, **14**: 797–806.

Mills, C.P.R. and Piggins, D.J. (1983) The release of reared salmon smolts (*Salmo salar*) into the Burrishoole system (Western Ireland) and their contribution to the rod and line fishery. *Fisheries Management*, **14**: 165–175.

Mills, C.P.R., Mahon, G.A.T. and Piggins, D.J. (1986) Influence of stock levels, fishing effort and environmental factors on angler's catches of Atlantic salmon, *Salmo salar* L., and sea trout, *Salmo trutta* L. *Aquaculture and Fisheries Management*, **17**: 289–297.

Murray, A.R. (1968) Smolt survival and adult utilization of Little Codroy River, Newfoundland, Atlantic salmon. *Journal of the Fisheries Research Board of Canada*, **25**: 2165–2218.

NASCO (1998) *Agreement on the Adoption of a Precautionary Approach: Report of the fifteenth annual meeting of the Council*. CNL(98)46. North Atlantic Salmon Conservation Organization, Edinburgh.

NASCO (2005) *Report of the twenty-second annual meeting of the Council*. CNL(05)50. North Atlantic Salmon Conservation Organization, Edinburgh.

Nelson, K. and Soulé, M. (1987) Genetical conservation of exploited fishes. In: N. Ryman and F. Utter (Ed.) *Population Genetics and Fishery Management*, pp. 345–368. University of Washington Press, Seattle, WA.

Nordqvist, O. (1924) Times of entering of the Atlantic salmon (*Salmo salar* L.) in the rivers. *Conseil Permanent International pour l'Exploration de la Mer, Rapports et Procés-Verbaux*, **33**: 1–58.

O'Connell, M.F. (2003) An examination of the use of angling data to estimate total returns of Atlantic salmon, *Salmo salar*, to two rivers in Newfoundland, Canada. *Fisheries Management and Ecology*, **10**: 201–208.

Olsen, E.M., Heino, M., Lilly, G.R., Morgan, M.J., Brattey, J., Ernande, B. and Dieckmann, U. (2004) Maturation trends indicative of rapid evolution preceded the collapse of northern cod. *Nature*, **428**: 932–935.

Ó Maoiléidigh, N., McGinnity, P., Prévost, E., Potter, E.C.E., Gargan, P., Crozier, W.W., Mills, P. and Roche, W. (2004) Application of pre-fishery abundance modelling and Bayesian hierarchical stock and recruitment analysis to the provision of precautionary catch advice for Irish salmon (*Salmo salar* L.) fisheries. *ICES Journal of Marine Science*, **61**: 1370–1378.

Pauly, D., Christensen, V., Dalsgaard, J., Froese, R. and Torres, F. (1998) Fishing down marine food webs. *Science*, **279**: 860–863.

Peet, R.F. and Pratt, J.D. (1972) *Distant and local exploitation of a Labrador Atlantic salmon population by commercial fisheries*. International Commission for the Northwest Atlantic Fisheries Research Document 72/82, Serial No. 2809.

Porter, T.R., Healey, M.F., O'Connell, M.F., Baum, E.T., Bielak, A.T. and Cote, Y. (1986) Implications of varying the sea age at maturity of Atlantic salmon (*Salmo salar*) on yield to the fisheries. *Canadian Special Publication in Fisheries and Aquatic Sciences*, **89**: 110–117.

Potter, E.C.E., MacLean, J., Wyatt, R.J. and Campbell, R.N.B. (2003) Managing the exploitation of migratory salmonids. *Fisheries Research*, **62**: 127–142.

Reddin, D.G. (1981) Estimation of fishing mortality for Atlantic salmon (*Salmo salar*) in Newfoundland and Labrador commercial fisheries. *ICES CM*, **1981/M:24**.

Ricker, W.E. (1981) Changes in the average size and average age of Pacific salmon. *Canadian Journal of Fisheries and Aquatic Sciences*, **38**: 1636–1656.

Riddell, B.E. (1986) Assessment of selective fishing on the age at maturity in Atlantic salmon (*Salmo salar*): a genetic perspective. *Canadian Special Publication in Fisheries and Aquatic Sciences*, **89**: 102–109.

Ritter, J.A. and Newbould, K. (1977) Relationship of parentage and smolt age to first maturity of Atlantic salmon (*Salmo salar*). *ICES CM*, **1977/M:32**.

Ryman, N., Utter, F. and Laikre, L. (1995) Protection of intraspecific biodiversity of exploited fishes. *Reviews in Fish Biology and Fisheries*, **5**: 417–446.

Sættem, L.M. (1995) Gytebestander av laks og sjøaure. En sammenstilling av registreringer fra ti vassdrag i Sogn og Fjordane fra 1960–94. Utredning for DN, 1995-7. Direktoratet for naturforvaltning, Trondheim, Norway. ('Spawning populations of Atlantic salmon and sea trout. A compilation of studies in ten rivers in Sogn og Fjordane county from 1960–94'; in Norwegian)

Sandhaugen, A.I. and Hansen, L.P. (2001) Exploitation of Atlantic salmon (*Salmo salar* L.) in the River Drammenselv. *NINA Fagrapport*, **51**: 1–44. (in Norwegian with English summary)

Saunders, R.L. (1969) Contribution of salmon from the North-west Miramichi River, New Brunswick, to various fisheries. *Journal of the Fisheries Research Board of Canada*, **26**: 269–278.

Scarnecchia, D.L., Ísaksson, Á. and White, S.E. (1989) Effects of oceanic variations and the West Greenland fishery on age at maturity of Icelandic west coast stocks of Atlantic salmon *(Salmo salar)*. *Canadian Journal of Fisheries and Aquatic Sciences*, **46**: 16–27.

Schaffer, W.M. (2004) Life histories, evolution, and salmonids. In: A.P. Hendry and S.C. Stearns (Ed.) *Evolution Illuminated: Salmon and their relatives*, pp. 20–51. Oxford University Press, Oxford.

Schaffer, W.M. and Elson, P.F. (1975) The adaptive significance of variations in life history among local populations of Atlantic salmon in North America. *Ecology*, **56**: 577–590.

Semple, J.R. and Cameron, J.D. (1990) *Biology, exploitation and escapement of Atlantic salmon* (Salmo salar), *Liscomb River, N.S.* Canadian Manuscript Report of Fisheries and Aquatic Sciences No. 2077.

Shearer, W.M. (1988) Relating catch records to stocks. In: D. Mills and D. Piggins (Ed.) *Atlantic Salmon: Planning for the future*, pp. 256–274. Croom Helm, London.

Shearer, W.M. (1992) *The Atlantic Salmon: Natural History, Exploitation and Future Management*. Fishing News Books, London.

Ståhl, G. (1987) Genetic population structure of Atlantic salmon. In: N. Ryman and F. Utter (Ed.) *Population Genetics and Fishery Management*, pp. 121–140. University of Washington Press, Seattle, WA.

Statens Dyrehelsetilsyn (2002) Fiske basert på fang og slipp-dyrevern. Letter dated 14th March 2002. Statens Dyrehelsetilsyn, Oslo. (in Norwegian)

Stewart, D.C., Smith, G.W. and Youngson, A.F. (2002) Tributary-specific variation in timing of return of adult Atlantic salmon (*Salmo salar*) to fresh water has a genetic component. *Canadian Journal of Fisheries and Aquatic Sciences*, **59**: 276–281.

Summers, D.W. (1995) Long-term changes in the sea-age at maturity and seasonal time of return of salmon, *Salmo salar* L., to Scottish rivers. *Fisheries Management and Ecology*, **2**: 147–156.

Taylor, E.B. (1991) A review of local adaptation in Salmonidae, with particular reference to Pacific and Atlantic salmon. *Aquaculture*, **98**: 185–207.

Thorpe, J.E. (1993) Impacts of fishing on genetic structure of salmonid populations. In: J.G. Cloud and G.H. Thorgaard (Ed.) *Genetic Conservation of Salmonid Fishes*, pp. 67–80. Plenum Press, New York.

Thorstad, E.B., Næsje, T.F., Fiske, P. and Finstad, B. (2003) Effects of hook and release on Atlantic salmon in the River Alta, northern Norway. *Fisheries Research*, **60**: 293–307.

Tufto, J. and Hindar, K. (2003) Effective size in management and conservation of subdivided populations. *Journal of Theoretical Biology*, **222**: 273–281.

Tufto, J., Engen, S. and Hindar, K. (1996) Inferring patterns of migration from gene frequencies under equilibrium conditions. *Genetics*, **144**: 1911–1921.

Verspoor, E. (1988) Identification of stocks in the Atlantic salmon. In: R.H. Stroud (Ed.) *Proceedings of the Symposium on Present and Future Atlantic Salmon Management*, pp. 37–46. Atlantic Salmon Federation, Ipswich, MA; National Coalition for Marine Conservation, Savannah, GA.

Wang, S., Hard, J.J. and Utter, F. (2002a) Salmonid inbreeding: a review. *Reviews in Fish Biology and Fisheries*, **11**: 301–319.

Wang, S., Hard, J.J. and Utter, F. (2002b) Genetic variation and fitness in salmonids. *Conservation Genetics*, **3**: 321–333.

Waples, R.S. (1998) Separating the wheat from the chaff: patterns of genetic differentiation in high gene flow species. *Journal of Heredity*, **89**: 438–450.

Waples, R.S. (2002) Definition and estimation of effective population size in the conservation of endangered species. In: S.R. Beissinger and D.R. McCullough (Ed.) *Population Viability Analysis*, pp. 147–168. University of Chicago Press, Chicago.

Warner, K. and Havey, K.A. (1985) *Life History, Ecology and Management of Maine Landlocked Salmon (Salmo salar)*. Maine Department of Inland Fisheries and Wildlife, Augusta, ME.

Webb, J.H. and Campbell, R.N.B. (2000) Patterns of run timing in adult Atlantic salmon returning to Scottish rivers: some new perspectives and management implications. In: F.G. Whoriskey and K.E. Whelan (Ed.) *Managing Wild Atlantic Salmon*, pp. 100–138. Atlantic Salmon Federation, St Andrews, New Brunswick, Canada.

Wohlfarth, G.W. (1986) Decline in natural fisheries: a genetic analysis and suggestion for recovery. *Canadian Journal of Fisheries and Aquatic Sciences*, **43**: 1298–1306.

WWF (2001) *The Status of Wild Atlantic Salmon: A river by river assessment*. World Wildlife Fund (WWF-US), Washington, DC.

Wright, S. (1969) *Evolution and the Genetics of Populations. Volume 2. The Theory of Gene Frequencies*. University of Chicago Press, Chicago.

Youngson, A.F., Jordan, W.C., Verspoor, E., McGinnity, P., Cross, T. and Ferguson, A. (2003) Management of salmonid fisheries in the British Isles: towards a practical approach based on population genetics. *Fisheries Research*, **62**: 193–209.

11 Stocking and Ranching

T. F. Cross[1], P. McGinnity[1], J. Coughlan, E. Dillane, A. Ferguson, M-L. Koljonen, N. Milner, P. O'Reilly and A. Vasemägi

Clockwise from top left: collecting broodfish on the River Scaur, south-west Scotland; stripping broodfish at Lairig on the River Shin, north-east Scotland; laying down ova in the hatchery at Contin on the River Connon; salmon ova with eyed embryos visible; planting out ova in the River Loth, north-east Scotland; salmon fry in a stream after planting out. (Photos credit: 1, 2, 5, E. Verspoor; 3, 4, D. Hay; 6, J. Pratt.)

[1] Joint first authors.

Various stocking and ranching programmes have generally been perceived by anglers and many fishery managers as first options when riverine populations of Atlantic salmon, *Salmo salar* L., decline. These programmes are often implemented regardless of the cause(s) of such declines or whether the causative agents have been removed. Past stocking and ranching records show that salmon have been extensively moved between catchments, often over long distances, or even between continents, on the presumed assumption that Atlantic salmon from all of these areas are functionally equivalent. Many of these efforts, however, have been unsuccessful, and genetic considerations have largely been ignored. In this chapter we consider four scenarios where stocking and ranching might be invoked and examine each from a genetic viewpoint, with particular regard to the wild population structure described in earlier chapters. These scenarios are (1) where salmon have been extirpated from a particular location (here termed reintroduction), (2) where natural freshwater populations still occur but in reduced numbers (termed rehabilitation), (3) where numbers are at natural carrying capacity but the management aspiration is to increase production (termed enhancement), and (4) where numbers are constrained by anthropogenic developments which cannot be removed (termed

Box 11.1 Lifetime success and performance characteristics under natural conditions of communally reared offspring of wild native Burrishoole, ranched native, and non-native Atlantic salmon from the adjacent Owenmore river.

The lifetime success and performance characteristics of communally reared offspring of wild native Burrishoole, ranched native, and non-native Atlantic salmon from the adjacent Owenmore river (subsequently designated: wild, ranched and Owenmore) were examined (McGinnity *et al.* 2004). The experiment was carried out in the Burrishoole river system and hatchery (western Ireland) and involved a similar protocol to that used for comparison of wild, farm and hybrid salmon (see Box 12.4 for details). The ranched fish used as parents were from the Burrishoole ranched stock, which was established from the wild Burrishoole population in 1964. A separate line has been maintained since then by adipose-fin clipping all ranched smolts and, in most years, using only fin-clipped adults for propagation. The mouth of the Owenmore is approximately 80 km north-west of the mouth of the Burrishoole system, although the two rivers have tributaries arising about 0.5 km apart on the same mountain. All broodstock were 1SW and all families were established by artificial fertilisation on the same day.

As a control on the field experiments, aliquots of each group were maintained in a hatchery tank until 8 months after hatching. No significant differences in survival among the three groups were found to this age in the hatchery. (However, given that total mortality was less than 10% under 'protected' hatchery conditions, there was little scope for detecting any differential survival.) These hatchery controls serve to demonstrate that all groups were potentially equally viable and that the differential survival apparent in the wild was the result of genetic and/or maternal differences.

Owenmore 0+ parr showed a substantial downstream migration, which was not shown by wild and ranched native parr. This appears to have been an active migration rather than competitive displacement as there was no difference in size among the three groups or between Owenmore emigrants and Owenmore parr remaining in the experimental river. Also as 1+ fish, Owenmore parr showed significantly lower emigration than wild parr. This early emigration may reflect an adaptation to the Owenmore river, where the main nursery habitat is downstream of the spawning area. There were no differences between wild and ranched fish in smolt output or adult return. However, both of these measures were significantly lower for the Owenmore group. A greater proportion of the Owenmore salmon was taken in the coastal nets compared to the return to the Burrishoole system. Ranched salmon showed a significantly greater male parr maturity, a greater proportion of 1+ smolts, a later time of year of return, and differences in sex ratio of returning adults compared to wild, which may have fitness implications under specific conditions. The overall lifetime success of the Owenmore group, from fertilised egg to returning adult (in the experimental stream and subsequently being ranched from the Burrishoole system), was only 35% of wild and ranched natives.

mitigation). In the first two of these scenarios it is presumed that the man-made constraints have been removed before stock rebuilding begins. Scenarios 3 and usually 4 involve ranching. Conservation hatcheries may also be involved in scenario 4.

11.1 Introduction

This chapter highlights the genetic considerations involved when considering deliberate releases of artificially spawned Atlantic salmon into the wild. Accidental introductions (as in farm escape incidents) are considered in Chapter 12. Deliberate release or stocking is generally termed as introduction (fish introduced into waters outside their historical distribution) or reintroduction (where fish are absent from their former distribution), or as rehabilitation or enhancement (where native salmon are present). It is argued in this chapter that the genetic implications for wild salmon populations from interactions with deliberately released reared salmon can be as detrimental as the interactions resulting from accidental escapes (Chilcote, 2003; McGinnity *et al.* 2004; Boxes 11.1 and 11.2). This is important in that stocking was previously invoked as the first management response to any situation where salmon numbers were considered to be depleted. It is argued here that, from a genetic viewpoint, stocking should not be used at all in certain situations or, in others, only if it can be used as a precisely targeted measure, forming part of an overall well-planned rehabilitation strategy. Thus the objective of the present chapter is to provide an assessment of risks, from a genetic perspective, to both wild populations (where these occur) and to the success of the stocking programme itself. We omit discussion of introductions into non-endemic areas, since these have enjoyed little success as a deliberate strategy for establishing self-sustaining populations. It is acknowledged that Atlantic salmon escape from sea cages in areas outside the native range, such as British Columbia, Chile and Tasmania, but these occurrences are also not considered in this chapter, though the possible consequences might, in some cases, pose considerable problems. Ranching, which generally refers to deliberate release of reared smolts, can be part of any of the stocking exercises outlined below.

11.2 Genetic characteristics of wild salmon populations

The genetic characteristics of wild populations and the delineation of appropriate management units have been described in earlier chapters. Each river system contains one or more populations, and preservation or restoration of this natural structure is considered paramount in any rehabilitation effort (see Waples 1991; Youngson *et al.* 2002, for detailed treatment of this area). These principles should guide management in any of the scenarios listed below.

11.3 Nature of strains reared for stocking and ranching

Salmon strains used for stocking and ranching have been investigated in many countries in Europe and North America using the molecular methods described in earlier chapters. In many of these studies, results are compared either with wild progenitor populations or with

> **Box 11.2** How stocking may lead to reduced salmon production.
>
> There is a major source of discrepancy in calculating production where extra contribution has been measured from stocked tagged fish. That is, it is assumed that if say 10% are tagged fish then this is 10% in addition to what would have been there without stocking. It does not take account that this 10% could be at the expense of an equal or even greater number of wild fish. Certainly the latter is the case for farm escaped salmon experiments (see Chapter 12).
>
> Stocked salmon have two main impacts on wild populations:
>
> (1) competition;
> (2) interbreeding.
>
> Often supplemental stocking is undertaken without adequate consideration as to why there are too few salmon present to start with. If reduced numbers are due to reduced habitat and food, then adding more fish will result in increased competition and lower survival overall, giving fewer fish at the end of the day than if stocking had not been carried out! Unfortunately, there is a belief in some quarters that adding more fish must be doing some good. However, gradually it is being appreciated that stocking may be one of the reasons for declining natural stocks, not a solution to the decline (Chilcote 2003).
>
> Since some stocked salmon do survive to maturity, in the next generation part of the wild production is converted to hybrids with these non-native fish. In terms of survival and other aspects of performance, such hybrids are likely to be intermediate between the native and non-native salmon (e.g. McGinnity *et al.* 2003). Since hybrids have reduced survival compared to wild fish, the conversion of part of the potential wild production to hybrids results in an overall reduction in survival in the population, giving fewer fish and lower recruitment of juveniles in the next generation. Repeated stocking results in a cumulative reduction in recruitment over generations, which could lead to extinction in vulnerable populations.

other nearby populations. These stocking strains can have a considerably different genetic composition compared with wild progenitors or nearest wild neighbours. In contrast to wild populations, allele frequencies, an index of genetic composition, are often observed to vary significantly between year classes of the same strain. The levels of genetic variability, measured as the number of alleles per locus, are also reduced in many stocked strains (Koljonen *et al.* 2002; Säisä *et al.* 2003). Sometimes lower levels of heterozygosity are also observed. Although the functional genetics of Atlantic salmon remain poorly understood, the differences between wild populations and stocking strains are often indicative of the reduced fitness of stocked strains in the wild.

11.4 Approach based on numbers of salmon present

Many publications (Meffe 1992; Cross 1999; Aprahamian *et al.* 2002; Youngson *et al.* 2002; Reisenbichler *et al.* 2003; Utter 2003) have focused on the genetic approaches to stocking and the consequences of this practice from both theoretical and practical perspectives, either with salmonid fish in general or for the genus *Salmo*. In this chapter we take a different approach, introducing commonly asked questions relating to specific stocking situations and addressing those questions for four stocking scenarios. A logical sequence is utilised to support management decision making. Generalised answers to each question are provided, though it is recognised that specific answers will depend on individual case circumstances. It is felt that such a practical approach will be of most use to fisheries managers and biologists, when considering

Table 11.1 The salmon augmentation scenarios addressed in the present chapter, based on the numbers of fish present in the target area of river and the feasibility of removing the source of constraint (pollution, barriers to upstream migration, etc.).

1   Stocking where population(s) extinct (**reintroduction**)
Aim: Re-establishment of healthy population(s) (ideally self-sustaining and at carrying capacity)

2   Stocking where there are extant native populations, at levels between very low numbers and numbers approaching carrying capacity (**rehabilitation**)
Aim: Increase in population size up to carrying capacity

3   Artificial production in excess of natural potential (**enhancement**)
Aim: Increase population size above natural carrying capacity to allow for increased harvest

4   Stocking for mitigation and gene banking (**mitigation**)
Aim: Compensatory fisheries production (sometimes a legal requirement) and/or protection of biodiversity

stocking as a management strategy. It is proposed that stocking questions can be divided into at least four different categories or scenarios on the basis of the distribution and proportion of wild fish present, ranging from zero to 100% of carrying capacity, and beyond (Table 11.1).

The first two of these scenarios relate to situations where problems limiting the natural productivity of the resource can be resolved, e.g. through water quality and habitat improvements, and fishery controls to reduce exploitation. A third scenario is where natural productivity is insufficient to maintain a fishery at the desired level. Lastly, scenario 4 describes situations where problems limiting or eliminating production are unlikely to be resolved in the short to medium term. An example of the latter case is a situation where the strategic economic importance of hydroelectric generating facilities transcends fisheries management considerations, but where there is a socio-economic requirement to maintain subsistence fisheries or angling.

With each scenario, we first provide a description of the typical circumstances and cite examples, where these are available. The ideal restoration objective for the first two scenarios (here termed reintroduction and rehabilitation) is self-sustaining populations with levels of genetic and phenotypic diversity typical of wild neighbouring populations (providing neighbouring populations are large and relatively pristine). Where adjacent riverine populations are of reduced size, levels of genetic variability typical of the region may have to be considered. However, we recognise that in some cases fishery managers may suggest opting for the less ideal circumstances of continued hatchery intervention to support the stock. We do not recommend this approach since it is likely to be detrimental to the native population(s) of the target river (where present), and also because of the possibility of gene flow from the reared salmon to the native populations of neighbouring rivers (Chilcote 2003; Reisenbichler *et al.* 2003).

A great deal of Atlantic salmon stocking has gone on throughout the last 50 years. These efforts were generally driven by a desire to 'improve' fishing. With this aim in mind salmon have been moved within and between catchments, regions and even continents (Galvin *et al.* 1996). In all of these efforts little or no consideration was given to the factors, including genetics, which may contribute to the success or failure of these activities. Advances in the application of genetics in fisheries management (see, for example, Ryman and Utter 1987; Carvalho and Pitcher 1995) have highlighted the importance of genetic considerations relating to stocking exercises. Another aspect shown by recent genetic surveys of wild populations previously exposed to deliberate release of non-native and/or hatchery reared native fish is that these introductions have had little or no long-term effect (Galvin *et al.* 1996; Hansen 2002) in terms of contributing genes to extant populations. However, this does not mean that stocking exercises have not and do not directly affect the genetic composition of these wild populations (see Box 11.2). As noted in Box 11.2, there may have been a transient indirect genetic effect with the depression of wild salmon production due to competitive ecological and subsequent reproductive interactions with reared conspecifics of reduced fitness. Genetic changes may also have occurred in recipient wild populations as a response to exposure to new or different forms of pathogens accompanying the stocked material (Bakke *et al.* 1990).

Armed with an emerging awareness of genetic risk factors associated with stocking, responsible management requires further information on the best genetic practices to utilise when considering the appropriateness of stocking, undertaking stocking and subsequently assessing the success of the exercise. Additionally, increased fish production can no longer be the sole consideration in reintroduction or enhancement scenarios. Conservation issues and the maintenance of biodiversity must also be taken into account in accordance with commitments under various international treaties and European Union legislation (see, for example, EU 1992). Furthermore, economic criteria in terms of initial capital outlay and ongoing sustainability costs (where needed) must be considered.

11.5 Scenario 1: Where salmon are extinct in a river (reintroduction)

Salmon have been extirpated from a large part of their natural range as a result of industrial pollution and human population pressure. Where extinction occurs within the native range of Atlantic salmon, and the causative factors have been removed, some recolonisation will occur naturally as a result of straying, for example, the recent reoccurrence of salmon in the River Mersey in England (Mawle and Milner 2003), River Dove in England (Milner *et al.* 2004) and in the River Selja in Estonia (Box 11.3). These salmon are likely to be strays from nearby rivers and may therefore be genetically similar to the original river stock. Over many generations, this recolonisation should result in the establishment of a new population that is genetically adapted to conditions in the recolonised river. Often, because natural recolonisation may be a protracted process, there is a desire to use stocking in an attempt to speed up the recolonisation process. If stocking is to be invoked, it is vital to introduce measures for protecting newly established natural populations, and to recognise that stocking may have detrimental effects on such populations (Reisenbichler *et al.* 2003), and possibly also on populations of neighbouring rivers.

Fishless systems of two basic types have been the focus of many attempts to reintroduce salmon:

- where river watersheds and neighbouring catchments are entirely bereft of salmon;
- where individual tributaries within a river catchment no longer sustain salmon.

The fishless tributary within a catchment situation is considered as part of rehabilitation (scenario 2). While the ultimate stated aim of reintroduction is to re-establish self-sustaining populations, in many cases managers appear to accept a partial resolution, some form of

Box 11.3 Natural recolonisation and reintroduction of salmon population in the River Selja, the Baltic Sea.

The River Selja, which flows over a limestone bedrock plateau to the Gulf of Finland, Baltic Sea, is one of the few rivers in Estonia where most of the potential spawning grounds can be reached by migrating salmon and trout. The first definite obstacle for upstream migratory fishes, including salmon, is a water reservoir dam, situated 34 km from the river mouth. The catchment area is 410 km^2 and annual runoff was 2.43 m^3/s during the period from 1932 to 1959. Historically, sea trout catches have outnumbered recorded Atlantic salmon catches in the River Selja. During the period from 1946 to 1952 annual catches of sea trout varied from 100 to 3105 kg, while recorded salmon catches ranged from 80 to 350 kg. Reproduction of wild salmon in the River Selja ceased in the early 1970s because of heavy pollution from the town of Rakvere, together with increased nutrient inflow from agricultural activities. Electrofishing surveys in the 1970s and 1980s failed to detect any salmon parr or smolts, indicating that native salmon population in this river most likely had been driven to extinction.

During the 1990s environmental conditions and water quality improved, although at present pollution still exists and current water quality in the River Selja is classified as 'poor' (class IV) according to the EU Water Framework Directive adopted in 1998 (Loigu *et al.* 2001). Nevertheless, small numbers of juvenile salmon parr (0+) were again detected in lower stretches close to the river mouth in 1995. To identify the origin of these fish, six microsatellite loci were genotyped and compared with genetic profile of six geographically closest potential donor populations by applying an assignment test (Vasemägi *et al.* 2001). The results of this study suggested that the initial strayers/recolonisers most likely originated from the geographically nearest (7 km) river Kunda wild population, despite the fact that hatchery releases outnumber the estimated wild production in the Gulf of Finland approximately 37 times (ICES 1998). Recolonisation from the geographically nearest population is consistent with the correlation between genetic and geographical distances (termed isolation by distance), and indicates that the spawners tend to stray and successfully reproduce in adjacent rivers.

Soon after the natural straying/recolonisation event, stocking of the river with salmon of multiple origin (River Neva strain, River Narva strain and mixture of fish caught in the sea near the mouth of River Selja) was started (in 1997) to support the formation of a self-sustaining salmon population in the River Selja. Microsatellite analysis of juvenile fish sampled in subsequent years (1999) indicated that interbreeding between initial recolonisers and hatchery individuals had taken place (Vasemägi *et al.* 2001). Nevertheless, despite intensive stocking, natural reproduction of salmon has been irregular in recent years and wild juvenile salmon parr (0+) have not been detected from 2002 to 2004. The main factors restricting the recovery of River Selja salmon population at present are probably the following: (1) poor water quality; (2) minimal runoff being too low; and (3) impact of illegal fishing of ascending spawners. Other potential negative factors include reduced sea survival of post-smolts, pathogens (e.g. the pathogenic myxosporean parasite *Chloromyxum truttae* has been found only in the River Selja) (Loigu *et al.* 2001) and M74 syndrome (Norrgren *et al.* 1993). Consequently, it has been more difficult for a viable self-sustaining population to become established after a natural recolonisation event than initially recognised. The River Selja case demonstrates that rapid recolonisation can naturally occur when geographically nearby wild population(s) exist, and also emphasises the importance of addressing both environmental and social problems when attempting to restore self-sustaining salmon population(s) in Estonia.

hatchery-based smolt release (ranching) programme being required to maintain the presence of salmon in these waters. Examples of such cases, where reintroductions are in progress but have not yet achieved self-sustainability because of a failure to eliminate environmental problems and ensure fish passage, include projects that have been undertaken in the rivers Lagan in Northern Ireland (Rosell and MacOscar 1997), Rhine in Germany and the Thames in England (Fig. 11.1).

Recent efforts to reintroduce self-sustaining populations of Atlantic salmon to Lake Ontario (North America) have also been unsuccessful (Stewart and Shaner 2002). The persistence of spawning populations of several non-native salmonids including rainbow trout (*Oncorhynchus mykiss*), brown trout (*Salmo trutta*), chinook salmon (*O. tshawytscha*) and coho salmon (*O. kisutch*), in Lake Ontario tributaries, suggests that river habitat quality may

Fig. 11.1 The River Kennet, a classic English chalk stream and the main tributary of the River Thames, is a former salmon river and a focus of stocking of salmon in the early 1980s in an attempt to restore salmon to the catchment; above, at Hungerford showing trout redds, and below at Kintbury. (Photo credit: J. Webb.)

not be an important issue in this instance (Scott *et al.* 2005). Non-native species of salmonids, however, may be impacting on Atlantic salmon restoration efforts in several ways. Firstly, chinook salmon and brown trout influence agonistic behaviour in Atlantic salmon juveniles, in such a way as to potentially impact net energy gain and long-term growth (Scott *et al.* 2005). It should be noted, however, that stocking efforts to date have been modest, and have utilised non-local salmon, including the anadromous LeHave strain from Nova Scotia. The Ontario Ministry of Natural Resources is now undertaking steps to scale up operations, and is considering several donor strains. The criteria being evaluated include phylogenetic similarity to the original Lake Ontario population, levels of within-population genetic diversity and ecological similarity of the habitat of the donor strain. For example, consideration is being given to populations of Atlantic salmon that coexist with alewife. The ultimate intention is to use several strains simultaneously, and microsatellite analysis and parentage assessment methods will be used to assess reproductive success.

Salmon managers who have fishless rivers in their care and intend reintroduction should ask the following questions:

- 'Will salmon re-establish themselves naturally and, if so, at what rate?'
- 'What is the most appropriate method of re-establishing salmon into a fishless system from the genetic viewpoint?'

The answer to the first question is almost certainly 'yes', but it may take considerable time. As stated above, probably the most genetically appropriate way to re-establish salmon in a river is to allow natural recolonisation (if wild salmon populations exist in nearby rivers). However, it is recognised that in the early generations of recolonisation effective population size (N_e) will be small. Natural recolonisation has been reported for the river Tyne in the north-east of England (Fig. 11.2). In the reappearance of large self-sustaining salmon populations in this river after almost a century of apparent absence, it was demonstrated that natural recolonisation was a far more important factor than hatchery releases (Milner et al. 2004; Box 11.4). (In fact, small numbers of salmon may have persisted in the River Tyne and the establishment of a hatchery was intended to mitigate the effects of a dam (Milner et al. 2004), so the situation may be more typical of scenarios 2 and 4: see below.)

Since natural recolonisation may not be possible at all in some cases (or in an acceptable time scale), here artificial intervention may be necessary. In these cases, fishery managers will require an answer to the second question. Unfortunately, the answer to this question can be complicated and is influenced by a number of factors including geographic location, size of potential donor populations and two specific genetic considerations:

(1) achieving the appropriate genetic architecture (Hard et al. 1999);
(2) maximising evolutionary potential by ensuring that genetic variability is comparable to other large self-sustaining salmon populations in the region.

In terms of geography and size of donor populations, two factors are of paramount importance. First, reintroduction should only be carried out using fish of the same major lineages (from rivers entering the western or eastern North Atlantic, or Baltic respectively; see earlier chapters). Given the molecular structure observed in wild populations (see earlier chapters), it is important to select donor strains from within a regional population group within these major lineages (see earlier chapters). Secondly, donor strains would ideally be selected from local rivers, which have similar ecological characteristics to the recipient river. However, in many cases (see Chapter 9), neighbouring rivers are suffering similar pressures to those that caused the extinction in the target river. Aside from further depleting stocks in these rivers by removing donor broodstock, levels of genetic variability may have been compromised there by severe population decline. With these considerations in mind, managers are advised to select donor stocks from the geographically closest, ecologically comparable and genetically 'healthy' populations. (Throughout this chapter, 'genetically healthy' is used to describe a large salmon population with levels of genetic variability comparable to pristine populations in a particular part of the species range.) Sometimes, genetic material from the wild population before

Fig. 11.2 The Kielder hatchery on the River Tyne (above) and the Kielder Burn (lower), an upper tributary into which juvenile salmon are stocked in a reintroduction exercise. (Photo credit: R. Bond.)

extirpation is available in a living gene bank (essentially a rearing station where every effort has been made to avoid selective effects, whether deliberate or inadvertent). Using such material, the aim would be to reconstitute a wild population.

Genetic architecture, in terms of co-adapted gene complexes (adaptation), is presently difficult to describe or measure using molecular genetics (but see Hansen *et al.* 2002). Hard *et al.* (1999), recognising this difficulty, suggest an alternative approach where donor populations are selected on the basis of matching quantitative life-history traits, such as age of maturity,

Box 11.4 The role of stocking in the recovery of the River Tyne salmon fisheries.

A recent study on the River Tyne, north-east England, has provided a comparison of the roles of natural processes and stocking in recovery of a salmon stock. The Tyne annual salmon rod catch increased from very low levels (zero in two years) in the 1950s to 2585 in 2002 (Fig. B11.4), now being the single biggest in England and Wales. The demise of the Tyne salmon stock was due to severe estuarine water quality decline, reaching a low point in the 1950s. Water quality (e.g. DO and NH_3) improved considerably following reduction in industrial activity and investment in effluent treatment and disposal between the 1960s and 1980s. Coincidentally, a stocking programme was started in 1979 to mitigate for the loss of 6–8% of salmon production area caused by construction of a dam and water supply reservoir in the upper Tyne. The increase in salmon stocks has been attributed entirely to the stocking programme in some accounts; however, this was contested in others and no formal assessment had been carried out until this study. The report considered three main sources of information: returns of micro-tagged stocked fish, the patterns of rod catch and effort, and a limited amount of juvenile survey data. Salmon eggs were obtained from Scottish rivers for the first 6 years, thereafter Tyne broodstock, annually replaced, has been used. The legally agreed annual mitigation requirement is for 100 000 0+ and 60 000 1+ salmon, but the annual stocking level has almost always greatly exceeded 160 000 (by on average 93%).

First hatchery returns were in 1980. Estimates of the long-term (1980–2000) weighted mean returns of stocked fish to the coast and river respectively were 0.6% (range 0.5–0.8%) and 0.3% (range 0.1–0.6%). Over the same time the weighted contributions to the north-east coast net fishery and the Tyne rod catch were 1.5% (range 1.2–2.0%) and 6% (range 3–14%) respectively. At their peak, between 1983 and 1986, annual stocking contributions to rod catch were between 22% and 42% (Fig. B11.4). The overall stocking contribution to total cumulative spawning escapement, accumulated over the start-up period up to 1986, was 20% (range 9–43%). Currently, the direct stocked fish returns contribute between 2% and 7% of the annual catch.

There was clear evidence of natural recovery pre-dating the start of the stocking programme. Historical river reports by fisheries staff reported spawning and smolt runs (sometimes notable for the resultant fish kills in the estuary) during the 1950s. Rod licence sales and catches began to increase at least 15 years before the first returns of stocked fish. By 1978, juvenile surveys showed that the upper reaches of the Tyne were colonised by salmon,

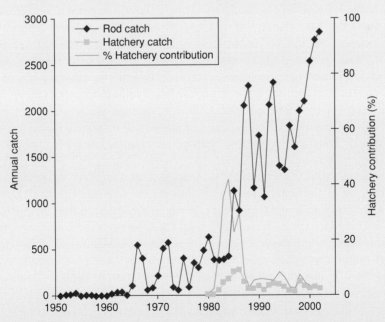

Fig. B11.4 Annual salmon rod catch on the River Tyne, UK (adjusted for licence return rate) and the estimated hatchery contribution to catch, partly based on micro-tag returns.

> **Box 11.4** *(cont'd)*
>
> with estimated smolt outputs of 0.6 to 4.7 100 m^{-2}. Sea trout *Salmo trutta* rod catch recovery occurred at rates similar to those of salmon, but with very little stocking, demonstrating that natural recovery had also occurred in another coexisting migratory salmonid species.
>
> It was not possible to detect a statistically significant impact of stocking on catches, but the most likely estimated contribution of stocked fish to the cumulative spawning escapement was 20% (9–40%), a level reached in 1986. There remains an unresolved difficulty in evaluating the long-term sustainable benefits of the stocking programme from the later generations of the progeny of hatchery-origin parents. No information on this contribution was available for the Tyne. However, making the conservative assumption of equal performance of hatchery and wild fish progeny, it could never have exceeded the 20% (9–40%) contribution. The role of natural processes in stock recovery is thought to be variable dependent upon (1) the starting level of the depleted local stock and (2) the availability of natural strays, which was high in the case of the Tyne. The overall conclusion is that that the recovery of the Tyne salmon stocks has been a mainly (80%, range 60–91%) natural process; but stocking is thought to have accelerated and stabilised this during a critical, early stage when water quality improvements were still partial and inconsistent between years.

smolt migration timing and spawning time. An illustration of the importance of quantitative variation in determining levels of performance is provided by Thorpe (1988). In attempting to regenerate salmon in a number of upland streams entering Loch Rannoch (part of the Tay river system in Scotland; Fig. 11.3), which became extinct following the construction of hydroelectric dams, fertilised eggs were taken from fish from lower down in the system below the Pitlochry dam and introduced into the streams. Smolt migration of stocked fish did not begin until after emigration of local populations was almost over, suggesting that the native populations of the Rannoch region had very different developmental timing from that of fish from lower down the Tay system, and that such downriver stocks were genetically inappropriate for stocking into upriver tributaries. In the event, no returning adults were ever recorded from these plantings (J. Thorpe, pers. comm.). From separate tracking studies on Loch Tummel, it is known that smolts move through the lochs at only a quarter of the speed of an inert object. So delays result in late entry to the sea, if indeed it happens at all. A more comprehensive list is given in Youngson *et al.* (2002). Important mediators of quantitative traits are the physical, chemical and biological characteristics of putative donor environments (items such as catchment physiography, hydrogeomorphology and pH, and the presence of predators and their abundance).

In reintroducing salmon into certain fishless situations, it may be necessary to decide between two or more alternative donor populations, or to decide to use all these simultaneously. Based on population genetics theory, it might be suggested that the former approach be adopted, since the latter would result in hybridisation between the donor populations, giving rise to possible outbreeding depression. This problem of hybridisation between strains will assume great importance when rehabilitation and enhancement are considered. However, there may be merit in introducing more than one population in fishless situations, in that it possibly provides for a greater chance of introducing the most suitable population. It is recognised that there will be possible negative consequences as described above (unless the recipient strains are managed separately and no interbreeding is allowed; Cross 1999), but it is possible that the population most adapted to the new environmental circumstances will eventually predominate (as has been shown in the Norwegian salmon breeding programme (Gjedrem 1999)

Fig. 11.3 Loch Rannoch and the upper Tummel River catchment. Clockwise from upper right: Dunalastair Dam and Fish Pass on the Tummel below Rannoch; the Tummel below Loch Rannoch; Loch Rannoch; Carie Burn which flows into the loch; the Gaur, the main upper tributary entering the loch, below Gaur Dam; the dried Gaur between the dam and the power station, where the river resumes normal flow. (Photos credit: E. Verspoor.)

where salmon from six rivers were initially mixed in equal numbers and where now, several generations and approximately 30 years later, one population, derived from a single river, predominates). (The definition of population used by Gjedrem (1999) is the same as that used throughout this chapter, i.e. a wild salmon population representing a river or part thereof.)

Table 11.2 Genetic considerations in selection of broodstock and rearing and release of progeny when stocking or ranching is to be evoked. Compiled from principles discussed in Cross (1999).

- Investigate wild population structure using molecular genetics (where salmon still exist)
- Use native strains, if possible
- Collect large random sample of broodstock
- Use 100+ spawners of equal sex ratio each year
- Use single-pair mating
- Avoid deliberate selection and attempt to minimise domestication selection
- Release equal numbers of each family
- Track family survival during rearing and after release using molecular methods
- Decide long-term broodstock strategy (new wild parents each generation or line breeding of returning reared salmon)
- Monitor exercise using appropriate molecular methods
- Avoid crossing strains

In selecting donor strains in fishless situations the possibility exists of using 'designer' strains (Cross and Rogan 1991; Wilkins *et al.* 1999) selected, for example, for a greater propensity to return as early-run multi-sea-winter salmon (so-called 'spring fish', preferred by anglers for their much greater size and by fishery owners because they promote angling during the spring months). However, the same cautions apply as when choosing other donor strains, i.e. these strains should originate from the same region and be relatively high in genetic variability, as demonstrated by a molecular genetic study (preferably using microsatellite DNA; see earlier chapters). There is also the need to collect a large random sample of broodstock (Table 11.2).

It is also assumed that a donor population will be selected on the basis of molecular genetic variability, which, to ensure long-term evolutionary potential, should be comparable to other healthy populations within the area. A difficulty arises if the only fish available from the nearest appropriate donor river are of relatively low genetic variability. Because of geographical proximity, this putative donor may have appropriate genetic architecture, but if used in reintroduction may have limited evolutionary potential and therefore may be deemed less suitable. Young (1999) invokes metapopulation theory (see earlier chapters) to assist the selection of suitable donor populations as part of their 'artificial recolonisation concept', but it must be recognised that traditional metapopulation models may not apply to Atlantic salmon (Fontaine *et al.* 1997), perhaps because of the extent of local adaptation in this species (also see Box 11.1).

Once a donor population(s) has been selected, it is assumed that best genetic practices will be utilised to maximise and maintain variability and composition, i.e. in terms of broodstock selection, utilisation of adequate numbers of broodstock, mating practices, and rearing and release of equal numbers of each family (Cross 1999; Miller and Kapuscinski 2003; Table 11.2). It should be noted that, in a fishless situation, stocking will have to be repeated annually for at least 4 years, utilising the same population(s), since the duration of the life cycle of the salmon, at least in the south of the range, is four or more years (see Chapter 2).

Since farmed fish are relatively cheap, in plentiful supply and easily obtainable, a fishery manager might be prompted to ask: 'Why not use farm strains?' In the case of farm strains, at least in Europe, the majority are of Norwegian origin and have been subject to both inadvertent and deliberate anthropogenic selection for several generations. It has been shown (McGinnity *et al.* 2003) that one of these strains has a far inferior performance in the wild compared with locally derived wild populations (for a more complete discussion refer to Chapter 12). As well as being unsuitable for reintroduction, these farm fish may inhibit natural recolonisation.

As stated above, the measure of success of the stocking programme is self-sustainability in subsequent generations as evaluated by juvenile surveys, adult counts and catch data. Molecular genetic monitoring is also recommended, particularly in a multiple stock situation where statistical methods such as mixed-stock analysis (see Chapter 9) can then be used to estimate the proportions of each donor population.

Specific recommendations

- The optimum way of reintroducing salmon into a fishless catchment, from a genetic viewpoint, is to alleviate the constraint to production and allow natural recolonisation to occur.
- Where the target catchment is geographically distant from healthy wild populations, it may be necessary to introduce suitable strains. If this is necessary, great care must be taken in the selection and subsequent treatment of putative donors, to avoid producing a suboptimal restored population or compromising the genetic potential of remnant wild populations in nearby rivers.
- Wild donor populations should not be sourced from another major genetic grouping, nor should farmed strains be used.
- A comprehensive molecular study should be undertaken of putative donor/neighbouring populations, and of the reintroduced strain(s) subsequent to introduction and annually thereafter.

11.6 Scenario 2: Where small to near optimal numbers of the local population(s) remain (rehabilitation)

Many riverine populations of salmon have suffered severe decline in numbers, mainly due to anthropogenic activities and, latterly, poor survival at sea. However, some fish can still inhabit these rivers, albeit at low population numbers (see Box 11.4). In this scenario, fishery managers often face a situation where small numbers of fish are not attaining carrying capacity (i.e. between less than 10 and numbers approaching carrying capacity). In some cases it has been presumed that salmon are extinct in a river but subsequent genetic studies have shown that a few natives remained (Nielsen *et al.* 2001). From a conservation standpoint, it is essential to determine whether fish remain before taking any ameliorative action involving stocking, since such a process might lead to extinction of an important remnant population. Typically, a reduction in population size can result in either population fragmentation (where residual fragments occur in optimal habitats within the original population distribution) or where a population continues to exist over its original distribution, but at a reduced level. The presumed aim of fishery managers in this type of situation is the rehabilitation of a self-sustaining population(s), and specifically to increase natural production.

Before considering rehabilitation by stocking, the first action must be to determine why the population is at a reduced level: whether due to habitat deterioration, pollution, overexploitation, and so on. Without identification and, where possible, correction of the factors limiting production there is little likelihood of the successful restoration of a self-sustaining population, and virtually no chance of increased natural production.

Where wild salmon production more nearly approaches carrying capacity, we suggest that no hatchery intervention should occur, i.e. the problem limiting production should be resolved and population expansion allowed to proceed on its own. The reason for this recommendation is that we perceive the risk of reduction in fitness as a result of interaction between wild and stocked fish to be unacceptably high (see Reisenbichler et al. 2003). Any stocking may have detrimental effects on population recovery (Chilcote 2003).

This natural population expansion back to optimal carrying capacity is likely to take a number of generations. We recognise, therefore, that there may be considerable pressure on fisheries managers to rehabilitate fish numbers and fish distribution more quickly. The two questions that a fishery manager must ask, if stocking is to be evoked as a rehabilitation strategy, are:

- 'What are the genetic considerations in attempting to rehabilitate river systems where an extant population or extant populations exist at reduced numbers?'
- 'Does the size of the extant population influence the genetic advice?'

Recent research has shown reduced fitness in stocked non-natives compared with the wild native salmon (see Box 11.1). From other research findings (McGinnity et al. 1997, 2003, 2004), it seems reasonable to expect that the fitness of F_1 hybrids will be intermediate between the progeny of native and stocked fish. Thus, any rehabilitation, enhancement or, in some cases, mitigation stocking must be recognised as potentially damaging to existing wild populations, i.e. having the potential to undermine the productive and evolutionary potential of the recipient river population(s).

Until recently, it was often argued that populations with effective numbers of spawners (N_e; see earlier chapters) being fewer than 50 (Wilcox 1986) were functionally extinct and that the principles described in a re-establishment scenario should be invoked. However, it now appears from studies in Denmark (Nielsen et al. 1999) that much lower numbers of naturally occurring fish can retain genetic composition (and possibly genetic architecture; Hard et al. 1999), biological function and productive potential of the original population. (Genetic composition and level of variability in a population prior to a decline in numbers can be assessed from a microsatellite DNA study of archival scales.) The consequence of these findings is that the concept of minimal viable population number is now difficult to define. It is therefore suggested that very small population remnants should be protected since these may have sufficient potential to allow population rehabilitation. Paradoxically, it is the most reduced populations that are most threatened by the effects of hybridisation with introduced fish.

Busack and Currens (1995) define extinction as a complete loss of all genetic information (though a more appropriate definition may differentiate between extinction in the wild, with retention of material in living gene banks, and complete loss). Extinction must be regarded as the ultimate genetic hazard, because once a population is gone, all the unique aspects of the diversity it contained are also lost. Evaluating the extinction risk for a population involves estimating the probability that a population will not persist for a given length of time. A great deal of progress has been made in developing quantitative models of population sensitivity. This general approach, referred to as population viability analysis (PVA), can be quite powerful, but uncertainties associated with parameter estimation and model choice have limited its usefulness to date (Jonsson et al. 1999).

Supportive breeding programmes offer opportunities for reducing extinction risk in remnant populations. Such programmes need careful planning and management, and extensive monitoring to ensure that genetic integrity is not compromised. Monitoring of genetic 'health' of the target population(s) and reared donors should be carried out before supportive breeding intervention. Such monitoring, using the most sensitive molecular systems available (currently microsatellites; see earlier chapters), should also be undertaken post-stocking on a regular basis (so that F_1 and F_2 hybrids between wild and reared salmon are monitored), to assess the effectiveness of intervention. Adherence to stringent genetic principles, as described for reintroduction scenarios for hatchery intervention (Table 11.2), is also necessary.

Supportive breeding programmes have been proposed as a way of rapidly boosting the productivity of depressed or fragmented populations, following environmental improvement. These programmes typically encompass three major paradigms (for a fuller discussion see Chapter 14).

(1) Captive broodstock programmes, where all, or a fraction, of the wild parental fish from an extant population are brought into a hatchery for breeding and the offspring are released into the natural habitat as fry, parr, smolt or adult life stages, where they mix and presumably interbreed with the wild fish.

(2) Juvenile capture programmes (eggs, fry, parr or smolts), where all, or a fraction, of juvenile fish from an extant population are brought into a hatchery and on-grown to maturity past some perceived bottleneck and the offspring subsequently released into natural habitat as eggs, fry, parr, smolts or adults, where they mix with wild fish if present.

(3) Kelt reconditioning programmes where all, or a fraction, of the kelts (post-spawned fish) from an extant population are reconditioned in fresh or seawater rearing facilities and held to sexual maturity, and their offspring subsequently released into the wild as eggs, fry, parr, smolts or adults (Davidson and Bielak 1993; Poole *et al.* 1994).

As stated above, supportive breeding programmes must be carefully planned and managed with regard to a number of factors, not the least of which are genetic considerations:

- Where broodstock are being removed for a rearing programme, natural populations must be large and remain so after broodstock removal (to avoid inbreeding in native populations).
- Ryman and Laikre (1991) have shown that a reduction in effective population size may result from preferentially augmenting that part of the gene pool taken into the hatchery. In other words, a large fraction of the individuals in the population post-supplementation will be derived from a relatively small number of founders. The result can be an artificial bottleneck that can lead to increased levels of inbreeding and erosion of genetic diversity (the so-called 'Ryman–Laikre effect').
- To supplement a population without correcting the constraint that caused the population to decline in the first place will put that population at greater risk than would be the case if it was left alone (Waples and Do 1994). If Ricker-type spawner recruit relationships apply and carrying capacity remains low, a large temporary increase in abundance could theoretically drive the population to extinction.

It has been suggested (Cross *et al.* 1998) that native stocking, as envisaged in supportive breeding programmes, is acceptable from a genetic viewpoint when environmental or overexploitation problems have been resolved. The presumption was that it was possible to maintain the genetic architecture and composition of the wild population during a short-term or well-managed longer-term culture phase. Several recent publications, particularly Reisenbichler *et al.* (2003), caution against this approach on the basis that any anthropogenic intervention will potentially change the genetic make-up of the cultured group (e.g. domestication selection). This can even be the case if inbreeding effects are minimised by the use of large numbers of broodstock (Cross 1999). Therefore, we now regard it as unlikely that salmon can be maintained in a rearing programme without changing their genetic composition in the way that will detrimentally affect fitness in the wild (Chilcote 2003).

Furthermore, the advice to use only native broodstock for stocking in this type of scenario has not historically been followed in sourcing parental material. For example, stocking in Northern Spain, using donor stocks from Iceland, Norway, Ireland and Scotland has been largely unsuccessful (see Chapter 7). In certain Norwegian rivers, strains resistant to *Gyrodactylus salaris* and carrying this ectoparasite were introduced from the Baltic, resulting in high mortality in susceptible Norwegian populations (Bakke *et al.* 1990).

Overall, the risks associated with stocking for remnant or small populations appear to be unacceptably high. Probably, the best option in this scenario is to correct the problem limiting productivity of the remnant population(s), then be patient and let nature take its course. However, it is recognised that in certain situations the small remnant population will fail to become established, leading to extinction, and that the principles outlined in reintroduction and/or mitigation situations will need to be invoked (with the possibility of stocking being involved). In the US, for example, the National Research Council's Committee on Atlantic salmon in Maine has recommended stocking in support of the severely declining populations of Atlantic salmon along the east coast. Specifically, primarily alevins of native origin are being released into six of eight rivers that comprise the distinct population segment (DPS), listed in 2000 as endangered under the Endangered Species Act (ESA). The committee has also prescribed 'urgently needed actions' (NRC 2004), including removal of dams and reduction of early mortality of smolts, as they enter the marine phase of their life cycle.

Specific recommendations
- Comprehensive surveying should be undertaken to determine whether any wild salmon remain in the target river.
- The previous recommendation of assuming a population functionally extinct if the effective population size (N_e) is less than 50 must now be regarded with caution, and smaller numbers of spawners might be used as the basis for rebuilding a population.
- The constraint to production must be identified and corrected, and then the first option should be to allow natural population increase to occur.
- A molecular study on archival scales, if available, should be undertaken, and then the present and expanding population monitored (on an regular basis) by the same methods.
- A supportive breeding programme to accelerate recovery should only be invoked as a last resort, since the genetic risks of reduced fitness progeny, through hybridisation between wild and reared salmon and other phenomena, are high.
- If stocking is considered to be the only available option, we suggest that this be undertaken with native progeny.
- Farmed strains of Atlantic salmon used for aquaculture should never be used for rehabilitation.

11.7 Scenario 3: Attempting to achieve productivity in excess of naturally constrained production (enhancement)

We see the objective of fisheries managers in this type of situation as being to fulfil a fisheries production function in either the freshwater or marine environments for commercial or sports fishery enhancement. This objective is usually addressed by implementing large-scale smolt ranching exercises, where the intention is to harvest all the returning fish, and smolt release programmes, where returning fish are free to spawn in the wild. In this scenario, a healthy, fully productive environment is assumed and that ongoing exploitation is biologically sustainable, e.g. in the Burrishoole River in the west of Ireland. It is interesting to note that in the Burrishoole the survival from smolt to grilse is consistently lower for the ranched strain (which was of native origin) than for wild fish (Table 11.3) (Salmon Research Agency of Ireland 1955–2003; Piggins and Mills 1985), and similar trends have been reported in many other instances. The presumption that the wild population(s) will continue to function

Table 11.3 Survival from smolt to grilse of wild and ranched smolts from Burrishoole system in each of 29 years, with average, illustrating the consistently higher survival of wild smolts. Compiled from annual reports of the Salmon Research Agency of Ireland (1955–98).

Year to sea	% survival Burrishoole wild smolts	% survival Burrishoole ranch smolts
1970	5.3	1.3
1971	11.0	6.7
1972	12.3	1.5
1973	8.7	1.1
1974	9.2	3.5
1975	6.3	1.6
1976	4.3	0.7
1977	6.3	1.0
1978	9.4	2.5
1979	7.8	1.6
1980	3.1	2.1
1981	5.4	1.3
1982	5.7	1.7
1983	3.4	0.5
1984	7.2	3.0
1985	8.1	3.7
1986	8.7	1.7
1987	12.0	3.5
1988	10.1	3.3
1989	3.5	2.4
1990	9.2	3.7
1991	9.5	2.4
1992	7.6	2.0
1993	9.5	3.4
1994	9.4	2.1
1995	6.8	3.0
1996	9.2	2.4
1997	8.2	4.5
1998	5.3	0.8
Average	7.7	2.4

optimally and maintain its genetic integrity accordingly (i.e. will continue to have positive biodiversity implications) may be challenged in these situations.

The most important genetic consideration is to protect the population(s) that may be affected by enhancement programmes. In this context, fishery managers must address the following question:

- 'What are the genetic factors to be considered in embarking on such programmes, particularly with regard to the necessity to protect the genetic integrity of local populations?'

In terms of protecting local populations, it is vitally important to prevent genetic interaction between the cultured strain and the wild population(s). Prevention of genetic interaction may be achieved by having full adult trapping facilities at the top of the tide and removing all cultured fish to secure broodstock holding tanks. If such interactions were to be allowed, the relative numbers of cultured and wild salmon would lead to the Ryman–Laikre effect and also substantially increased hybridisation (McGinnity *et al.* 2003). The latter would be even more damaging if non-native cultured fish were being used in the smolt release programme.

It is recognised that smolt release programmes may be 'open' in the sense that broodstock are taken from returning adults both cultured and wild *de novo* in each generation, or 'closed' programmes, where the ranched strain is line bred from ranch returns. The former situation is often referred to as broodstock 'mining' and might be genetically damaging to the wild population(s) in terms of reducing numbers of natural spawners over time. If this broodstock 'mining' effort is concentrated in a specific location within a catchment (often necessary for practical reasons related to catching the fish), it runs the risk of eliminating a biologically important element of population structure.

In the specific case where native broodstock have been used to found the cultured strain, it is often argued that constantly refreshing the broodstock, by taking new individuals from the wild in each generation, introduces 'new blood'. Stated more formally, this practice is designed to minimise hatchery effects. It is also suggested that in these circumstances the cultured strain will be functionally equivalent to the wild populations and therefore breeding between these two groups will have no detrimental genetic effect. However, the genetic risks for the wild population(s) can outweigh any potential benefits of this practice, because of the Ryman–Laikre effect described earlier.

Another potential strategy for eliminating hybridisation between wild and cultured populations would be to release all-female triploid smolts (see Box 12.6 for description of technology involved). This approach has been tried experimentally (Wilkins *et al.* 2001) with some success, in that all-female triploid salmon were sterile but returned to coastal waters. It requires more development before it might be used as a wide-scale strategy. To maintain such a strategy over several years, it is necessary to continue to collect broodstock from the wild as the majority of returning adults will be sterile or of low fertility.

Large-scale smolt release programmes (assuming a terminal fishery and/or line breeding) also present a risk to wild salmon populations in adjacent rivers due to strays entering and successfully breeding with native populations in these rivers. Between-river straying (implying gene flow) is stock specific and is likely to be higher where non-native strains are used in the smolt release programme (Hard and Heard 1997). Adipose fin clipping in addition to microtagging or genetic marking (see Chapter 6) would allow an evaluation of the magnitude of this occurrence, which might have severe genetic consequences for salmon of affected adjacent rivers (i.e. hybridisation effects).

Another consequence of large-scale smolt release programmes is that the presence in the ocean of large numbers of ranched fish will encourage heavy commercial exploitation in the ocean or rod exploitation in fresh water (referred to in North America as 'enhancement overfishing'). Such heavy exploitation will have a disproportionate detrimental effect on wild populations in the target rivers and adjacent rivers, since they are often numerically less abundant. Specifically, the genetic effect of this activity will be to reduce the potential number of wild spawners in particular cases to dangerously low levels, where inbreeding effects can become profound. Small wild populations are particularly vulnerable and may become extinct. Note, however, that contrary results have been reported from Denmark (Nielsen *et al.* 1999, see above).

Another circumstance where there is potential for increasing the productivity of a population, but on a much smaller scale, is by redistributing native salmon as eyed eggs or unfed fry into areas that would not normally be accessible to migrating fish (e.g. above impassable waterfalls). However, while this practice carries minimum risk for the genetics of the wild extant salmon population, the risk to congeneric species such as brown trout, through indirect genetic effects, because of increased competition or the introduction of disease, should be considered.

Specific recommendations
- Every effort (including the production of sterile reared salmon) must be made to avoid interbreeding between reared salmon and the wild population(s).
- The genetic structure of both the wild population(s) and reared strains should be monitored regularly at a molecular level while augmentation continues, in order to detect any interbreeding.
- The risks of reared fish entering nearby rivers and interbreeding with the natives should be considered.
- The dangers of such an exercise enabling enhanced commercial fishing in nearby coastal waters to the detriment of the wild recipient river population(s) and those of nearby rivers must be appreciated.
- Released smolts should be micro-tagged so that levels of escapement into the river proper or straying to local catchments can be assessed.

11.8 Scenario 4: Mitigation programmes and conservation hatcheries to counter irreversible loss of natural production (mitigation)

The mitigation scenario is one where there is irreversible or at least long-term possibility of a catastrophic effect on production, i.e. the river or a portion of the river no longer has the capacity to maintain a self-sustaining population or populations. This sort of scenario usually relates to the damming of rivers, where the ecological functioning of the habitat above the dam is compromised to a lesser or greater extent (see Box 11.5 on the Baltic situation), or sometimes where natural phenomena such as volcanic eruptions have destroyed spawning areas. (An example of the latter is the Ranga in Iceland.) Usually, because of larger economic and political considerations, there is no possibility of removing dams and restoring rivers to their pre-impoundment state. Thus, fisheries managers will have to consider alternative measures to restore production. It is interesting to note however that in the US state of Maine, the Federal endangered species legislation has been invoked allowing inter alia for the removal of dams, in an attempt to protect Atlantic salmon.

Box 11.5 Atlantic salmon in the Baltic Sea: a mixture of wild and hatchery-reared fish.

The majority of wild Atlantic salmon populations in the Baltic Sea area have been lost through the destruction of habitats suitable for reproduction, by power plant construction since the 1940s. Of the 90 original salmon riverstocks (the term 'stock' refers to the salmon of a single river catchment) in the Baltic Sea, only about 36 remain (ICES 2005). Their combined production capacity of 3.5 million smolts amounts to about one-third of the original estimated production capacity of 8–10 million smolts in the whole Baltic Sea drainage area (Lindroth 1965). In Finland, for instance, more than 70% of the total original smolt production capacity of about 3 million smolts was irreversibly destroyed when damming closed the largest Baltic Sea salmon rivers (Kemijoki, 700 000 smolts; Iijoki, 300 000 smolts; Oulujoki; 450 000 smolts; Kokemäenjoki, 300 000 smolts; Kymijoki 250 000 smolts; Sjöblom *et al.* 1974). The trend was similar in other countries around the Baltic Sea, when large rivers such as the Luleälven, Indalsälven, Ångermanälven and Dalälven in Sweden, the Daugava in Latvia and the Vistula in Poland were closed. In more recent years, the intensive offshore fishery and the degradation of river habitats have caused a further decline in the remaining wild stocks.

In the 1960s and 1970s, extensive hatchery rearing and release programmes were launched in Finland and Sweden. These programmes, most of which were based on the original stocks of the salmon rivers, aimed to compensate for the lost smolt production, to maintain the fisheries, and to preserve the genetic resources of the salmon stocks of the dammed rivers. Most current smolt release programmes in Finland and Sweden are based on water court decisions regarding compensation for closed rivers, and amending these decisions would be a difficult process, both legally and politically.

Over the last 15 years (1990–2004), the amount of annually released smolts has ranged from 4.5 to 5.9 million within the whole Baltic Sea. Finnish releases have accounted, on average, for 34% and Swedish for 30% of the total smolt release. The remaining 22% has consisted of releases from Denmark, Estonia, Latvia, Lithuania, Poland and Russia, with wild production comprising an average of 14% (ICES 2005). The production of hatchery smolts has been based either on captive hatchery broodstock, which is the preferred method in Finland, or on the catching of feral spawners from river mouths, which is the main method used in Estonia, Latvia, Poland and Sweden.

The total number of released smolts has been very large compared with the level of natural production. From 1987 to 2004 the proportion of wild production ranged from 3% to 25%, with an increasing trend (Fig. B11.5). Stockings helped to sustain salmon catches in offshore and coastal fisheries at virtually the same high level of ~ 3000 tonnes a year from 1972 to 2004. An unwanted side effect of this achievement was the very high fishing pressure (enhancement overfishing) on the remaining wild stocks. In the 1970s and 1980s, all the naturally reproducing stocks of Atlantic salmon in the Baltic Sea area were regarded as threatened, and by the end of the 1980s several wild riverine stocks in the northern Baltic Sea were close to extinction (Romakkaniemi *et al.* 2003).

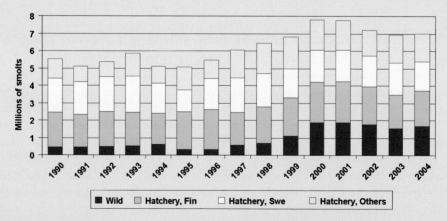

Fig. B11.5 Annual smolt production of wild and hatchery fish in the Baltic Sea drainage area. Fin = Finland, Swe = Sweden.

Box 11.5 *(cont'd)*

Since 1991, the annual salmon catch at sea has been limited by the International Baltic Sea Fishery Commission (IBSFC), which now imposes a quota (a total allowable catch: TAC) on this fishery. For 2004, the TAC was 495 000 fish, corresponding to a total catch of about 2300 tonnes. The production of wild stocks has clearly increased since 2000, partly as a result of the TAC regulation and the stricter Finnish and Swedish coastal fishing restrictions in the Gulf of Bothnia since 1996, and partly due to the exceptionally strong year classes of spawners (Fig. B11.5). Nowadays the large wild salmon stocks, e.g. those of the Tornionjoki and Kalixälven, have reached from 40% to 60% of their estimated maximum smolt production capacity, and all wild stocks together are producing approximately 48% of their currently estimated production capacity of about 3.5 million smolts. For some unknown reason, post-smolt mortality has increased among hatchery fish, and thus since 2002 more than half of the total offshore catch has derived from wild stocks, which is a completely new situation in the Baltic Sea fishery (Koljonen 2004; ICES 2005).

The contemporary Atlantic salmon gene pool in the Baltic Sea catchment is only a fraction of what it was before human impact, many populations having become extinct. A part of the genetic material of the destroyed river populations has been maintained in reared strains derived from native populations. The genetic diversity of these strains is slightly lower than that of the large contemporary wild populations, but with captive breeding it has been possible to maintain some of the genetic diversity that would otherwise have been irrevocably lost, and also to maintain the level of genetic diversity of these hatchery strains, at least at a higher level than that in the smallest wild populations (Koljonen *et al.* 1999, 2002). The genetic diversity of reared strains used as broodstock has remained at a relatively high level, even when compared with the level of diversity in the corresponding wild populations prior to dam construction (as assessed by molecular genetic screening of archival scales). In reared strains of wild origin, the decrease in allelic richness has ranged from 1.5% to 2.5% per generation and the maximum loss in mean heterozygosity has been less than 1% per generation since the damming of the rivers (Säisä *et al.* 2003).

Hydroelectric dams can be built at the top of the tide, effectively eliminating all salmon production, or in varying positions upriver where production continues in tributaries and the main channel downstream of the dam. (Downstream production may be adversely affected by a change in water temperature regime, for example.) One view is that areas above dams are no longer capable of maintaining self-sustaining populations and mitigation is required to maintain historical fisheries production levels and/or to provide for the protection of adaptive genetic potential of the population(s) that previously lived above these dams. While it is recognised that fish passes allow upstream migration over many dams and access to the affected spawning grounds, the hydro-morphological changes that occur in the river channels following impoundment (formation of lakes that can slow migration and aggregate predators in smolt migration corridors) and the destructive effect of the turbines, may, de facto, lead to the drastic reduction or elimination of salmon production in these areas.

Where there is minimal possibility of natural salmon production in affected rivers, the maintenance of salmon runs (at least as far as the dam) becomes chiefly a ranching exercise. It is recognised that the extensive ranching that takes place from rivers entering the Baltic has a commercial function in that it supports a marine fishery (see Box 11.5). However, the initial aim was mitigation of production lost by damming of rivers for electricity generation (Kangur and Wahlberg 2001; Koljonen 2001). Dams for hydroelectric power occur to a lesser extent in other parts of the range, but may have major effects on rivers on which they occur. The case of the St John River in New Brunswick, eastern Canada (Fig. 11.4) is described in Box 11.6.

We do not discuss commercial ocean ranching, where returning fish are harvested, usually at or near the mouth of the river of origin, and which is an extensive practice with Pacific salmon (genus *Oncorhynchus*), since this is not a widespread practice in the North Atlantic.

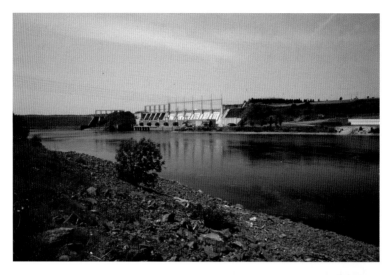

Fig. 11.4 The Mactaquac Dam (above) and an angler playing a salmon (below) on St John River, Canada. (Photo credit: J. Webb.)

(Ranching to enhance a marine fishery, on the other hand, is widespread in the Baltic; see Box 11.5.) There have been efforts to initiate large-scale commercial ranching of Atlantic salmon in Iceland, but these efforts have not been economically successful. Our major reason for excluding discussion of commercial ranching here is that it is more akin to salmon farming (see Chapter 12) in that deliberate selection is involved in production (Jonasson *et al.* 1997).

In mitigation exercises, interaction with wild populations should be minimised as noted in Chapter 12 for escaped farm salmon. The genetic considerations are in general similar to those described for enhancement scenarios and in some ways the questions fishery managers should ask are also similar:

Box 11.6 Compensating for the loss of natural Atlantic salmon production in the St John River, New Brunswick.

The St John River, New Brunswick, Canada, is the largest river east of the St Lawrence, and occupies a watershed of approximately 54 500 km^2. Around the middle of the twentieth century, several dams were constructed on the main stem and upper tributaries, including the Tobique Narrows Dam in 1953, the Beechwood Dam in 1957 and the Mactaquac Dam in 1967 (Fig. B11.6). Although reasonably efficient upstream fish passage exists at all three facilities, safe downstream fish passage is lacking, and cumulative turbine mortality for outmigrating smolts at all three dams has been estimated to be approximately 50% (Carr 2001). Mitigation stocking was undertaken to compensate for the loss of spawning habitat and increased mortality of downstream migrating smolts. Under a legal agreement between the New Brunswick Electric Power Commission (NBEPC) and the Department of Fisheries and Oceans (DFO), Canada, NBEPC was obliged to construct fish culture facilities immediately below the Mactaquac dam, and DFO to carry out fish culture and associated stocking activities (fish collection, captive rearing, spawning, fish release, and so on). Throughout the 1970s, 1980s and 1990s, the compensation programme centred around smolt production; several hundred wild returning salmon of local origin, captured in fish traps at Mactaquac, were spawned each autumn, and approximately 300 000 smolts released each spring.

In the mid to late 1970s, over 8000 wild 1SW and as many wild MSW salmon passed through the fishway at Mactaquac en route to upstream spawning habitat. Over 10 000 1SW and 3000 MSW hatchery salmon also returned to Mactaquac in some years during this period (Jones *et al.* 2004). However, salmon runs, particularly the wild MSW component, began to decline in the early 1980s, and helped precipitate closures of commercial fisheries in 1984 throughout the entire watershed. Continuing declines and persistent failures to meet conservation requirements resulted in the complete cessation of all remaining legal fisheries (aboriginal and sports) by 1998. In both 2002 and 2003, fewer than 1000 MSW and 2500 1SW salmon (wild and hatchery) returned to Mactaquac and the upper St John watershed (Jones *et al.* 2004). In addition to mortality associated with

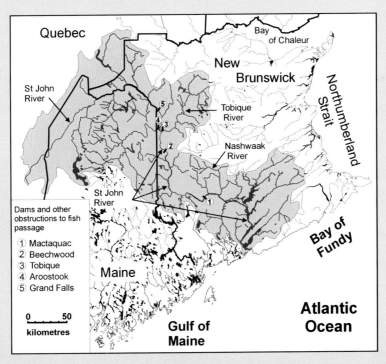

Fig. B11.6 The St John River (blue) and associated watershed (grey), including several major tributaries. Location of dams and other obstructions to fish passage are also indicated.

> **Box 11.6** (cont'd)
>
> downstream fish passage, other potential contributing factors to the long-term decline in numbers include changes in water body morphology, increased predation on juveniles from fish species introduced into the river system (Marshall *et al.* 2000), interaction with aquaculture salmon in the estuaries and in river habitat, and possible reduced fitness associated with the captive spawning and rearing programme at Mactaquac. However, recent marked increases in marine mortality of St John River smolts (Jones *et al.* 2004), reported to be impacting many southern and mid-latitude river populations in North America (ICES 2003), is also likely to be a key factor in the decline of St John River salmon.
>
> Today, the primary objective of the compensation programme, developed by the Department of Fisheries and Oceans and the St John River Salmon Management Advisory Committee (SJRSMAC), is to conserve St John River salmon. Rather than sequestering a substantial portion of the now markedly reduced number of returning wild salmon for use in the production of smolts, emphasis has shifted to the capture of wild pre-smolts from the Tobique River tributary (Fig. B11.6), rearing in captivity through to maturity, and subsequent return to the Tobique for natural spawning. In addition to minimising the loss of wild spawning salmon to the upper St John, this adult release strategy offers the following advantages over the production of smolts: (1) possible benefits of mate selection and the resulting increase in fitness of offspring in the next generation (reviewed in Jennions and Petrie 1997), (2) exposure of early life-history stages to natural selection, (3) maintenance of potential genetic differences among salmon from different tributaries, and (4) avoidance of high mortality during the marine phase of their life cycle. Disadvantages include genetic and behavioural changes brought about by domestication during the late juvenile–adult phase of their life cycle, possible loss of genetic variation if relatively few salmon spawn successfully, and possible competition for spawning habitat between released and wild returning adults (though much of the existing spawning habitat is currently unused due to the low numbers of returning adults).
>
> The first pre-smolts (approximately 2600 individuals) were captured in 2001 using rotary screw traps located on the upper and lower Tobique River, and released as adults into the same river in 2003 and 2004 to spawn. The programme is ongoing, with a similar number of pre-smolts captured, and approximately 1000 to 2000 adults released each year into the Tobique River. If successful, the programme will be expanded to other major tributaries above the Mactaquac Dam.
>
> A combination of electrofishing, chemical (canthaxanthin) marking and genetic tagging methods are being used to assess the success of the programme, measured as the proportion of juveniles produced by released versus wild returning salmon and the number of adults contributing to subsequent generations of salmon. Other extensive evaluations, using microsatellite markers and parentage determination methods, are currently under way to assess the success of this adult release strategy in the Maritimes.
>
> A small number of captive-reared adults, from the same Tobique River programme, are also used for broodstock in the production of 300 000 0+ parr and up to 60 000 smolts for research purposes; parr are released into the tributaries from which their parents were obtained, and smolts into the St John river below the Mactaquac Dam.

- 'What are the likely genetic considerations pertinent to mitigation programmes?'
- 'Can conservation hatcheries (or cryopreserved gene banks, see Chapter 14) help to maintain native salmon in the ranching programme?'
- 'What are the genetic considerations associated with such conservation hatcheries?'

Two specific situations where mitigation might be invoked are described in Boxes 11.7 and 11.8.

Another catastrophic event requiring drastic response, where hatchery intervention (living gene bank) (Cross and Rogan 1991) or cryopreserved gene banks are required, is where an entire stream-resident salmon cohort is affected by *Gyrodactylus salaris* (Gausen and Moen 1991) or where increasing acidity prevents successful breeding (as in areas in southern Norway (Hesthagen and Hansen 1991) and eastern Canada affected by air pollution). Here,

Box 11.7 Mitigation programme started simultaneously with the introduction of the constraint to salmon production.

In these circumstances ranching with native salmon commences in tandem with the building of the dam, thus there is the possibility of conserving all aspects of genetic legacy of affected populations. Downstream populations are endangered only if the newly established hatchery population is changed genetically. So ideally the conservation hatchery should be a rearing facility that is designed to breed and propagate a stock of fish with genetic resources equivalent to the native stock (at risk). Flagg and Nash (1999) suggest that conservation hatcheries should provide fish with minimal genetic divergence from their natural counterparts to maintain long-term adaptive traits.

Box 11.8 Genetic considerations when the ranching programme begins some years after the construction of the dam and when a large proportion or all of the upstream genetic legacy has been extirpated.

In this situation it is recommended that a downstream tributary, if holding sufficient numbers of surplus salmon, be used as a hatchery donor. If salmon from downstream tributaries are reduced in number, a strain from an external source might be considered, thus disregarding genetic legacy in order to fulfil a fisheries production function. This will possibly be to the detriment of downstream remnant populations because of hybridisation. Where there is a dam located at the top of tide the latter approach is necessary (see also discussion of scenario 1).

In the past, it was often common practice for hydroelectric companies, in setting up and maintaining ranching strains, to mix remaining native salmon with externally sourced non-native fish. Recent advances in molecular biology (Nielsen *et al.* 2001) make it possible to identify native remnants within often quite heterogeneous assemblages of native and non-native fish. Thus native or near-native salmon with the correct genetic legacy might be retrieved and used as broodstock with a greater potential for use in any subsequent rebuilding of the wild stock. The same molecular genetic analyses and good genetic practices in broodstock selection and management, and in rearing and release of progeny, as outlined in enhancement scenarios, are also required in these situations (see Table 11.2).

early pre-emptive action is required to retrieve sufficient fish to protect the unique adaptive genetic legacy of the population(s). Genetic principles must be applied in planning and executing such endeavours in order to achieve these goals. Use of captive breeding to arrest the further decline and possible extinction of small populations is discussed in Chapter 14.

Specific recommendations
- The genetic risks of ranching exercises for the wild population(s) within the target river and in neighbouring catchments should be assessed.
- The molecular genetic structure of both the wild population(s) and reared strains should be monitored regularly, while augmentation continues.
- The dangers of mitigation ranching enabling enhanced angling within the river (or increased commercial exploitation at sea, or in the estuary or fresh water) to the detriment of the wild population(s) must be appreciated.
- Every effort should be made to preserve the original genetic composition while utilising gene banking, using either cryopreservation or hatcheries.
- Micro-tagging, in conjunction with adipose fin clipping, of ranched smolts will help to identify the extent of straying and possible genetic exchange with neighbouring rivers or tributaries.

11.9 Summary and conclusions

Many rivers have a history of stocking and ranching with Atlantic salmon, and it is not clear at this stage to what extent the original within- and between-population genetic diversity has been affected. Other rivers have not been stocked, but stocking is currently being considered due to low productivity, linked to poor sea survival. As emphasised above, stocking may have a detrimental effect on wild populations (where these still occur) within the target catchment (see Box 11.2), and also on wild salmon in neighbouring catchments. Therefore, it is likely that past stocking and ranching efforts may have had similar detrimental effects, the extent of which might have been influenced by factors such as relative numbers of introduced and wild salmon, and how much lower the fitness of introduced fish was relative to wild salmon. While it is impossible to quantify these effects after the event, every effort should be made to avoid such effects in the future. Stocking and ranching may be as serious (or more serious) a threat to the fitness of wild populations as farm escapes, and this concept must be paramount in risk assessment of any such exercises (see Chilcote 2003; McGinnity *et al.* 2004).

11.10 Management recommendations

- In any situation where salmon rehabilitation is being considered, the aims need to be defined, for example (1) to protect biodiversity, (2) to produce a self-sustaining population, or (3) to boost commercial fishing or angling catches. A risk assessment of the various rehabilitation options should then be undertaken by a team which includes an experienced fish geneticist.
- The current status of the population(s) within the target catchment, both in terms of numbers and genetic structure, must be evaluated using the most up-to-date molecular and statistical methods. This is especially important for those catchments where the population(s) appears to be under threat at present. Molecular genetic analysis of such a population(s) will facilitate informed management decisions in the future. In addition, evidence of local adaptation must be sought, using a combination of genetic, ecological, behavioural and physiographic information.
- Constraints to production, where present, should be removed, and natural recolonisation or population increase allowed to proceed.
- From a genetic viewpoint, stocking and ranching should be considered as a last resort in rehabilitation exercises, rather than the first option, as heretofore.
- Only where natural processes have failed, or in situations where they will not or cannot achieve rehabilitation, should stocking or ranching be invoked (see above). In most such cases, it is best to use native broodstock from a strain that has spent a minimum amount of time in the hatchery, and all possible measures to avoid changing genetic composition and to minimise inbreeding have been taken.
- There must also be a means of determining progress towards achieving rehabilitation aims at appropriate milestones in the programme, i.e. monitoring and evaluation. Broodstock should be typed genetically in each generation. This has two advantages: it facilitates post-stocking evaluation of the success of the stocking and it allows the identification of siblings and thus the avoidance of inbreeding in the hatchery.
- Transfer of strains for rehabilitation from outside a particular major population grouping should not be allowed, nor should farmed strains be used for stocking or ranching.

- An alternative approach to decision making in relation to stocking is advocated in the recently published *NASCO Guidelines for Stocking Atlantic Salmon* (CNL(04)54, Annex 4), where the degree of intervention recommended is dependent on how pristine or otherwise is the salmon habitat of a particular river. NASCO's *Guidelines on the Use of Stock Rebuilding Programmes* (CNL(04)55) provide guidance on compliance assessment, evaluation of the problem, development of a management plan and monitoring and evaluation of progress. In addition, to assist its parties in applying the precautionary approach, NASCO has developed a *Decision Structure for Management of North Atlantic Salmon Fisheries* (CNL31.332) and a *Plan of Action for the Protection and Restoration of Atlantic Salmon Habitat* (CNL(01)51). It is recommended that these documents be consulted to determine whether stocking is an appropriate management response to a perceived problem. Because there are both similarities and differences between the NASCO approach and that advocated in this chapter, we would recommend that both be considered in formulating management responses.

References

Aprahamian, M.W., Martin Smith, K., McGinnity, P., McKelvey, S. and Taylor, J. (2002) Restocking of salmonids: opportunities and limitations. *Fisheries Research*, **62**: 211–227.

Bakke, T.A., Jansen, P.A. and Hansen, L.P. (1990) Differences in the host resistance of Atlantic salmon, *Salmo salar* L., stocks to the monogenean, *Gyrodactylus salaris*, Malmberg, 1957. *Journal of Fish Biology*, **30**: 713–721.

Busack, C.A. and Currens, K.P. (1995) Genetic risks and hazards in hatchery operations: fundamental concepts and issues. *American Fisheries Society Symposium*, **15**: 71–80.

Carr, J. (2001) A review of downstream movements of juvenile Atlantic salmon (*Salmo salar*) in the dam-impacted Saint John River drainage. *Canadian Manuscript Report in Fisheries and Aquatic Science*, **2573**.

Carvalho, G.R. and Pitcher, T.J. (Ed.) (1995) *Molecular Genetics in Fisheries*. Chapman & Hall, London.

Chilcote, M.W. (2003) Relationship between natural productivity and the frequency of wild fish in mixed spawning populations of wild and hatchery steelhead (*Oncorhynchus mykiss*). *Canadian Journal of Fisheries and Aquatic Sciences*, **60**: 1057–1067.

Cross, T.F. (1999) Genetic considerations in stock enhancement and sea ranching of invertebrates, marine fish and salmonids. In: B.R. Howell, E. Moksness and T. Svasand (Ed.) *Stock Enhancement and Sea Ranching*, pp. 37–48. Fishing News Books, Oxford.

Cross, T.F. and Rogan, E. (1991) *The feasibility of developing and utilising gene banks for sea trout (Salmo trutta) conservation*. National Rivers Authority, Fisheries Technical Report No. 4.

Cross, T., McGinnity, P. and Galvin, P. (1998) Genetic considerations in stocking Atlantic salmon, *Salmo salar*. In: I.G. Cowx (Ed.) *Stocking and Introductions of Fish*, pp. 355–370. Fishing News Books, Oxford.

Davidson, K. and Bielak, A.T. (1993) New enhancement strategies in action. In: D.H. Mills (Ed.) *Salmon in the Sea and New Enhancement Techniques*, pp. 299–320. Fishing News Books, Oxford.

EU (1992) *EU Habitats Directive: European Council Directive 92/43/EEC of 21 May 1992, on the conservation of natural habitats and of wild fauna and flora*. OJ L 206/7: 22.7.92.

Flagg, T.A. and Nash, C.E. (1999) *A conceptual framework for conservation hatchery strategies for Pacific salmonids*. US Department of Commerce, NOAA Technical Memo. NMFS-NWFSC-38.

Fontaine, P.M., Dodson, J.J., Bernatchez, L. and Slettan, A. (1997) A genetic test of metapopulation structure in Atlantic salmon (*Salmo salar*) using microsatellites. *Canadian Journal of Fisheries and Aquatic Sciences*, **54**: 2434–2442.

Galvin, P., Taggart, J., Ferguson, A., O'Farrell, M. and Cross, T.F. (1996) Population genetics of Atlantic salmon (*Salmo salar*) in the river Shannon system in Ireland: an appraisal using minisatellite (VNTR) probes. *Canadian Journal of Fisheries and Aquatic Sciences*, **53**: 1933–1942.

Gausen, D. and Moen, V. (1991) Large-scale escapes of farmed Atlantic salmon (*Salmo salar*) into Norwegian rivers threaten natural populations. *Canadian Journal of Fisheries and Aquatic Sciences*, **48**: 426–428.

Gjedrem, T. (1999) Genetic improvement of cold water fish species. *Aquaculture Research*, **31**: 25–33.

Hansen, M.M. (2002) Estimating the long-term effects of stocking domesticated trout into wild brown trout (*Salmo trutta*) populations: an approach using microsatellite DNA analysis of historical and contemporary samples. *Molecular Ecology*, **11**: 1003–1015.

Hansen, M.M., Ruzzante, D.E., Nielsen, E.E., Bekkevold, D. and Mensberg, K.D. (2002) Long-term effective population sizes, temporal stability of genetic composition and potential for local adaptation in anadromous brown trout (*Salmo trutta*) populations. *Molecular Ecology*, **11**: 2523–2535.

Hard, J.J. and Heard, W.R. (1997) Analysis of straying variation in Alaskan hatchery chinook salmon (*Oncorhynchus tshawytscha*) following transplantation. *Canadian Journal of Fisheries and Aquatic Sciences*, **56**: 578–589.

Hard, J.J., Winans, G.A. and Richardson, J.C. (1999) Phenotypic and genetic architecture of juvenile morphometry in chinook salmon. *The American Genetic Association*, **90**: 597–606.

Hesthagen, T. and Hansen, L.P. (1991) Estimates of annual loss of Atlantic salmon, *Salmo salar* L., in Norway due to acidification. *Aquaculture and Fisheries Management*, **22**: 85–91.

ICES (1998) *Report of the Baltic and Salmon Trout Assessment Working Group*. International Council for the Exploration of the Sea. Helsinki, Finland, 16–24 April 1998. *ICES CM* **1998/ACFM:17**.

ICES (2003) *Report of the Working Group on North Atlantic Salmon*. International Council for the Exploration of the Sea, ICES Headquarters, Copenhagen. *ICES CM* **2003/ACFM:19**.

ICES (2005) *Report of the Working Group on Assessment of Baltic Salmon and Trout*. International Council for the Exploration of the Sea, Advisory Group on Fishery Management. Helsinki, Finland. 5–14 April 2005. *ICES CM* **2005/ACFM:18**.

Jennions, M.D. and Petrie, M. (1997) Variation in mate choice and mating preferences: a review of causes and consequences. *Biological Reviews*, **72**: 283–327.

Jonasson, J., Gjerde, B. and Gjedrem, T. (1997) Genetic parameters for return rate and body weight in sea ranched Atlantic salmon. *Aquaculture*, **154**: 219–231.

Jones, R.A., Anderson, L. and Goff, T. (2004) Assessments of Atlantic salmon stocks in southwest New Brunswick, an update to 2003. *CSAS Res. Doc.* 2004/019.

Jonsson, B., Waples, R.S. and Friedland, K.D. (1999) Extinction considerations for diadromous fishes. *ICES Journal of Marine Science*, **56**: 405–409.

Kangur, M. and Wahlberg, B. (Ed.) (2001) *Present and Potential Production of Salmon in Estonian Rivers*. Estonian Academy Publishers, Tallinn.

Koljonen, M-L. (2001) Conservation goals and fisheries management units for Atlantic salmon in the Baltic Sea area. *Journal of Fish Biology*, **59** (Supplement A): 269–288.

Koljonen, M-L. (2004) Changes in stock composition of annual Atlantic salmon catches in the Baltic Sea on basis of DNA-microsatellite data and Bayesian estimation. *ICES CM* **2004/Stock Identification Methods/EE:08**.

Koljonen, M-L., Jansson, H., Paaver, T., Vasin, O. and Koskiniemi, J. (1999) Phylogeographic lineages and differentiation pattern of Atlantic salmon in the Baltic Sea with management implications. *Canadian Journal of Fisheries and Aquatic Sciences*, **56**: 1766–1780.

Koljonen, M-L., Tähtinen, J., Säisä, M. and Koskiniemi, J. (2002) Maintenance of genetic diversity of Atlantic salmon (*Salmo salar*) by captive breeding programmes and the geographic distribution of microsatellite variation. *Aquaculture*, **212**: 69–93.

Lindroth, A. (1965) The Baltic salmon stock. *Mitteilungen. Internationale Vereinigung für Theoretische und Angewandte Limnologie*, **13**: 163–192.

Loigu, N., Leisk, Ü., Hannus, M. and Blinova, I. (2001) Water quality. In: B. Wahlberg and M. Kangur (Ed.) *Present and Potential Production of Salmon in Estonian Rivers*, pp. 17–31. Estonian Academy Publishers, Tallinn.

Marshall, T.L., Jones, R.A. and Anderson, L. (2000) *Assessment of Atlantic salmon stocks in southwest New Brunswick, 1999*. DFO CSAS Research Document 2000/010.

Mawle, G.W. and Milner, N.J. (2003) The return of salmon to cleaner rivers: England and Wales. In: D. H. Mills (Ed.) *Salmon at the Edge*, pp. 186–199. Blackwell Science, Oxford.

McGinnity, P., Stone, C., Taggart, J.B., Cooke, D., Cotter, D., Hynes, R., McCamley, C., Cross, T. and Ferguson, A. (1997) Genetic impact of escaped farm Atlantic salmon (*Salmo salar* L.) on native population: use of DNA profiling to assess freshwater performance of wild, farm and hybrid progeny in a natural environment. *ICES Journal of Marine Sciences*, 54: 998–1008.

McGinnity, P., Prodöhl, P., Ferguson, A., Hynes, R., Ó Maoiléidigh, N., Baker, N., Cotter, D., O'Hea, B., Cooke, D., Rogan, G., Taggart, J. and Cross, T. (2003) Fitness reduction and potential extinction of wild populations of Atlantic salmon, *Salmo salar*, as a result of interactions with escaped farm salmon. *Proceedings of the Royal Society, Series B*, 270: 2443–2450.

McGinnity, P., Prodöhl, P., Ó Maoiléidigh, N., Hynes, R., Cotter, D., Baker, N., O'Hea, B. and Ferguson, A. (2004) Differential lifetime success and performance of native and non-native Atlantic salmon examined under communal natural conditions. *Journal of Fish Biology*, 65 (Supplement A): 173–187.

Meffe, G.K. (1992) Techno-arrogance and halfway strategies: salmon hatcheries on the Pacific coast of North America. *Conservation Biology*, 6: 350–354.

Miller, L.M. and Kapuscinski, A.R. (2003) Genetic guidelines for hatchery supplementation programs. In: E.M. Hallerman (Ed.) *Population Genetics: Principles and applications for fisheries scientists*, pp. 329–356. American Fisheries Society, Bethesda, MD.

Milner, N.J., Russell, I.C., Aprahamian, M., Inverarity, R., Shelley, J. and Rippon, P. (2004) *The role of stocking in the recovery of the River Tyne salmon fisheries*. Environment Agency, Fisheries Technical Report No. 2004/1.

Nielsen, E.E., Hansen, M.M. and Loeschke, V. (1999) Genetic variation in time and space: microsatellite analyses of extinct and extant populations of Atlantic salmon. *Evolution*, 53: 261–268.

Nielsen, E.E., Hansen, M.M. and Bach, L.A. (2001) Looking for a needle in a haystack: discovery of indigenous Atlantic salmon (*Salmo salar* L.) in stocked populations. *Conservation Genetics*, 2: 219–232.

Norrgren, L., Andersson, T., Bergqvist, P-A. and Björklund, I. (1993) Chemical, physiological and morphological studies of feral Baltic salmon (*Salmo salar*) suffering from abnormal fry mortality. *Environmental Toxicology and Chemistry*, 12: 2065–2075.

NRC (2004) *Atlantic salmon in Maine*. National Research Council Committee on Atlantic salmon in Maine. The National Academies Press, Washington, DC.

Piggins, D.J. and Mills, C.P.R. (1985) Comparative aspects of the biology of naturally produced and hatchery reared Atlantic salmon smolts (*Salmo salar* L.). *Aquaculture*, 45: 321–334.

Poole, W.R., Dillane, M.G. and Whelan, K.F. (1994) Artificial reconditioning of wild sea trout, *Salmo trutta*, as an enhancement option: initial results on growth and spawning success. *Fisheries Management and Ecology*, 1: 179–192.

Reisenbichler, R.R., Utter, F.M. and Krueger, C.C. (2003) Genetic concepts and uncertainties in restoring fish populations and species. In: R.C. Wissmar and P.A. Bisson (Ed.) *Strategies for Restoring River Ecosystems: Sources of variability and uncertainty in natural and managed systems*, pp. 149–183. American Fisheries Society, Bethesda, MD.

Romakkaniemi, A., Perä, I., Karlsson, L., Jutila, E., Carlson, U. and Pakarinen, T. (2003) Development of wild Atlantic salmon stocks in the rivers of the northern Baltic Sea in response to management measures. *ICES Journal of Marine Science*, 60: 1–14.

Rosell, R. and MacOscar, K.C. (1997) The Lagan salmon restoration experiment: towards a self sustaining population. *Proceeding of the Institute of Fisheries Management Annual Study Course*, 28: 70–86.

Ryman, N. and Laikre, L. (1991) Effect of supportive breeding on the genetically effective population size. *Conservation Biology*, **3**: 325–329.

Ryman, N., and Utter, F.M. (Ed.) (1987) *Population Genetics and Fishery Management* (2nd edn). University of Washington Press, Seattle, WA.

Säisä, M., Koljonen, M-L. and Tähtinen, J. (2003) Genetic changes in Atlantic salmon stocks since historical times and the effective population sizes of the long-term captive breeding programmes. *Conservation Genetics*, **4**: 613–627.

Salmon Research Agency of Ireland (1955–2003) Annual reports of Salmon Research Agency of Ireland. **Nos. 1–49**.

Scott, R.J., Poos, M.S., Noakes, G. and Beamish, F.W.H. (2005) Effects of exotic salmonids on juvenile Atlantic salmon behaviour. *Ecology of Freshwater Fish*, **14**: 282–288.

Sjöblom, V., Tuunainen, P., Toivonen, J., Westman, K., Sumari, O., Simola, O. and Salojärvi, K. (1974) Itämeren ja Belttien kalastusta ja elollisten luonnonvarojen säilyttämistä koskevan yleissopimuksen perusteella Suomen osalle tuleva lohen istutusvelvollisuus. (The salmon stocking estimated for Finland in accordance with the convention on fishing and conservation of the living resources in the Baltic Sea and the Belts). Finnish Game and Fisheries Research Institute. *Tiedonantoja*, **2**: 22–52.

Stewart, T.J. and Shaner, T. (2002) Lake Ontario salmonid introductions 1970 to 1999: stocking, fishery and fish community influences. In: *Lake Ontario Fish Communities and Fisheries: 2001 Annual Report of the Lake Ontario Management Unit*, pp. 12.1–12.10. Queen's Printer for Ontario, Picton, Canada.

Thorpe, J.E. (1988) Salmon enhancement: stock discreteness and choice of material for stocking. In: D. Mills and D. Piggins (Ed.) *Atlantic Salmon: Planning for the future*, pp. 373–388. Croom Helm, London.

Utter, F.M. (2003) Genetic impacts of fish introductions. In: E.M. Hallerman (Ed.) *Population Genetics: Principles and applications for fisheries scientists*, pp. 357–378. American Fisheries Society, Bethesda, MD.

Vasemägi, A., Gross, R., Paaver, T., Kangur, M., Nilsson, J. and Eriksson L-O. (2001) Identification of the origin of Atlantic salmon (*Salmo salar* L.) population in a recently recolonised river in the Baltic Sea. *Molecular Ecology*, **10**: 2877–2882.

Waples, R.S. (1991) Genetic interactions between hatchery and wild salmonids: lessons from the Pacific Northwest. *Canadian Journal of Fisheries and Aquatic Sciences*, **48**: 124–133.

Waples, R.S. and Do, C. (1994) Genetic risks associated with supplementation of Pacific salmonids: captive broodstock programs. *Canadian Journal of Fisheries and Aquatic Sciences*, **51** (Supplement 1): 310–329.

Wilcox, B.A. (1986) Extinction models and conservation. *Trends in Ecology and Evolution*, **1**: 46–48.

Wilkins, N.P., O'Farrell, M. and O'Connor, W. (1999) Parteen salmon breeding programme. In: K.F. Whelan and F.S. O'Muircheartaigh (Ed.) *Managing Ireland's Spring Salmon Stocks: The options*, pp. 41–42. Salmon Research Agency and Central Fisheries Board, Newport, Ireland.

Wilkins, N.P., Cotter, D. and O'Maoileidigh, N. (2001) Ocean migration and recaptures of tagged, triploid, mixed-sex and all-female Atlantic salmon (*Salmo salar* L.) released from rivers in Ireland. *Genetica*, **111**: 197–212.

Young, K.A. (1999) Managing the decline of Pacific salmon: metapopulation theory and artificial recolonisation as ecological mitigation. *Canadian Journal of Fisheries and Aquatic Sciences*, **56**: 1700–1706.

Youngson, A.F., Jordan, W.C., Verspoor, E., McGinnity, P., Cross, T.F. and Ferguson, A. (2002) Management of salmonid fisheries in the British Isles: towards a practical approach based on population genetics. *Fisheries Research*, **62**: 193–209.

12 Farm Escapes

A. Ferguson, I. A. Fleming, K. Hindar, Ø. Skaala, P. McGinnity, T. Cross and P. Prodöhl

Upper: River Polla, north-west Scotland. (Photo credit: E. Verspoor.) Lower: wild (upper) and farm (lower) salmon netted in the River Polla in 1989, the main distinguishing feature being the misshapen and eroded fins on the farm fish. (Photo credit: D. Hay.)

The farming of Atlantic salmon, *Salmo salar*, is arguably the most public and controversial issue facing managers and policy makers concerned with wild Atlantic salmon. From a genetic perspective, the concerns raised are to some extent the same in their basic nature as those associated with stocking, something the reader will probably note. However, in so far as stocking is a management tool and deliberate, while farm escapes are an accidental side effect of the farming industry, the two are somewhat different in respect of socio-political context and associated issues. Because of this they require different management and policy responses. Given the management focus of this book, it is more appropriate that the two issues are treated separately.

12.1 Introduction

Farming of Atlantic salmon commenced in Norway in 1969, when Mowi a/s in Bergen and the Grøntvedt Brothers in Hitra put salmon smolts into holding facilities in the sea. The successful harvest two years later triggered considerable interest in salmon farming. Over the past 30 years the industry has expanded exponentially and in 2003 the global production of farmed Atlantic salmon exceeded 1.1 million tonnes, with 761 752 t in the North Atlantic (ASF 2004). The major producers (and approximate production) were Chile (450 000 t), Norway (430 000 t), Scotland (140 000 t), Faroes (45 000 t), Iceland (35 000 t), Ireland (25 000 t) and North America (east coast) (55 000 t). This farm production is about 400 times the wild Atlantic salmon catch in Europe and North America.

Atlantic salmon farming mimics the natural life cycle by rearing juveniles in fresh water to the smolt stage and then transferring these smolts to net cages in the sea for ongrowing to marketable size. Farm fish can escape from containment during both these freshwater and marine life stages (Stokesbury and Lacroix 1997). Juvenile rearing farms are often situated adjacent to rivers from which they take and discharge water (Fig. 12.1). Without adequate screening on the outlet, escapes of juveniles can occur into the river. In some cases juvenile rearing involves the use of net cages in freshwater lakes (Fig. 12.2). However, in Norway the rule is that discharge from smolt-rearing units is directly to the sea, generally into the estuaries of small rivers without salmon populations.

The physical nature of marine net cages is such that escapes from confinement inevitably occur (Fig. 12.3). Although improvements in cage design and husbandry have resulted in proportionally less escapement, the increasing scale of the industry means that substantial numbers of fish still escape.

Concern has arisen as to the potential detrimental genetic changes that may occur in wild populations as a result of escaped farm salmon entering rivers and interacting with wild populations, particularly given the endangered status of many populations (WWF 2001). This concern has been expressed since the 1980s (Hansen *et al.* 1991). A review of the literature on the genetic effects following releases of non-native salmonid populations suggested that fitness losses in wild populations must be expected due to interbreeding with escaped farm salmon (Hindar *et al.* 1991). Two broad conclusions were drawn from that review:

- The genetic effects of (intentionally or accidentally) released salmonids on natural populations are typically unpredictable; they vary from no detectable effect to complete introgression or displacement.

Fig. 12.1 Smolt-rearing unit discharging into adjacent river where wild Atlantic salmon are present. (a) Discharge point. (b) Overview of unit. (Photos credit: A. Ferguson.)

- Where genetic effects on performance traits have been detected, they appear always to be negative in comparison with the unaffected native populations. For example, reduced total population sizes have been observed following introductions of exogenous populations, and also reduced performance in a number of traits which can explain such population declines (e.g. lower survival in fresh and sea water).

This chapter reviews the information available on genetic changes in wild populations of Atlantic salmon as a result of farm escapes and the consequent changes in fitness (i.e. survival + reproduction), performance characteristics, and genetic diversity of wild populations.

Fig. 12.2 Smolt rearing cages in a freshwater lake that is part of a major Atlantic salmon river catchment. (Photo credit: A. Ferguson.)

Fig. 12.3 Farm salmon cages in sea lough. (Photo credit: A. Ferguson.)

12.2 Magnitude of farm salmon escapes

12.2.1 Identifying escaped farm salmon

Clearly the extent of genetic change in wild populations is related to the scale of farm escapes, as well as the ability of these escapes to survive and breed. In order to determine the extent of escapes, it is necessary to be able to distinguish farm salmon from their wild counterparts. This has been successfully accomplished based on differences in external morphology (Lund *et al.* 1989, Fleming *et al.* 1994), growth patterns in scales and otoliths (Lund and Hansen 1991; Hindar and L'Abée-Lund 1992), and various chemicals (e.g. carotenoid pigments, Lura and Sægrov 1991a,b). However, the longer fish have been in the wild since escape, the more difficult it is to use such characters to distinguish them from wild fish. Consequently, these methods can only be used to make minimum estimates of the occurrence of escaped farm salmon in the sea and rivers. Pigments, of maternal origin, in eggs and early juvenile stages can also be used to identify offspring of recently escaped farm fish (Craik and Harvey 1986; Lura and Sægrov 1991a,b; Webb *et al.* 1993a). In some specific circumstances it has been possible to use allozyme, mitochondrial DNA (mtDNA) and variable number of tandem repeat (VNTR, i.e. microsatellites and minisatellites) genetic markers (see Chapter 4) to distinguish farm, wild and hybrid salmon, especially when the farm fish were not of local origin (e.g. Crozier 1993; Clifford *et al.* 1998a,b; Skaala *et al.* 2004).

12.2.2 Escapes from sea cages

Escapes from marine net cages occur during routine handling operations such as net changing and due to damage to the nets as a result of collision by boats, and especially due to storm damage causing construction failure. The exact number of farm salmon escaping is poorly known. Adult fish escaping during large-scale failure of net pens are usually registered, and their numbers may be considerable. In Norway, for example, more than 1 million Atlantic salmon (sub-adults and adults) escaped from net pens during storms in the winter of 1988–89, and again in the winter of 1991–92. Official estimates of total numbers of fish escaping from Norwegian net pens during later years ranged from 250 000 to 650 000 (NOU 1999), although these are probably underestimates. High numbers of escaped farm salmon in recent years have also been recorded elsewhere, including Scotland, Ireland and eastern Canada. In 2001 about 1 million salmon escaped in Scotland of which 400 000 were in a single incident. In one of the largest incidents some 600 000 salmon escaped during a storm accident in spring 2002 in the Faroes (ASF 2004). A similar number also escaped in the west of Scotland during storms in January 2005 (BBC 2005).

We know less about the number of fish escaping during the daily handling of fish, but it has been estimated that their total number from all aquaculture facilities may be as large as the large-scale accidents in a few net pens. In addition, little is known of the extent of smolt escapes in the period immediately after transfer to sea cages, where in some cases the net size allows smaller smolts to escape. Thus currently it is likely that at least 2 million salmon escape each year in the North Atlantic (McGinnity *et al.* 2003). That is around 50% of the total prefishery abundance of wild salmon in the area, which is estimated at 4 million fish (ASF 2004).

Escaped farm Atlantic salmon made up 20–40% of the salmon in the fisheries off the Faroes in the 1990s (Hansen *et al.* 1999), and a similar percentage of spawners overall in Norwegian rivers, although in some of these latter rivers up to 80% of the spawners were of farm origin (Fiske *et al.* 2001). In the Hardanger fjord, on the western coast of Norway, 86% of the salmon catch in 2003 consisted of escaped farm fish (WWF-Norway 2005). Other salmon-producing countries have had high percentages of farm fish in the spawning populations in some rivers (Gudjonsson 1991; Carr *et al.* 1997; Youngson *et al.* 1997), but not at the same high level as observed in many rivers in Norway. On the east coast of North America, escaped salmon outnumbered wild fish by as much as 10 to 1 in some rivers. For example, after massive escapes in south-west New Brunswick, Canada, in 1994, 1200 farm salmon were counted entering the Magaguadavic River compared to 137 wild fish (ASF 2004). Escaped farm salmon of at least partial European origin (as determined by DNA analyses, see Chapter 5) have also been found in the Magaguadavic River even though only farm salmon of local origin are licensed for use in Canada (ASF 2005).

It has been estimated that over 396 000 Atlantic salmon escaped into the Pacific Ocean from farms in British Columbia, Canada, from 1991 to 2001 (Gaudet 2002). Over 595 000 fish were accidentally released from fish farms in Washington State, USA, from 1996 to 1998 (Noakes *et al.* 2000). Escaped farm Atlantic salmon have been found as far north as the Bering Sea (Brodeur and Busby 1998). Atlantic salmon are not native to the Pacific Ocean and these escapements will thus not impact on any wild Atlantic salmon stocks. However, their impact on local native Pacific salmon (*Oncorhynchus* spp.) populations, as an invasive species, can be equally detrimental and requires careful consideration and monitoring (Naylor *et al.* 2001, 2005; Nielsen *et al.* 2003).

12.2.3 Juvenile escapes

Little is known of the extent of parr escapes into rivers from juvenile rearing units and the threat from such escapes is generally insufficiently recognised. Clifford *et al.* (1998a) demonstrated that substantial escapes had occurred in a river in north-west Ireland and that some of these fish had been able to complete the life cycle and home to a section of the river adjacent to the farm outlet. On the basis of multivariate scale analysis, Stokesbury *et al.* (2001) found that in 1996, 1997 and 1998, 36%, 59% and 43%, respectively, of parr in the Magaguadavic River were direct escapees from commercial hatcheries. Some of these escaped juveniles were later found to return to the river as adults, although survival at sea was considerably less than that of the wild salmon (Lacroix and Stokesbury 2004). In those countries where juvenile rearing units discharge to wild salmon rivers and especially where net-cage rearing takes place on freshwater lakes in such river catchments, the extent of juvenile escapes may be substantial, although to date this has not been adequately studied.

12.3 Genetic differences between wild and farm salmon

12.3.1 Founder effects

In addition to the magnitude of escapes and spawning, the impact of escaped farm fish will also depend on the degree of genetic differentiation from wild stocks. Genetic differentiation can occur due to both founding effects and subsequent changes in culture of the farm strain as a result of domestication. Given that there is extensive genetic differentiation among wild populations (see Chapter 5), a farm strain derived from a non-local river will be genetically different irrespective of any further changes. Even when derived from the same river as a wild stock, the farm strain may differ due to a founder effect such as small number of broodstock being used or broodstock being taken from a temporal or maturity component of the wild stock. After the farm strain is established genetic changes can occur in the culture environment through intentional (Gjedrem *et al.* 1991) and unintentional domestication selection (Fleming and Einum 1997; Johnsson *et al.* 2001; Fleming *et al.* 2002; Huntingford 2004). As well as these directional changes, random changes can occur through genetic drift leading to a loss of genetic variation (Mjølnerød *et al.* 1997; Clifford *et al.* 1998a,b; Norris *et al.* 1999; Skaala *et al.* 2004, 2005), especially as some farm stocks have involved low numbers of broodstock in the founding or subsequent generations.

12.3.2 Differences due to domestication

Artificial selection experiments were initiated in Norway by Harald Skjervold at the Agricultural University of Norway, and by Gunnar Nævdal of the Institute of Marine Research, Bergen (Nævdal *et al.* 1975; Gjedrem *et al.* 1991). Two main breeding programmes were established for Atlantic salmon in Norway (Box 12.1). The first was the Mowi strain established in the late 1960s. For the second, broodstock were sampled from one Swedish and 40 Norwegian rivers, between 1971 and 1974, and used to set up four separate strains, initially at Sunndalsøra and later duplicated at, and replaced by, Kyrksæterøra (Gjøen and Bentsen 1997).

Much of salmon farming elsewhere in Europe makes use of these Norwegian strains, although further differentiation has occurred as a result of directional and inadvertent changes subsequent to import. Strains of local origin have also been established in Iceland and Scotland. In Eastern Canada the principal aquaculture strain is based on salmon from the

---

Box 12.1 Origin of farm salmon strains.

Sunndalsøra and Kyrksæterøra, Norway

Although rivers from most of the Norwegian coast were included in establishing these farm strains, there was a dominance in the material by stocks from the Møre and Romsdal and the Trøndelag region, the middle part of the Norwegian coast. The intention was to obtain 12 females and 4 males from each river stock, but for practical reasons, the numbers of sires and dams from many rivers were lower than the requested numbers. This effective population size (N_e) of 12 or less would have resulted in a substantial founding bottleneck. Selection was carried out at two stages. Before individual marking and communal rearing, families with low survival and low body weight were culled. Selection was then carried out after two years in the sea based on body weight and low grilse component from the 1980 year class onwards. Each strain was stripped every four years, and in 1985 at generation 4, the remaining contribution from the various river stocks was evaluated. Somewhat surprisingly, it was shown that a low number of river stocks dominated the gene pool. In strain 1, River Namsen and River Surna constituted 70% and 12% of the material respectively, while in strain 2 an unknown mixture of farmed fish constituted 70% and River Driva about 22% of the material. In strains 3 and 4, mixtures of farm strains constituted close to 100%. Thus, the contribution of the majority of the initial 40 wild stocks was strongly reduced already after four generations. This has been explained by genetic differences among river stocks in traits such as growth rate in fresh water and body weight at slaughter, and by high representation of some stocks among the founders (Gjøen and Bentsen 1997; Gjedrem *et al.* 1991).

Mowi, Norway

Beside the four major strains referred to above, which provide 70–90% of the total number of eggs used in Norwegian salmon farming, the Mowi strain is also of considerable importance. The Mowi strain was established from Norwegian west coast rivers in the 1960s with a major contribution of material from River Bolstad in the Vosso watercourse (County Hordaland), River Årøy (County Sogn and Fjordane) and possibly a contribution from salmon in the Maurangerfjord area (Rita Brokstad, Marine Harvest, pers.comm.). The Vosso and Årøy salmon are famous for their large size and late maturity.

Other farm strains

Local strains have been established in Scotland, Iceland, Canada and the USA. In some cases these local strains have been crossed with Norwegian farm strains.

Table 12.1 Comparison of genetic variation in domesticated and wild salmon using molecular markers.

Marker	Observation	Reference
Allozyme loci	Reduction in heterozygosity in 1st generation farm strains in Canada	Verspoor (1988)
Allozyme loci	Reduction in genetic variability in some, but not all, studied farm strains in Ireland	Cross & NiChallanain (1991)
Allozyme loci	Temporal genetic changes in farm strains, differences between farm strains and wild source, directional change in MEP-2* in farm strains in Scotland	Youngson et al. (1991)
Allozyme loci	A 14% reduction in number of alleles in a farm strain in Norway	Mjølnerød et al. (1997)
Allozyme loci	Reduction in mean number of alleles, percentage polymorphic loci and mean heterozygosity in farm strains in Norway	Skaala et al. (2005)
Minisatellites	An Irish farm strain of Norwegian Mowi origin had only 56% of alleles and 53% of mean heterozygosity compared to wild local populations in small rivers	Clifford (1996) Clifford et al. (1998a,b)
Microsatellites	Between 52% and 80% of the alleles present at 15 loci in a domesticated Irish strain of Mowi origin compared to wild salmon	Norris et al. (1999)
Microsatellites	Strong reductions in number of alleles at 12 loci in domesticated strains. On average 58% of alleles in wild salmon retained in the farm strains in Norway	Skaala et al. (2004)

St John River, which were introduced into a breeding programme at St Andrews. In the Eastern USA, principally Maine, the farm fish were initially derived from crosses between European fish (Scottish Landcatch) and St John River fish. These fish have subsequently been crossed with Penobscot River (Maine, USA) fish. Baum (1998) estimated that there is a European genetic influence in ~ 30–50% of the production of farm Atlantic salmon in Maine. It is now mandated legally, however, that farm salmon stocks used in Maine should not have significant European ancestry.

12.3.3 Genetic marker differences between wild and farm salmon

A number of studies have been conducted since the late 1980s to compare domesticated and wild salmon by assessing genetic variability at protein coding loci, and more recently mtDNA and nuclear VNTR loci (Table 12.1). Protein studies (Verspoor 1988; Cross and NiChallanain 1991; Youngson et al. 1991; Mjølnerød et al. 1997) have shown genetic differentiation of farm strains from their wild origin populations and, more significantly, reductions in genetic variability in farm strains both in terms of number of alleles and mean heterozygosity. Skaala et al. (2004, 2005) compared the broodstocks of the five major Norwegian farm strains with four major wild populations in Norway at eight polymorphic enzyme coding loci. The genetic distance between one farm strain and its source populations in the River Namsen and River Surna was about 10 times higher than that observed between three wild populations from the River Namsen, River Vosso and River Surna. Mean F_{ST}, a statistical measure of genetic differentiation (see Chapter 4) over eight polymorphic allozyme loci was 0.161 among the domesticated strains, with high values at MEP-2* (0.340), TPI-3*

(0.115) and *MDH-2** (0.111), compared to a much lower F_{ST} estimate of 0.021 among the four wild stocks representing salmon from the west coast up to the north of Norway. The mean number of alleles was about 12% lower in farm strains than in wild stocks, percentage polymorphic loci was 14% lower in farm strains, and mean heterozygosity was about 17% lower in farm strains than in wild stocks. This further demonstrates that a rapid and significant differentiation has taken place in Norwegian farm salmon, and that the existing farm strains have differentiated significantly from their origin. Thus domestication and founder effects, in many cases, have a much greater impact on the differentiation of farm salmon from wild populations compared to non-native origin (McGinnity *et al.* 2004).

Several studies have employed VNTR loci, which are, in general, more highly variable than allozymes, and have demonstrated that farm salmon have even greater reductions in genetic variability than shown by protein studies. Clifford (1996) and Clifford *et al.* (1998a,b) found that an Irish farm strain of Mowi origin had 56% of the number of alleles and 53% of the mean heterozygosity over three minisatellite loci compared with local wild populations in small rivers. Norris *et al.* (1999), examining later cohorts of the same strain with 15 microsatellites, found between 52% and 80% of the alleles present in wild salmon. Skaala *et al.* (2004), using 12 microsatellite loci to compare domesticated and wild strains in Norway, found strong reductions in the number of alleles at all loci, with on average 58% of the allelic variability observed in wild populations retained in the farm strains. A direct comparison of allelic variability between a wild source population and its domesticated derivative showed that 50% of the alleles in the wild stock were retained in the domesticated strain. Also, the genetic differentiation observed between a specific strain and its wild founder stock was two to six times higher than the genetic differentiation observed among wild salmon stocks.

12.3.4 Phenotypic differences between wild and farm salmon

Most phenotypic traits are controlled by multiple gene loci and are a product of both genes and environment (Box 12.2). Given the several orders of magnitude higher fecundity of salmon compared to traditional farm livestock, much higher levels of selection can be applied. Domestication has thus been prevalent in cultured salmonids as a result of such breeding programmes, which have selected directly for a variety of traits sought by farmers, including growth rate, body size, survival, delayed maturity, stress tolerance, temperature tolerance, disease resistance, flesh quality and egg production (reviewed in Fleming 1995) (Table 12.2).

Due to the moderate heritability for growth, and the active selection for growth in breeding programmes (e.g. Gjedrem *et al.* 1991; Glebe 1998), it is not surprising that farm salmon outgrow wild salmon in artificial culture (Einum and Fleming 1997; Thodesen *et al.* 1999; Fleming *et al.* 2002) and this also carries over in the wild (Einum and Fleming 1997; McGinnity *et al.* 1997, 2003; Fleming *et al.* 2000). However, until recently surprisingly little was known about the mechanisms underlying growth and the impact of selection on these mechanisms. A recent study has shown that there is a direct link between domestication selection for growth and its endocrine regulation, where individuals with more active endocrine regulatory components (e.g. growth hormone production) will be targeted in breeding programmes (Fleming *et al.* 2002). Thus there can be unintentional changes associated with such directed selection.

Such changes caused by domestication are common in cultured salmon and frequently include alteration of fitness-related traits such as survival, deformity, feed conversion rate,

Box 12.2 Phenotypic/quantitative traits and selective breeding in farm salmon.

Most whole fish phenotypic traits such as growth, survival, disease resistance and age of maturity are quantitative traits. That is, they are controlled by a number of genes (often five to 20+) and are also partly influenced by the environment. The extent to which phenotypic variation within a population is determined by genetic variability is known as the broad-sense heritability (h_b^2) or simply heritability. More widely used is narrow-sense heritability (h_n^2), which is a measure of the proportion of phenotypic variance due to additive gene action and which ranges from 0 (trait influenced only by environment variability) to 1 (purely genetically determined). Most traits of interest show h^2 values from 0.2 to 0.8, although the actual value refers only to specific environmental conditions.

The phenotype of any trait with a heritability above 0 can be changed in a farm strain by artificial selection (e.g. Bentsen 1994). This can involve either mass selection where the best-performing individuals are used as broodstock, or family selection where the best families are used. The latter is particularly appropriate where heritabilities are low or where the phenotype of a trait cannot be determined without killing the individual, e.g. disease resistance.

Table 12.2 Comparison of phenotypic traits in domesticated and wild salmon.

Trait	Observation	Reference
Growth rate	Domesticated salmon parr outgrow wild salmon	Einum & Fleming (1997)
Growth rate	Domesticated salmon parr outgrow wild salmon in natural habitat	McGinnity *et al.* (1997)
Aggression	Domesticated salmon parr more aggressive than wild salmon	Einum & Fleming (1997)
Domination	Domesticated salmon parr dominate wild salmon	Einum & Fleming (1997)
Predator response	Time elapsed before reappearance after exposure to predator model shorter in domesticated salmon than in wild	Einum & Fleming (1997)
Predator response	Domesticated salmon parr had lower heart rate and less pronounced flight and heart responses to a model predator at attack	Johnsson *et al.* (2001)
Growth hormone	Individuals with high levels of growth hormone are targeted during domestication. Higher levels of growth hormone in domesticated than in wild salmon	Fleming *et al.* (2002)

spawning time, morphology, aggression, egg viability and production, risk-taking behaviour and growth hormone production (reviewed in Fleming 1995; Fleming *et al.* 2002). For instance, not only may selection for growth rate alter growth hormone regulatory components, but it may also affect fish behaviour. Growth hormone treatment has been shown to increase appetite (Johnsson and Björnsson 1994; Jönsson *et al.* 1996), aggression and activity (Jönsson *et al.* 1998), the tendency to forage under risk of predation (Johnsson *et al.* 1996; Jönsson *et al.* 1996) and dominance (Martin-Smith *et al.* 2004).

Not surprisingly then, farm salmon show behavioural and physiological differences (genetically based) related to differences in growth rate, including differences in activity, aggression, dominance, risk taking and cardiac responses to a perceived threat (Einum and Fleming 1997; Fleming and Einum 1997; Johnsson *et al.* 2001; Fleming *et al.* 2002) (Figs 12.4, 12.5 and 12.6). Farming thus generates rapid genetic change, resulting in distinct differences relative to wild fish, so much so, that Atlantic salmon can be considered one species with two biologies (Gross 1998). These changes to fitness-related traits, as a result of domestication, have important implications for the survival of farm salmon in nature, and for their interactions with wild salmon.

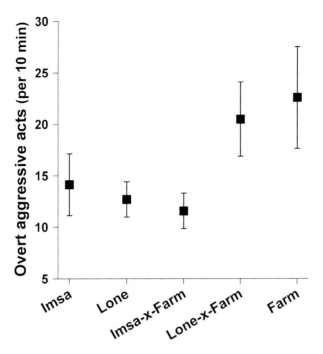

Fig. 12.4 Differences in overt aggression among native (Rivers Imsa and Lone, Norway), farm (AquaGen, Norway) and hybrid (Imsa × Farm, Lone × Farm) age 0+ juveniles. Data are means ± S.E. Redrawn from Einum and Fleming (1997).

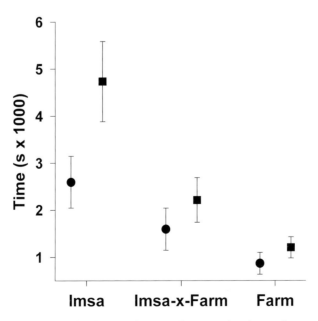

Fig. 12.5 Differences in response to a simulated predator attack measured as time until reappearance from cover (●) and time until remaining out for a minute or more (■). Data are means ± S.E. Redrawn from Einum and Fleming (1997).

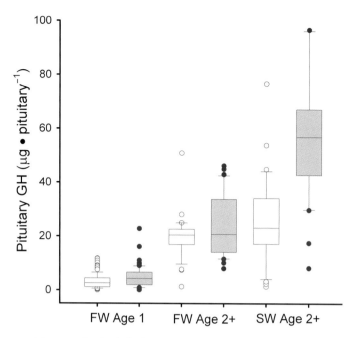

Fig. 12.6 Pituitary growth hormone levels of wild (open) and farm (shaded) Atlantic salmon at ages 1 and 2+ in fresh water (FW) or seawater (SW). The data are medians (lines within boxes), 25th to 75th percentiles (boxes), 10th and 90th percentiles (whiskers) and outlying points (open and solid circles). Source: Fleming *et al*. (2002).

12.4 Potential impact of farm escapes on wild populations

12.4.1 Fate of adult escapees

Farm salmon escapes in the marine environment occur as smolts, post-smolts and adults. When these escapees begin to mature they show a strong tendency to migrate into rivers in the vicinity of the site of escape (Hansen and Jonsson 1991; Youngson *et al*. 1997; Whoriskey and Carr 2001; Bridger *et al*. 2001). However, the survival and dispersal of farm salmon depend on the life stage and the time of year they escape. Autumn and winter escapes of farm salmon are associated with high mortality and wide dispersal, except for maturing fish entering fresh water directly following escape. Spring and summer escapes seem to survive better and disperse to rivers in the neighbourhood of the marine location (Hansen *et al*. 1987; Hansen and Jonsson 1989, 1991). Farm fish tagged in the feeding areas off the Faroes have poorer survival than wild fish tagged in the same area (Hansen and Jacobsen 2003). Farm escapees seem to approach the coast and enter rivers later in the season than wild fish, many of them after the angling season (Fiske *et al*. 2001). There is a significant correlation between the intensity of fish farming in an area (estimated as density of farms, or total numbers of smolts put into net pens) and the occurrence of escaped farm fish in the rivers (Lund *et al*. 1991). In western Scotland in the period 1990–2001, Butler and Watt (2003) found a mean contribution of farm escapes to the salmon rod catch of 9% in rivers with salmon farms compared to 2% in rivers without farms. Due to the wide dispersal of escaped farm fish, no river can be regarded as having no farm fish. For example, following an escape from a farm in north-east Ireland in August 2002 escaped fish were found in a number of rivers in England and Wales up to 250 km away, on the other side of the Irish Sea, in areas where salmon farms are absent (Milner and Evans 2003).

12.4.2 Juvenile escapes

Juvenile stages of farm fish escaping into freshwater locations have a migratory behaviour that is more similar to wild fish. Generally, the homing precision of adults released as freshwater juveniles or as smolts in rivers is much higher than that for fish escaping or being released at marine sites, without any connection with a river (Hansen and Jonsson 1994; Hansen and Quinn 1998). However, the straying rate of farm fish is still higher than that of wild fish migrating in the same river, and the straying rate of the major farm strains is higher than the straying rate of farm fish developed from the local population (Jonsson *et al.* 2002).

12.4.3 Indirect genetic effects of farm escapes

Escaped farm salmon can have an impact on the genetic composition and thus on fitness and characteristics of wild populations both indirectly and directly (Waples 1991; Youngson and Verspoor 1998). Indirect genetic effects occur due to behavioural, ecological and disease interactions with the wild population. These interactions may reduce the success of wild fish thereby reducing the effective population size of the wild population and increasing genetic drift. This results in the well-known phenomenon of inbreeding depression (Wang *et al.* 2002a,b) due to an increase in the number of homozygotes for deleterious recessive alleles as well as the loss of heterozygous advantage (detailed in Chapter 8). Interaction of farm salmon with wild fish may also result in changes in selection pressures in natural populations through differential impacts on particular size, life history, geographical, or temporal components of the wild stock.

It has been observed that farm females may destroy the redds of wild salmon in nature (Lura and Sægrov 1991b; Webb *et al.* 1991). As noted above, escaped farm salmon may enter rivers and attempt to spawn later in the season when wild fish have already spawned. Superimposition of redds is common among salmonids, especially when the density of spawners is high. Late-spawning individuals may dig up the eggs of early-spawning fish, and thereby lower the latter's reproductive success. Thus, even when farm salmon have low spawning success, they can reduce the success of local wild fish.

Cultured salmon released into the native range of the species should in general be expected to behave similarly to the wild species. As summarised in section 12.3, however, genetic changes in captivity may mean that cultured fish behave differently from their wild counterparts. Salmonid fishes compete for food and space in fresh water (Chapman 1966). Body size and territoriality are often good predictors of competitive ability in streams. Therefore, it seems that the most important changes in cultured salmonids are increased growth performance and increased aggressiveness. These changes may give cultured fish a competitive advantage at particular life stages and in favourable environments (e.g. high food availability and few predators), while at other times, such as in low-food and/or high-risk environments, wild fish are probably favoured.

The potential for competition is significant, as diet and habitat choice of farm and hybrid juveniles overlap highly with wild conspecifics (Einum and Fleming 1997; McGinnity *et al.*

1997; Fleming *et al.* 2000). Territorial and social dominance behaviour in salmonids, as a result of interactions between species or between cultured and wild fish, can affect both mortality and growth (e.g. Fausch and White 1986; Einum and Fleming 1997). The intensity and form of intraspecific competition may be altered when salmon from different populations (e.g. wild and farm) that have not co-evolved interact, resulting in deleterious consequences as suggested by Fausch (1988) for interspecific competition. Displacement of native fish by larger, more aggressive farm fish can result in shifts in habitat use and increased mortality (McGinnity *et al.* 1997, 2003; Fleming *et al.* 2000).

In addition to competition for space and territories, in some situations the rapid growth rates of farm and hybrid juveniles relative to wild juveniles may increase early maturation rates, and result in increased mating competition among early maturing male parr. Furthermore, the farm and hybrid juveniles are likely to have a size advantage and to show behavioural differences (e.g. aggression) that increase breeding success, and thus genetic introgression (Garant *et al.* 2003). However, McGinnity *et al.* (1997, 2003) found that farm fish of another strain, under natural conditions, had substantially lower parr maturity, with hybrids being intermediate, presumably as a result of selection against parr maturity in this strain (see also Fleming and Einum 1997).

Fish farming operates in open aquatic systems, and there is a possibility that disease organisms may be transferred from farm to wild fish, or from wild to farm and back to wild at higher densities. Moreover, transport of fish in aquaculture across vast distances makes it likely that novel combinations of fish species and pathogen/parasite communities are encountered. Epidemics arising from the transmission of pathogens and parasites can result from a single farm fish being transported to an otherwise inaccessible location, and can occur even without any fish being released or escaping from captivity due to disease-causing organisms being carried in the water. This is not new to fish farming, however, as transport of fish and releases have taken place on a large scale for more than a century. It remains clear, nonetheless, that reduction and loss of salmon populations have followed disease outbreaks that can be linked to salmon aquaculture (Bakke and Harris 1998).

The most prominent candidate in this respect is a freshwater parasite, *Gyrodactylus salaris*, which was introduced into Norway as a result of sea ranching activities and subsequently spread by aquaculture. It has led to population reductions, and near extinctions, in several Norwegian rivers for more than 25 years (Johnsen and Jensen 1986, 1991; Malmberg 1989). Other such problems include the bacterium, *Aeromonas salmonicida salmonicida*, which has led to furunculosis epidemics in a few rivers (Johnsen and Jensen 1994), the marine salmon lice (*Lepeophtheirus salmonis, Caligus* sp.) which have been a problem in marine aquaculture since the mid-1970s (Brandal and Egidius 1979) and for wild brown trout *Salmo trutta* and Atlantic salmon since 1989 (Tully 1992; Tully *et al.* 1993a,b; Tully and Whelan 1993; Birkeland 1996; Finstad *et al.* 2000; Heuch *et al.* 2005), and viral diseases such as infectious salmon anaemia (ISA) and infectious pancreatic necrosis (IPN). ISA was detected in Norwegian fish farms in 1984, and subsequently in Canada (1996) and Scotland (1999). In 1999, the Atlantic Salmon Federation reported that ISA had been detected in escaped farm and wild adult Atlantic salmon in a Canadian river. IPN now affects 60–70% of salmon farms in Scotland with 28 escape incidents, involving an estimated 500 000 fish, coming from farms infected with the disease (Staniford 2002). Both ISA and IPN have caused significant mortalities on fish farms and are therefore almost certain to cause increased mortality in the wild, particularly under conditions of environmental stress. It is very difficult to study directly the impacts of such diseases under natural conditions, where natural mortality is much higher and where deliberate introduction of disease to a river would be unethical.

In a common garden experiment, De Eyto *et al.* (in press) brought the progeny of Atlantic salmon from a wild river with no history of disease to another river with a long history of disease outbreaks associated with salmon ranching (a case of bringing the fish to the disease rather than the disease to the fish). The study revealed that MHC class II alpha genes, crucial to the immune response of salmon, were under selection in the wild. They concluded that diseases originating from aquaculture could indirectly be an important mechanism of evolutionary change in wild salmon populations and could have negative consequences for the long-term persistence of the species in the wild.

12.4.4 Direct genetic effects of farm escapes

Direct effects occur due to hybridisation of farm with wild salmon and gene flow from farm to wild salmon through backcrossing of these hybrids (introgression) in subsequent generations. Two main types of genetic change can occur. The first is a change in the level of genetic variability, and the second is a change in the frequency and type of alleles present. Such genetic changes will only be important if the extent and nature of genetic variability is important for survival and recruitment (i.e. fitness) of wild populations. Contrary to some statements in the literature, it does not, however, require that there are adaptive differences among wild populations but only that, as a result of genetic changes in farm fish during domestication, hybrids between wild and farm salmon have lower fitness than wild fish. As noted above, domestication has a greater impact on genetic differentiation than non-native origin. However, the extent of fitness reduction will be increased due to local adaptive differentiation (see Chapter 7). Genetic changes due to hybridisation and introgression may also change the characteristics of a population even if there are no obvious changes in fitness. Characteristics such as age and timing of adult return are important for angling exploitation and alteration of such characteristics may have economic consequences irrespective of whether it impacts on the survival and recruitment of that population.

When interbreeding between genetically different populations results in a reduction in fitness relative to both parental genotypes, it is often referred to as 'outbreeding depression'. Several genetic mechanisms may be responsible for outbreeding depression including the breakdown in co-adapted gene complexes (Templeton 1986). Outbreeding depression may occur in the first hybrid generation, or among their offspring (Lynch 1996). While the degree of fitness loss seems to depend on how distant a cross is (i.e. the extent of genetic differentiation between the parents), quantitative data are largely lacking on the frequency and severity of outbreeding depression in animals (Frankham 1995).

The timing of salmon runs into rivers has generally evolved to be in synchrony with natural flow or temperature regimes that favour success of the wild stock (reviewed in Fleming 1996, 1998). A change in the timing of river ascent could therefore have potentially serious implications if it introduced the population to adverse conditions. Farm salmon often enter rivers later than wild salmon (Lund *et al.* 1991; Carr *et al.* 1997; Fiske *et al.* 2001), and this appears to be related to a lack of juvenile experience with the river resulting in delayed entry (Jonsson *et al.* 1990; Fleming *et al.* 1997). Asynchrony in run timing between wild and farm fish could result in a high degree of type-assortative mating, where interbreeding between farm and wild fish is uncommon (Webb *et al.* 1991). In addition, spawning time likely influences embryo and fry survival, because the thermal regime that embryos experience during development largely determines when they hatch and emerge from the gravel as fry (e.g. Crisp 1981; Jensen *et al.* 1991).

This timing appears adapted to ensure hatching and initial feeding occurs when it is optimal for the offspring (Brannon 1987; Heggberget 1988; Quinn *et al.* 2000). Delayed river entry by farm fish, however, does not necessarily translate into delayed spawning relative to wild fish (Lura and Sægrov 1993; Fleming *et al.* 1996). Rather, spawning time may be more a reflection of genetic differences between populations due to adaptation to climatic and river temperature regimes (reviewed in Fleming 1996). For example, in southern Norway escaped farmed salmon often appear to spawn before wild salmon, while the opposite occurs further north. This appears to reflect the origin of most Norwegian farm salmon from rivers in the mid-western regions of the country.

The migratory behaviour of escaped farm salmon upon river entry appears quite variable among regions and seldom matches exactly that of the local wild salmon. In Norway, farm salmon have been shown to migrate upriver as quickly as wild salmon and distribute themselves further upriver (Økland *et al.* 1995; Heggberget *et al.* 1996; Thorstad *et al.* 1998). The farm females are also less stationary in a particular section of river during the breeding season than are wild females (Økland *et al.* 1995). This latter observation may reflect lack of juvenile experience within the river (Jonsson *et al.* 1990) and/or inferior competitive ability (Fleming *et al.* 1996). Spawning site will dictate the environment embryos and, subsequently, emerging offspring will experience as fry, and hence their growth and survival (reviewed in Chapman 1988). In contrast to the observations from Norway, escaped farm salmon in the River Polla, Scotland, were distributed further downstream than wild salmon during spawning (Webb *et al.* 1991, 1993a). This, however, does not appear to be a general pattern as considerable overlap exists in the spawning distribution of farm and wild fish within other Scottish rivers (Webb *et al.* 1993b) and Norwegian rivers (Fleming *et al.* 2000). Clifford *et al.* (1998a) found variability among years in the occurrence of the offspring of farm salmon in different parts of a river in north-west Ireland.

12.5 Breeding of escaped farm salmon in the wild

12.5.1 Evidence for breeding of escaped farm salmon in the wild

While indirect genetic effects can result from the mere presence of escaped farm fish, or their 'pure' offspring, in a river, direct genetic effects require that interbreeding takes place between farm and wild. Evidence of breeding of escaped farm fish can be obtained by several methods, such as direct observation of spawning of farm fish in the wild, identification of eggs deposited by farm females, and by genetic markers identifying the parentage of a group of fish or individuals. The history of such observations has to a large extent followed the development of techniques that could make inferences about wild or farm origin of naturally produced salmon.

Farm fish have been observed to spawn among wild fish in small rivers that received large inputs of escaped farm fish (Webb *et al.* 1991). That they actually could leave viable offspring was demonstrated by analysis of pigments in eggs and alevins, based on a comparison of optical isomers of astaxanthin which differ between cultured and wild salmon through differences in their diet (Lura and Sægrov 1991a,b). In areas where farm fish were fed canthaxanthin, which is not a natural pigment of salmon (Craik and Harvey 1986), this pigment could distinguish farm offspring from wild well into the start-feeding period (Webb *et al.* 1993a).

In Scotland, a survey in 16 rivers found canthaxanthin to be present in 14 of them, and the overall egg contribution of farm females was 5.1% (Webb *et al.* 1993b). In Norway, investigations in seven rivers demonstrated farm female contribution to the eggs in five of them (Lura and Sægrov 1991b; Lura 1995) and at proportions varying from zero up to 38%. As it is not possible to demonstrate synthetic astaxanthin in farm fish that escaped a year or more before spawning, the farm female contribution may be higher than observed in the eggs (Lura and Økland 1984). Adjustment of observations in one Norwegian river, the Vosso, suggested farm female contribution of up to 80% or more of the eggs (Sægrov *et al.* 1997). Pigments have also been used to identify eggs laid by escaped farm females in a Canadian river on the Atlantic coast (Carr *et al.* 1997).

The pigment-based techniques demonstrated that farm females left fertilised and viable offspring under natural conditions in rivers, but did not disclose whether these were farm × farm or farm × wild crosses. The same type of observation relates to the evidence from mtDNA studies, where the occurrence of mtDNA haplotypes only found in farm females among juveniles demonstrate that they have contributed to the natural spawning. Such a method was used by Clifford *et al.* (1998a) to demonstrate that escaped farm females left offspring in two Irish rivers. They showed that farm female spawning was highly heterogeneous within each river with up to 70% at some sites and complete absence in others. In addition, these authors used a bi-parentally inherited minisatellite locus to demonstrate the presence of pure farm offspring in the rivers and also the breeding of farm males with wild females in a different part of one river from that in which farm female spawning took place. Clifford *et al.* (1998b), using the same two markers, showed that farm fish escaping into a river at the juvenile stage completed the life cycle in the wild to return to that river to breed and interbreed with wild fish.

Other genetic demonstrations of farm contribution to wild populations come from the observation that farm and wild adults entering the same stream differ in allele frequencies at protein-coding loci, and that in the offspring generation change occurred in the direction of farm fish. If the alleles recorded are found in both parental groups, other factors than the successful spawning of farm fish can explain this observation. However, when the farm escapes have alleles that are not found in wild fish, these have been used to demonstrate farm contribution of alleles to the wild population (Crozier 1993, 2000).

Finally, direct evidence that farm escapes breed in the wild comes from observations of Atlantic salmon offspring in areas where the only Atlantic salmon present are farm escapes. In British Columbia, Volpe *et al.* (2000) showed that Atlantic salmon had bred successfully in rivers otherwise occupied by wild Pacific salmon (*Oncorhynchus* spp.).

12.5.2 *Differences in breeding behaviour of farm and wild salmon*

Good estimates of farm male contribution to successful spawning are limited to experimental studies (Fleming *et al.* 1996, 1997, 2000). These have shown that farm males on average perform less well in the wild than farm females, but that they are capable of siring offspring and can have moderate success in the absence of wild males. The reproductive traits of escaped farm fish reflect the environmental effects of artificial rearing, domestication (i.e. intentional and unintentional genetic effects of artificial rearing) and the stock's non-local genetic origin.

As such, farm adults show altered expressions of morphological characters important during breeding, such as secondary sexual characters (Fleming *et al.* 1994; Hard *et al.* 2000).

Such reduced expressions of secondary sexual characteristics are known to have negative consequences for natural breeding success (e.g. Fleming and Gross 1994). Moreover, the breeding behaviour of farm fish has been similarly affected. Evidence from semi-natural breeding experiments indicates that the breeding performance (measured from mate and territory acquisition to egg deposition and fertilisation to egg survival) of farm salmon can be significantly inferior to that of wild salmon (Fleming *et al.* 1996, 2000; see also Berejikian *et al.* 1997, 2001). Farm females show less appropriate breeding behaviour, construct fewer nests, are less efficient at nest covering, and incur greater nest destruction, retention of unspawned eggs and egg mortality than wild females.

The breeding behaviour of farm males appears even more strongly affected by artificial rearing than that of females, reflecting the greater intensity of selection on male competitive ability during this period. Farm males tend to be less dominant, court females less actively and partake in fewer spawnings than wild males. Reproductive inferiority relative to wild conspecifics appears to be a pattern common to cultured salmonids, having also been observed to a lesser degree in sea-ranched/hatchery fish (Leider *et al.* 1990; Jonsson *et al.* 1991; Fleming and Gross 1993; Fleming *et al.* 1997). Greater secondary male, probably mature parr, contribution was found in redds of ranched females compared to wild females (Thompson *et al.* 1998).

The breeding performance of farm salmon appears to be positively related to the length of the period from escape until spawning, i.e. fish that escape early show better performance than those that escape shortly before spawning (Fleming *et al.* 1996, 1997). In addition to such environmental effects of farming on breeding performance, there are also likely to be genetic effects associated with the number of generations the fish have been cultured, though this has never been tested directly. The competitive inferiority of farm fish during spawning is likely to be density/sex ratio-dependent, as has been observed in sea-ranched/hatchery fish (Fleming and Gross 1993; Fleming *et al.* 1997). The competitive and reproductive performance of cultured fish appears to decline as the intensity of competition for breeding resources increases (e.g. density increases or the sex ratio becomes more male-biased). This suggests that 'healthy', dense spawning populations of wild fish will be more resistant to intrusions by farm fish than populations already in a poor state.

The competitive weakness of farm fish also appears to be sex-biased, i.e. farm females perform significantly better than farm males (Fleming *et al.* 1996, 2000; Berejikian *et al.* 1997, 2001). This suggests that much of the initial gene flow that will occur between farm and wild populations will involve farm females and wild males. There is also evidence that the culturing of Atlantic salmon alters female egg traits, resulting in the production of more, but smaller eggs (Jonsson *et al.* 1996; Fleming *et al.* 2000; see also Heath *et al.* 2003), which may affect subsequent juvenile survival (Einum and Fleming 2000a).

12.5.3 Increased hybridisation with brown trout as a result of farm escapes

The rate of interspecific hybridisation between Atlantic salmon and brown trout is increasing in Scotland (Youngson *et al.* 1993) and Norway (Hindar and Balstad 1994), and shows associations with the presence of escaped farm salmon. The average proportion of interspecific hybrids in north-western Europe is low (1%) (Matthews *et al.* 2000) but reaches 7.5% in some rivers (NINR 1997). Hybrids survive well but rarely reproduce (NINR 1997), and thus may lower the productivity of local populations and in very rare cases lead to introgression of genetic material from one species into the other. Direct evidence that farm salmon may increase rates of interspecific hybridisation comes from the Imsa experiment (see section 12.6.1, below) where, among the 3.2% of the offspring that were hybrids, the salmon parent was significantly more often farm than native (Hindar and Fleming 2005).

12.6 Experimental studies of the impact of farm escapes

Molecular marker studies have shown that escaped farm salmon breed in the wild and that changes in the genetic composition of wild populations have taken place. However, such studies do not provide information on the extent of fitness-related genetic differences. Currently such information can only come from 'common garden' experiments where wild, farm and hybrid salmon are reared under communal natural environmental conditions. Molecular markers have an important role in such studies in enabling identification of the groups and families used. The results of two major experiments have been published to date (McGinnity *et al.* 1997, 2003; Fleming *et al.* 2000). These experiments simulated escaped farm salmon entering a river and give quantitative information on the lifetime success of farm salmon and hybrids.

12.6.1 Imsa experiment

A study was undertaken in the Imsa River, a small Norwegian Atlantic salmon river, to quantify the lifetime success (adult to adult) and interactions resulting from farm salmon invading a native population (Fleming *et al.* 2000) (See Box 12.3 for experimental details).

In the river, farm and native adults had similar migration patterns and nesting locations, though farm females spawned before native females. Both types of males began courting females shortly after release; however, native males were more active doing so and retained less of their testes unspawned. The findings from the artificial arena experiment (Box 12.3; see also Fleming *et al.* 1996) paralleled those from the river, indicating that farm males were competitively and reproductively inferior, obtaining fewer spawnings and having 24% of the breeding success (i.e. number of live embryos parented) of native males. Farm females also

---

Box 12.3 Details of Imsa experiment.

A total of 22 farm salmon (fifth generation) derived from Norway's national breeding programme (Gjedrem *et al.* 1991) and reared locally and 17 native salmon were radiotagged and released above a fish trap in the River Imsa, south-western Norway. The fish were sexually mature and had been selected such that all farm salmon were homozygous for the muscle enzyme locus *MEP-2** (**125/*125*) and all native salmon homozygous for the **100* allele. The selection reflected the background allele frequencies of the farm and native populations. Using radiotag signals, the positions of all fish were determined daily throughout the spawning season, with a few exceptions. Visual observations were made of fish stationed in spawning areas in order to record activity.

In parallel with the release experiment, farm and native salmon were introduced into a semi-natural spawning arena where their breeding performance could be closely monitored by direct observation and video 24 hours/day (see Fleming *et al.* 1996). The arena was excavated in the spring, all nests recovered and the number of live (eyed) and dead eggs recorded. Nests were assigned to females using spawning records and egg-size date.

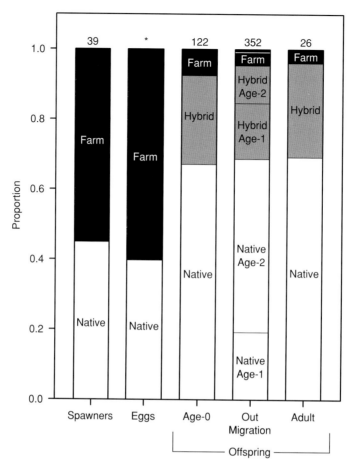

Fig. 12.7 Summary of Imsa results. Changes in the proportional constitution of the Atlantic salmon population in the River Imsa following the release of native and farm spawners. The number above each bar represents either the total population size (spawners and adult offspring) or the sample size examined at each life stage (age −0 and out-migration). Two age groups of out-migrants existed, age −1 and −2, and are stacked on top of each other for each offspring type. *Potential egg deposition was 19 443 for native females and 29 388 for farm females. Solid bars indicate farm offspring, open bars indicate native offspring and hatched bars indicate hybrid offspring. Source: Fleming et al. (2000).

showed a reproductive inferiority (e.g. fewer nests, lower egg survival), achieving just 32% of the breeding success of native females.

During September and October the following year, offspring from the spawnings in the river were sampled by electrofishing the Imsa. The proportion of farm to native genotypes had shifted dramatically from that at release (56% farm) to this stage, with farm genotypes now comprising slightly less than 20% of the population (Fig. 12.7). Moreover, most of the farm genetic representation was in the form of hybrid offspring between farm females and wild males (identified by restriction fragment length polymorphism (RFLP) analysis of mitochondrial DNA). Pure farm offspring comprised less than 8% of the 0+ parr.

Based on the breeding success in the arena experiment, the early survival of farm genotypes was estimated to be 70% that of native genotypes. Thereafter, there was no significant evidence of differential freshwater survival, as farm genotypes composed 18% of the smolt population (Fig. 12.7). There were, however, indications of resource competition in fresh water, as there was considerable diet overlap among native, farm and hybrid offspring. Moreover, the total production of smolts was 28% below that expected based on the potential egg deposition and the 16-year stock–recruitment relationship for the Imsa (Jonsson *et al.* 1998). For native females, smolt production was 31–32% below that expected in the absence of farm females. This effect may reflect competitive asymmetries as native parr were smaller than farm and hybrid parr due to differences in growth rate and parental spawning dates. There were also indications of displacement of native parr further upstream.

As smolts, the offspring types showed distinct behavioural and life-history differences. Farm smolts descended earlier and at a younger age than native smolts, with hybrids being intermediate. Hybrid smolts were also longer and heavier than native smolts, while farm smolts weighed less for a given length than their counterparts. Despite these differences, there were no indications of differential marine survival to maturity (Fig. 12.7). All adult recaptures were made in the coastal fishery or Imsa, and no fish were reported straying into other rivers. The mean age at maturity of hybrid salmon (3.4 years) was significantly less than that of native salmon (4.2 years) because of differences in their age at smolting and poor survival of native age 1 smolts.

12.6.2 Burrishoole experiment

The experiment, comprising three cohorts (1993, 1994, 1998) of Atlantic salmon, was undertaken in the Burrishoole system in western Ireland (Fig. 12.8). This involved multiple families of the following seven groups: native wild (all cohorts); farm (all cohorts); F_1 hybrid wild × farm (male and female reciprocal groups – 93 and 94 cohorts); F_2 hybrid wild × farm (98 cohort); BC_1 backcrosses to wild (98 cohort); and BC_1 backcross to farm (98 cohort). As the aim of the experiment was to look at genetic differences, without the confusion of behavioural differences, eggs and milt were stripped from mature adults and artificially fertilised. (See Box 12.4 for experimental details and McGinnity *et al.* 2003 for further details.)

Box 12.4 Details of Burrishoole experiment.

This experiment, which took 10 years to complete, examined, for the first time, multiple families of both first- and second-generation hybrids between wild and farm salmon in the freshwater and marine life-history phases. By carrying out the experiments under common environment conditions, the effects of environmental variation are eliminated and thus any differences found are the result of differing genetic make-up.

The experiment was undertaken in the Burrishoole system in western Ireland. This system consists of a freshwater lake (Lough Feeagh), connected to Lough Furnace, a tidal brackish lough, by two outlet channels with permanent smolt and adult trapping facilities ('sea entry traps'), and a number of afferent rivers. One of these rivers (Srahrevagh – ~ 7250 m² of juvenile salmonid habitat) was used for the freshwater stages of the experiment and was equipped with a further trap capable of capturing all downstream juvenile migrants and upstream adults (hereafter referred to as the 'experiment river' and 'experiment trap'). Three cohorts were involved with juveniles hatching in 1993, 1994 and 1998.

Native wild Burrishoole salmon of one sea-winter maturity (1SW) and 2SW farm salmon (Norwegian Mowi origin) were used for the 98 cohort, whereas 3SW and 4SW farm fish were used for the earlier cohorts. Returning F_1 hybrid Atlantic salmon (2SW), which had been ranched from the 94 cohort, were captured at the Burrishoole traps from August to November 1997 and used to produce the F_2 hybrids and BC_1 backcrosses (see Table B12.4 for details). A muscle tissue specimen from each parent was retained for DNA profiling. Fertilised eggs were incubated in the

Box 12.4 (cont'd)

hatchery on the Burrishoole system until the developmental stage when eyes were visible ('eyed eggs'), with cumulative mortality being recorded daily. At this stage, live eggs were counted accurately, families mixed, and planted out in the experiment river in artificial redds constructed according to Donaghy and Verspoor (2002). Aliquots of eggs from each family were retained in the hatchery in a communal tank with additional eggs of the 98 cohort (except F_2 hybrid) being reared in separate group tanks. The communally reared hatchery parr were sampled at approximately 11 months age. Juveniles (0+) were sampled from the experiment river by electrofishing and the experiment trap was inspected daily from April post-emergence to the end of the potential 3+ smolt run.

Since insufficient adult returns would have been obtained from the smolts produced in the experiment river, the marine phase of the life cycle of all cohorts (with the exception of F_2 hybrids) was examined by ranching, i.e. smolts were reared in the hatchery and released to sea to complete the life cycle. Prior to release, smolts were tagged with coded wire microtags. Communally reared smolts from the 93 and 94 cohorts were each given a single code whereas each group of the 98 cohort was reared in a separate tank and could be assigned a unique group code. Smolts of the 93, 94 and 98 cohorts were released to Lough Furnace on 3 May 1994, 3 May 1995 and 29 April 1999, respectively. Microtags and tissue specimens were recovered from returning adult salmon taken by angling in Lough Furnace and at the sea entry upstream traps on the Burrishoole system. For the 98 cohort, additional returning fish were also obtained from the commercial net fisheries around the Irish coast through the National Microtag Recovery Programme (Wilkins *et al.* 2001). The relationship between fecundity (F) and weight (W, g) of returning females was estimated using the formula given by Mangel (1996): $F=cW^k$, where $c=4.832$ and $k=0.8697$.

Sampled individuals, except returning adults of the 98 cohort (identifiable from microtags), were identified by microsatellite profiling involving six minisatellite or microsatellite loci. Progeny were identified to family and group parentage using the FAP program (J.B. Taggart, unpubl.). Overall 96.7% of individuals were unambiguously assigned to a single group.

As relative number of groups in some samples is determined by both survival and migration, this is referred to as representation. Differences in survival and representation, relative to the wild group, were tested using G-tests incorporating Williams' correction (Sokal and Rohlf 1995), and were expressed relative to a wild value of 1.0. Length data did not meet the requirements for parametric analyses and were analysed using the Kruskal–Wallis non-parametric one-way ANOVA and, if this showed significant overall heterogeneity, unplanned pairwise comparisons were carried out using Dunn's multiple comparison test.

Table B12.4 Experimental groups of Atlantic salmon in the 1993, 1994 and 1998 cohorts of the Burrishoole experiment.

Group	Cohort	Code	#♀	#♂	Families*	Eyed eggs to river[†]	Smolts to sea[‡]
Burrishoole wild	93	Wild93	6	6	6	5 273	1 842
Farm	93	Farm93	15	15	15	14 997	1 722
F_1 hybrid: Wild ♀ × Farm ♂	93	F_1HyW93	6	6	6	5 886	1 962
F_1 hybrid: Farm ♀ × Wild ♂	93	F_1HyF93	8	8	8	8 659	1 914
Burrishoole wild	94	Wild94	11	11	11	10 537	854
Farm	94	Farm94	11	11	11	10 537	1 138
F_1 hybrid: Wild ♀ × Farm ♂	94	F_1HyW94	11	11	11	10 537	1 211
F_1 hybrid: Farm ♀ × Wild ♂	94	F_1HyF94	11	11	11	10 537	1 028
Burrishoole wild	98	Wild98	8	5	12 (24 sea)	8 787	2 544
Farm	98	Farm98	6	9	33	9 832	9 131
F_2 hybrid: F_1 hybrid × F_1 hybrid	98	F_2Hy	15	2	26	8 337	0
BC_1 wild backcross: F_1 hybrid × Wild	98	BC_1W	15	5	45	9 549	5 661
BC_1 farm backcross: F_1 hybrid × Farm	98	BC_1F	15	5	45	9 928	7 297

* The number of families.
[†] The number of eggs planted out in the experiment river.
[‡] The number of microtagged smolts released to sea.

Fig. 12.8 The Burrishoole experiment. (a) Experiment river with typical Atlantic salmon spawning habitat. (b) Experiment trap with interchangeable screens, capable of capturing 0+ parr onwards. (c) Egg incubation in river. (d) Filling egg wallets. (Photos credit: A. Ferguson.)

Fertilised eggs were incubated to the eyed stage in the hatchery with cumulative mortalities being recorded. The highest egg mortality occurred in the F_2 hybrid group (median 68%), which was significantly higher than all other groups (e.g. wild 3%). Since the backcrosses, which used aliquots of the same eggs as F_2 hybrids, showed significantly lower mortality (8%) this high F_2 hybrid mortality is not due to maternal or egg quality effects and most likely reflects outbreeding depression.

Aliquots of each family were maintained in a hatchery tank until 11 months as a control on the field experiments. No significant differences in survival among groups were found to this age. However, given that total mortality was less than 10% under 'protected' hatchery conditions, there was little opportunity for detectable differential survival. These hatchery controls

Fig. 12.8 (cont'd)

serve to demonstrate that all groups were potentially equally viable and that the differential survival apparent in the wild was the result of genetic and/or maternal differences.

Farm salmon showed significantly lower representation than wild in the samples of 0+ parr of all three cohorts from the experiment river at the end of the first summer; 'hybrids' (i.e. F_1 and F_2 hybrids and BC_1 backcrosses) were intermediate or not significantly different from wild fish (Fig. 12.9). During the period from May 0+ to September 1+ (i.e. second year), the highest proportion of emigrant parr, taken in the experiment trap, was from the wild group and the lowest from the farm group, with 'hybrids' intermediate in representation (in all three cohorts). In the river 0+ parr, it was found that farm parr were largest in size, wild parr smallest and 'hybrids' intermediate, as expected from the selection of farm strains for increased growth rate (Box 12.2). Thus, downstream migration was inversely proportional to parr size, and proportional to cohort density over the three cohorts, indicating competitive displacement

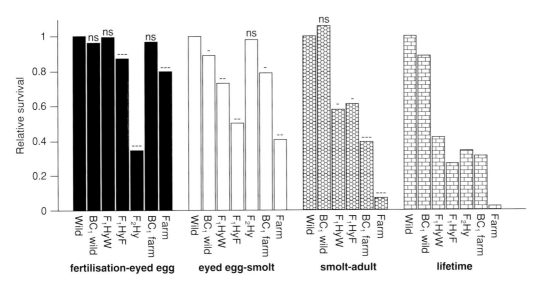

Fig. 12.9 Summary of Burrishoole experiment results. See Table B12.4 for details of groups. Survival at each of the three stages and overall lifetime success is shown relative to a wild value of 1.0. Mean values over cohorts are given where available. Significance of pairwise differences to wild (chi square) is indicated above bar: ns non-significant, – indicates significantly less; < 0.05–0.01; ––/< 0.01–0.001; –––/< 0.001. Only significant values are used in the calculation of lifetime success.

of wild parr by the larger farm and 'hybrid' fish. Although displaced wild parr were found to survive downstream under the experimental conditions used, such survival would not occur if suitable unoccupied habitat is not available. This would be the case, for example, when a river is at parr carrying capacity or where the spawning area debouches directly to sea, as may be typical for escaped farm salmon spawning in some circumstances.

Smolt output was assessed in two ways. First, as the actual numbers of migrants taken in the experiment trap, which assumes that emigrant parr do not survive, i.e. the river is at its parr habitat carrying capacity. In spite of displacement of the wild parr, the farm group produced significantly fewer smolts for the 93 and 94 cohorts but was not significantly different for the 98 cohort. The 'hybrids' had variable representation among cohorts due to differential emigration as a result of different planting densities. The second estimate of smolt output assumed that emigrant parr have the same survival downstream as parr of the equivalent group remaining in the experimental river. This scenario is equivalent to the intrusion of farm salmon into a river with parr habitat in excess of that required by the wild population. With the exception of the F_1 wild mother hybrid group of the 93 cohort and the F_2 hybrid group, all groups had significantly lower smolt production relative to wild. Again farm salmon had consistently the lowest smolt production relative to wild in all three cohorts (34%, 34%, 55%) (Fig. 12.9).

Adult salmon returned from sea after one and two sea-winters (1SW and 2SW). In the 1SW returns, all groups, except the BC_1 backcross to wild, showed a significantly lower return relative to wild. In the 2SW returns, all groups, except farm of the 98 cohort, showed a proportionately greater return. However, the Burrishoole population is primarily a 1SW stock and the wild 2SW return was only 2.5% of the total return. Farm salmon have been bred for late maturity, a trait with high heritability under such conditions (Jónasson *et al.* 1997).

Overall the farm group showed a 0.3% return compared with 8% for wild smolts. Egg deposition is likely to be the limiting factor in salmon recruitment, and taking account of the differential egg production of 1SW and 2SW females (Mangel 1996) shows that total potential egg deposition was significantly lower than wild for all groups except BC_1 backcross to wild (Fig. 12.9). Overall the concordance of the results in the three cohorts considerably increases confidence in the findings.

12.7 Discussion of genetic implications of farm escapes

Escaped farm salmon have multiple effects on wild populations (Table 12.3) and repeated escapes result in cumulative effects over generations. In the Imsa experiment, the lifetime reproductive success of the farm fish was 16% that of the native salmon. This effect comes on top of the apparent effects of intraspecific interactions on productivity (e.g. smolt production). The > 30% depression in smolt production observed was the second largest in 16 years of records (Jonsson *et al.* 1998). Moreover, it occurred despite the absence of competition from older salmon cohorts and during a period that was favourable for smolt production, at least at a broad scale.

In the Burrishoole experiment, the product of survival at the different life-history stages (i.e. survival from fertilisation to reproducing adult) provides a quantitative measure of overall lifetime success which, by taking account of differential egg production between 1SW and 2SW females, can be equated to potential fitness. In the experimental conditions where emigrants survived, farm salmon had a lifetime success of 2% relative to wild fish (Fig. 12.9). This increased to 4% if it is assumed that displaced wild parr did not survive. The 'hybrids' showed intermediate fitness and decreased in survival in the rank order: BC_1 backcross to wild (89%); F_1 hybrid (wild mother); F_2 hybrid (but marine stage not measured for this group); BC_1 backcross to farm; F_1 hybrid (farm mother) (27%). This is as expected from additive genetic variation for survival with a maternal component also in F_1 hybrids, the latter primarily influencing early developmental stages.

Successful breeding and interbreeding by farm salmon will generate pure farm and hybrid (farm × wild) offspring that will compete directly with wild offspring the next generation. These latter genetic and ecological interactions may profoundly affect the productivity of wild populations. The offspring of farm escapees, both pure and hybrid, were observed to incur higher mortality during the earliest life stages but were similar thereafter until the smolt stage (Einum and Fleming 1997; McGinnity *et al.* 1997, 2003; Fleming *et al.* 2000). It is during

Table 12.3 Summary of main impacts of escaped farm salmon on natural populations.

- Conversion of part of wild production to hybrids, which have lower survival, thus reducing population fitness
- Reduction of wild parr survival and smolt production due to competition, thus reducing overall fitness
- Introduction of diseases and parasites reducing wild survival and reducing effective population size and fitness
- Changes in extent of male parr maturity, smolt age, age of maturity, run-timing, and other life history characteristics
- Homogenisation of genetic differences among populations
- Reduction in within-population genetic variability
- Effects are cumulative over generations due to repeated escapes

early life that overall mortality of salmon is high and thus there is ample opportunity for differential survival resulting from maternal effects (e.g. egg size, and emergence time and location; Einum and Fleming 2000a,b) and genetic differences in behavioural adaptations.

There are clear and consistent differences in behaviour of farm and wild juveniles, such as predator avoidance behaviour (see section 12.3), that would contribute to such a result. Farm juveniles also typically outgrow wild juveniles, even in nature (Einum and Fleming 1997; McGinnity *et al.* 1997, 2003; Fleming *et al.* 2000), reflecting the directed domestication selection for growth and, as a result, its endocrine regulatory components (see section 12.3). Observations of increased food consumption and conversion, increased aggression and decreased response to predation risk in the farm juveniles are consistent with the effects of increased growth regulatory activity. These findings suggest that there may be fewer (i.e. relative to the number of adult spawners) but larger farm than wild juveniles. First-generation hybrid juveniles are phenotypically intermediate between the two forms (Einum and Fleming 1997; McGinnity *et al.* 1997; Fleming *et al.* 2000). Farm offspring and hybrids also showed much higher mortality in the marine phase in the Burrishoole experiment (McGinnity *et al.* 2003) but not in the Imsa one (Fleming *et al.* 2000). This difference is likely to be due to different farm strains being used as well as to different natural conditions in Ireland and Norway.

Where escaped farm salmon enter a river, in many situations, production of F_1 hybrids rather than pure farm offspring is the outcome (section 12.5.2). Thus the usual outcome is for part of the potential wild juvenile recruitment to be converted to 'hybrids'. These 'hybrids' can result in an increase in 2SW salmon in rivers that naturally are primarily 1SW producers, which may be desirable from an angling perspective in such rivers by increasing the number of larger fish. However, given their reduced lifetime success, 'hybrids' do not compensate for the loss of wild recruitment resulting in a decrease in fitness in the population. Wild recruitment can be further decreased due to competitive displacement.

The overall extent of reduction in fitness in the wild population, as a result of both interbreeding and competition, depends on a number of factors including availability of unoccupied juvenile habitat and relative numbers of wild, farm and hybrid salmon. However, given the proportion of hybrids found following farm salmon intrusions and in experiments, together with proportions of farm salmon entering rivers, a reduction in juvenile recruitment of 15–30%, in the first generation, could be expected based on the results of the McGinnity *et al.* (1997, 2003) studies. This may be within the range of natural variability for strong wild populations (i.e. high adult production relative to replacement requirements). However, such reductions would bring about further decline in populations on the verge of self-sustainability. Since farm escapes are repetitive, often resulting in annual intrusions in some rivers, such reductions in fitness are cumulative and potentially lead to an extinction vortex.

Hybridisation and introgression due to backcrossing will result in gene flow from farm to wild. One-way gene flow from cultured to wild populations is a potent evolutionary force (Hedrick 1983). The recipient, wild population is eventually composed of individuals that have all descended from the immigrants, and this situation may be approached rapidly for selectively neutral loci/traits. As only a few farm strains are used throughout the industry, this gene flow will reduce the natural inter-population heterogeneity found in Atlantic salmon, reducing the adaptive potential of the species. Under some simplifying assumptions, it is possible to use estimates of the level of gene flow into the wild population to calculate the rate at which the wild population is becoming similar to the immigrants.

Box 12.5 Transgenic or GMO salmon.

In addition to farmed Atlantic salmon from breeding programmes (using quantitative genetics techniques to, for example, increase growth rate and delay sexual maturity) (Box 12.2), transgenic or GMO salmon are becoming available commercially. The term genetic modification might be used to refer to traditional breeding programmes, or to manipulation of sex or ploidy, but here the narrow definition, transgenesis, is used.

Transgenesis refers to the insertion, using the techniques of genetic engineering, of one or more copies of a specific gene into the recently fertilised egg of a donor individual, resulting in a salmon referred to as 'transgenic' or more commonly, as a 'genetically modified organism' (GMO). The rationale of this practice in salmon farming is to attain greatly enhanced growth (Hew and Fletcher 1992; Sin 1997; Fletcher *et al.* 2004) or tolerance of lower water temperatures due to production of antifreeze proteins (Fletcher *et al.* 1988).

In transgenesis it is necessary to synthesise a so-called gene construct that consists of a promoter region to switch on the gene, upstream of the target gene, and then the gene itself. Multiple copies of this construct are then inserted into fertilised eggs, where cell division has not yet begun. The traditional method of insertion was by microinjection into individual eggs, but other methods such as electroporation (use of electricity to pass the gene construct through the cell membrane) are now available. A transgenic Atlantic salmon has been produced by a North American company, containing a growth hormone gene from coho salmon with an ocean pout antifreeze promoter region, and is available commercially although not yet licensed for use. Higher levels of growth hormone are present in the treated fish and because the liver is targeted (in addition to the pituitary which produces growth hormone in untreated fish), the hormone is produced throughout the year instead of production reducing or switching off in winter. The result is greatly enhanced and sustained growth.

This process might appear ideal for the aquaculture industry but environmental concerns about transgenic salmon being released, albeit inadvertently into the wild, have inhibited uptake. For example, there are worries about transgenic escapes interbreeding with wild salmon. The fitness of such progeny has not been demonstrated experimentally but is likely to be at least as reduced as progeny between wild salmon and non-GMO farm escapes (McGinnity *et al.* 2003). It has been suggested that sterility (using processes as described in Box 12.6) be induced in transgenics to prevent such interbreeding. Thus, if transgenic salmon are sterile, direct genetic effects through interbreeding with wild fish would be eliminated, and the effects of escaped transgenic fish would be determined by ecological interactions. It is also anticipated that the production of transgenic salmon will only be allowed in escape-proof locations, in which case the combined physical and biological containment is likely to have a lower impact on wild salmon than the current aquaculture practice based on reproductively viable salmon.

However, even if sterile fish are used there is still environmental concern because of suggestions that escapes might remain in the ocean and, because of their large size, act as effective predators. There are also worries about customer resistance against excessive genetic modification of what is perceived as a green product and also consumer concerns about potential health problems. Finally, there are different legislative approaches to GMO salmon in different regions. Field experiments, similar to those described in section 12.6, should be undertaken with transgenics to determine the relative freshwater performance of such strains compared with wild populations. It seems likely that transgenic salmon will be utilised somewhere in the industry in the near future (if indeed this is not already occurring), so scientific research and discussion must continue so that informed management decisions can be made to protect wild populations. NASCO (2004) has developed preliminary guidelines for the use of transgenic salmonids. These state that transgenic salmonids should be confined to secure, self-contained, land-based facilities.

For transgenic fish sterility may also be targeted by other means than triploidy (Box 12.6), for example by gene technology itself. Reversible sterility is (at least in theory) possible to achieve by gene knockout of crucial hormone-encoding genes via homologous recombination, or alternatively, knock down the expression of such a gene via ribozyme or antisense technologies (see Uzbekova *et al.* 2000; Maclean *et al.* 2002). Targets for such an approach are the gonadotropin releasing hormone (GnRH) and luteinising hormone which are crucial to gonad formation in fishes. For a full review of recent advances in the production of sterile transgenic fish see the report by the Committee on the Biological Confinement of Genetically Engineered Organisms, National Research Council (NRC 2004).

Average proportions of escaped salmon in the North Atlantic have been reported at 20–40% (e.g. Fiske and Lund 1999). Based on the reproductive success calculated from the experiment of Fleming *et al.* (2000), where 55% intrusion of farm spawners left 19% 'farm' genes at the spawning of their offspring, this corresponds to a gene flow at $m \sim 0.06$. In that case, the half-life of the difference between farm and wild salmon would be of the order of 10 generations (see Hedrick 1983). The same conclusion has been reached by modelling the fate of the total effective population size of wild Atlantic salmon receiving immigrants from the farm component, where the total effective population size is reduced to that of the farm fish in a little more than 10 generations (Tufto and Hindar 2003).

Genetically modified Atlantic salmon (Box 12.5) are poised to be one of the first transgenic animals farmed for human consumption (Stokstad 2002). Such salmon would be expected to result in at least the same genetic effects as non-modified ones, both with respect to changes in genetic structure and with respect to fitness. Thus the knowledge gained from studies of farm salmon escapes provides a good foundation for assessing the risks of genetically modified salmon. However, two new aspects relate specifically to transgenic fish. First, the hybrids produced by crosses between transgenic and wild fish will be hemizygous for the transgene. The strength of natural selection against (or for) the new trait will depend on the expression of this trait in hemizygotes relative to homozygotes. A well-established single-locus theory can be used to predict the fate of the transgenes at various levels of gene flow from transgenic to wild fish, and at various selection regimes and dominance relationships (e.g. Muir and Howard 1999, 2001). When immigration rates into natural populations are very high, the recipient populations may be swamped by the inflowing genes irrespective of the strength of selection (Haldane 1932).

12.8 How can the genetic impact of farm escapes be reduced?

A major reason for the high level of escapes is probably the variable standards used for marine fish farm constructions and the lack of adequate certification in some countries. Damage has occurred to sea cages during conditions that these cages were meant to withstand and in Norway construction failure is a most important factor in salmon escapes (WWF-Norway 2005). As well as poor technology standards, there is also a lack of standards, in some cases, relating to routine husbandry such as changing of nets and transfer of fish. For example, boats that carry fish, feed or goods to the marine farms may use open propellers with no means of protecting against potential entanglement in, and damage to the nets. Poor management and training are thus also major causes of escapes (WWF-Norway 2005). A major requirement to reduce escapes is legal enforcement of better standards for construction and operation of fish farms in all countries. While some advances towards better reporting of escapes and stricter technical requirements for fish farms are occurring, policies to prevent or mitigate escapes remain weak in most salmon-farming regions (Naylor *et al.* 2005). In Norway, new technical standards for floating fish farm installations came into force in April 2004, although it will be 2008 before existing equipment is required to be converted to the new standards (WWF-Norway 2005).

Hindar (1992) has argued that the most important measure for marine aquaculture is to base it on closed culture, where the possibility for escape is eliminated and where inflowing and outflowing water are controlled. Land-based technology currently seems too expensive to be able to compete with open-water net pens. A major problem for land-based fish farms is converting the large volume of a sea cage, which results from a depth of over 30 m, to an equivalent area on land where depth is constrained. This would require enormous land areas and some regard this option as non-viable from this perspective alone. It may be better for technologists to develop closed-wall fish farms that are sea based (Karlsen 1993). These would probably need to be located in sheltered localities, but could pump water from exposed localities with little extra cost.

An important measure would be to establish coastal protection zones (nature reserves or national parks) that prevent fish farms being sited close to important river populations and migratory routes (e.g. NOU 1999), which has been done on a limited basis in some countries (WWF-Norway 2005). Exactly what distances should be chosen is not clear, but both the transmission of disease to wild fish and the spread of farm escapes into rivers appear to decrease with distance from fish farms.

Selective harvesting of farm escapes can be developed based on knowledge about migratory differences between cultured and wild fish. It has proven difficult to recapture escaped fish if they escape during extreme weather conditions. However, as half of the escapes occur independently of bad weather conditions (NOU 1999), video surveillance could be used to determine when recapture fisheries need to be executed.

Restrictions on transport of cultured fish have already been implemented. Diseases have nevertheless spread rapidly in aquaculture and some diseases, especially viral ones, are still poorly understood in wild fish (Bakke and Harris 1998). Strict regionalisation of aquaculture should therefore proceed rapidly.

An important measure for reducing gene flow would be to base aquaculture on sterile (all-female) fish, which can be produced at a large scale through simple, inexpensive technology (Box 12.6). Cotter *et al.* (2000) and Wilkins *et al.* (2001) suggested that the reduced return of triploid salmon to the coast and to fresh water, together with their inability to reproduce demonstrated the potential of the use of such fish to eliminate the genetic impact and reduce the freshwater ecological impact of escaped farm salmon. While this measure cannot alleviate problems related to competition in the sea and transmission of parasites and pathogens, it would be an efficient and rapid way of reducing genetic problems. Sterility should at any rate be used while developing a technology that targets full containment. However, the industry cites problems with triploids, e.g. increased mortality at low oxygen tensions as during bath treatment for sea lice.

Increasingly other methods for producing sterility are becoming available through gene manipulation (see Box 12.5). It could be argued that use of genetically modified salmon that were 100% sterile would be preferable to the current use of reproductively competent fish. Thus, if transgenic salmon are sterile, direct genetic effects through interbreeding with wild fish would be eliminated, and the effects of escaped transgenic fish would be determined by ecological interactions.

Gene banks can be used as a last resort to conserve threatened wild populations. For Norwegian Atlantic salmon populations, threatened and numerically weak populations have been targeted in a programme combining cryopreservation of sperm and rearing of wild populations. As virtually no population is unaffected by farm escapes, it has been suggested that emphasis should also be directed towards protecting the strong populations, because these may harbour a greater amount of the genetic diversity of the species (Mork *et al.* 1999). This approach, unfortunately, externalises costs and places it on the public.

> **Box 12.6** All-female triploid production and use in farming.
>
> Triploidy in salmonid fish, which was first used extensively in rainbow trout *Oncorhynchus mykiss* rearing, is achieved by preventing the disjunction of the second polar body from the newly fertilised egg. This process, which is undertaken by temperature or high-pressure shocking at a particular time after fertilisation, results in a fish with two copies of the maternal chromosome complement, instead of one from each sex as in normal diploid individuals. The process results in complete sterility in females, but not in males. Thus all female embryos are produced for triploidisation by using sex-reversed females as the functional 'male' parent in the previous generation. The latter is achieved by first feeding female fry with a diet containing a particular concentration of male hormone, resulting in functional if misshapen sperm-producing testes. Therefore a two-generation process is involved. It should be noted that the second-generation triploid fish are not hormonally treated. (Direct hormonal treatment of marketed food fish contravenes legislation in some countries.)
>
> Complete sterility was required by the rainbow trout industry since the aim was to produce large fish without the diminution of somatic growth and accompanying development of secondary sexual characteristics, which occur with sexual maturation. This process is now an integral part of rainbow trout culture, particularly since this species matures sexually at a small body size.
>
> In the case of Atlantic salmon farming, the triploid process was first invoked to prevent grilse maturation in sea cages (allowing year-round sales, instead of sales being confined to a few months between achievement of adequate body size for market and maturation of grilse). However, alternative methods of addressing the 'grilse problem' were made available by dietary manipulation and selective breeding, which in addition to fast growth, selected for late maturity and also identified late-maturing riverine populations as sources of broodstock. There were also the worries that triploidy reduced viability in Atlantic salmon and might cause adverse customer perception (being seen as untoward genetic manipulation). Since the interaction of farm escapes with wild conspecifics emerged as a problem, it has been suggested that the use of triploid and sterile all-female farmed strains be reconsidered. Sterility would obviously eliminate the direct genetic impact of farm escapes on wild populations, and other impacts might also be reduced since ranching experiments Cotter *et al.* (2000) have shown substantially reduced river returns of triploid salmon, when compared with diploids.

Some of the measures that provide opportunities for coexistence between cultured and wild fish are initially costly to the industry. But maintenance of genetic diversity in wild populations may be crucial in the long run, both for wild populations and for cultured strains (Hindar *et al.* 1991). It remains to be seen whether the long-term costs will be lower if effective measures to protect native populations are ignored rather than taken seriously.

12.9 Summary and conclusions

- Since its origin in 1969, salmon farming in the North Atlantic has increased production to ~ 761 752 tonnes in 2003, with Norway and Scotland being the major producers.
- Currently, around two million salmon escape from farms each year in the North Atlantic, which is equivalent to about 50% of the wild prefishery abundance of salmon in this ocean. Escaped farm salmon comprise some 20–40% of the salmon in some North Atlantic areas and rivers and up to 80% in some Norwegian rivers. Farm salmon parr also escape from juvenile rearing units and freshwater cages but the extent of this has been poorly studied and its impact is probably considerably underestimated.
- Farm salmon are genetically different from wild stocks due to geographical origin, founding effects, and as a result of deliberate and accidental selection, and genetic drift, during domestication. Many farm salmon differences can be related to selection for faster growth

and later maturity together with inadvertent changes affecting survival, deformity, feed conversion rate, spawning time, morphology, aggression, egg viability, egg production and risk-taking behaviour.
- Escaped farm salmon enter rivers generally adjacent to the site of escape but sometimes at considerable distances. These fish have been shown to breed, and interbreed with wild fish, although the greater reproductive success of farm females relative to males, and differences in behaviour, mean that more hybrids are produced than pure farm offspring.
- Farm salmon have both indirect and direct genetic effects on wild populations. Indirect genetic effects occur due to behavioural, ecological and disease interactions, thereby reducing the effective population size of the wild population and increasing genetic drift. In particular, competition with farm fish and hybrids, which are larger, can reduce wild smolt production. Direct genetic effects occur due to interbreeding with wild fish and backcrossing in subsequent generations.
- Farm salmon offspring and hybrids show substantially reduced lifetime success with poorer survival in the early juvenile stages and again in the sea. This results in a loss of fitness (reduced recruitment) in individual wild populations. Since farm escapes are regular occurrences, such reductions in fitness are cumulative and potentially lead to an extinction vortex in weak populations (i.e. populations on the verge of self-sustainability).
- Hybridisation and introgression can change the performance characteristics in wild populations with, for example, an increase in MSW salmon in otherwise predominantly grilse populations, which may be desirable from an angling perspective in such rivers. However, given their reduced lifetime success, hybrids do not compensate for the loss of wild recruitment resulting in a decrease in fitness in the population.
- Domestication would appear to be the main reason for the genetic differences between farm and wild populations. Thus detrimental genetic impacts would occur even if all wild populations were genetically identical. The results are therefore pertinent to other farm salmonids and to cultured marine fish species even though the latter have generally lower levels of natural population differentiation.
- Hybridisation and introgression due to backcrossing will result in gene flow from farm to wild. As only a few farm strains are used throughout the industry, this gene flow will reduce the natural inter-population heterogeneity found in Atlantic salmon, thereby reducing the adaptive potential of the species.
- Genetically modified (transgenic) salmon would be expected to result in the same genetic effects as non-modified ones, both with respect to changes in genetic structure and with respect to fitness. However, the negative impact on fitness could be even greater, although to date appropriate experiments have not been undertaken.
- Potential measures for reducing the genetic impact of farm escapes include improved mandatory standards for the construction and operation of fish farms, closed culture, use of sterile fish and other bioconfinement measures, coastal protection zones, regionalisation of farming, post-escape harvesting, and gene banks to conserve natural genetic diversity.

12.10 Management recommendations

- The Guidelines on Containment of Farm Salmon, developed by the North Atlantic salmon farming industry and the North Atlantic Salmon Conservation Organization (NASCO

2001), should be the *minimum* standard for the construction and operation of fish farms. Research into further improving both technological and operation standards should be undertaken.
- Smolt rearing units should not outflow into salmon rivers and cages should not be placed in lakes in such catchments (as already required in Norway).
- Marine cages should not be situated within 30 km of major salmon rivers.
- Where escapes occur, appropriate recovery plans and resources should be available for immediate deployment.
- Triploids should be used until full containment is possible although further investigations into the use of triploids and other bioconfinement methods are urgently required.
- If it is intended to introduce sterile transgenic salmon in the industry in the future, research, along the lines described in section 12.6, should be undertaken, prior to permission being granted, to determine the ecological impact that such fish may have on wild populations. If transgenic salmon are introduced to the industry in the future only sterile fish should be used (Box 12.6). Note should be taken of the initial guidelines for action on transgenic salmonids that have been developed by NASCO (2004); however, these will need elaboration should permission be granted for the use of such fish.

Acknowledgements

The authors gratefully acknowledge the comments and assistance from the editors and referees, which considerably improved the final text.

Further reading

Fleming, I.A., Hindar, K., Mjølnerød, I.B., Jonsson, B., Balstad, T. and Lamberg, A. (2000) Lifetime success and interactions of farm salmon invading a native population. *Proceedings of the Royal Society of London, Series B*, **267**: 1517–1523. (Published online, DOI: 10.1098/rspb.2000.1173)

Garant, D., Fleming, I.A., Einum, S. and Bernatchez, L. (2003) Alternative male life-history tactics as potential vehicles for speeding introgression of farm salmon traits into wild populations. *Ecology Letters*, **6**: 541–549.

Hindar, K., Fleming, I.A., McGinnity, P. and Diserud, O. (2006) Genetic and ecological effects of salmon farming on wild salmon: modelling from experimental results. *ICES Journal of Marine Science*, **63**: 1234–1247.

McGinnity, P., Stone, C., Taggart, J.B., Cooke, D., Cotter, D., Hynes, R., McCamley, C., Cross, T. and Ferguson, A. (1997) Genetic impact of escaped farmed Atlantic salmon (*Salmo salar* L.) on native populations: use of DNA profiling to assess freshwater performance of wild, farmed, and hybrid progeny in a natural river environment. *ICES Journal of Marine Science*, **54**: 998–1008.

McGinnity, P., Prodöhl, P., Ferguson, A., Hynes, R., Ó Maoiléidigh, N., Baker, N., Cotter, D., O'Hea, B., Cooke, D., Rogan, G., Taggart, J. and Cross, T. (2003) Fitness reduction and potential extinction of wild populations of Atlantic salmon *Salmo salar* as a result of interactions with escaped farm salmon. *Proceedings of the Royal Society of London, Series B*, **270**: 2443–2450. (Published online, DOI: 10.1098/rspb.2003.2520)

NASCO (2001) *Guidelines on containment of farm salmon* CNL(01)53. (Available at: http://www.nasco.int/pdf/nasco_cnl_04_54.pdf)

Naylor, R., Hindar, K., Fleming, I.A., Goldburg, R., Mangel, M., Williams, S., Volpe, J., Whoriskey, F., Eagle, J. and Kelso, D. (2005) Fugitive salmon: assessing risks of escaped fish from aquaculture. *BioScience*, **55**: 427–437.

WWF-Norway (2005) *On the run: escaped farmed fish in Norwegian waters.* (Available at: http://www.wwf.no/core/pd/wwf_escaped_farmed_fish_2005.pdf)

References

ASF (2004) *Atlantic Salmon Aquaculture: A primer.* Atlantic Salmon Federation. (Available: http://www.asf.ca/Aquaculture/ASFaquaculture2004.pdf)

ASF (2005) *Backgrounder on European Strains in Magaguadavic River.* Atlantic Salmon Federation. (Available: http://www.asf.ca/backgrounder/europeanstrains.pdf)

Bakke, T.A. and Harris, P.D. (1998) Diseases and parasites in wild Atlantic salmon (*Salmo salar*) populations. *Canadian Journal of Fisheries and Aquatic Sciences*, **55** (Supplement 1): 247–266.

Baum, E.T. (1998) *Maine Atlantic Salmon: A national treasure.* Atlantic Salmon Unlimited, Hermon, ME.

BBC (2005) Salmon storm escape figure given. (http://news.bbc.co.uk/1/hi/scotland/4287407.stm)

Bentsen, H.B. (1994) Genetic effects of selection on polygenic traits with examples from Atlantic salmon. *Salmo salar* L. *Aquaculture and Fisheries Management*, **25**: 89–102.

Berejikian, B.A., Tezak, E.P., Schroder, S.L., Knudsen, C.M. and Hard, J.J. (1997) Reproductive behavioural interactions between wild and captively reared coho salmon (*Oncorhynchus kisutch*). *ICES Journal of Marine Science*, **54**: 1040–1050.

Berejikian, B.A., Tezak, E.P. and Schroder, S.L. (2001). Reproductive behavior and breeding success of captively reared chinook salmon. *North American Journal of Fisheries Management*, **21**: 255–260.

Birkeland, K. (1996) Consequences of premature return by sea trout (*Salmo trutta* L.) infested with the salmon louse (*Lepeophtheirus salmonis* Krøyer): migration, growth and mortality. *Canadian Journal of Fisheries and Aquatic Sciences*, **53**: 2808–2813.

Brandal, P.O. and Egidius, E. (1979) Treatment of salmon lice (*Lepeophtheirus salmonis* Krøyer) with Neguvon: description of method and equipment. *Aquaculture*, **18**: 183–188.

Brannon, E.L. (1987) Mechanisms stabilizing salmonid fry emergence timing. *Canadian Special Publications in Fisheries and Aquatic Sciences*, **96**: 120–124.

Bridger, C.J., Booth, R.K., McKinley, R.S. and Scruton, D.A. (2001) Site fidelity and dispersal patterns of domestic triploid steelhead trout (*Oncorhynchus mykiss* Walbaum) released to the wild. *ICES Journal of Marine Science*, **58**: 510–516.

Brodeur, R.D. and Busby, M.S. (1998) Occurrence of an Atlantic salmon in the Bering Sea. *Alaska Fisheries Research Bulletin*, **5**: 64–66.

Butler, J.R.A. and Watt, J. (2003) Assessing and managing the impacts of marine salmon farms on wild Atlantic salmon in western Scotland: identifying priority rivers for conservation. In: D. Mills (Ed.) *Salmon at the Edge*, pp. 93–118. Blackwell Publishing, Oxford.

Carr, J.W., Anderson, J.M., Whoriskey, F.G. and Dilworth, T. (1997) The occurrence and spawning of cultured Atlantic salmon (*Salmo salar*) in a Canadian river. *ICES Journal of Marine Science*, **54**: 1064–1073.

Chapman, D.W. (1966) Food and space as regulators of salmonid populations in streams. *American Naturalist*, **100**: 345–357.

Chapman, D.W. (1988) Critical review of variables used to define effects of fines in redds of large salmonids. *Transactions of the American Fisheries Society*, **117**: 1–21.

Clifford, S.L. (1996) *The genetic impact on native Atlantic salmon populations resulting from the escape of farmed salmon.* PhD Thesis, Queen's University, Belfast.

Clifford, S.L., McGinnity, P. and Ferguson, A. (1998a) Genetic changes in Atlantic salmon (*Salmo salar*) populations of northwest Irish rivers resulting from escapes of adult farm salmon. *Canadian Journal of Fisheries and Aquatic Sciences*, **55**: 358–363.

Clifford, S.L., McGinnity, P. and Ferguson, A. (1998b) Genetic changes in an Atlantic salmon population resulting from escaped juvenile farm salmon. *Journal of Fish Biology*, **52**: 118–127.

Cotter, D., O'Donovan, V., Ó Maoiléidigh, N., Rogan, G., Roche, N. and Wilkins, N.P. (2000) An evaluation of the use of triploid Atlantic salmon (*Salmo salar* L.) in minimising the impact of escaped farmed salmon on wild populations. *Aquaculture*, **186**: 61–75.

Craik, J.C.A. and Harvey, S.M. (1986) The carotenoids of eggs of wild and farmed Atlantic salmon and their changes during development. *Journal of Fish Biology*, **29**: 549–565.

Crisp, D.T. (1981) A desk study of the relationship between temperature and hatching time for the eggs of five species of salmonid fishes. *Freshwater Biology*, **11**: 361–368.

Cross, T. and NiChallanain, D. (1991) Genetic characterisation of Atlantic salmon (*Salmo salar*) lines farmed in Ireland. *Aquaculture*, **98**: 209–216.

Crozier, W.W. (1993) Evidence of genetic interaction between escaped farmed salmon and wild Atlantic salmon (*Salmo salar* L.) in a Northern Irish River. *Aquaculture*, **113**: 19–29.

Crozier, W.W. (2000) Escaped farmed salmon, *Salmo salar* L., in the Glenarm River, Northern Ireland: genetic status of the wild population 7 years on. *Fisheries Management and Ecology*, **7**: 437–446.

De Eyto, E., McGinnity, P., Consuegra, S., Coughlan, J., Tufto, J., Farrell, K., Jordan, W.C., Cross, T., Megens, H-J. and Stet, R. (in press) Natural selection acts on Atlantic salmon MHC variability in the wild. *Proceeding of the Royal Society of London, Series B*.

Donaghy, M.J. and Verspoor, E. (2002) A new design of instream incubator for planting out and monitoring Atlantic salmon eggs. *North American Journal of Fisheries Management*, **20**: 521–527.

Einum, S. and Fleming, I.A. (1997) Genetic divergence and interactions in the wild among native, farmed and hybrid Atlantic salmon. *Journal of Fish Biology*, **50**: 634–651.

Einum, S. and Fleming, I.A. (2000a) Selection against late emergence and small offspring in Atlantic salmon (*Salmo salar*). *Evolution*, **54**: 628–639.

Einum, S. and Fleming, I.A. (2000b) Highly fecund mothers sacrifice offspring survival to maximise fitness. *Nature*, **405**: 565–567.

Fausch, K.D. (1988) Tests of competition between native and introduced salmonids in streams: what have we learned? *Canadian Journal of Fisheries and Aquatic Sciences*, **45**: 2238–2246.

Fausch, K.D. and White, R.J. (1986) Competition among juveniles of coho salmon, brook trout, and brown trout in a laboratory stream, and implications for Great Lakes tributaries. *Transactions of the American Fisheries Society*, **115**: 363–381.

Finstad, B., Bjørn, P.A. and Hvidsten, N.A. (2000) Laboratory and field investigations of salmon lice (*Lepeophtheirus salmonis* Krøyer) infestation on Atlantic salmon (*Salmo salar* L.) post-smolts. *Aquaculture Research*, **31**: 795–803.

Fiske, P. and Lund, R.A. (1999) Escapes of reared salmon in coastal and riverine fisheries in the period 1989–1998. *NINA Oppdragsmelding*, **603**: 1–23. (in Norwegian with an English abstract)

Fiske, P., Lund, R.A., Østborg, G.M. and Frøystad, L. (2001) Escapes of reared salmon in coastal and riverine fisheries in the period 1989–2000. *NINA Oppdragsmelding*, **704**: 1–26. (in Norwegian with an English abstract)

Fleming, I.A. (1995) Reproductive success and the genetic threat of cultured fish to wild populations. In: D.P. Philipp, J.M. Epifanio, J.E. Marsden and J.E. Claussen (Ed.) *Protection of Aquatic Biodiversity*, pp. 117–135. Proceedings of the World Fisheries Congress, Theme 3. Oxford and IBH Publishing, New Delhi, India.

Fleming, I.A. (1996) Reproductive strategies of Atlantic salmon: ecology and evolution. *Reviews in Fish Biology and Fisheries*, **6**: 379–416.

Fleming, I.A. (1998) Pattern and variability in the breeding system of Atlantic salmon (*Salmo salar*), with comparison to other salmonids. *Canadian Journal of Fisheries and Aquatic Sciences*, **55** (Supplement 1): 59–76.

Fleming, I.A. and Einum, S. (1997) Experimental tests of genetic divergence of farmed from wild Atlantic salmon due to domestication. *ICES Journal of Marine Science*, **54**: 1051–1063.

Fleming, I.A. and Gross, M.R. (1993). Breeding success of hatchery and wild coho salmon (*Oncorhynchus kisutch*) in competition. *Ecological Applications*, **3**: 230–245.

Fleming, I.A. and Gross, M.R. (1994) Breeding competition in a Pacific salmon (coho: *Oncorhynchus kisutch*): measures of natural and sexual selection. *Evolution*, **48**: 637–657.

Fleming, I.A., Jonsson, B. and Gross, M.R. (1994) Phenotypic divergence of sea-ranched, farmed, and wild salmon. *Canadian Journal of Fisheries and Aquatic Sciences*, **51**: 2808–2824.

Fleming, I.A., Jonsson, B., Gross, M.R. and Lamberg, A. (1996) An experimental study of the reproductive behaviour and success of farmed and wild Atlantic salmon (*Salmo salar*). *Journal of Applied Ecology*, **33**: 893–905.

Fleming, I.A., Lamberg, A. and Jonsson, B. (1997) Effects of early experience on reproductive performance of Atlantic salmon. *Behavioral Ecology*, **8**: 470–480.

Fleming, I.A., Hindar, K., Mjølnerød, I.B., Jonsson, B., Balstad, T. and Lamberg, A. (2000) Lifetime success and interactions of farm salmon invading a native population. *Proceedings of the Royal Society of London, Series B*, **267**: 1517–1523. (Published online, DOI: 10.1098/rspb.2000.1173)

Fleming, I.A., Agustsson, T., Finstad, B., Johnsson, J.I. and Björnsson, B.T. (2002) Effects of domestication on growth physiology and endocrinology of Atlantic salmon (*Salmo salar*). *Canadian Journal of Fisheries and Aquatic Sciences*, **59**: 1323–1330.

Fletcher, G.L., Shears, M.A., King, M.J., Davies, P.L. and Hew, C.L. (1988) Evidence for antifreeze protein gene transfer in Atlantic salmon *Salmo salar*. *Canadian Journal of Fisheries and Aquatic Sciences*, **45**: 352–357.

Fletcher, G.L., Shears, M.A., Yaskowiak, E.S., King, M.J. and Goddard, S.V. (2004). Gene transfer: potential to enhance the genome of Atlantic salmon for aquaculture. *Australian Journal of Experimental Agriculture*, **44**: 1095–1100.

Frankham, R. (1995) Inbreeding and extinction: a threshold effect. *Conservation Biology*, **9**: 792–799.

Garant, D., Fleming, I.A., Einum, S. and Bernatchez, L. (2003) Alternative male life-history tactics as potential vehicles for speeding introgression of farm salmon traits into wild populations. *Ecology Letters*, **6**: 541–549.

Gaudet, D. (2002) *Atlantic salmon*. Alaska Department of Fish and Game Technical Report. (http://www.state.ak.us/local/akpages/FISH.GAME/geninfo/special/AS/docs/AS_white2002.pdf)

Gjedrem, T., Gjøen, H.M. and Gjerde, B. (1991) Genetic origin of Norwegian farmed salmon. *Aquaculture*, **98**: 41–50.

Gjøen, H.M. and Bentsen, H.B. (1997) Past, present, and future of genetic improvement in salmon aquaculture. *ICES Journal of Marine Science*, **54**: 1009–1014.

Glebe, B.D. (1998) *East coast salmon aquaculture breeding programs: history and future*. Department of Fisheries and Oceans Canadian Stock Assessment Secretariat Research Document 98/157.

Gross, M.R. (1998) One species with two biologies: Atlantic salmon (*Salmo salar*) in the wild and in aquaculture. *Canadian Journal of Fisheries and Aquatic Sciences*, **55** (Supplement 1): 131–144.

Gudjonsson, S. (1991) Occurrence of reared salmon in natural salmon rivers in Iceland. *Aquaculture*, **98**: 133–142.

Haldane, J.B.S. (1932) *The Causes of Evolution*. Longmans, Green and Co., London. (Reprinted 1990 Princeton University Press, Princeton, NJ.)

Hansen, L.P. and Jacobsen, J.A. (2003) Origin and migration of wild and escaped farmed Atlantic salmon, *Salmo salar* L., in oceanic areas north of the Faroe Islands. *ICES Journal of Marine Science*, **60**: 110–119.

Hansen, L.P. and Jonsson, B. (1989) Salmon ranching experiments in the River Imsa: effect of timing of Atlantic salmon (*Salmo salar*) smolt migration on survival to adults. *Aquaculture*, **82**: 367–373.

Hansen, L.P. and Jonsson, B. (1991) The effect of timing of Atlantic salmon smolt and post-smolt release on the distribution of adult return. *Aquaculture*, 98: 61–71.

Hansen, L.P. and Jonsson, B. (1994) Homing of Atlantic salmon: effects of juvenile learning on transplanted post-spawners. *Animal Behaviour*, 47: 220–222.

Hansen, L.P. and Quinn, T.P. (1998) The marine phase of the Atlantic salmon (*Salmo salar*) life cycle, with comparisons to Pacific salmon. *Canadian Journal of Fisheries and Aquatic Sciences*, 55 (Supplement 1): 104–118.

Hansen, L.P., Døving, K.B. and Jonsson, B. (1987) Migration of farmed adult Atlantic salmon with and without olfactory sense, released on the Norwegian coast. *Journal of Fish Biology*, 30: 713–721.

Hansen, L.P., Håstein, T., Nævdal, G., Saunders, R.L. and Thorpe, J.E. (Ed.) (1991) Interactions between cultured and wild Atlantic salmon. *Aquaculture*, 98: 1–324.

Hansen, L.P., Jacobsen, J.A. and Lund, R.A. (1999) The incidence of escaped farmed Atlantic salmon, *Salmo salar* L., in the Faroese fishery and estimates of catches of wild salmon. *ICES Journal of Marine Science*, 56: 200–206.

Hard, J.J., Berejikian, B.A., Tezak, E.P., Schroder, S.L., Knudsen, C.M. and Parker, L.T. (2000) Evidence for morphometric differentiation of wild and captively reared adult coho salmon: a geometric analysis. *Environmental Biology of Fish*, 58: 61–73.

Heath, D.D., Heath, J.W., Bryden, C.A., Johnson, R.M. and Fox, C.W. (2003) Rapid evolution of egg size in captive salmon. *Science*, 299: 1738–1739.

Hedrick, P.W. (1983) *Genetics of Populations*. Science Books International, Boston, MA.

Heggberget, T.G. (1988) Timing of spawning in Norwegian Atlantic salmon (*Salmo salar*). *Canadian Journal of Fisheries and Aquatic Sciences*, 10: 1681–1698.

Heggberget, T.G., Økland, F. and Ugedal, O. (1996) Prespawning migratory behaviour of wild and farmed Atlantic salmon (*Salmo salar*) in a north Norwegian river. *Aquaculture Research*, 27: 313–322.

Heuch, P.A., Bjørn, P.A., Finstad, B., Holst, J.C., Asplin, L. and Nilsen, F. (2005) A review of the Norwegian 'National Action Plan against Salmon Lice on Salmonids': the effect on wild salmonids. *Aquaculture*, 246: 79–92.

Hew, C.L. and Fletcher, G.L. (Ed.) (1992) *Transgenic Fish*. World Scientific Publishing, Singapore.

Hindar, K. (1992) Conservation and sustainable use of Atlantic salmon. In: O.T. Sandlund, K. Hindar and A.H.D. Brown (Ed.) *Conservation of Biodiversity for Sustainable Development*, pp. 168–185. Scandinavian University Press, Oslo, and Oxford University Press, Oxford.

Hindar, K. and L'Abée-Lund, J.H. (1992) Identification of hatchery-reared and wild Atlantic salmon juveniles based on examination of otoliths. *Aquaculture and Fisheries Management*, 23: 235–241.

Hindar, K. and Balstad, T. (1994) Salmonid culture and interspecific hybridization. *Conservation Biology*, 8: 881–882.

Hindar, K. and Fleming, I.A. (2005) Behavioral and genetic interactions between escaped farm and wild Atlantic salmon. In: T.M. Bert (Ed.) *Ecological and Genetic Implications of Aquaculture Activities*. Kluwer Academic Publishers, Dordrecht.

Hindar, K., Ryman, N. and Utter, F. (1991) Genetic effects of cultured fish on natural fish populations. *Canadian Journal of Fisheries and Aquatic Sciences*, 48: 945–957.

Huntingford, F.A. (2004) Implications of domestication and rearing conditions for the behaviour of cultivated fishes. *Journal of Fish Biology*, 65 (Supplement A): 122–142.

Jensen, A.J., Johnsen, B.O. and Saksgård, L. (1991) Temperature requirements in Atlantic salmon (*Salmo salar*), brown trout (*Salmo trutta*), and Arctic charr (*Salvelinus alpinus*) from hatching to initial feeding compared with geographic distribution. *Canadian Journal of Fisheries and Aquatic Sciences*, 46: 786–789.

Johnsen, B.O. and Jensen, A.J. (1986) Infestations of Atlantic salmon, *Salmo salar*, by *Gyrodactylus salaris* in Norwegian rivers. *Journal of Fish Biology*, 29: 233–241.

Johnsen, B.O. and Jensen, A.J. (1991) The *Gyrodactylus* story in Norway. *Aquaculture*, 98: 289–302.

Johnsen, B.O. and Jensen, A.J. (1994) The spread of furunculosis in salmonids in Norwegian rivers. *Journal of Fish Biology*, **45**: 47–55.

Johnsson, J.I. and Björnsson, B.T. (1994) Growth hormone increases growth rate, appetite and dominance in juvenile rainbow trout, *Oncorhynchus mykiss*. *Animal Behaviour*, **48**: 177–186.

Johnsson, J.I., Petersson, E., Jönsson, E., Björnsson, B.T. and Järvi, T. (1996) Domestication and growth hormone alter antipredator behaviour and growth patterns in juvenile brown trout, *Salmo trutta*. *Canadian Journal of Fisheries and Aquatic Sciences*, **53**: 1546–1554.

Johnsson, J.I., Höjesjö, J. and Fleming, I.A. (2001) Behavioural and heart rate response to predation risk in wild and domesticated Atlantic salmon. *Canadian Journal of Fisheries and Aquatic Sciences*, **58**: 788–794.

Jónasson, J., Gjerde, B. and Gjedrem, T. (1997) Genetic parameters for return rate and body weight in sea-ranched Atlantic salmon. *Aquaculture*, **154**: 219–231.

Jonsson, B., Jonsson, N. and Hansen, L.P. (1990) Does juvenile river experience affect migration and spawning of adult Atlantic salmon? *Behavioural and Ecological Sociobiology*, **26**: 225–230.

Jonsson, B., Jonsson, N. and Hansen, L.P. (1991) Differences in life history and migration behaviour between wild and hatchery-reared Atlantic salmon in nature. *Aquaculture*, **98**: 69–78.

Jonsson, B., Jonsson, N. and Hansen, L.P. (2002) Atlantic salmon straying from the River Imsa. *Journal of Fish Biology*, **62**: 641–657.

Jönsson E., Johnsson, J.I. and Björnsson, B.T. (1996) Growth hormone increases predation exposure of rainbow trout. *Proceedings of the Royal Society of London, Series B*, **263**: 647–651.

Jönsson E., Johnsson, J.I. and Björnsson, B.T. (1998) Growth hormone increases aggressive behavior in juvenile rainbow trout. *Hormones and Behaviour*, **33**: 9–15.

Jonsson, N., Jonsson, B. and Fleming, I.A. (1996) Does early growth cause a phenotypic plastic response in egg production of Atlantic salmon? *Functional Ecology*, **10**: 89–96.

Jonsson, N., Jonsson, B. and Hansen, L.P. (1998) The relative role of density-dependent and density-independent survival in the life cycle of Atlantic salmon *Salmo salar*. *Journal of Animal Ecology*, **67**: 751–762.

Karlsen, L. (1993) Developments in salmon aquaculture technology. In: K. Heen, R.R. Monaghan and F. Utter (Ed.) *Salmon Aquaculture*, pp. 59–82. Fishing News Books, Oxford.

Lacroix, G.L. and Stokesbury, M.J.W. (2004) Adult return of farmed Atlantic salmon escaped as juveniles into freshwater. *Transactions of the American Fisheries Society*, **133**: 484–490.

Leider, S.A., Hulett, P.L., Loch, J.J. and Chilcote, M.W. (1990) Electrophoretic comparison of the reproductive success of naturally spawning transplanted and wild steelhead trout through the returning adult stage. *Aquaculture*, **88**: 239–252.

Lund, R.A. and Hansen, L.P. (1991) Identification of wild and reared Atlantic salmon, *Salmo salar* L., using scale characters. *Aquaculture and Fisheries Management*, **22**: 499–508.

Lund, R.A., Hansen, L.P. and Järvi, T. (1989) Identification of reared and wild salmon by external morphology, size of fins and scale characteristics. *NINA Research Report* **1**. (in Norwegian, with English summary)

Lund, R.A., Økland, F. and Hansen, L.P. (1991) Farmed Atlantic salmon (*Salmo salar* L.) in fisheries and rivers in Norway. *Aquaculture*, **98**: 143–150.

Lura, H. (1995) *Domesticated female Atlantic salmon in the wild: spawning success and contribution to local populations*. Dr. Scient. Thesis, University of Bergen.

Lura, H. and Økland, F. (1984) Content of synthetic astaxanthin in escaped farmed Atlantic salmon, *Salmo salar* L., ascending Norwegian rivers. *Fisheries Management and Ecology*, **1**: 205–216.

Lura, H. and Sægrov, H. (1991a) Documentation of successful spawning of escaped farmed female Atlantic salmon, *Salmo salar*, in Norwegian rivers. *Aquaculture*, **98**: 151–159.

Lura, H. and Sægrov, H. (1991b) A method of separating offspring from farmed and wild Atlantic salmon (*Salmo salar*) based on different ratios of optical isomers of astaxanthin. *Canadian Journal of Fisheries and Aquatic Sciences*, **48**: 429–433.

Lura, H. and Sægrov, H. (1993) Timing of spawning in cultured and wild Atlantic salmon (*Salmo salar*) in the River Vosso, Norway. *Ecology of Freshwater Fish*, **2**: 167–172.

Lynch, M. (1996) A quantitative-genetic perspective on conservation issues. In: J.C. Avise and J.L. Hamrick (Ed.) *Conservation Genetics: Case histories from nature*, pp. 471–501. Chapman & Hall, New York.

McGinnity, P., Stone, C., Taggart, J.B., Cooke, D., Cotter, D., Hynes, R., McCamley, C., Cross, T. and Ferguson, A. (1997) Genetic impact of escaped farmed Atlantic salmon (*Salmo salar* L.) on native populations: use of DNA profiling to assess freshwater performance of wild, farmed, and hybrid progeny in a natural river environment. *ICES Journal of Marine Science*, **54**: 998–1008.

McGinnity, P., Prodöhl, P., Ferguson, A., Hynes, R., Ó Maoiléidigh, N., Baker, N., Cotter, D., O'Hea, B., Cooke, D., Rogan, G., Taggart, J. and Cross, T. (2003) Fitness reduction and potential extinction of wild populations of Atlantic salmon *Salmo salar* as a result of interactions with escaped farm salmon. *Proceedings of the Royal Society of London, Series B*, **270**: 2443–2450. (Published online, DOI: 10.1098/rspb.2003.2520)

McGinnity, P., Prodöhl, P., Ó Maoiléidigh, N., Hynes, R., Cotter, D., Baker, N., O'Hea, B. and Ferguson, A. (2004) Differential lifetime success and performance of native and non-native Atlantic salmon examined under communal natural conditions. *Journal of Fish Biology* **65** (Supplement A), 173–187.

Maclean, N., Rahman, M.A., Sohm, F., Hwang, G., Iyengar, A., Ayad, H., Smith, A. and Farahmand, H. (2002) Transgenic tilapia and the tilapia genome. *Gene*, **295**: 265–277.

Malmberg, G. (1989) Salmonid transports, culturing and *Gyrodactylus* infections in Scandinavia. In: O.N. Bauer (Ed.) *Parasites of Freshwater Fishes of North-West Europe*, pp. 88–104. Petrozavodsk, Nauka.

Mangel, M. (1996) Computing expected reproductive success of female Atlantic salmon as a function of smolt size. *Journal of Fish Biology*, **49**: 877–892.

Martin-Smith, K.M., Armstrong, J.D., Johnsson, J.I. and Björnsson, B.T. (2004) Growth hormone increases growth and dominance of wild juvenile Atlantic salmon without affecting space use. *Journal of Fish Biology*, **65** (Supplement A): 156–172.

Matthews, M.A., Poole, W.R., Thompson, C.E., McKillen, J., Ferguson, A., Hindar, K. and Whelan, K.F. (2000) Incidence of hybridization between Atlantic salmon, *Salmo salar* L., and brown trout, *Salmo trutta* L., in Ireland. *Fisheries Management and Ecology*, **7**: 337–347.

Milner, N.J. and Evans, R. (2003) The incidence of escaped Irish farmed salmon in English and Welsh rivers. *Fisheries Management and Ecology*, **10**: 403–406.

Mjølnerød, I.B., Refseth, U.H., Karlsen, E., Balstad, T., Jakobsen, K.S. and Hindar, K. (1997) Genetic differences between two wild and one farmed population of Atlantic salmon (*Salmo salar*) revealed by three classes of genetic markers. *Hereditas*, **127**: 239–248.

Mork, J.H.B., Bentsen, H., Hindar, K. and Skaala, Ø. (1999) Genetiske interaksjoner mellom oppdrettslaks og vill laks, pp. 181–200. In: *Til laks åt alle kan ingen gjera? Norges offentlige utredninger 1999: 9*, Statens forvaltningstjeneste, Oslo. ('Genetic interactions between farmed Atlantic salmon and wild salmon', in Norwegian)

Muir, W.M. and Howard, R.D. (1999) Possible ecological risks of transgenic organism release when transgenes affect mating success: sexual selection and the Trojan gene hypothesis. *Proceedings of the National Academy of Sciences*, **96**: 13853–13856.

Muir, W.M. and Howard, R.D. (2001) Fitness components and ecological risk of transgenic release: a model using Japanese medaka (*Oryzias latipes*). *American Naturalist*, **158**: 1–16.

NASCO (2001) *Guidelines on containment of farm salmon* CNL(01)53. (Available at: http://www.nasco.int/pdf/nasco_cnl_04_54.pdf)

NASCO (2004) *NASCO Guidelines for action on transgenic salmonids* CNL(04)41. (Available at: http://www.nasco.int/pdf/nasco_cnl_04_54.pdf)

Naylor, R.L., Williams, S.L. and Strong, D.R. (2001) Aquaculture: a gateway for exotic species. *Science*, **294**: 1655–1656.

Naylor, R., Hindar, K., Fleming, I.A., Goldburg, R., Mangel, M., Williams, S., Volpe, J., Whoriskey, F., Eagle, J. and Kelso, D. (2005) Fugitive salmon: assessing risks of escaped fish from aquaculture. *BioScience*, **55**: 427–437.

Nævdal, G., Holm, M., Møller, D. and Østhus, O.D. (1975) Experiments with selective breeding of Atlantic salmon. *International Council for the Exploration of the Seas* **1975/M-15**.

Nielsen, J.L., Williams, I., Sage, G.K. and Zimmerman, C.E. (2003) The importance of genetic verification for differentiation of Atlantic salmon in north Pacific waters. *Journal of Fish Biology*, **62**: 871–878.

NINR (1997) Hybridisation between escaped farmed Atlantic salmon (*Salmo salar*) and brown trout (*Salmo trutta*): frequency, distribution, behavioural mechanisms and effects on fitness. *Final Report, Commission of the European Community Contract AIR3 CT94 2484*. K. Hindar (coordinator). Norwegian Institute for Nature Research, Trondheim.

Noakes, D.J., Beamish, R.J. and Kent, M.L. (2000) On the decline of Pacific salmon and speculative links to salmon farming in British Columbia. *Aquaculture*, **183**: 363–386.

Norris, A.T., Bradley, D.G. and Cunningham, E.P. (1999) Microsatellite genetic variation between and within farmed and wild Atlantic salmon (*Salmo salar*) populations. *Aquaculture*, **180**: 247–264.

NOU (1999) *Til laks åt alle kan ingen gjera? Norges offentlige utredninger* 1999: 9, Statens forvaltningstjeneste, Oslo. (in Norwegian, with English summary)

NRC (2004) *Biological confinement of genetically engineered organisms*. National Academy Press, Washington, DC. (Also available at: http://www.nap.edu)

Økland, F., Heggberget, T.G. and Jonsson, B. (1995) Migratory behaviour of wild and farmed Atlantic salmon (*Salmo salar*) during spawning. *Journal of Fish Biology*, **46**: 1–7.

Quinn, T.P., Unwin, M.J. and Kinnison, M.T. (2000) Evolution of temporal isolation in the wild: genetic divergence in timing of migration and breeding by introduced chinook salmon populations. *Evolution*, **54**: 1372–1385.

Sægrov, H., Hindar, K., Kålås, S. and Lura, H. (1997) Escaped farmed Atlantic salmon replaces the original salmon stock in the River Vosso. *ICES Journal of Marine Science*, **54**: 1166–1172.

Sin, F.Y.T. (1997) Transgenic fish. *Reviews in Fish Biology and Fisheries*, 7: 417–441.

Skaala, Ø., Høyheim, B., Glover, K. and Dahle, G. (2004) Microsatellite analysis in domesticated and wild Atlantic salmon (*Salmo salar* L.): allelic diversity and identification of individuals. *Aquaculture*, **240**: 131–143.

Skaala, Ø., Taggart, J.B. and Gunnes, K. (2005) Genetic differences between five major domesticated strains of Atlantic salmon and wild salmon. *Journal of Fish Biology*, **67** (Supplement A): 118–128.

Sokal, R.R. and Rohlf, F.J. (1995) *Biometry* (3rd edn). Freeman & Co, New York.

Staniford, D. (2002) *Sea cage fish farming: an evaluation of environmental and public health aspects (the five fundamental flaws of sea cage fish farming)*. (Available at: http://www.salmonfarmmonitor.org/stanifordpaper.doc)

Stokesbury, M.J. and Lacroix, G.L. (1997) High incidence of hatchery origin Atlantic salmon in the smolt output of a Canadian river. *ICES Journal of Marine Science*, **54**: 1974–1081.

Stokesbury, M.J.W., Lacroix, G.L., Price, E.L., Knox, D. and Dadswell, M.J. (2001) Identification by scale analysis of farmed Atlantic salmon juveniles in southwestern New Brunswick rivers. *Transactions of the American Fisheries Society*, **130**: 815–822.

Stokstad, E. (2002) Engineered fish: friend or foe of the environment? *Science*, **297**: 1797–1799.

Templeton, A.R. (1986) Coadaptation and outbreeding depression. In: M.E. Soulé (Ed.) *Conservation Biology: The science of scarcity and diversity*, pp. 105–116. Sinauer, Sunderland, MA.

Thodesen, J., Grisdale-Helland, B., Helland, S.J. and Gjerde, B. (1999) Feed intake, growth and feed utilization of offspring from wild and selected Atlantic salmon (*Salmo salar*). *Aquaculture*, **180**: 237–246.

Thompson, C.E., Poole, W.R., Matthews, M.A. and Ferguson, A. (1998) Comparison, using minisatellite DNA profiling, of secondary male contribution in the fertilisation of wild and ranched Atlantic salmon (*Salmo salar*) ova. *Canadian Journal of Fisheries and Aquatic Sciences*, **55**: 2011–2018.

Thorstad, E.B., Heggberget, T.G. and Økland, F. (1998) Migratory behaviour of adult wild and escaped farmed Atlantic salmon, *Salmo salar* L., before, during and after spawning in a Norwegian river. *Aquaculture Research*, **29**: 419–428.

Tufto, J. and Hindar, K. (2003) Effective size in management and conservation of subdivided populations. *Journal of Theoretical Biology*, **222**: 273–281.

Tully, O. (1992) Predicting infestation parameters and impacts of caligid copepods in wild and cultured fish populations. *Invertebrate Reproductive and Development*, **22**: 91–102.

Tully, O. and Whelan, K.R. (1993) Production of nauplii of *Lepeophtheirus salmonis* (Krøyer) (Copepoda: Caligidae) from farmed and wild salmon and its relation to the infestation of wild sea trout (*Salmo trutta* L.) off the west coast of Ireland in 1991. *Fisheries Research*, **17**: 187–200.

Tully, O., Poole, W.R. and Whelan, K.R. (1993a) Infestation parameters for *Lepeophtheirus salmonis* (Krøyer) (Copepoda: Caligidae) parasitic on sea trout, *Salmo trutta* L., off the west coast of Ireland during 1990 and 1991. *Aquaculture and Fisheries Management*, **24**: 545–555.

Tully, O., Poole, W.R., Whelan, K.R. and Merigoux, S. (1993b) Parameters and possible causes of epizootics of *Lepeophtheirus salmonis* (Krøyer) infesting sea trout (*Salmo trutta* L.) off the west coast of Ireland. In: G.A. Boxshall and D. Defaye (Ed.) *Pathogens of Wild and Farmed Fish: Sea lice*, pp. 202–213. Ellis Horwood, New York.

Uzbekova, S., Chyb, J., Ferrière, F., Bailhache, T., Prunet, P., Alestrom, P. and Breton, B. (2000) Transgenic rainbow trout expressed sGnRH-antisense RNA under the control of sGnRH promoter of Atlantic salmon. *Journal of Molecular Endocrinology*, **25**: 337–350.

Verspoor, E. (1988) Reduced genetic variability in first-generation hatchery populations of Atlantic salmon (*Salmo salar*). *Canadian Journal of Fisheries and Aquatic Sciences*, **45**: 686–1690.

Volpe, J.P., Taylor, E.B., Rimmer, D.W. and Glickman, B.W. (2000) Natural reproduction of aquaculture escaped Atlantic salmon (*Salmo salar*) in a coastal British Columbia river. *Conservation Biology*, **14**: 899–903.

Wang, S., Hard, J.J. and Utter, F. (2002a) Salmonid inbreeding: a review. *Reviews in Fish Biology and Fisheries*, **11**: 301–319.

Wang, S., Hard, J.J. and Utter, F. (2002b) Genetic variation and fitness in salmonids. *Conservation Genetics*, **3**: 321–333.

Waples, R.S. (1991) Genetic interactions between hatchery and wild salmonids: lessons from the Pacific Northwest. *Canadian Journal of Fisheries and Aquatic Sciences*, **48** (Supplement 1): 124–133.

Webb, J.H., Hay, D.W., Cunningham, P.D. and Youngson, A.F. (1991) The spawning behaviour of escaped farmed and wild adult Atlantic salmon (*Salmo salar* L.) in a northern Scottish river. *Aquaculture*, **98**: 97–110.

Webb, J.H., McLaren, I.S., Donaghy, M.J. and Youngson, A.F. (1993a) Spawning of farmed Atlantic salmon, *Salmo salar* L., in the second year after their escape. *Aquaculture and Fisheries Management*, **24**: 557–561.

Webb, J.H., Youngson, A.F., Thompson, C.E., Hay, D.W., Donaghy, M.J. and McLaren, I.S. (1993b) Spawning of escaped farmed Atlantic salmon, *Salmo salar* L., in western and northern Scottish rivers: egg deposition by females. *Aquaculture and Fisheries Management*, **24**: 663–670.

Whoriskey, F.G. and Carr, J.W. (2001) Returns of transplanted adult, escaped, cultured Atlantic salmon to the Magaguadavic River, New Brunswick. *ICES Journal of Marine Science*, **58**: 504–509.

Wilkins, N.P., Cotter, D. and Ó Maoiléidigh, N. (2001) Ocean migration and recaptures of tagged, triploid, mixed-sex and all-female Atlantic salmon (*Salmo salar* L.) released from rivers in Ireland. *Genetica*, **25**: 1–16.

WWF (2001) *The status of wild Atlantic salmon: a river by river assessment*. (Available at: http://www.wwf.org.uk/filelibrary/pdf/atlanticsalmon.pdf)

WWF-Norway (2005) *On the run: escaped farmed fish in Norwegian waters*. (Available at: http://www.wwf.no/core/pd/wwf_escaped_farmed_fish_2005.pdf)

Youngson, A.F. and Verspoor, E. (1998) Interactions between wild and introduced Atlantic salmon (Salmo salar). *Canadian Journal of Fisheries and Aquatic Sciences*, **55** (Supplement 1): 153–160.

Youngson, A.F., Martin, S.A.M., Jordan, W.C. and Verspoor, E. (1991) Genetic protein variation in Atlantic salmon in Scotland: comparison of wild and farmed fish. *Aquaculture*, **98**: 231–242.

Youngson, A.F., Webb, J.H., Thompson, C.E. and Knox, D. (1993) Spawning of escaped farmed Atlantic salmon (*Salmo salar*): hybridisation of females with brown trout (*Salmo trutta*). *Canadian Journal of Fisheries and Aquatic Sciences*, **50**: 1986–1990.

Youngson, A.F., Webb, J.H., MacLean, J.C. and Whyte, B.M. (1997) Frequency of occurrence of reared Atlantic salmon in Scottish salmon fisheries. *ICES Journal of Marine Science*, **54**: 1216–1220.

13 Genetics and Habitat Management

E. Verspoor, C. García de Leániz and P. McGinnity

Upper: exposed gravels on the upper Garry, a tributary of the River Tay in Scotland, due to water diversion as part of a hydroelectric generation scheme, resulting in a loss of tens of thousands of square metres of salmon habitat. (Photo credit: Crown copyright FRS, reproduced with the permission of FRS Freshwater Laboratory, Pitlochry, photo by D. Hay.) Lower: Loch Faskally on the River Tummel, a tributary of the River Tay, above Pitlochry Dam, whose formation resulted in the loss and fragmentation of existing juvenile and spawning habitat in the river. (Photo credit: E. Verspoor.)

Habitat modification and destruction underlies much of the worldwide loss of genetic diversity, both among and within species. The losses in respect of Atlantic salmon, *Salmo salar*, are no exception and, along with stocking, habitat restoration and enhancement have been the main targets of species management for more than a century. However, while the last two decades have witnessed an increased awareness of the genetic issues raised by stocking (Chapter 11), appreciation of the potential genetic effects of habitat modification and destruction has developed only slowly. Research into this question is limited, both generally and with regard to the Atlantic salmon specifically. Yet what is known makes it clear that there are genetic issues which need to be considered in habitat management.

13.1 Introduction

Over the last two centuries, humans have had a serious impact on the environment of the Atlantic salmon and, thereby, on the abundance and biological character of the species in many of its native rivers (Netboy 1968; Parrish *et al.* 1998; WWF 2001; Lackey 2002). Some environmental changes have clearly exceeded the inherent capacity of the species to cope, at least without human assistance (Fig. 13.1) though, other changes, from the point of view of the Atlantic salmon at least, have been more positive (Fig. 13.2). To address the negative impacts of habitat change and modification, humans have engaged in activities aimed at maintaining and improving Atlantic salmon habitat. At the same time, habitat modification has also been aimed at increasing abundance, or access to fish, at least some of which is of questionable value at best as regards its effect on local fisheries abundance (Fig. 13.3).

Atlantic salmon habitat, in the general sense, refers to environmental space which has the required physical, chemical and biological conditions for the species to survive and thrive

Fig. 13.1 The hydroelectric dam built on the River Tummel in Pitlochry, Scotland, represents one of the most common examples of human-caused changes to the environment of the Atlantic salmon. (Photo credit: E. Verspoor.)

Fig. 13.2 A functioning fish ladder on the River Fleet in north-east Scotland in the late 1800s, built to allow anadromous salmon to surmount an impassable falls and colonise the upper river. (Photo credit: E. Verspoor.)

Fig. 13.3 The gabions, galvanised wire boxes filled with stones, used to form an instream dam and the retaining walls of the outlet for an artificial pool in a small Scottish river, constructed in an attempt to 'improve' fishing; already limited spawning and juvenile habitat has been destroyed, habitat fragmented, and the upstream movement of adults made more difficult. (Photo credit: E. Verspoor.)

throughout its life cycle. However, the use of the term is often qualified to be more specific, recognising that the habitat needs of the species vary with life-history stage, e.g. parr or fry habitat (Fig. 13.4) or, even more specifically, winter parr habitat. The term habitat is also frequently qualified by words such as good, poor or marginal, reflecting the notion that it is not absolutely suitable, or not, but that environmental needs can be met with varying degrees

Fig. 13.4 An example of typical habitat for Atlantic salmon fry and parr (inset photos) in the River North Esk, north-east Scotland. (Photo credits: main, E. Verspoor; insets Crown copyright FRS, reproduced with the permission of FRS Freshwater Laboratory, Pitlochry.)

of success in terms of supporting fish. The nature of what constitutes habitat has been explored in detail by Hall *et al.* (1997).

The basic habitat requirements of the Atlantic salmon in fresh water are reasonably well understood, at least relative to those of brown trout, *Salmo trutta*, and other fish species (Chapter 2). In contrast, its marine habitat requirements are largely an enigma, and defining the nature of marine habitat for Atlantic salmon remains a major research challenge. However, perhaps equally poorly understood is the extent to which the genetic character of the Atlantic salmon is important to its habitat requirements, and how the former is affected by changes in the latter. What is known is that the genetic character of the species is relevant to its abundance and character, and genetic character and habitat needs are inextricably linked (Chapters 5 and 7). Thus habitat changes and habitat management can in many cases be expected to have genetic consequences which need to be understood to limit or avoid detriment to local species abundance and character.

The linkage of genetics and habitat is complex, and the impacts of habitat changes and modifications will often be difficult to predict. The task of understanding the impacts of habitat change is made harder by the fact that Atlantic salmon habitat is naturally characterised by day-to-day, seasonal and inter-annual variation. This includes key variables such as temperature, stream flow, density of competitors and presence of disease vectors (see Chapters 2 and 7). This often involves diurnal, yearly or decadal fluctuations linked to natural climatic variations, and varies from river to river and region to region. Some variation involves long-term fluctuations that are effectively directional in management time scales. At the same time, aspects of habitat variation, both within and among years, will also be random and, where outside the range of the variation normally seen, will involve catastrophic events such as 100-year floods or volcanic eruptions.

Fig. 13.5 The Torridon River in the highlands of Scotland, showing a classic example of a postglacial landscape, with drumlins (small, oval, hummocky hills formed from the detritus of a retreating glacier) prominent in the background. (Photo credit: D. Hay.)

Natural environmental change, to some degree, is an essential condition for a given physical space to provide a suitable habitat. For example, annual spates in salmon rivers are important in clearing fine sediments from spawning gravels (Hicks *et al.* 1991) or providing the volumes of waters needed for salmon to ascend waterfalls on their spawning migrations. Seasonal fluctuations will also drive trophic dynamics and sustain the interactions of predators and prey. As such, these variations will be integral to the observed distribution, character and abundance and something on which populations will often depend for their survival and productivity.

Environmental changes and fluctuations on larger time scales are also the norm for species such as the Atlantic salmon, given their association with regions which have been alternately habitable and covered by glaciers (Fig. 13.5). The repeated retreat and advance of the ice sheets throughout the Pleistocene (Bernatchez and Wilson 1998) are part of natural historical environmental change which involves a continual creation and destruction of habitat. Indeed, over its history, the species as a whole has undoubtedly been engaged in a long-term metapopulation dynamic (Chapter 5), where populations have variously gone extinct or served as sources of colonisers establishing new populations as deglaciated areas become suitable. Dealing with a habitat distribution, which over the long term is dynamic, is something to which the species as a whole has had to adapt to over the course of its historical development. Thus it is likely to have conditioned the current linkage in the species between genetics and habitat.

Where problems arise, as pointed out in Chapter 7, is when human-induced habitat change exceeds the scale and tempo of natural habitat change and the heritable capacity of an individual population, or the species as a whole, to adapt. In such cases, habitat change can generally be expected to have a negative impact on the local character and abundance of Atlantic salmon. The genetic effects leading to negative impacts need to be understood and addressed, just as much as non-genetic effects, if habitat management initiatives are to be successful in their conservation, restoration or enhancement objectives.

13.2 Genetic issues

Each species, conditioned by its genetic character, has habitat requirements distinct from every other species. Unfortunately, the monotypic view of species which is widely held, but is clearly invalid (Chapter 5), has led to the general perception that each species and its habitat needs are the same from one locality to the next. In principle at least, this is unlikely to be true for species such as the Atlantic salmon, which are highly structured into reproductively, independently evolved, locally adapted, and spatially distinct populations. Rather there can be expected to be differences, sometimes obvious but often subtle, in habitat requirements, associated with differences in their evolutionary history and their geographical location. However, the implications of adaptive population structuring for habitat management remain to be properly considered in species management, something which reflects the slow development of scientific insight into population structuring and, particularly, into local population adaptation (Ashley et al. 2003).

In general, the habitat needs of different populations within a species might be expected to be more similar to each other than they are to other species but, with respect to certain habitat parameters at least, this may not always be the case. What constitutes suitable habitat for one population will, to the extent that populations are adaptively differentiated, be unsuitable habitat for another (e.g. Donaghy and Verspoor 1997; Nemeth et al. 2003). Thus in habitat management for Atlantic salmon it is important to avoid a typological mindset which sees habitat requirements as being the same wherever one goes in the species range, and as necessarily the same for all populations.

The most obvious example in the Atlantic salmon of population-specific habitat requirements is the difference in the needs of anadromous and freshwater resident populations. The marine environment provides important habitat for the former but not the latter, something which is due to evolved genetic differences in the migratory behaviour between the two population types. Differences are also seen within these two population types. Some freshwater resident populations utilise only rivers, such as those found in the Long Beach River in Newfoundland (Fig. 13.6; Gibson et al. 1996), some only lakes (Fig. 13.7; Verspoor and Cole 1989, 2005) and others use both types of habitat, i.e. are adfluvial (Warner and Havey 1985; Kazakov 1992).

The same is also the case, though less obviously, with regard to anadromous salmon populations. For example, in most parts of the species range there is no use of lake habitat by anadromous populations for juvenile feeding, even though lakes may be present. In contrast, in many rivers on the island of Newfoundland, anadromous juveniles commonly migrate from rivers into lakes where they become resident and feed for one or more years before migrating to sea (Hutchings 1986; Ryan 1986).

Anadromous populations, if habitat is also viewed in a geographical context, are known to vary in the types of marine habitat they utilise (see Chapter 2), though the full extent of this variation remains to be established. For example, for anadromous Baltic populations, which are behaviourally restricted in their marine phase to the Baltic Sea (Box 2.2), the North Atlantic clearly does not represent suitable habitat. Yet for anadromous populations in Atlantic drainages elsewhere in Europe it clearly does while the Baltic Sea does not.

This is also seen among populations exploiting each of these marine areas, something illustrated by the apparent restriction of the marine phase of most inner Bay of Fundy populations to the Bay of Fundy and the adjacent Gulf of Maine (Box 2.1). Potentially, at least,

Fig. 13.6 The mouth of the small river at Long Beach, Cape Race on the Avalon Peninsula of the island of Newfoundland, which contains a stream-resident population of small Atlantic salmon (~ 15 cm at maturity). The shingle bank which normally blocks the mouth of the river had been breached by unusual storm conditions shortly before the picture was taken. (Photo credit: E. Verspoor.)

Fig. 13.7 Little Gull Lake, in the upper reaches of the Gander River on the island of Newfoundland, which contains a lake-resident population of small Atlantic salmon (~ 24 cm at maturity). These cohabit with juvenile parr and smolts, from a genetically distinct anadromous population which spawns in inlet and outlet streams of the lake, and uses the lake for feeding.

differences in migrations and feeding areas will also be linked to differences in other habitat conditions to which the populations are adaptively suited. For example, the Baltic Sea provides a relatively low-salinity environment compared to the Atlantic Ocean. The Baltic Sea can also be expected to be different with regard to other factors such as temperature and available prey, to which populations using the sea are likely to be to some extent heritably adjusted.

More subtle differences in the habitat requirements among salmon populations have come to light from detailed studies of requirements in relation to specific environmental variables. For example, rivers characterised by flushes of acid water may provide suitable habitat for some populations of Atlantic salmon but not for others (e.g. Donaghy and Verspoor 1997). This specific habitat need, a part of their local adaptation, will be a contributory factor in the widespread failure of transplantations of salmon to non-native rivers (e.g. Burger *et al.* 2000; Chapter 11) which, at a superficial level at least, appear to provide suitable habitat.

In addition to most populations probably having specific habitat needs, it may also be the case that the same habitat is used by different populations. However, if so, the nature and timing of use is likely to differ. Overlapping use is commonly seen with respect to anadromous populations within rivers during migratory phases of the life cycle, given most large salmon rivers are likely to contain multiple populations (Chapter 5). For example, this is supported by early genetic work on the Miramichi (Møller 1970) and more recent work on the Varzuga (Primmer *et al.* 2006). It also occurs in the marine environment; however, the full extent to which this is the case is far from clear (Chapter 2). A striking, but perhaps less general, example is in rivers with sympatric freshwater resident and anadromous populations such as in Little Gull Lake (Verspoor and Cole 1989, 2005). Little Gull Lake, in the upper part of the Gander River in Newfoundland (Fig. 13.7), provides for the habitat needs over the whole of the life cycle of the resident population but only appears to be exploited by anadromous salmon during the later juvenile phase and as a migratory route for spawners going upstream.

Population-specific habitat needs, and overlapping use of habitats by several populations, may or may not be relevant to management in a given context, depending on the habitat issue and the populations involved. For example, in rivers affected by hydroelectric dams, timings of compensation flows, needed for optimal upstream and downstream passage of smolts and adults, may vary for different populations in a system. Certainly, it is known that there are differences in the timing of smolt and adult runs in some rivers (e.g. Stewart *et al.* 2002). Another example is where a population utilises lake as well as river habitat for feeding and growth in the juvenile phase. Such populations, which occur in Newfoundland (Hutchings 1986; Ryan 1986) as well as in the Western Isles of Scotland, mean that juvenile habitat management in these situations must encompass both lakes and rivers. In other areas of the species range, where lakes are only important in the migration of smolts and adults, the impact of environmental changes to lakes will be more constrained. Thus the existence of differences in the habitat requirements of populations reinforces the need for a population-centred approach to Atlantic salmon management (Youngson *et al.* 2003).

The full extent of differences in the habitat requirements across Atlantic salmon populations remains to be established. However, at a more general level, it is clear from population genetics theory and the results of empirical studies of a range of other species that simple changes to the quantity, distribution and quality of habitat can have important effects on population structuring. Through these effects, it might be expected that in many situations there will be an impact on local species abundance and character. The precise nature of the impact will depend on the specific biology of the populations involved (e.g. anadromous or non-anadromous, lake-resident, stream-resident or adfluvial) and the distribution of the different types of habitat that they use at different life-history stages. The general types of habitat changes and their genetic effects are considered below.

13.2.1 Habitat reduction

The amount of habitat available, at one or more stages of the Atlantic salmon's life cycle, will be a factor limiting the number of individuals which can be supported in a given locality. If this limiting habitat is reduced, local abundance and the size of effected populations can be expected to decrease and, in many situations, entire populations may be lost. Reductions may lead to the loss of the populations, if the habitat remaining cannot support a population of sufficient size to cope with fluctuations in abundance caused by other biological factors.

Habitat reductions within rivers have occurred in many parts of the species range. Common causes are the construction of weirs and dams, road culverts, and localised severe chronic pollution. Entire populations may be lost when river migration of returning anadromous salmon is obstructed by dams or weirs, or by severe pollution near the mouth of river systems, as most river systems contain more than one distinct genetic population (Chapter 5). Dams and pollution in the lower reaches of rivers such as the Thames (England), Seine (France), Rhine (Germany) and Connecticut (US) (Netboy 1968) have in each case resulted in the loss of many unique genetic populations. Obstructions higher up in many other rivers, particularly in regions such as Spain, the Baltic and the north-eastern US, will have resulted in a partial or total loss of local populations. Non-anadromous populations have also been affected in situations where dams have interfered with the return migration of fish from lakes, where they feed, to natal rivers, where they spawn, e.g. Lake Ladoga in Russia and Lake Vänern in Sweden (Kazakov 1992). Such losses of entire populations reduces overall genetic diversity in the river system or region affected as well as in the species as a whole.

Where affected populations do not go extinct but are only reduced in size, genetic diversity may be lost within them due to increased inbreeding and genetic drift (see Chapter 8). Small inbred populations, and those subjected to recurring bottlenecks, are particularly at risk of losing genetic variation due to random loss or fixation of alleles. In such situations, the influence of genetic drift outweighs the effects of natural selection and adaptive genes begin to act as neutral genes (i.e. their frequencies are determined by genetic drift), further restricting the capacity of the populations to adapt (Lande 1988, Day et al. 2003). Evolutionary theory predicts that, in small populations, the main diversifying force is genetic drift (Lande 1988) and that local adaptations are favoured in large and stable populations (Adkinson 1995). For natural selection to operate at maximum efficiency, salmon populations need to be maintained above a certain minimum size, though determining what this size might be is not an easy task. Recent studies by Koskinen et al. (2002) suggest that evolutionary change can still occur in small populations over time frames of 80–120 years.

Reduced populations will have smaller, more impoverished gene pools, which may have a more limited capacity to adapt and an increased risk of extinction, i.e. they become more vulnerable to environmental change. An increasing body of research shows that genetic diversity in host populations plays an important role in buffering populations against disease epidemics, such as those associated with parasites (Altizer et al. 2003). In Atlantic salmon, this may be important with regard to the ability of populations from outside the Baltic Sea drainages, to handle infestations of the monogenean ectoparasite *Gyrodactylus salaris*, where this endemic Baltic parasite is introduced (Bakke 1991; Johnsen and Jensen 1991).

Reductions in population numbers caused by habitat loss at one life-history stage may also potentially lead to directed genetic changes in the population gene pool where there is density-dependent selection. Density-dependent selection is expected but the study of density-dependent selection presents major practical obstacles (Endler 1986) and, to date, there have been few studies which have addressed this issue, even for species more amenable to study than Atlantic salmon. Certainly it is known, for example, that growth rate is influenced by the density of conspecific individuals (Gibson 1993) as well as by genetic factors (Tave 1993), so selective changes caused at one life-history stage by habitat reductions at another life-history stage must be seen as highly likely.

Reductions in spawning habitat may also have a further impact. The types of habitat used by Atlantic salmon are in many situations similar to those used by brown trout. Thus, where spawning habitat is reduced, levels of interspecific mating may be increased dramatically. This will be particularly acute in rivers where the spawning of the species overlaps temporally, there is a high incidence of freshwater male maturation in salmon, and the sizes of mature fish of the two species are similar. Such interactions result in an increase in the incidence of hybridisation (e.g. Jansson *et al*. 1991; Jansson and Ost 1997) and wastage of gametes, which may in turn further reduce population size. The increase in hybrids may also have further impacts in some contexts. For example, recent studies suggest that brown trout–Atlantic salmon hybrids can survive infestation by the monogenean ectoparasite, *Gyrodactylus salaris* (Bazilchuk 2004), thus providing a potential reservoir for the parasite in regions of the species range where populations are not resistant to this pest (see Chapter 7).

13.2.2 Habitat fragmentation

Habitats may fragment when they are changed or reduced. This can also have genetic effects which impact local species abundance and character. In some situations, fragmentation may lead to the subdivision of an existing population into several smaller populations, or the greater isolation of existing populations, though further potential scenarios can be envisaged. In such cases, not only are the combined populations likely to be smaller than the original, but disproportionate reductions in levels of recruitment, as normal demographic processes such as those involving metapopulation dynamics are disrupted, are possible.

Some of the genetic consequences of fragmenting habitat have recently been explored by Meldgaard *et al*. (2003) in relation to the grayling, *Thymallus thymallus*, in the River Skjern, Denmark. Over the last 50 years, populations in the river have decreased and at the same time many weirs with fish ladders were built. Unfortunately, while adequate for brown trout, they were complete barriers to grayling movement. Meldgaard *et al*. compared the genetic characteristics of the grayling stocks of 60 years ago (using DNA extracted from historical scale collections) with modern fish. While no overall loss of genetic variation in the river was found, they found evidence of reduced genetic exchange among fish in different parts of the river, which increased with the number of separating weirs rather than with geographical distance. This suggested that movements of breeding adults had been restricted and that either gene flow had been reduced between existing populations, or that the previously existing single population had fragmented into multiple populations. In the latter case, given that the new populations would then be much smaller in size, these would be more likely to suffer loss of genetic variation due to genetic drift and reduced fitness due to inbreeding, as already discussed.

The increased isolation of populations may also have implications for the longer-term presence of Atlantic salmon in a river or region. Natural levels of gene flow among Atlantic salmon populations appear to be highly constrained, in situations where populations are healthy and fill available local habitat. However, where habitat opens up, it is clear that natural recolonisation occurs (Vasemägi *et al.* 2001; Doughty and Gardiner 2003). Furthermore, the straying of fish from natal rivers in anadromous salmon appears to be a function of geographic distance (e.g. Reisenbichler 1988). Therefore, the likelihood and speed of recolonisation of recovered or restored rivers or tributaries might be expected to be negatively affected as the distribution of salmon both among and within rivers becomes more fragmented. Indeed, as populations become increasingly isolated as the range of a species becomes more and more fragmented, the whole historical process of recolonisation of rivers following extinction can be expected to be disrupted (Hanski 1997).

13.2.3 Habitat expansion

Atlantic salmon habitat can also be expanded in certain situations. This may involve a restoration of habitat where weirs or dams built in rivers are removed, or an expansion into habitat when natural barriers to movement are removed. The latter may happen when a fish ladder is constructed to allow salmon to surmount a previously impassable falls, an impassable falls is leveled, or a hydroelectric development floods an area used by lake-based populations. The latter may involve the linking of habitat in what may initially have been different watersheds.

In many management initiatives, anadromous fish habitat is expanded by the removal of old weirs and dams built in rivers as part of local industrial or navigation works. Where the expanded habitat is contiguous, the consequences of such removals should be the expansion and numerical enhancement of existing downstream populations. Where appended habitat is disconnected, the removal of a barrier could lead to a new population being re-established, something which may represent the restoration of a lost element of a river's historical population diversity. In either case, the consequences of the habitat expansion can be viewed positively for a population in so far as they involve an increase in its size, reducing genetic losses and increasing population viability, or restoring historical levels of population diversity.

In situations where existing natural circumstances are changed, the consequences of habitat expansion may be less positive or even potentially negative. Where river systems contain multiple anadromous populations, attempts to increase access to new freshwater habitat may risk disrupting existing population structuring. This would be a concern, for example, in a river where distinct anadromous populations exist above and below a partially passable falls. In this situation, the removal of the falls is likely to lead to increased population mixing and, ultimately, may result in the merging of the populations. If so, outbreeding depression (see Chapter 8) would be expected from the reproductive mixing of different genetic populations, leading to an overall reduction in abundance. At the same time, if the populations involved have distinct traits such as timing of the return of their adults to rivers, these may be lost in the new hybrid population which is formed, potentially affecting fisheries exploitation, energetics, ecological interactions with other species and, thereby, overall fitness.

The same situation may exist where there are impassable falls and distinct resident and anadromous populations exist above and below the falls, as occurs in the River Namsen in Norway and in many river systems in Newfoundland and Labrador. If falls are removed or circumvented by fish passes in these situations, the result is likely to be competitive displacement and increased hybridisation above the falls with native resident populations. This will be particularly likely if the removal of the barrier is linked to supplemental stocking of anadromous fish above the falls. A very likely consequence of the removal of the barrier is the loss at least of the native resident population and its replacement by a less productive and potentially unviable hybrid population. It may also lead to changes in the character and productivity of the anadromous populations downstream, though changes in abundance may be masked by increased numbers of fish resulting from supplemental stocking.

Genetic concerns also exist where impassable falls are breached to allow access by downstream anadromous salmon even if there are no native salmon populations present in the newly accessed habitat. In most cases, in Europe, such areas will have resident populations of salmonids such as brown trout with which salmon can hybridise naturally. Work done to date suggests that such hybridisation is likely to increase where the two species are newly mixed (Verspoor and Hammar 1991). There is also likely to be increased competition for food and space between the two species during the juvenile stage which may lead to a depressed abundance of the native trout, resulting in a loss of genetic diversity. The impacts may also extend beyond those on closely related species. For example, a consequence of the introduction of fish into previously fishless lakes has been the severe depression, or even loss, of other species such as newts and cladocera, due to predation and changes in the community structure which result (e.g. Pilliod and Peterson 2001). Svärdson (1979) described the effect of introductions of species of whitefish (*Coregonus*) and consequent extinction of the *Salvelinus alpinus* complex in Scandinavia. These issues have, however, been little studied in situations where habitat for either resident or anadromous Atlantic salmon has been expanded.

13.2.4 *Habitat degradation*

Changes in habitat quality involve alterations of physical, chemical and biological character of habitat. In principle this encompasses all aspects of habitat including predators, among which humans can be included. However, in the management context, human exploitation is seen as a distinct issue and has been dealt with separately (see Chapter 10). Changes in habitat quality which exceed those seen naturally, from the point of view of native populations, represent habitat degradation. Hence they can be expected to lead to maladaptation and reduction in population abundance, as well as altered selective pressures and genotype–environment interactions. Given that habitat is seldom spatially homogeneous, in many cases habitat degradation will go hand in hand with reductions in the overall amount of habitat available to a population.

Natural selection may be expected to have moulded the genes of Atlantic salmon to produce individuals better capable of surviving under the historical environmental conditions experienced by each population. Changes in habitat quality can be expected to often lead to changes in the selective pressures acting on populations. If so, the result will be that the affected populations become relatively more poorly adapted, i.e. maladapted. Furthermore, such changes can also often be expected to affect the way the genotype interacts with the environment and alter the phenotypes arising from the genotypes generated by a population's gene pool. This will lead to changes in population character as well as reducing abundance though a lower mean fitness of individuals.

Compared to other freshwater fishes, Atlantic salmon show a relatively narrow habitat breadth, and have relatively distinct habitat requirements (Klemetsen *et al.* 2003; Chapter 2). Furthermore, based on what is known (Chapter 7), it is likely that habitat requirements for individual populations are likely to be even narrower, and the historical habitat experienced by a population should be assumed to be the one under which it will perform best. Loss of fitness and eventual extinction can be expected if the environment is allowed to go 'out of bounds', either outside the species' absolute habitat requirements, or the population's more restricted adaptive landscape (Fig. B7.6d). In general the habitat changes that may be expected to be most damaging are those that take place during the critical times for survival, depending on the relative roles of density-dependent and density-independent factors on the survival of each population (Jonsson *et al.* 1998). A well studied case of the impact of habitat changes, as well as other factors, is that of the spring chinook salmon in the south Umpqua River in Oregon (Ratner *et al.* 1997). Here the population decline is associated with temperature increases and reduced juvenile habitat caused by the removal of riparian vegetation, siltation of spawning areas due to logging-associated bank erosion, and increased removal of spawning gravel and debris due to the increased winter water flows.

Maladaptation and loss of fitness may also occur if the environment changes more quickly than a population is able to adapt to genetically, something which will depend on generation time (Chapter 7). This is true even if the magnitude of the environmental change is a relatively small one, well within the tolerance limits for the population. Examples of potentially detrimental rapid environmental change include many anthropocentric disturbances such as deforestation, impoundment and stream regulation, siltation, point-source pollution, blockage of migratory routes (Frissell 1993), or parasite introductions (Johnsen and Jensen 1991). Together these may constitute 'ecological traps' (sensu Schlaepfer *et al.* 2002).

There is little direct evidence from studies of Atlantic salmon regarding the time frames of adaptation with regard to different types of habitat change. Studies of other species suggest that with regard to some changes, genetic adjustments can sometimes be rapid i.e. within decades rather than centuries or millennia. In a study of heavy metal pollution, increased fitness was observed after only a single generation in an experimental population of tilapia, *Oreochromis niloticus* (Cuvin-Aralar and Aralar 1993). In a study of mummichogs, *Fundulus heteroclitus*, tolerance to dioxin-like compounds was found to be inherited and tolerant populations were found to be indigenous to sites with elevated sediment PCB (polychlorinated biphenyl) concentrations (Nacci *et al.* 2002). However, in the case of contamination of the marsh habitat of the darter goby (*Gobionellus boleosoma*) by toxic polyaromatic hydrocarbons (PAH), despite a long history of exposure, there was no evidence of genetic adaptation (Klerks *et al.* 1997).

13.2.5 Loss of biodiversity

There is little doubt that the habitat changes experienced by Atlantic salmon have been profound and widespread (Netboy 1968; Parrish *et al.* 1998; WWF 2001), though the full extent of this is probably not completely known. However, the actual genetic impacts on the Atlantic salmon from habitat changes, associated with dam construction and pollution, have not been documented and have seldom been considered. Yet, from the insight which has been gained into the genetics of the species over the last few decades, it is not difficult to surmise that the impacts have been wide ranging and substantive. Most easily appreciated is the demise of a large number of distinct genetic populations and, by inference, the loss of a substantial and unique component of intraspecific biodiversity. This is not widely appreciated and concern is generally focused on the losses at the taxonomic species level rather than on the loss of unique populations (Frissell 1993). Both represent losses of biodiversity that are irreversible in management time frames, and indigenous populations and intraspecific evolutionary groups are potentially incipient species.

As documented in Chapters 5 and 7, the species is highly structured into large numbers of phylogenetically and adaptively distinct populations, not only among but also within river systems. As such, where salmon have become extinct in rivers and tributaries, the extinctions will mean the loss of unique components of the species' overall genetic diversity. The scale of the loss can be appreciated by looking at habitat changes in four illustrative cases from across the species range (Box 13.1). However, the impacts of habitat are likely to extend well beyond the loss of population diversity: for example, habitat loss has also increased local range fragmentation which has the potential to disrupt normal regional metapopulation dynamics, slowing down natural recolonisation processes once river habitat is restored. The extent to which habitat change in the marine environment may be driving genetic changes and, thereby, changes in the character and abundance of runs of anadromous salmon can only be guessed. Much more research is needed before the genetic implications of habitat change will be fully appreciated. Yet what we do know means that they will be found to be extremely important.

13.2.6 Global climate change

Not all environmental changes will be rapid and obvious. Some changes, such as global warming, may be relatively gradual. Global climatic change has the potential for altering the adaptive genetic response of aquatic organisms (Carpenter *et al.* 1992), including that of salmonids. Climatic records indicate that average global temperatures have increased over recent decades in a highly anomalous trend (IPCC 2001); more generally correlated changes in seasonal weather patterns have also occurred. For Europe and the North Atlantic, the North Atlantic Oscillation (NAO) is regarded as a proxy for changes in climatic processes affecting both freshwater and marine environments. Although some difficulty is associated with correlating recent trends, the changes probably involve absolute values for two obviously relevant climate parameters, temperature and rainfall, and measures of within- and between-year variation for both.

In the case of fresh water, evidence from the few time series that are available indicates a warming trend (Webb 1996). In the Girnock Burn, a tributary of the Aberdeenshire Dee in Scotland, average annual temperatures in the spring period, which are critical for seasonal growth (Juanes *et al.* 2000) and for smolt migration, have increased by about 2°C since the mid-1960s. These changes were attributed to reduced trends for snow pack accumulation and ablation (Langan *et al.* 2001).

Changes have also occurred in the surface features of the North Atlantic Ocean. In general, these trends have been towards warming in the eastern and cooling in the north-western part of this ocean (Dickson and Turrell 1999). In the latter case, the Labrador Sea and adjacent areas to the north are the known location for substantial numbers of sub-adult salmon, originating both from eastern North America and from Europe. Given evidence for declining trends in marine survival rates for fish of both groups (Reddin *et al.* 1999; Youngson *et al.* 2002), several studies have supported the hypothesis that ocean climate and marine performance are linked and that recent climatic effects are unfavourable for salmon (reviewed by Friedland 1998).

A recent paper by Boylan and Adams (2006) showed that, over the period 1875–2001, below a critical NAO index level, the index correlated closely with the number of adult

Box 13.1 The unrecorded loss of biodiversity: a damming legacy of unknown proportions.

Case 1

The Baltic Sea historically had natural runs of anadromous salmon in 80–120 of its rivers (Box 2.2). However, due to pollution, habitat disturbance and the damming of rivers for electricity generation, there are now wild salmon populations in only 31 (IBSFC/HELCOM 1999). Furthermore, even in many rivers which still have salmon, their distribution is much more restricted than before. The numbers of distinct genetic populations lost is unknown but is likely to number in the hundreds, given the size and complexity of the river systems and tributaries where salmon have gone extinct. The reduction in the salmon distribution in rivers draining into the Gulf of Bothnia, the northwestern arm of the Baltic, is shown in Fig. B13.1.1; similar or even more severe losses have occurred in other parts of the Baltic Sea.

Case 2

Salmon in rivers at the southern limits of the species range, in both Europe and North America, have been particularly impacted by habitat change due to human activities. On the Iberian Peninsula, the species probably inhabited all accessible and suitable rivers flowing into the North Atlantic and the Bay of Biscay, from the River Duero in Portugal to the Bidasoa on the border with France (Fig. B13.1.2). As the region was unglaciated, populations are likely to have existed in these rivers during much of the last 100 000 or more years. Certainly, salmon vertebrae, the discards of some prehistoric caveman's meal, have been found which have been dated at 40 000 years BP (Consuegra *et al.* 2002). With the advent of the industrial revolution, however, the species has been lost from many rivers and, even in those where it still remains, it is confined to a small fraction of the catchment in which it was formerly found (Fig. B13.1.2). Dams, along with pollution in the lower reaches of the rivers, have been the main cause (Netboy 1968). As in the Baltic, the loss of unique populations potentially extends into the hundreds.

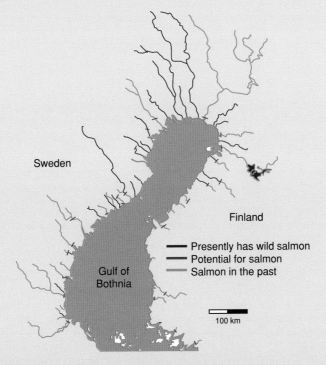

Fig. B13.1.1 Map showing the former, potential and present distribution of wild salmon in rivers draining into the Gulf of Bothnia of the Baltic Sea and the position of major hydroelectric dams. Based on information in IBSFC/HELCOM (1999).

Box 13.1 *(cont'd)*

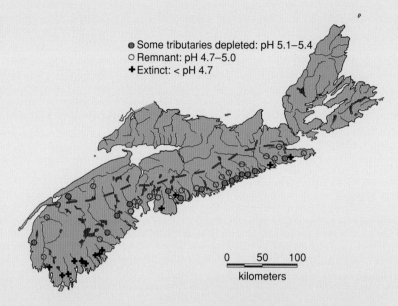

Fig. B13.1.2 Map showing the number of rivers in the Southern Uplands of Nova Scotia, Canada (dashed red line), whose salmon populations, due to problems with low pH, are either extinct, for which only remnant numbers remain, or have tributaries with depleted numbers. Based on data obtained from http://www.asf.ca/Communications/2002/mar/acidrain.html#anchor243992.

Case 3

Atlantic salmon populations in a number of parts of their range have been severely impacted by acid rain which has lowered pH levels to lethal or near-lethal levels. One of the most severely affected regions has been the southern upland region of Nova Scotia in eastern Canada. It has been suggested that it may take the best part of the current century for habitat in the area to be restored to a point where it can once again sustain healthy salmon populations (Watt and Hinks 1999). The acidification of the region's rivers has resulted in the loss of salmon from 14 rivers, the severe reduction of salmon in a further 20, and severe reductions in the tributaries of a further 16 (Fig. B13.1.3). Again the loss of genetic diversity through the extinction of unique river populations is unknown. It is likely to extend into the many tens and to have encompassed severe losses in genetic diversity within remaining remnant populations. It is also likely that the salmon in these rivers represent a unique phylogenetic group within the main Western Atlantic lineage, based on the occurrence in the region of a unique mitochondrial DNA lineage (Verspoor *et al.* 2005).

Case 4

Resident, non-anadromous populations represent a substantial component of intraspecific biodiversity and an important component of the biological community in many North Atlantic river systems. This is particularly the case in the Karelia–Lake Ladoga–Lake Onega region of northern Europe and across much of the species' North American range, though particularly in Newfoundland and Labrador (Chapter 2). In those areas where they occur, most lakes and sub-catchments are likely to contain separate, genetically distinct populations (Chapter 5). In many parts of the species range, with resident or adfluvial salmon, large areas have been flooded due to dams constructed for hydroelectric developments. One example is in central Labrador, where hundreds of lakes and streams, many likely to have contained populations of non-anadromous salmon (McCrimmon and Gots 1979; Verspoor, pers. obs.), have been joined and covered by formation of the Smallwood Reservoir (Fig. B13.1.4). This

Box 13.1 (cont'd)

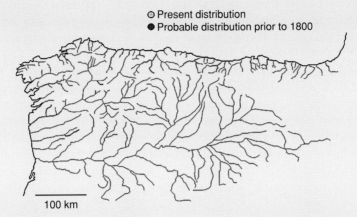

Fig. B13.1.3 Map showing the present and most probable former distribution of Atlantic salmon on the Iberian Peninsula. Based on C. García de Leániz (unpubl.).

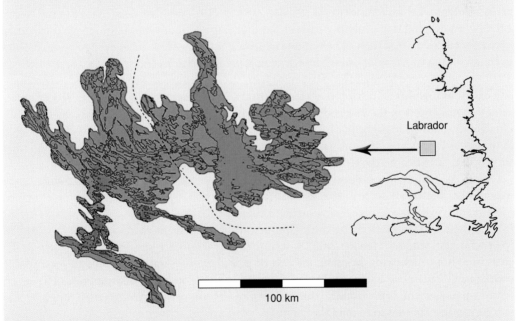

Fig. B13.1.4 Map showing the lake systems in the area now covered by the Smallwood Reservoir in central Labrador. Dotted line shows the former division between the Hamilton and Nauscaupi River drainages, whose headwater areas are now merged by formation of the reservoir.

reservoir was constructed in the 1960s above the impassable (75 m high) Churchill Falls on the Hamilton River and covers 6527 km^2, at an altitude of 472 m on the remote Labrador Plateau, near the Quebec border. The genetic impact of this development on species diversity is unknown but it is highly likely that hundreds of distinct populations have merged or been lost. Based on the analysis of mitochondrial DNA (Verspoor, unpubl.), the populations in this region also have a unique genetic history.

salmon entering the River Foyle. When the index was above this critical level, there was a higher probability that the population of adult salmon would be below its long-term mean. Furthermore, in eight of the ten years up to 2001, the index exceeded its critical level, a situation which independent climate models indicate is likely to become increasingly common, indicating that the abundance of salmon in the Foyle will decrease in the future.

Considering the thermal niche of Atlantic salmon (Jonsson et al. 2001), and given the preponderant influence of water temperature on salmonid growth and life history (e.g. McCarthy and Houlihan 1997), it is likely that a trend towards higher temperatures would certainly be accompanied by genetic changes at a variety of levels. In assessing what the likely impacts might be, it is important to consider that, throughout their history, Atlantic salmon have experienced a continually shifting distribution of suitable habitat due to landscape and climate changes linked to the advance and retreat of glacial ice sheets. For example, much of the modern species range was uninhabitable due to ice at the time of the last glacial maximum (LGM) ~ 18 000 years BP (see Fig. 5.12). Since this time, there has been an overall northward expansion and shift in the distribution of habitat suitable for the species as a whole.

The climatic fluctuations and general change since the LGM have had impacts on habitat which, in turn, have had clear genetic consequences, for example, the development of resident populations. During the Younger Dryas glacial period ~ 11 000 years ago, there was a long period of depressed sea temperatures (Wilson et al. 2000). This historical change in the North Atlantic marine habitat probably caused 'landlocking' of many populations established in newly deglaciated areas such as Newfoundland and Norway. This is believed to have resulted in the evolution of many of today's resident salmon populations (Berg 1985), some of which do not go to sea even though they now have free access (Verspoor and Cole 1989). The capacity exists, at least for some of today's populations, to undergo such changes. In recent times, salmon populations in the rivers of Ungava Bay have periodically lost access to marine habitat. This has been due to the occasional failure of water temperatures in the Bay, lethally cold in the winter, to warm sufficiently in the summer to allow fish to migrate to the North Atlantic for feeding, or back into the rivers for spawning (Power 1969).

One of the main consequences of global climatic changes will be a northward shift in the overall distribution of species such as the Atlantic salmon. Species establish in new regions more readily than they evolve a new range of climate tolerances (Davis and Shaw 2001). This is supported by observations of contemporary situations (Root et al. 2003) and historical (paleoclimatic) studies (Hewitt 2000). Given that global climate has fluctuated widely in the past 3 million years, with dramatic periods of global cooling and warming, an inescapable consequence for most living organisms is great changes in their distributions. However, within management time frames, the genetically determined environmental range of the species can be expected to remain the same.

As part of the climate response, within rivers in the more southerly parts of the range, populations may be lost from warmer tributaries; in more northern parts, tributaries previously too cold, such as those fed by glacial melt waters, may be colonised and new populations established. Within this overall response, there are likely to be genetic effects on the distribution of variation within and among its constituent populations. The result of the range shift will be that total genetic diversity within species will be reduced for a number of reasons. One will be due to the loss of older, established and more diverse southern populations; these may be replaced by new populations in the north but, established by limited numbers of founders, they can be expected to be less variable. Populations at the warming southern edge of the species' distribution will have no more capacity to adapt with no possibility of adaptive gene flow from populations adapted to warmer conditions. Examples of persistence through repeated periods of unfavourable climate are documented in the fossil record but the record of extirpations (Davis and Shaw 2001) suggests that limits to adaptation are greatest during periods of rapid change, such as predicted for the future. Thus populations at the warmer margins will be lost while the species range expands at the colder end of the existing distribution.

At the centre of a species' distribution, most local populations should be able to compensate for the climate changes they experience through a selective adjustment of their genetic character. Populations contain different genetic types, whose proportions are determined by gene frequencies, gene linkage and selection processes. In each population, this distribution can be expected to be optimised for the mean recent historical environmental conditions encountered, though this may be perturbed if there is gene flow. In so far as the variation is adaptive, this genetic distribution will be selectively adjusted toward some new, more optimal, distribution as climate change occurs. However, where climate change exceeds the rate at which populations are able to adjust, maladaptation will occur, in part because the basis of the genotype–environment interactions is likely to change. For any given genotype, there will be a range of potential phenotypic outcomes, which influence fitness, and which depend on environmental conditions. The effects of temperature may not only operate on the differential survival or reproductive success of individuals but may also influence the success of different genetic types of gametes at the time of fertilisation (e.g. Danzmann and Ferguson 1988). Change in selective pressures may lead to changes in the distribution of adaptive genetic variation such as that at the *MEP-2** protein locus (e.g. Verspoor and Jordan 1989).

There are further qualifications on the selective response. If adaptation depends on change in multiple traits, as seems likely, genetic interdependence among the traits may retard the evolutionary response (Rodriguez-Trelles and Rodriguez 1998) and rapid adaptive responses may have very large genetic and fitness costs. Furthermore, if climate change is too rapid then the adjustment may not be quick enough and significant depression of population abundance may still occur. This could lead to extinctions where populations are already depressed or small, though their habitat may be recolonised in the longer term if they are part of a regional metapopulation complex.

Atlantic salmon abundance, as judged from catch records, has varied markedly in the historical past (Summers 1993; Lajus *et al.* 2001; Youngson *et al.* 2002). Over recent decades, marine mortality appears to have affected population components differentially (Youngson *et al.* 2002) and selection may therefore have been involved. In the case of sockeye salmon, inferential studies have linked climatic variation with major fluctuations in abundance (Finney *et al.* 2002), suggesting that increased and relaxed selection may alternate over long periods.

At the northern edge of the distribution, migration is expected to gradually lead to the introduction of adaptive genetic variation from central parts of the distribution needed to adapt to warmer conditions, which will allow these populations to respond to the new conditions. Again, however, this may not occur sufficiently quickly where climate change is rapid relative to the dispersal capacity of the species in question. If population abundance is depressed by overexploitation or habitat degradation the ability to adjust may be compromised (Gomulkiewicz and Holt 1995). Furthermore, in previously isolated populations increased gene flow may initially give rise to outbreeding depression (Kirkpatrick and Barton 1997) prior to there being a fitness increase due to natural selection on the new spectrum of genetic variation. Genetic diversity may also be lost because genetically unique, isolated populations in the north of the species' distribution are swamped by new gene flow and integrated into the advancing centre of the distribution.

General reductions in abundance are expected across the range, due to increased maladaptation, increasing losses of genetic variability due to drift, which will only very slowly be restored by mutational and other evolutionary processes. However, if other anthropogenic factors have already reduced abundance and increased fragmentation, colonisation processes may be hindered. This will compound the other causes of biodiversity reductions.

Each population will respond differently depending on its specific genetic character and its location within the overall species range. For example, smaller populations with less variation may respond less successfully than larger populations which are more variable. Lost biodiversity will only gradually be restored as new fringe populations at the new northern limit are established and evolutionary processes such as mutation and recombination generate new variation (Chapter 3). Some fisheries should expand at the species' northern distributional limit with the colonisation of emerging suitable habitats. However, on the whole, a decline in fisheries abundance is likely, as long a climate change is in progress and in the short to medium term after it stops.

To understand the impact on biodiversity requires understanding not only of the genetics and demography of small populations but also their ecology and evolutionary biology. Lande and Shannon (1996) show that in constant or predictable environments genetic variance for fitness traits reduces population mean fitness, increasing the risk of extinction. However, in unpredictable, variable environments genetic variance appears to be essential for adaptive evolution and population persistence.

That the current rate of climatic change is likely to have many of these impacts seems likely. Based on current climate change projections for the twenty-first century, range shifts at rates of 300–500 km per century will be needed to keep progress with environmental shifts (Davis and Shaw 2001; also see Parmesan and Yohe 2003). Yet commonly observed range shifts in the recent past have been in the order of 20–40 km per century and even the most exceptional migration rates documented in the fossil record are less than 200 km per century.

13.3 Summary and conclusions

- Habitat, environmental space where an organism does or could live for part or all of its life cycle, is a simple concept but difficult to qualify and quantify in practice.
- The freshwater habitat requirements of Atlantic salmon compared to other species are reasonably well understood; relatively little is known about its marine habitat requirements and the full extent of differences in the habitat requirements across Atlantic salmon populations remain to be established.
- As each species, conditioned by its genetic character, has habitat requirements distinct from every other species, so does each distinct genetic population, to a greater or lesser degree; what constitutes suitable habitat for one population will, to the extent that populations are adaptively differentiated, be less suitable habitat for another, reinforcing the need for a population-centred approach to Atlantic salmon management.
- Freshwater habitat destruction and degradation have been one of the main factors underlying historical declines and extinctions of Atlantic salmon across the species range and in

many cases can be expected to have genetic consequences which need to be understood to limit or avoid detriment to local species abundance and character.
- Natural environmental change, to some degree, is an essential condition for a given physical space to provide a suitable habitat; problems arise when human-induced habitat change exceeds the scale and tempo of natural habitat change, and exceeds the heritable capacity of an individual population, or the species as a whole, to adapt.
- The actual genetic impacts of habitat changes have not been studied in Atlantic salmon but population genetics theory and studies of a range of other species show that reductions in the amount of habitat available to a population can reduce population abundance and genetic diversity; changes in habitat distribution can lead to changes in gene flow and population structure; linking habitats can lead to populations merging, and expanding habitat to hybridisation and introgression; habitat degradation will result in a loss of local adaptation, change selective pressures and gene frequencies, reduce population size and increase the loss of gene diversity by genetic drift.
- The trend towards higher temperatures in respect to global climatic change will be accompanied by genetic changes at a variety of levels, including the loss of southern populations and the expansion of northern populations, with an overall decrease in intraspecific biodiversity.

13.4 Management recommendations

- Implement population-centred management practices which are sensitive to the potentially different habitat needs of populations.
- Maintain and protect the natural habitat conditions of populations, including both natural habitat heterogeneity and environmental fluctuations.
- Operate a presumption against the enhancement of natural historical conditions and only allow such where there is a high probability of a positive outcome.
- Avoid reducing habitat area, fragmenting habitat, removing natural barriers between existing habitats, or degrading habitat quality.
- Historical habitat conditions should be restored where habitats have changed and numbers remain depressed, indicating populations have been unable to adapt to the change.
- In circumstances where it is clear that changes have been in place long enough to allow adaptive genetic adjustments, full or rapid habitat restoration may be counterproductive; whether such situations exist is unclear but they are likely to be uncommon.
- Existing habitat changes need not be reversed where the abundance of populations remains unaffected or has recovered to historical levels.
- Management action to restore normal historical conditions is reasonable where natural catastrophic events have perturbed habitat; catastrophic changes will in most cases lie outside a population's evolved capacity to cope.

Acknowledgements

We are grateful to Alan Youngson for reading and commenting on an early draft of the manuscript.

Further reading

Lackey, Robert T. (2002) Salmon recovery: learning from successes and failures. *Northwest Science*, **76**: 356–360.

Young, A.G. and Clarke, G.M. (Ed.) (2000) *Genetics, Demography and Viability of Fragmented Populations, Conservation Biology Series*, No. 4. Cambridge University Press, Cambridge.

References

Adkison, M.D. (1995) Population differentiation in Pacific salmon: local adaptation, genetic drift, or the environment? *Canadian Journal of Fisheries and Aquatic Sciences*, **52**: 2762–2777.

Altizer, S., Harvell, C.D. and Friedle, E. (2003) Rapid evolutionary dynamics and disease threats to biodiversity. *Trends in Ecology Evolution*, **18**: 589–596.

Ashley, M.V., Willson, M.F., Pergams, O.R.W., O'Dowd, D.J., Gende, S.M. and Brown, J.S. (2003) Evolutionary enlightened management. *Biological Conservation*, **111**: 115–123.

Bakke, T. (1991) A review of the inter- and intraspecific variability in salmonid hosts to laboratory infections with *Gyrodactylus salaris* Malmberg. *Aquaculture*, **98**: 303–310.

Bazilchuk, N. (2004) Salmon crisis in Norway. *New Scientist*, **181**: 14.

Berg, O.K. (1985) The formation of non-anadromous populations of Atlantic salmon, *Salmo salar* L., in Europe. *Journal of Fish Biology*, **27**: 805–811.

Bernatchez, L. and Wilson, C.C. (1998) Comparative phylogeography of Nearctic and Palearctic fishes. *Molecular Ecology*, **7**: 431–452.

Boylan, P. and Adams, C.E. (2006) The influence of broad-scale climatic phenomena on long-term trends in Atlantic salmon population size: an example from the River Foyle, Ireland. *Journal of Fish Biology*, **68**: 276–283.

Burger, C.V., Scribner, K.T., Spearman, W.J., Swanton, C.O. and Campton, D.E. (2000) Genetic contribution of three introduced life history forms of sockeye salmon to colonization of Frazer Lake, Alaska. *Canadian Journal of Fisheries and Aquatic Sciences*, **57**: 2096–2111.

Carpenter, S.R., Fisher, S.G., Grimm, N.B. and Kitchell, J.F. (1992) Global change and freshwater ecosystems. *Annual Review of Ecology and Systematics*, **23**: 119–140.

Consuegra, S., García de Leániz, C., Serdio, A., Gonzalez Morales, M., Straus, L.G., Knox, D. and Verspoor, E. (2002) Mitochondrial DNA variation in Pleistocene and Modern Atlantic salmon from the Iberian Glacial refugium. *Molecular Ecology*, **11**: 2037–2048.

Cuvin-Aralar, M.L. and Aralar, E.V. (1993) Effects of long-term exposure to a mixture of cadmium, zinc, and inorganic mercury on two strains of Tilapia *Oreochromis niloticus* (L.). *Bulletin of Environmental Toxicology*, **50**: 891–897.

Danzmann, R.G. and Ferguson, M.M. (1988) Temperature-dependent genotypic selection and embryonic survival of rainbow trout. *Biochemical Genetics*, **26**: 69–81.

Davis, M.B. and Shaw, R.G. (2001) Range shifts and adaptive responses to Quaternary climate change. *Science*, **292**: 673–678.

Day, S.B., Bryant, E.H. and Meffert, L.M. (2003) The influence of variable rates of inbreeding on fitness, environmental responsiveness, and evolutionary potential. *Evolution*, **57**: 1314–1324.

Dickson, R.R. and Turrell, W.R. (1999) The NAO: the dominant atmospheric process affecting oceanic variability in home, middle and distant waters of European Atlantic salmon. In: D. Mills (Ed.) *The Ocean Life of Atlantic Salmon*s, pp. 92–115. Fishing News Books, Oxford.

Donaghy, M.J. and Verspoor, E. (1997) Egg survival and timing of hatch in two Scottish Atlantic salmon stocks. *Journal of Fish Biology*, **51**: 211–214.

Doughty, R. and Gardiner, R. (2003) The return of salmon to cleaner rivers: a Scottish perspective. In: D. Mills (Ed.) *Salmon at the Edge*, pp. 175–185. Blackwell Science, Oxford.

Endler, J.A. (1986) *Natural Selection in the Wild*. Princeton University Press, Princeton, NJ.

Finney, B.P., Gregory-Eaves, I., Douglas, M.S.V. and Smol, J.P. (2002) Fisheries productivity in the northeastern Pacific Ocean over the past 2200 years. *Nature*, **416**: 729–733.

Friedland, K.D. (1998) Ocean climate influences on critical Atlantic salmon (*Salmo salar*) life history events. *Canadian Journal of Fisheries and Aquatic Sciences*, **55** (Supplement 1): 119–130.

Frissell, C.A. (1993) Topology of extinction and endangerment of native fishes in the Pacific Northwest and California. *Conservation Biology*, **7**: 342–354.

Gibson, R.J. (1993) The Atlantic salmon in fresh water: spawning, rearing and production. *Reviews in Fish Biology and Fisheries*, **3**: 39–73.

Gibson, R.J., Williams, D.D., McGowan, C. and Davidson, W.S. (1996) The ecology of dwarf fluvial Atlantic salmon, *Salmo salar* L., cohabiting with brook trout, *Salvelinus fontinalis* (Mitchill), in southeastern Newfoundland, Canada. *Polish Archives for Hydrobiology*, **43**: 145–166.

Gomulkiewicz, R. and Holt, R.D. (1995) When does evolution by natural selection prevent extinction? *Evolution*, **49**: 201–207.

Hall, L.S., Krausman, P.R. and Morrison, M.L. (1997) The habitat concept and a plea for standard terminology. *Wildlife Society Bulletin*, **25**: 173–182.

Hanski, I. (1997) Metapopulation dynamics: from concepts and observations to predictive models. In: I.A. Hanski and M.E. Gilpin (Ed.) *Metapopulation Biology*, pp. 69–91. Academic Press, San Diego, CA.

Hewitt, G.M. (2000) The genetic legacy of the quaternary ice ages. *Nature*, **405**: 907–913.

Hicks, B.J., Beschta, R.L. and Harr, R.D. (1991) Long-term changes in stream flow following logging in western Oregon and associated fisheries implications. *Water Resources Bulletin*, **27**: 217–226.

Hutchings, J.A. (1986) Lakeward migrations of juvenile Atlantic salmon (*Salmo salar*). *Canadian Journal of Fisheries and Aquatic Sciences*, **43**: 732–741.

IBSFC/HELCOM (1999) *Baltic Salmon Rivers: Status in the late 1990s as reported by the countries of the Baltic Region*. The Swedish Environmental Protection Agency/Swedish National Fisheries Board, Stockholm/Drottingholm.

IPCC (2001) *Climate change 2001. Third Assessment Report of the Intergovernmental panel on Climate Change*. IPCC (WG I and II), Cambridge University Press, Cambridge.

Jansson, H. and Ost, T. (1997) Hybridization between Atlantic salmon (*Salmo salar*) and brown trout (*S. trutta*) in a restored section of the River Dalalven, Sweden. *Canadian Journal of Fisheries and Aquatic Sciences*, **54**: 2033–2039.

Jansson, H.I., Holmgren, K.W. and Andersson, T. (1991) High frequency of natural hybrids between Atlantic salmon, *Salmo salar* L., and brown trout, *Salmo trutta* L. in a Swedish river. *Journal of Fish Biology*, **39**: 343–348.

Johnsen, B.O. and Jensen, A.J. (1991) The Gyrodactylus story in Norway. *Aquaculture*, **98**: 289–302.

Jonsson, N., Jonsson, B. and Hansen, L.P. (1998) The relative role of density-dependent and density-independent survival in the life cycle of Atlantic salmon *Salmo salar*. *Journal of Animal Ecology*, **67**: 751–762.

Jonsson, B., Forseth, T., Jensen, A.J. and Naesje, T.F. (2001) Thermal performance of juvenile Atlantic salmon, *Salmo salar* L. *Functional Ecology*, **15**: 701–711.

Juanes, F., Letcher, B. and Gries, G. (2000) Ecology of stream fish: insights gained from individual-based approach to juvenile Atlantic salmon. *Ecology of Freshwater Fish*, **9**: 65–73.

Kazakov, R.V. (1992) Distribution of Atlantic salmon, *Salmo salar* L., in freshwater bodies of Europe. *Aquaculture and Fisheries Management*, **23**: 461–475.

Kirkpatrick, M. and Barton, N.H. (1997) Evolution of a species' range. *American Naturalist*, **150**: 1–23.

Klemetsen, A., Amundsen, P-A., Dempson, J.B., Jonsson, B., Jonsson, N., O'Connell, M.F. and Mortensen, E. (2003) Atlantic salmon *Salmo salar* L., brown trout *Salmo trutta* L. and Arctic charr *Salvelinus alpinus* (L.): a review of aspects of their life histories. *Ecology of Freshwater Fish*, **12**: 1–59.

Klerks, P.L., Leberg, P.L., Lance, R.F., McMillin, D.J. and Means, J.C. (1997) Lack of development of pollutant-resistance or genetic differentiation in darter gobies (*Gobionellus boleosoma*) inhabiting a produced-water discharge site. *Marine Environmental Research*, **44**: 377–395.

Koskinen, M.T., Haugen, T.O. and Primmer, C.R. (2002) Contemporary fisherian life-history evolution in small salmonid populations. *Nature*, **419**: 826–830.

Lackey, R.T. (2002) Salmon recovery: learning from successes and failures. *Northwest Science*, **76**: 356–360.

Lajus, J., Alekseeva, Y., Davydov, R., Dmitrieva, Z., Kraikovski, A., Lajus, D., Lapin, V., Mokievsky, V., Yurchenko, A. and Alexandrov, D. (2001) Status and potential of historical and ecological studies on Russian fisheries in the White and Barents Seas: the case of the Atlantic salmon (*Salmo salar*). *Research in Maritime History*, **21**: 67–96.

Lande, R. (1988) Genetics and demography in biological conservation. *Science*, **241**: 1455–1460.

Lande, R. and Shannon, S. (1996) The role of genetic variation in adaptation and population persistence in a changing environment. *Evolution*, **50**: 434–437.

Langan, S.J., Johnston, L., Donaghy, M.J., Youngson, A.F., Hay, D.W. and Soulsby, C. (2001) Variation in river water temperatures in an upland stream over a thirty year period. *Science of the Total Environment*, **265**: 199–211.

McCarthy, I.D. and Houlihan, D.F. (1997) The effect of temperature on protein synthesis in fish: the possible consequences for wild Atlantic salmon (*Salmo salar* L.) stocks in Europe as a result of global warming. In: C.M. Wood and G. McDonald (Ed.) *Global Warming: Implications for freshwater and marine fish*, SEB Seminar Series no. 61, pp. 51–77. Cambridge University Press, Cambridge.

McCrimmon, H.R. and Gots, B.L. (1979) World distribution of Atlantic salmon, *Salmo salar*. *Journal of the Fisheries Research Board of Canada*, **36**: 422–457.

Meldgaard, T., Nielsen, E.E. and Loeschcke, V. (2003) Fragmentation by weirs in a riverine system: a study of genetic variation in time and space among populations of European grayling (*Thymallus thymallus*) in a Danish river system. *Conservation Genetics*, **14**: 735–747.

Møller, D (1970) Transferrin polymorphism in Atlantic salmon (*Salmo salar*). *Journal of the Fisheries Research Board of Canada*, **27**: 1617–1625.

Nacci, D.E., Champlin, D., Coiro, L., McKinney, R. and Jayaraman, S. (2002) Predicting the occurrence of genetic adaptation to dioxinlike compounds in populations of the estuarine fish *Fundulus heteroclitus*. *Environmental Toxicology and Chemistry*, **21**: 1525–1532.

Nemeth, M.J., Krueger, C.C. and Josephson, D.C. (2003) Rheotactic response of two strains of juvenile landlocked Atlantic salmon: implications for population restoration. *Transactions of the American Fisheries Society*, **132**: 904–912.

Netboy, A. (1968) *The Atlantic Salmon: A vanishing species?* Faber & Faber, London.

Parmesan, C. and Yohe, G. (2003) A globally coherent fingerprint of climate change impacts across natural systems. *Nature*, **421**: 37–42.

Parrish, D.L., Behnke, R.J., Gephard, S.R., McCormick, S.D. and Reeves, G.H. (1998) Why aren't there more Atlantic salmon (*Salmo salar*)? *Canadian Journal of Fisheries and Aquatic Sciences*, **55** (Supplement 1): 281–287.

Pilliod, D.S. and Peterson, C.R. (2001) Local and landscape effects of introduced trout on amphibians in historically fishless watersheds. *Ecosystems*, **4**: 322–333.

Power, G. (1969) The salmon of Ungava Bay. *Arctic Institute of North America Technical Paper*, **22**: 1–73.

Primmer, C.R., Veselov, A.J., Zubchenko, A., Poututkin, A., Bakhmet, I. and Koskinen, M.T. (2006) Isolation by distance within a river system: genetic population structuring of Atlantic salmon, *Salmo salar*, in tributaries of the Varzuga River in northwest Russia. *Molecular Ecology*, **15**: 653–666.

Ratner, S., Lande, R. and Roper, B.B. (1997) Population viability analysis of spring chinook salmon in the south Umpqua River, Oregon. *Conservation Biology*, **11**: 879–889.

Reddin, D.G., Helbig, J., Thomas, A., Whitehouse, B.G. and Friedland, K.D. (1999) Survival of Atlantic salmon (*Salmo salar* L.) related to marine climate. In: D. Mills (Ed.) *The Ocean Life of Atlantic Salmon*, pp. 88–91. Fishing News Books, Oxford.

Reisenbichler, R.R. (1988) Relation between distance transferred from natal stream and recovery rate for hatchery coho salmon. *North American Journal of Fisheries Management*, **8**: 172–174.

Rodriguez-Trelles, F. and Rodriguez, M.A. (1998) Rapid micro-evolution and loss of chromosomal diversity in Drosophila in response to climate warming. *Evolutionary Ecology*, **12**: 829–838.

Root, T.L., Price, J.T., Hall, K.R., Schneider, S.H., Rosenzweig, C. and Pounds, J.A. (2003) Fingerprints of global warming on wild animals and plants. *Nature*, **421**: 57–60.

Ryan, P.M. (1986) Lake use by wild anadromous Atlantic salmon, *Salmo salar*, as an index of subsequent adult abundance. *Canadian Journal of Fisheries and Aquatic Sciences*, **43**: 2–11.

Schlaepfer, M.A., Runge, M.C. and Sherman, P.W. (2002) Ecological and evolutionary traps. *Trends in Ecology and Evolution*, **17**: 474–480.

Stewart, D.C., Smith, G.W. and Youngson, A.F. (2002) Tributary-specific variation in timing of return of adult Atlantic salmon (*Salmo salar*) to fresh water has a genetic component. *Canadian Journal of Fisheries and Aquatic Sciences*, **59**: 276–281.

Summers, D.W. (1993) Scottish salmon: the relevance of studies of historical catch data. In: T.C. Smout (Ed.) *Scotland Since Prehistory*, pp. 98–112. Scottish Cultural Press, Aberdeen, UK.

Svärdson, G. (1979) Speciation of Scandinavian Coregonus. *Report of the Institute of Freshwater Research*, Drottningholm, **57**.

Tave, D. (1993) *Genetics for Fish Hatchery Managers* (2nd edn). Van Nostrand Reinhold, New York.

Vasemägi, A., Gross, R., Paaver, T., Kangur, M., Nilsson, J. and Eriksson, L.O. (2001) Identification of the origin of an Atlantic salmon (*Salmo salar*) population in a recently colonized river in the Baltic Sea. *Molecular Ecology*, **10**: 2877–2882.

Verspoor, E. and Cole, L.C. (1989) Genetically distinct sympatric populations of resident and anadromous Atlantic salmon *Salmo salar*. *Canadian Journal of Zoology*, **67**: 1453–1461.

Verspoor, E. and Cole, L.C. (2005) Genetic evidence for lacustrine spawning of the non-anadromous Atlantic salmon *Salmo salar* L. population of Little Gull Lake, Newfoundland. *Journal of Fish Biology*, **67** (Supplement A): 200–205.

Verspoor, E. and Hammar, J. (1991) Introgressive hybridization in fishes: the biochemical evidence. *Journal of Fish Biology*, **39** (Supplement A): 309–334.

Verspoor, E. and Jordan, W.C. (1989) Genetic variation at the Me-2 locus in the Atlantic salmon within and between rivers: evidence for its selective maintenance. *Journal of Fish Biology*, **35** (Supplement A): 205–213.

Verspoor, E., O'Sullivan, M., Arnold, A.M., Knox, D., Curry, A., Lacroix, G. and Amiro, P. (2005) The nature and distribution of genetic variation at the mitochondrial ND1 gene of the Atlantic salmon (*Salmo salar* L.) within and among rivers associated with the Bay of Fundy and the Southern Uplands of Nova Scotia. *FRS Research Services Internal Report* no. 18/05.

Warner, K. and Havey, K.A. (1985) *Life history, ecology and management of Maine landlocked salmon (Salmo salar)*. Maine Department of Inland Fisheries and Wildlife, ME.

Watt, W. and Hinks, L. (1999) Acid rain devastation. *Atlantic Salmon Journal*, **48**: 33–38.

Webb, B.W. (1996) Trends in stream and river temperature. *Hydrological Processes*, **10**: 205–226.

Wilson, R.C.L., Drury, S.A. and Chapman, J.L. (2000) *The Great Ice Age: Climate change and life*. Routledge, London.

WWF (2001) *The Status of Wild Atlantic Salmon: A river by river assessment.* Available at http://www.panda.org/news_facts/publications/general/index.cfm)

Youngson, A.F., MacLean, J.C. and Fryer, R.J. (2002) Rod catch trends for early-running MSW salmon in Scottish rivers: divergence among stock components. *ICES Journal of Marine Science*, **59**: 836–849.

Youngson, A.F., Jordan, W.C., Verspoor, E., Cross, T.F. and Ferguson, A. (2003) Management of salmonid fisheries in the British Isles: towards a practical approach based on population genetics. *Fisheries Research*, **62**: 193–209.

14 Live Gene Banking of Endangered Populations of Atlantic Salmon

P. O'Reilly and R. Doyle

The Norwegian Atlantic Salmon Living Gene Bank at Haukvik in south central Norway, operated by the Directorate for Nature Management. Clockwise from top left: aerial photo of facility, egg incubation unit, interior rearing tanks, growing family of juveniles, maturing adults, and external tanks for keeping mature, repeat spawners. (Photos credit: aerial view A. Haukvik; others E. Verspoor.)

The window of opportunity for conserving the remaining native wild anadromous Atlantic salmon, *Salmo salar*, in many watersheds in the south of the species range, where runs have been either extirpated or reduced to very small numbers, is rapidly closing. Elevated levels of marine mortality are accelerating the decline of many river runs, compounding previous impacts of acid precipitation, river obstructions and other human disturbances. The geographic isolation of many remaining viable populations makes the recolonisation of extirpated rivers via natural strays, when conditions improve, unlikely in any reasonable time frame, and there is no certainty that it will be possible to restore populations using salmon from remaining distant populations. Marine survival of stocked fish may decrease with increasing coastal distance between the recipient and source rivers (Chapter 11; Reisenbichler 1988); there are likely to be heritable differences in marine migration routes among populations. Furthermore, northerly populations may be maladapted to conditions in southern rivers (Chapter 5). In some instances, the live gene banking of remaining populations, until conditions improve, would seem prudent. This chapter describes the use of live gene banks to conserve some of the last populations of salmon in the lower latitudes of their native range in North America.

14.1 Introduction

Controversy surrounds the use of captive breeding and rearing of Atlantic salmon in the recovery of numerically depressed populations, or in the restoration of extirpated wild populations. A major criticism is that such programmes may fail in their objective of contributing to the natural productivity of wild populations (reviewed in Fleming and Petersson 2001; see also Reisenbichler *et al.* 2003). Lingering habitat quality problems no doubt contribute to these observations. However, a growing body of research (Chilcote *et al.* 1986; Campton *et al.* 1991; Fleming and Gross 1993; Reisenbichler and Brown 1995) indicates that reduced fitness of hatchery stocked salmon in the wild may be another factor in the limited success of some restoration or recovery programmes. These are discouraging findings for live gene bank (LGB) programmes that, by necessity, involve multiple generations of captive rearing to shelter populations at imminent risk of local extirpation. However, there is a notable gap between some practices that have been utilised in the past, and those achievable today. Recent research highlights several factors likely to contribute to the poor success of earlier restoration efforts, and offers insight into measures that may mitigate some unwanted side effects of captive rearing (Table 14.1).

The negative effects of small population size and of captive rearing on the fitness of salmon in the wild have been discussed in Chapters 9 and 12, respectively, but have focused largely on programmes of short duration or scenarios where a large portion of the total population remains in the wild. In contrast, LGB programmes will often involve the captive rearing and breeding of all, or the vast majority, of the remaining population for an extended period of time. This presents unique challenges, and requires innovative approaches and solutions.

14.1.1 Genetic concerns associated with the long-term captive rearing of salmonids

Long term capture rearing may negatively affect the long-term persistence and adaptability of captive populations upon their return to the wild by bringing about a number of genetic changes,

Table 14.1 Examples of management practices that may be improved upon to increase the likelihood of success of salmonid captive breeding and rearing programmes.

Management practices that may contribute to poor programme success	Threat	Steps to mitigate	Reference or source of additional information*
Use of donor stock from non-local source	• Outbreeding depression • Loss of adaptation to native conditions	• Use only locally obtained broodstock	reviewed in Fleming & Petersson (2001), see also Reisenbichler (1988)
Ongoing homogenisation of salmon from different reaches/tributaries in medium or large river systems	• Loss of adaptation to environmental conditions on finer spatial scales	• Consideration of tributary-specific broodstock collection/stocking strategies	Reisenbichler et al. (2003)
Nonvolitional release of fry, and lack of conditioning to river conditions (temperature, etc.)	• Increased early mortality	• Use of egg baskets or streamside incubators • Allow gradual adjustment to stream conditions prior to release	Miller & Kapuscinski (2003)
Suboptimal release of juveniles (site, method, distribution, time of day, time of year)	• Increased early mortality • Increased late mortality • Increased competition with wild juveniles • Reduced or inappropriate imprinting	• Disperse release • Where possible, release into suitable habitat currently unoccupied by wild juveniles • Avoid releasing larger, later-stage hatchery juveniles into habitat occupied by smaller, wild individuals • Select release sites and times based on historical or recent observations of habitat preferences for similar-aged wild salmon	Miller & Kapuscinski (2003) Cuenco et al. (1993)
Physiological, behavioural and morphological conditioning to captive conditions (non-genetic changes in response to captive conditions)	• Increased mortality throughout the lifecycle • Reduced breeding success	• Use of egg baskets, stream side incubators • Early release of juveniles • Adjustment of rearing practices (feeding, current, number of individuals per volume) • Naturalisation of captive environment	Maynard et al. (1995) Maynard et al. (1996)
Intentional selection (large size, early spawners)	• Loss of adaptation to natural conditions	• Broodstock (parr, smolt or adult) should be taken so as to be representative of the population with respect to timing, size, age, sex ratio, etc. • Selection of spawners should be done at random, without intentional bias for size or other phenotypic traits (early maturity) desired by the hatchery manager	reviewed in Miller & Kapuscinski (2003)

Table 14.1 (cont'd)

Management practices that may contribute to poor programme success	Threat	Steps to mitigate	Reference or source of additional information*
Unintentional selection for captive conditions	Adaptation to captive conditions	• Use of broodstock obtained from the wild • Early release of offspring	Reisenbichler *et al.* (2003) Miller & Kapuscinski (2003) but see also Reisenbichler *et al.* (2003) Maynard *et al.* (1995)
		• Modification of management regime (feeding, spawning, etc.) • Modification of hatchery environment (Examples) • Equalisation of family size • Use of cryopreserved sperm • Limit supplementation programme to relatively few years to allow adaptation to local wild conditions	Maynard *et al.* (1995) Maynard *et al.* (1996) Woodworth *et al.* (2002) this chapter Reisenbichler *et al.* (2003)
Use of small numbers of breeders and other departures from safe broodstock management practices	Inbreeding and loss of genetic variation Change due to drift	• Use of an appropriate number of individuals recently obtained from the wild, enough to minimise inbreeding and to adequately represent the genetic composition of the original population (removing too many individuals, particularly adults, is also ill-advised, and may deplete the natural population and accelerate genetic change. Numbers recommended by various authors range from 50 to 200, depending on several factors, and should be evaluated on a case by case basis)	Kincaid (1983) Allendorf & Ryman (1987) Ryman & Laikre (1991) Hedrick *et al.* (2000)
		• Rotational/systematic line crossing • Maximising effective population size by equalising the reproductive output of spawners • Avoidance of mating between sibs • Use of cryopreserved sperm obtained from individuals prior to dramatic population declines	Kincaid (1977); Kincaid (1983) Wang (1997); Hedrick *et al.* (2000) Shaklee *et al.* (1995) Sonesson *et al.* (2002); Wedekind & Müller (2004)

Assigned matings	Reduction in the fitness of individuals in the next generation and, potentially, in the long term, due to the absence of sexual selection and related benefits	Allow some individuals to select their own mates, at least on an experimental basis Accept reproductive skews that are linked to heritable viability	Wedekind & Müller (2004)
	Uninformed management actions, failure to adopt practices that may substantially increase programme success	When multiple alternative approaches are available with different advantages and disadvantages, attempt informed, well thought-out experiments to test different approaches	Fleming & Petersson (2001)
Lack of analysis of programme efficiency in the context of (1) juvenile survival, (2) marine survival, (3) breeding success, (4) reproductive success, and (5) contribution to the spawning population, and subsequent failure to adapt programme based on results observed		Use chemical, physical or genetic markers to differentiate management practices, or to isolate treatments by region or river Quantify success using increasingly stringent criteria (contribution to wild population > fitness of released offspring > breeding success of released offspring > rate of adult return > freshwater survival > number of juveniles released)	Fleming & Petersson (2001)
Lack of monitoring of life-history characteristics in pre- and post-supplemented populations, and between supplemented and adjacent non-supplemented populations		Monitoring divergence in life-history characteristics between the original wild population and the supplemented population	Hard (1995)

* Reference and associate readings may pertain to one or more adjacent columns.

Table 14.2 Probabilities of losing an allele due to genetic drift alone for four allelic frequencies, over one, four and ten generations, given different numbers of effective breeders (N_e) – based on Tave (1993).

N_e	Probability of loss over a single generation				Probability of loss over four generations				Probability of loss over ten generations			
	Allele frequency				Allele frequency				Allele frequency			
	0.5	0.2	0.1	0.01	0.5	0.2	0.1	0.01	0.5	0.2	0.1	0.01
2	0.0625	0.4096	0.6561	0.9606	0.2275	0.8785	0.9860	0.9999	0.4755	0.9949	0.9999	1
5	0.001	0.1074	0.3487	0.9043	0.0039	0.3651	0.8200	0.9999	0.0098	0.6788	0.9863	1
10	0	0.0115	0.1216	0.8179	0	0.0453	0.4046	0.9989	0.0001	0.1095	0.7265	1
15	0	0.0012	0.0424	0.7397	0	0.0049	0.1591	0.9954	0	0.0123	0.3515	0.9999
20	0	0.0001	0.0148	0.669	0	0.0005	0.0578	0.9880	0	0.0013	0.1383	0.9999
30	0	0	0.0018	0.5471	0	0	0.0072	0.9580	0	0	0.0179	0.9997
50	0	0	0	0.3660	0	0	0.0001	0.8385	0	0	0.0003	0.9895
100	0	0	0	0.1340	0	0	0	0.4375	0	0	0	0.7627
200	0	0	0	0.0180	0	0	0	0.0699	0	0	0	0.1657
400	0	0	0	0.0003	0	0	0	0.0013	0	0	0	0.0032

N_e Effective population size.

including (1) inbreeding, (2) loss of genetic variation, (3) adaptation to captive rearing conditions and (4) the accumulation of deleterious alleles (Woodworth et al. 2002). Inbreeding can occur if crosses are conducted among relatives, or if mates are selected based on higher than average phenotypic resemblance where phenotypic resemblance is based on heritable traits (Hallerman 2003a). Inbreeding will also result if the pool of available mates is small, as a high proportion of mates will be related (Crow and Kimura 1970). It is widely understood that, in general, as the degree of relatedness of the parents increases, so do the negative effects of inbreeding on the growth, survival and reproduction of their offspring (Ryman 1970; Kincaid 1976a,b, 1983; but see also Wang et al. 2002). What needs to be appreciated is that in small isolated populations inbreeding may accumulate through time, as can the magnitude of the negative effects on the survival and growth of individuals in subsequent generations. For example, after only three generations, levels of inbreeding in a randomly mating population of 10 males and 10 females, with equal family size, will be greater than that observed in a single generation of full-sib matings!

Loss of genetic variation, in addition to being associated with the accumulation of inbreeding, may also negatively impact population viability by limiting the ability of populations to adapt to new challenges. Alleles present in an original population at a frequency of less than 10% are at high risk of being lost when the effective number of breeders is in the low tens (Table 14.2). Low-frequency alleles may be very important for the long-term persistence of populations.

Long-term adaptation to captive rearing and the resulting decrease in wild fitness is poorly understood, and likely to be very complicated. Most investigations endeavour to quantify impacts of domestication on wild fitness, and few attempt to understand the mechanisms behind the reduced survival and breeding success often reported.

The accumulation of deleterious mutations, so-called 'mutation meltdown', was proposed by Lynch et al. (1995). When the effective population size (N_e) is < 50 individuals, the tendency of drift to fix even moderately deleterious alleles increases relative to the purging ability of selection (Whitlock 2000). This differs from inbreeding depression, due to the hypothesised expression of deleterious recessive alleles across the genome (see Chapter 9), in that mutation meltdown involves the accumulation of new mutations, whereas inbreeding depression is due to expression of existing recessive mutations.

14.1.2 Impact of long-term genetic changes on captive populations

The long generation time of salmonids places limits on the feasibility of empirical studies of the long-term effects of captive rearing on population viability in the wild. However, such studies have been conducted on other smaller-sized, short-lived species, with important take-home messages for Atlantic salmon LGB programmes. The effects of the accumulation of inbreeding can be seen in the study of Woodworth *et al.* (2002) of small, captively managed populations of fruit fly, *Drosophila* (Fig. 14.1). In this experiment, replicated populations of 25, 50, 100, 250 and 500 individuals were maintained in captivity for up to 50 generations, and then exposed to simulated 'wild' environments (i.e. harsher environments), and reproductive fitness evaluated. The smallest populations of 25 individuals experienced the sharpest and highest rates of decline in 'wild' fitness, most likely due, according to the authors, to inbreeding depression. Surprisingly, the second highest rates of decline in 'wild' fitness were observed in the largest ($n = 500$) populations. The authors present six lines of evidence to implicate adaptation to captive conditions as the cause of the rapid genetic deterioration of the larger populations, and conclude 'genetic adaptation of endangered species in captivity may be of much greater concern than hitherto recognised'. These results, and a growing literature demonstrating reduced wild fitness of hatchery-reared salmon relative to wild salmon (see above and

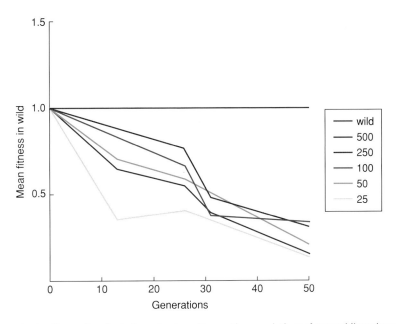

Fig. 14.1 Reproductive fitness of replicated, randomly mating captive populations of *Drosophila melanogaster* of 25, 50, 100, 250 and 500 individuals, upon exposure to simulated 'wild' conditions. A wild control population was included for comparison purposes. (Image credit: modified from Woodworth *et al.* 2002.)

Chapter 12), indicate that minimising inbreeding and adaptation to captive conditions (domestication selection) need to be key goals in salmon LGB programmes. Maintenance of genetic variation throughout the captive phase must also be an important goal, to ensure that captive populations can adapt to conditions encountered in the wild upon their release, and to future unknown changes in the natural environment.

14.2 Live gene banking of inner Bay of Fundy Atlantic salmon: a case study

The following case study, involving inner Bay of Fundy Atlantic salmon, is intended to provide an example of how one might design and implement an LGB programme for salmonid populations at imminent risk of extirpation. The Bay of Fundy is located between the Province of New Brunswick to the north-east and the Province of Nova Scotia to the southwest. The hydraulic resonance of the Bay of Fundy–Gulf of Maine system, and the bounding by the nearby and exceedingly steep continental shelf (Fig. 14.2), results in world record high tides. Strong currents associated with these high tides resuspend bottom nutrients, promoting exceptionally high biological productivity. Several larger species of fishes, whales and seabirds, in fact, migrate hundreds to thousands of kilometres to the area to feed on small invertebrates and fishes.

Wild anadromous Atlantic salmon inhabiting the drainages of the Bay of Fundy, eastward of (but not including) the St John and Annapolis rivers (Fig. 14.3), are characterised by unique life-history traits (see Chapter 2) and are commonly referred to as inner Bay of Fundy (iBoF) salmon. Whereas tagged multi-sea-winter salmon from most North American rivers are typically recovered in the Labrador Sea off Greenland (Mills 1989), iBoF salmon have, with rare exception, only been recovered from Bay of Fundy–Gulf of Maine waters (Jessop 1976). Consistent with the hypothesis of local migration is the observation that inner Bay salmon, unlike those from the outer Bay (and elsewhere along the Atlantic coast in Canada and the US), exhibit a high incidence of maturation after one winter at sea and a high incidence of repeat spawning (Ducharme 1969; Amiro 1987; Amiro and Jefferson 1996). An extensive analysis of geographic patterns of sequence variation at the mitochondrial *ND1* locus (see Chapter 5) also suggests that salmon of the iBoF, particularly those of the Minas Basin, may have had an evolutionary history distinct from that of salmon in the outer Bay and elsewhere (Verspoor *et al.* 2002). In this study, the mtDNA clade 1–3, consisting of two mtDNA haplotypes, was observed at moderate to high frequencies in multiple populations from the Minas Basin, but not elsewhere in the species' distribution.

Inner Bay of Fundy salmon may also be distinct from others in the region in experiencing particularly high rates of decline, beginning in the 1980s, that are most likely due to dramatic increases in marine mortality. Using multiple index models, Gibson and Amiro (2003) and Gibson *et al.* (2003a) estimated a 90% probability that the 5-year mean population sizes in the Stewiacke and Big Salmon rivers have declined by more than 99.8% and 94.7%, respectively, during the last 30 years. Also, juvenile Atlantic salmon were not detected in 25 of 35 non-LGB supported iBoF rivers electrofished in 2002 (Gibson *et al.* 2003b), indicating that declines are not limited to these two rivers.

In May 2001, iBoF salmon were listed as endangered by the Committee On the Status of Endangered Wildlife In Canada (COSEWIC). Under the new Canadian Species At Risk Act

Fig. 14.2 The unique bathymetry and orientation of the Bay of Fundy–Gulf of Maine contributes to tides in excess of 15 metres in portions of the Bay. (Image credit: Gulf of Maine Council on the Marine Environment.)

(SARA), a recovery team must be formed and a National Recovery Strategy document drafted for all COSEWIC-designated endangered species, identifying steps required to re-establish wild self-sustaining populations. Given the high unknown but likely marine source of mortality and very low numbers of adult returns, the iBoF recovery team, which included representatives of national and provincial governments, universities, aboriginal communities, non-government organisations and local industries, identified the use of LGBs (captive breeding and rearing programmes) to achieve the short-term objective of harbouring and protecting residual populations, for the eventual long-term goal of restoring self-sustaining populations to the inner Bay.

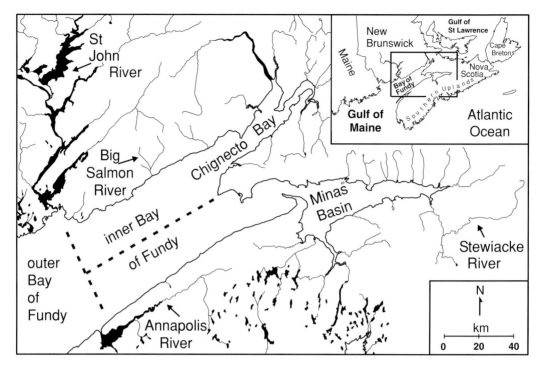

Fig. 14.3 Locations of primary inner Bay of Fundy live gene bank source populations and selected reference rivers.

14.2.1 Collection of founder broodstock

Founder broodstock were obtained from the primary LGB rivers, the Big Salmon and Stewiacke, in the spring and early summer of 1998 and subsequent years as numbers permitted (Fig. 14.4, Table 14.3). Collections consisted exclusively of wild parr captured from multiple sites by electrofishing. In the first collection year, 268 individuals were obtained from the Big Salmon River and 401 from the Stewiacke River (Table 14.4). Similar numbers of wild parr were collected by electrofishing from both rivers in 1999, 2000 and 2001. In 2002, parr abundance in the Stewiacke was such that very few individuals could be recovered.

Concerns that broodstock collected as parr may be descended from relatively few families that may not reflect the variability of the adult populations were mitigated by (1) collecting from multiple well-spaced sites, (2) collecting over multiple years and (3) assessing kinship among individuals (discussed below) prior to spawning. By collecting parr instead of adults, the danger of selecting on the basis of run timing was minimised. However, selection may have inadvertently occurred for other traits, including preference for particular sites more likely to be electrofished and late smolting; parr that persist longer in the river are more likely to be sampled than parr that smolt and leave the river earlier.

Smolt collections commenced in 2003 in the Big Salmon River and are ongoing. The advantages of collecting smolts include (1) greater geographical coverage of the river basin, as

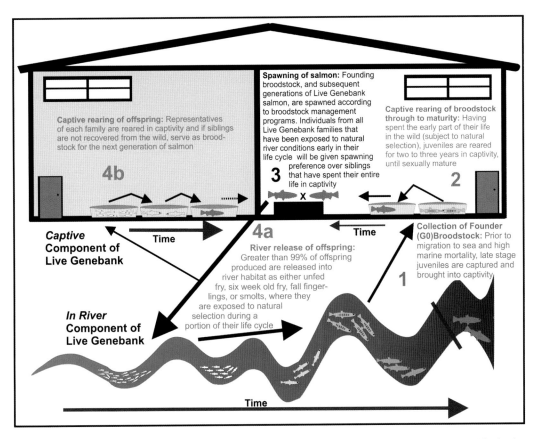

Fig. 14.4 Schematic depicting the inner Bay of Fundy live gene banking programme, including 'captive' and 'in river' components.

smolt wheels are located near the mouth of the river, and (2) greater exposure of juveniles to wild-river conditions prior to their collection for captive rearing (discussed below).

14.2.2 Captive rearing of broodstock

Parr from the Stewiacke and Big Salmon rivers captured in the autumn of 1998–2002 were reared in captivity, primarily at Mactaquac (Fig. 14.5) and Coldbrook (Fig. 14.6) biodiversity facilities. Most juveniles matured after two years of captive rearing, though some were not ready to spawn until the third or fourth year. Juveniles from the Stewiacke River were also reared at the Mersey biodiversity facility in south-west Nova Scotia; maintenance of representatives of most families at an additional site minimised risks of catastrophic failures (disease, pump failure, and so on) extirpating captive populations. Security at Coldbrook and Mactaquac sites is maintained by restricting access and through the use of intrusion alarms. Risks are further reduced by maintaining many families, as juveniles, in formerly extirpated rivers of the iBoF.

Table 14.3 Schedule of inner Bay of Fundy live gene bank operations.

1 Collection of founder (G0) broodstock	2 Captive rearing through to maturity	3 Tissue sampling, DNA fingerprinting, and pedigreeing G0 broodstock	3 Spawning of G0 founder broodstock*	4 Retention/ river release of G1 fry, G1 parr	4 River release of G1 smolt	1 Capture of wild exposed G1 fry and parr as late parr or smolt**	2 Captive rearing of wild exposed G1 parr/smolt through to maturity	Possible adult return and wild spawning of G1 released as smolts***	Tissue sampling, fingerprinting, pedigreeing captive-reared G1	3 Spawning of G1 salmon; priority given to wild-exposed individuals+	4 Retention/ river release of G2 fry, G2 parr
1998	1998–2000	2000	2000	2001	2002/**2003**	**2003**	**2003–2005**	**2003**/2004	2004	2005	2006
1999	1999–2001	2001	2001	2002	**2003**/2004	2004	2004–2006	2004/2005	2005	2006	2007
2000	2000–2002	2002	2002	**2003**	2004/2005	2005	2005–2007	2005/2006	2006	2007	2008
2001	2001–**2003**	**2003**	**2003**	2004	2005/2006	2006	2006–2008	2006/2007	2007	2008	2009
2002	2002–2004	2004	2004	2005	2006/2007	2007	2007–2009	2007/2008	2008	2009	2010
2003	**2003**–2005	2005	2005	2006	2007/2008	2008	2008–2010	2008/2009	2009	2010	2011

* Some individuals will not mature and spawn for an additional one to two years; rare founding broodstock that do spawn in year one may be spawned a second or third time, in subsequent years.

** Some fast-growing wild-exposed fry/parr may be captured the previous year as large parr, or in the two following years as slow-developing parr or smolts.

*** Some smolts may return the following year, having spent 2 years at sea (G1 sibs released as fry/parr may also return to spawn in following years).

+ Captive-exposed salmon will mature 1–2 years prior to their wild-exposed sibs. Only captive-reared individuals from very rare families will be spawned for inclusion in the live gene bank prior to the possible recovery of wild-exposed sibs. The remaining salmon will be spawned, but their offspring will not be incorporated into the LGB programme.

Table 14.4 Number and stage of collection of wild salmon (founders) recruited into the primary live gene bank (LGB) programmes. Number of crosses performed in a given year is also indicated.

Primary Live Gene Bank	Year	Number of juveniles recruited into primary iBoF LGBs	Stage collected	Possible origins of collected juveniles	Number of crosses performed
Big Salmon	1998	268	parr	wild	n/a
Big Salmon	1999	216	parr	wild	n/a
Big Salmon	2000	313	parr	wild	89
Big Salmon	2001	304	parr	wild	125
Big Salmon	2002	454	parr	wild/LGB*	129
Big Salmon	2003	n/a	smolt	wild/LGB	n/a
Stewiacke	1998	401	parr	wild	n/a
Stewiacke	1999	189	parr	wild	n/a
Stewiacke	2000	232	parr	wild	175
Stewiacke	2001	201	parr	wild	108
Stewiacke	2002	4	parr	wild/LGB*	35
Stewiacke	2003	n/a	n/a	wild/LGB	n/a

LGB* In order to maximise exposure of released juveniles to natural conditions, smaller individuals are avoided when sampling parr from the Stewiacke and Big Salmon rivers. Therefore, in the year 2002, most parr will likely be of wild origin.

Fig. 14.5 Mactaquac Biodiversity Facility, in the foreground, is situated on the St John River, 19 km upstream from the city of Fredericton, New Brunswick. (Photo credit: Department of Fisheries and Oceans, Canada.)

14.2.3 Spawning

Prior to spawning, broodstock from the Big Salmon and Stewiake rivers were DNA fingerprinted or genotyped at seven to nine tetranucleotide microsatellite loci (Box 14.1). This information was then used to group individuals into full- and half-sib families, and to place individuals into river-specific pedigrees (see Box 14.2 and Chapter 6). Full-sib families generally consisted of offspring descended from a specific male–female pair, whereas the half-sib grouping included full-sibs, half-sibs (siblings with single common female or male parent) and first cousins (Smith *et al.* 2001). Founder broodstock collections were also screened for brown trout (*Salmo trutta*) or brown trout–Atlantic salmon hybrids, and for St John River and

Fig. 14.6 Coldbrook Biodiversity Facility, as viewed from the air, is located near the town of Kentville, Nova Scotia. (Photo credit: Department of Fisheries and Oceans, Canada.)

European aquaculture ancestry by comparison of microsatellite genotypes to existing database information for these groups.

In the first spawning year (2000), the primary goal of the breeding programme was to capture as much as possible of the genetic diversity remaining in the wild populations and to minimise inbreeding in the following generation. For each LGB founding female, unacceptable founding males (potential full or half-sibs) and acceptable males (not identified as full or half-sibs) were identified. Ready-to-spawn males and females were sorted from other mature salmon and crossed with acceptable males. The first spawning occurred in late October and early November for Stewiacke and Big Salmon rivers broodstock, respectively. Spawning continued until all mature females were mated, ending in December for Stewiacke and late November for Big Salmon River salmon. In more than 95% of the crosses, single females were

---

Box 14.1 Genotyping accuracy: implications of errors and minimising mistakes.

Microsatellite genotype analyses involve multiple steps (Fig. B14.1a) and even a very low rate of error per operation will result in genotype information incorrectly assigned to a particular sample. Certain errors, including the inadvertent skipping of wells, unintentional duplication of samples into adjacent wells, inversion of strips of tubes, incorrect orientation of PCR plates, and so on, are a particular concern. Such mistakes could result in many or all genotype profiles being incorrectly matched to a particular sample, resulting in the incorrect pedigree placement of a large portion of LGB salmon *and* their subsequent offspring.

In the schematic of a typical high-throughput microsatellite analysis utilising a 96-well microplate format (Fig. B14.1b), red-coloured wells contain cross-gel standards to ensure that alleles of a particular size, but occurring in different individuals and analysed on different days, are assigned the same values. All other coloured wells represent duplicated samples (e.g. wells 1 and 87 contain tissue from the same individual, as do wells 8 and 88, and so on). Most procedural errors, from tissue sampling through to the uploading of data, that involve multiple consecutive samples of eight or more individuals will be identified using this approach.

DNA fingerprinting analyses can be carried out for as little as $US15 per sample by some commercial laboratories. While adhering to sampling recommendations of the laboratory used, the suggested checks and sample duplications should be considered. Ideally, analyses should be carried out by a laboratory without knowledge of *which* samples are duplicated, to ensure that bias is not affecting genotype assignment.

Box 14.1 *(cont'd)*

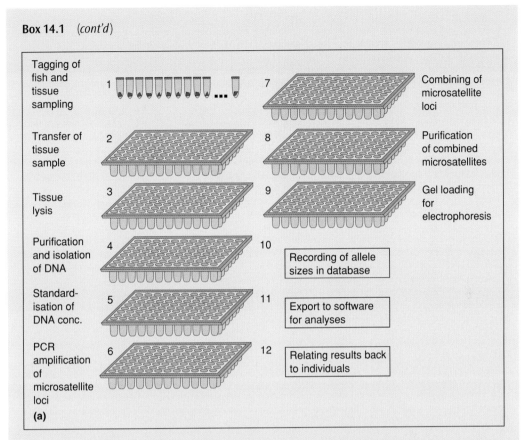

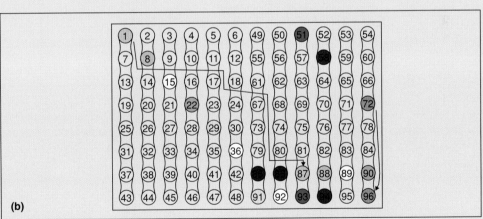

Fig. B14.1 (a) Procedural steps in laboratory analyses of microsatellites. (b) Strategic duplication of samples to minimise errors.

Box 14.2 The role of parentage and kinship analyses in the inner Bay of Fundy live gene bank programme.

Most broodstock management programmes that maximise effective population size by minimising mean kinship utilise pedigree information from as many prior generations as possible to determine average kinship among contributing individuals. Relationships in subsequent generations are tracked by tagging offspring of parents of known prescribed matings. Founding broodstock from the wild are generally assumed to be unrelated unless there is reason to suspect otherwise. In Atlantic salmon, a highly fecund species, founder individuals may be descended from relatively few parents, especially in small, declining populations. For example, genetic analyses of parr electrofished from multiple sites from the Big Salmon River demonstrate that up to 40% of the 268 juveniles sampled belonged to five Kingroups (Herbinger *et al.*, 2006). Assessment of relatedness among individuals in founding populations, especially when the option exists for collecting additional families, is very important in minimising inbreeding in subsequent generations. In the iBoF LGB programme (Fig. B14.2), ancestry or relatedness among the founding generation LGB salmon was estimated by the kinship method of Smith *et al.* (2001) (Chapter 6), which identifies siblings from microsatellite data without parental genotype information, and by visual inspection of genotype information of similar full-sib families.

Relatedness analyses in the founding broodstock are an ongoing process. Each year, additional fish are obtained from the wild, analysed in the context of wild parr recovered the previous year, and placed into their respective families. Kinship assignments are continuously re-evaluated as information from additional microsatellite loci becomes available and as new individuals are added to full-sib groupings. As fish are added, the number of clustered full-sibs grows. This means that, for an increasing number of full-sib families, all four parental alleles are known for an increasing number of loci. As such, even full-sibs that by chance inherit mostly different alleles at multiple loci will be able to be correctly assigned.

Implementing a purely pedigree-based broodstock management programme for Atlantic salmon presents additional challenges: offspring are too small to be physically tagged, and the number of crosses too numerous to keep families separate by rearing in different tanks, making it difficult to determine the ancestry of communally reared individuals. Here, family lineages are tracked by physically tagging communally reared offspring once they reach a certain size, then reconstructing parentage using microsatellite genotype analysis and parentage determination methods.

Juveniles recovered from the wild in 2002, and in subsequent years, will be compared to the founder generation parents using parentage analysis; those matching specified crosses will be placed into the appropriate tier of their respective population-specific pedigrees. Individuals not matching any sets of parents will be identified as new founders, analysed for kinship in the context of all previously collected founder broodstock according to Smith *et al.* (2001), and placed into their respective population-specific pedigrees.

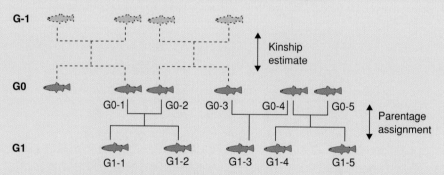

G-1 Wild salmon that spawned in natural river habitat, to produce the G0 generation of salmon

G0   Salmon captured from the wild as parr or smolt, and subsequently brought into captivity as the LGB founder broodstock

G1   The first generation of LGB salmon that were produced in captivity, that may or may not have spent some of their life cycle in the wild

Fig. B14.2 Atlantic salmon pedigrees in the inner Bay of Fundy LGB programme are estimated using both kinship and parentage assignment analyses.

Box 14.3 Minimisation of mean kinship and other broodstock management programmes.

In small captive populations, drift (see Chapters 4 and 9) will be the primary cause of loss of genetic variation (Falconer and Mackay 1996; Fernandez and Caballero 2001; Fernandez *et al.* 2001); the smaller the number of breeders, the greater the drift and the greater the likelihood of losing alleles, particularly rare ones. This is because, when producing one generation from an earlier generation, not all alleles are passed on to offspring and not all individuals successfully breed, resulting in chance allele frequency changes over time (Hallerman 2003a). Retention of genetic variation can be maximised by increasing the number of breeders, equalising sex ratios, equalising family size, and minimising variation in population size over time (Foose *et al.* 1986; Wang 1997).

With pedigree information, more efficient methods for conserving genetic variation are available (Ballou and Lacy 1995; Fernandez and Toro 1999; Caballero and Toro 2000). Individuals that are less related to others in the population are more likely to carry low-frequency alleles. By preferentially breeding such individuals (Fig. B14.3), the likelihood of losing rare or lower frequency alleles is reduced and gene diversity is maximised (Fernandez *et al.* 2001).

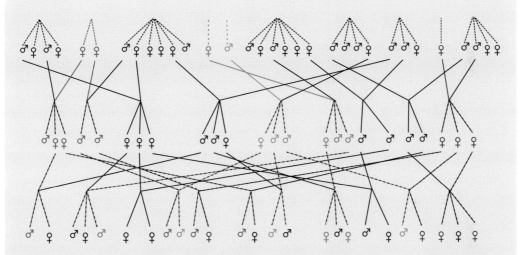

Fig. B14.3 Schematic illustrating how in the inner Bay of Fundy LGB programme individuals from uncommon families are spawned preferentially in order to minimise the loss of rare alleles.

mated with single males; because of an excess of mature females, some Stewiacke River males were spawned with two females in 2000, and some Big Salmon females with mature parr.

In subsequent years attempts were made to reduce the long-term accumulation of inbreeding and loss of genetic variation by attempting to minimise mean kinship (Ballou and Lacy 1995; Doyle *et al.* 2001). Typically, this would be accomplished by following one of several pedigree-based broodstock management programmes (Box 14.3); these methods can maintain 90% or more of the genetic variation present in the founding broodstock for over 100 years of captive propagation. Implementation of such a programme requires pedigree information and the ability to conduct prescribed pairwise matings between specific males and females.

Since broodstock were collected from the wild, extensive pedigree records were of course not available. Therefore, estimates of relatedness going back a single generation were made and used as a surrogate for pedigree data (Doyle *et al.* 2001). Pairwise matings were not, at that time, logistically possible. Hence, a modified programme that was to evolve as more information became available and as the ability to conduct pairwise matings became feasible was developed. In 2001 and 2002, maturing salmon were DNA fingerprinted, and full and half-sib relatedness assessed as above, but in the context of river-specific samples from earlier years. All eligible spawners were ranked according to (1) the number of previous successful spawnings and (2) the number of previous successful spawnings attained by all of an individual's full-sibs; a spawning was deemed successful if 85% or more of the offspring from a particular cross survived through to the grouping of families for communal rearing. In the year 2002, for example, individuals that had spawned twice previously were placed into the lowest of three bins, those that had spawned once before were placed into the middle bin, and those that had never spawned into the top bin. Each salmon was then ranked within each of the three bins according to the previous spawning success of their sibs. Individuals that had spawned twice previously were rarely spawned a third time. Whether individuals that had spawned once previously were spawned in the following year depended on how many of their sibs had contributed before. All individuals in the top bin, from family sizes of three or fewer, were spawned, regardless of the contribution of their sibs; this decision was based on (1) uncertainty in our ability to recover families early in the programme and (2) reduced statistical confidence in kinship groupings of three or fewer individuals (Smith *et al.* 2001). In instances where family size was four or greater, the first two sibs that had not mated previously were spawned. Individuals were partitioned into several spawning priority categories, depending on their spawning rank so obtained. Most crosses were made within spawning priority categories. Individuals in the lowest category were not spawned within the context of their river-specific LGB programmes, although many were mated to others in the lowest category, and their offspring released into other extirpated rivers of the iBoF.

As the next generation of LGB salmon are assigned to their respective parents, three generations of pedigree information, including the generation for which relatedness was inferred from microsatellite genotype information (Box 14.2), will become available. Also, steps are now being taken to facilitate prescribed pairwise matings at Coldbrook and Mactaquac. Within a few years, it is hoped that a true minimisation of mean kinship breeding programme will be implemented for iBoF salmon. To achieve this, the genetic value of all founders will be determined as outlined by Ballou and Lacy (1995) (Fig. 14.7). Mean kinship, or the average relatedness between a specific potential breeder and all other potential breeders, including itself, will be determined using the coancestry coefficient, f_{ij}. This is the probability that alleles drawn randomly from individuals i and j at a locus are identical by descent (Hartl and Clark 1989); the f_{ij} for individual G1-1 and itself is 0.5. To estimate the f_{ij} for G1-1 and G1-2, all common ancestors in the pedigree must be identified. In this example, the only ancestors in common are the immediate parents, G0-2 and G0-3, so the f_{ij} between these two individuals is that which one would expect for full-sibs, 0.25.

When additional common ancestors exist (e.g. the sharing of a great grandparent by G0-2 and G0-3), f_{ij} can be easily calculated from pedigree information using path analysis (see Hallerman 2003a). Next, f_{ij} is computed for G1-1 and G1-3, G1-1 and G1-4 and G1-1 and G1-5. All f_{ij} values are then summed, and divided by N. This is the mean kinship for individual G1-1. The same procedure will be repeated for all individuals in the population. Next, the males and females will be ranked, and the male and female with the lowest mean kinship values that are not related at the full or half-sib level will be selected for mating, producing a single breeder for the next generation. These two parents will be returned to the pool of

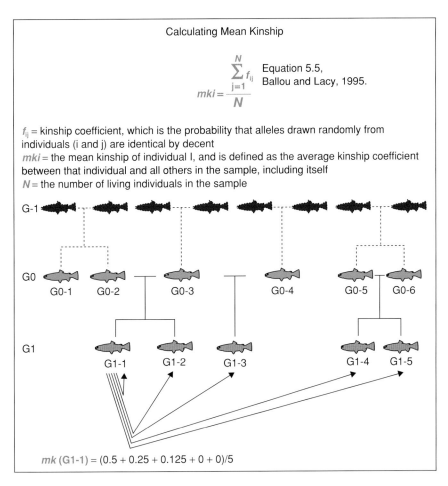

Fig. 14.7 Estimating the genetic value or mean kinship of individuals, information required to carry out breeding programmes designed to minimise global co-ancestry.

candidate breeders, and mean kinship recalculated, incorporating the new single offspring produced, and the highest ranked male and female selected once again. This could be the same male and female selected earlier, in which case the pair would be assigned to produce a second breeder for the next generation. Once the capacity of the rearing facility is met, the process is terminated. The result of the procedure is (1) a list of which males and females are to be mated, (2) designation of which female is to be mated with which male, and (3) a list of how many offspring to produce from each paired mating. Other breeding programmes have been recently developed with different advantages and disadvantages compared to the Ballou and Lacy method employed here (Box 14.4).

14.2.4 Captive rearing and river release of offspring

After spawning, salmon eggs are placed into cross-specific trays and allowed to water harden. Within 2–3 months, eyed eggs are transferred to incubators, where they hatch in several weeks.

Box 14.4 Additional potential implications of mean kinship broodstock management programmes for the fitness of salmon in the wild.

Broodstock management programmes should also be concerned with maintaining long-term wild fitness for intended future reintroduction as well as maintaining genetic variation. Conditions in captivity are relatively benign compared to the wild, allowing individuals with mild or moderately deleterious alleles to persist and breed successfully, so that deleterious alleles may accumulate over time. Furthermore, as captive populations are typically small, drift can more easily increase and fix deleterious alleles despite their selective disadvantage.

Fernandez and Caballero (2001) simulated the long-term fitness of populations of various sizes under the Ballou and Lacy (1995) method with avoidance of crosses between full- or half-sibs (Box 14.2), their own two-step method, and random mating. Under a model of many mutations of little effect, both minimisation of mean kinship approaches either increased population fitness or were similar to that expected under a random mating strategy, depending on the population size. Under a model of few mutations of large effect, fitness was maintained by all three methods as the strong selection purged all mutations. Both controlled mating approaches minimised loss of genetic variation and the accumulation of inbreeding over random mating. Thus their analysis suggests that the concern that balancing founder contributions may inadvertently preserve deleterious alleles and lower population fitness (Meffert 1999) may not be justified.

Broodstock mating programmes such as those outlined may also reduce the loss of wild fitness to domestication selection, relative to programmes where mating is random. For example, Heath *et al.* (2003) report that under conventional supportive breeding programmes egg size in chinook salmon declined dramatically in a few generations, with likely negative consequences on the wild fitness of salmon in natural river habitat due to expected lower early survival of juveniles. The authors hypothesise that change in egg size may be due to reduced natural selection for large eggs, and selection in the hatchery for high fecundity. Presumably, in benign hatchery environments small eggs have a very good chance of surviving.

When selecting the next generation of broodstock, typically from a mixed pool of offspring from multiple crosses, large families would be at numerical advantage relative to small families. Under a minimisation of kinship programme, this advantage is removed. Furthermore, these strategies may even increase frequencies of genes (e.g. genes promoting large egg size) that might be rare in conventional multigenerational captive breeding programmes where broodstock are selected at random.

Another potential concern of programmes that endeavour to minimise mean kinship is that they circumvent mate choice. Sexual selection may contribute to the production of genotypes that increase survival, viability or male attractiveness (Jennions and Petrie 1997). For example, Landry *et al.* (2001) found that Atlantic salmon choose their mates in a way which maximised MHC diversity. This could increase offspring resistance to parasites and pathogens, though Langefors *et al.* (2001) found that MHC class II heterozygotes did not exhibit higher resistance than homozygotes to the proteobacterium *Aeromonas salmonicida*, the causative agent of furunculosis. Lohm *et al.* (2002), however, demonstrated experimentally that some alleles confer greater resistance than others to *A. salmonicida*. If different MHC alleles confer resistance to different pathogens, increased heterozygosity may be generally advantageous where fish are exposed to many different pathogens over the course of their life.

Loss of fitness in the next generation due to the absence of mate choice among parents, and a resulting suboptimal combination of alleles in the offspring, may not be a high priority for LGB programmes, where the primary objectives are (1) retaining genetic variation, and (2) maximising fitness in the future generation intended for release into the wild, providing survival in the meantime is not too severely impacted. Allowing mate choice over multiple generations in small captive populations risks losing more variation at the MHC and other loci than controlled mating strategies outlined above.

The circumvention of mate choice may cause other cumulative genetic changes, however, such as increases in deleterious mutations that may erode fitness. Wedekind and Müller (2004) recommend that when excessive losses of genetic variation can be avoided, strategies that utilise predictors of heritable viability be considered. Where phenotypic attributes (e.g. condition factor, breeding ornamentation, fluctuating asymmetry, and so on) are good predictors of offspring survival, 'superior' parents could be allowed to contribute more offspring to the next generation. However, given the complexities and uncertainties of captive breeding and rearing programmes, it is recommended that LGB programmes first carry out experiments to assess the benefits of mate choice, and strategies using predictors of heritable viability, on the survival and breeding success of salmon. Based on results and published findings, where appropriate, adjustments to broodstock management plans can then be made.

Less than 1% of the production of juveniles was sampled from each family and placed into each of two separate incubators, then reared until large enough to tag and tissue sample. Shortly after, all were DNA fingerprinted, assigned to their respective parents using simple compatibility analysis, and placed into population-specific pedigrees (Box 14.2). Several representatives from each family were reared in tanks as the captive component of the LGB (Fig. 14.4).

The majority (> 99%) of production was released into native Big Salmon or Stewiacke river habitat, as either unfed fry, 6-week-old fry, autumn fingerlings, or 1-year-old smolts. Unfed and 6-week-old fry were released upstream into the smaller creeks and tributaries, autumn fingerlings further downstream into larger streams, and smolts further downstream still; downstream release of older larger individuals was done to minimise negative effects (competition, cannibalism) of larger fish on smaller fish.

Little is known about the relative importance of captive rearing at different life-history stages (fertilisation to hatching, hatching to first feeding, and so on) on survival in freshwater or marine habitats. Even less is known about how important captive rearing at these different stages is on subsequent breeding success in the wild. Given that mortality rates of fry in native river habitat are highest (and considerably so) soon after emergence (Henderson and Letcher 2003) and that the greatest differential in mortality between captive and wild exposed fry is likely during this period, early captive rearing may be an important factor in the reduced wild fitness of stocked salmon often reported. Although traditional means of stocking fry may result in little or no increase in the number of returning adults, the objective of the iBoF LGB programme is not the maximum production of outward migrating smolts, but rather the maintenance of wild fitness, so long as late-stage parr or smolts from most families can be recovered from river habitat. Therefore, emphasis was placed on release of early-stage salmon, and we are encouraging experimentation with egg baskets and stream-side incubators.

14.2.5 Ongoing founder broodstock collection and recovery of wild-exposed live gene bank salmon

Starting in 2002, parr collections include offspring of LGB salmon (spawned in captivity) and, possibly, residual wild salmon spawning in river habitat; although LGB fry were released prior to the sampling of juveniles in 2001, selection of larger parr should have precluded all LGB salmon from this collection year. As discussed above, all captured juveniles are brought into captivity, tagged, and tissue sampled. Each individual will be DNA fingerprinted and its multilocus genotype compared to possible LGB parental crosses. When parent–offspring matches are found, individuals will be placed into population-specific pedigrees, under their respective parents (Box 14.2).

Individuals that do not match any of the possible LGB parents, and which are likely to have descended from native wild salmon, will be analysed using kinship reconstruction methods (without parental genotype information) in the context of wild parr collected in earlier years. These individuals will be incorporated into population-specific pedigrees as new founders, either into existing families or into new families. LGB smolts released in 2002 could return to natural river habitat to spawn in 2003. Therefore, in 2005 (the first year that their offspring would be large enough to recover in ongoing sample collections) all captured individuals will also be subjected to grandparentage analysis, as described by Letcher and King (2001); this analysis traces individuals back two generations, in this case to their LGB parents spawned in captivity.

All individuals will be placed into population-specific pedigrees as described above. In future years, when captive-spawned, wild-exposed salmon that have been recovered from the wild mature, spawning priority (within families) will be given to those individuals with maximum number of months or years of wild exposure. Only when representatives of specific families could not be recovered from the wild will siblings reared in captivity be spawned.

14.3 Conservation and management of small remnant populations of Atlantic salmon

14.3.1 Prioritising rivers for conservation measures

Often, as in the case of iBoF salmon, several remnant populations, within a previously defined conservation unit or within a given management jurisdiction, will remain for consideration for LGB measures. Given the objective of retaining the maximum genetic diversity to ensure the long-term adaptive potential of the species, how should remaining populations be prioritised for conservation efforts? A number of criteria, biological and otherwise, should be considered. Is there clear evidence that the population is at imminent risk of extirpation, or reduction to very small numbers of individuals where small population effects are likely? Captive rearing, particularly in this context where the entire population is managed in captivity, is especially risky. Have other alternative measures, such as freshwater restoration, improvement of dam passage, cessation of fishing, and so on, that might recover the population in question been considered? What is the cause of the population decline and what is the likelihood of ever recovering the native population? However, one should bear in mind that there may be other reasons to consider LGB measures than the restoration of the original native population: restoration of runs in other nearby rivers, provision of broodstock for future sea ranching programmes, and maintenance of genetic variation for future aquaculture programmes (for selective breeding purposes or to increase the likelihood of resistance to new diseases).

The effective population size, expected accumulation of inbreeding and likelihood of persistence in captivity through to release should also be considered for each candidate population, as should the probability of continuance of the population upon release to the wild due to the above genetic factors. Several studies of *Drosophila* and other insects indicate that inbreeding can markedly reduce population fitness and the likelihood of a population persisting over time compared to larger control populations, both in captivity (Armbruster *et al.* 2000; Bijlsma *et al.* 2000; Woodworth *et al.* 2002) and in the wild or in stressful environments (Saccheri *et al.* 1998; Bijlsma *et al.* 2000; Woodworth *et al.* 2002). Assuming that populations are to be managed in isolation (but see below), and all else being equal, very small bottlenecked populations should probably be prioritised below larger populations because (1) of decreased likelihood of persistence upon release to the wild and (2) they may contribute less to global species diversity (discussed below).

Information on patterns of within- and among-population genetic variation may also be useful in ranking populations for conservation priority. One approach is to assess the phylogeny or relatedness among populations, using molecular genetic markers, and then to remove one population at a time, noting how much the removal of each reduces the overall tree length (Weitzman 1992) (Fig. 14.8). In this example, removal of population B results in

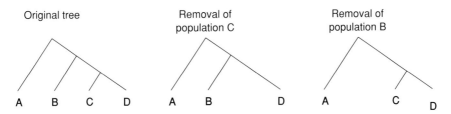

Fig. 14.8 Maximising retention of **among**-population variation by selective pruning of individual population samples. Molecular genetic markers (typically, mtDNA) are used to reconstruct the phylogeny of populations or subspecies. In this example, one of four populations is removed, and the overall tree length assessed. The population that results in the smallest reduction in the tree length (C) likely contributes least to the total genetic variance, and is prioritised below all others (Weitzman 1992).

the greatest reduction in tree length (the greatest loss of among-population diversity), and thus B is prioritised above remaining populations. This approach, however, does not consider within-population genetic diversity. Petit *et al.* (1998) and Caballero and Toro (2002) develop methods for analysing genetic diversity within and among populations, and present procedures for maximising global diversity within a given metapopulation or species, based on patterns of variation observed at neutral molecular genetic markers. Conservation efforts, though, should be concerned with both the maintenance of molecular genetic variation and genetically based phenotypic variation in quantitative traits, as both contribute to intraspecific diversity (Crandall *et al.* 2000). Since evidence for correlation between molecular variation and heritability of important fitness-related traits is at present ambiguous, Wang *et al.* (2002) recommend the maintenance of genetic and phenotypic variation within and among populations. Populations could be ranked in accordance with what they contribute to both genetic and phenotypic variation. Utter *et al.* (1993) suggest that molecular genetic analyses be used to first identify ancestral relationships among populations, and that adaptively distinct populations (based on life-history and ecological information) within these lineages would be considered for units of conservation.

Information considered in the prioritisation and management of salmon from rivers of the iBoF is given in Table 14.5. The two primary LGB rivers, the Stewiacke (Fig. 14.9) and the Big Salmon (Fig. 14.10), drain into the two geographically distinct regions of the inner Bay – the Minas Basin and the Chignecto Bay. The Stewiacke and Big Salmon populations are also representative of the two genetically distinct assemblages of populations identified by Verspoor *et al.* (2002): those of the Minas Basin, exhibiting, at high frequency, a distinct mtDNA clade not found elsewhere in the species distribution, and those of the Chignecto Bay. These two populations were also among the last to decline, and likely harboured the largest number of returning adults at (or just prior to) the time of collection of LGB founder individuals (juveniles). The increased census size, and likely increased effective population size, is reflected in the higher levels of genetic variation compared to other rivers of the inner Bay.

14.3.2 Should very small populations be combined or managed separately?

When several geographically proximate, remnant populations remain, decisions must be made to pool populations (to arrange crosses between individuals from different populations) or to keep populations separate (to arrange crosses between individuals within populations). On the

Table 14.5 Information being considered in the prioritisation and management of salmon from rivers of the inner Bay of Fundy.

Population and year of sample collection	Basin	MtDNA clade 1–3+	Primary life-history characteristics	Primary responsible agency	Standardised number of microsatellite alleles (variance)	Gene diversity
Primary live gene bank (sample year)						
Big Salmon R. (2000)	C	no	L	DFO	11.35 (197.04)	0.848
Stewiacke R. (2000)	M	yes	L	DFO	10.61 (187.86)	0.833
Additional live gene banks (under way or under consideration)						
Gaspereau R. (1999–2001)	M	yes	D	DFO	8.48 (188.38)	0.756
Great Village R. (2000)	M	yes	L	DFO	8.68 (202.35)	0.807
Economy R. (2000)	M	yes	L	DFO	6.81 (135.72)	0.738
Harrington R. (2002)	M	n/a	L	DFO	n/a	n/a
Black R. (2000)	C	no	L	DFO	n/a	n/a
Irish R. (2000)	C	no	L	DFO	n/a	n/a
Upper Salmon R. (2001, 2002)	C	n/a	L	EC	9.27 (140.45)	0.791
St John River reference population (2001)		no	D	DFO	11.67	0.850

C = Chignecto Basin.
M = Minas Basin.
MtDNA clade 1–3+ = Two related mtDNA haplotypes found at high frequency in multiple Minas Basin rivers that have not been observed outside the inner Bay of Fundy (Verspoor *et al.* 2002).
L = Local migration, high 1-sea-winter component, high incidence of multiple repeat spawning.
D = Distant migration, one-and multi-sea winter components.
DFO = Department of Fisheries and Oceans, Canada.
EC = Environment Canada.
Standardised number of alleles = Estimated by standardising to the smallest sample size observed ($N = 42$) using resampling procedures.
Gene diversity, also referred to as effective heterozygosity, was estimated according to Nei (1973).

one hand, considerable evidence exists to support the hypothesis that salmon exhibit variation in adaptive traits at various spatial scales, including among tributaries within rivers (Chapter 7, Reisenbichler *et al.* 2003). Large-scale, uncontrolled transfers of individuals between genetically differentiated populations may be a serious threat to salmon populations (Utter 2003) in potentially causing loss of local adaptation and outbreeding depression, defined as the loss of fitness due to disruption of co-adapted gene complexes (Hallerman 2003b). It should also be borne in mind, though, that drift is expected to be the more important evolutionary force when selection coefficients (s) are less than $1/N_e$ (effective population size) (Li 1978), and likely predominate in salmon populations that have been reduced to very few effective breeders for several generations. Given the ratio of the effective to observed population size commonly observed for salmonid species in the wild of approximately 0.1–0.2 (Hedrick *et al.* 1995), it is hard to imagine how populations that have remained small (fewer than 50 or so individuals) for several generations may have remained adapted to local environments. On

Fig. 14.9 The Stewiacke River, located in the Province of Nova Scotia, drains an area of approximately 619 km². Much of the gradient, however, is quite low, resulting in approximately 2 700 000 m² of suitable juvenile habitat, and an estimated carrying capacity of 237 000 juvenile salmon. (Photo credit: Department of Fisheries and Oceans, Canada.)

Fig. 14.10 The Big Salmon River, located in the Province of New Brunswick, is a much smaller and simpler system, but because gradients over much of the river are moderate, over 900 000 m² of streambed is suitable salmon habitat, capable of supporting an estimated 200 000 juveniles. (Photo credit: Department of Fisheries and Oceans, Canada.)

the other hand, the general consensus among salmon geneticists is that inbreeding is a tangible and serious threat to population fitness, although there is still debate as to the mechanisms of inbreeding depression (dominance versus overdominance; see Chapter 9).

When populations are sufficiently large, and are not at risk of loss of fitness due to small-population effects, they should be managed independently. Random mating populations consisting of 100 individuals (given equal numbers of males and females) are expected to accumulate inbreeding at 0.5% per generation (Kincaid 1983; Tave 1993), and are probably at minimal risk of short-term inbreeding depression. The more cautious minimum number of individuals, recommended by Allendorf and Ryman (1987) as 200 individuals, is consistent with the findings of Bryant *et al.* (1999) of loss of population fitness below 200 individuals. Similar low levels of inbreeding can be achieved in somewhat smaller populations by equalising family size (Wang 1997) or by implementing pedigree-based broodstock management, discussed above.

When remaining populations consist of a handful of individuals, knowledge-based adaptive management strategies should be considered. First, the risks of inbreeding can be estimated by calculating levels of accumulated inbreeding expected, given a range of conditions (skewed ratios of males to females, variation in family size, and so on; see Gall 1987). Potential implications of inbreeding at various levels on growth and survival in captivity and in the wild can be appreciated from controlled experiments involving rainbow trout, *Oncorhynchus mykiss*, maintained in captivity (Kincaid 1976a,b, 1983; Gjerde *et al*. 1983; see also the review by Wang *et al*. 2002 and Chapter 9) and Atlantic salmon in the wild (Ryman 1970). However, lineage and environment interactions have been observed in studies of inbreeding depression in other organisms, where high levels of inbreeding depression for a given level of inbreeding are observed for some families but not others and under some environments but not others (Pray *et al*. 1994; Pray and Goodnight 1995), so care must be taken not to infer too much from existing studies of inbreeding depression in salmonids. Recent studies of inbreeding and other small-population effects on population fitness and survival in other species (see above) should also be considered.

The effects of potential outbreeding depression in captivity can be monitored and evaluated by comparing relative growth and survival of potentially outbred progeny with progeny resulting from crosses performed within both parental populations. Of course, individuals from all three treatments should be reared in a common environment. Family identification may be determined using genotyping and traditional tagging methods (Box 14.2). Adjustments in mating strategies could be made based on the above observations, and on ongoing results obtained from studies of the consequences of inbreeding and outbreeding observed in other organisms; because of the shorter generation times of some of these other species evaluated, and lower maintenance costs, such studies can incorporate information from many more generations, treatments (number of individuals, with and without migration, and so on) and population replicates then would be possible for similar analyses of salmonids.

Potential loss of local adaptations might be further mitigated by keeping populations largely reproductively isolated, but allowing a very small number of migrants among populations (Woodworth *et al*. 2002). Bryant *et al*. (1999) found that one migrant per generation was sufficient, in the long run, to minimise declines in embryo viability (attributed to inbreeding depression) in captive populations of housefly, *Musca domestica*. Several recent studies involving greater prairie chickens, *Tympanuchus cupido* (Westemeier *et al*. 1998), Scandinavian grey wolves, *Canis lupis* (Vila *et al*. 2003), and adders, *Vipera berus* (Madsen *et al*. 1999), also indicate that similarly low levels of immigration can rescue declining natural populations from negative genetic effects of small population size.

Finally, it must be acknowledged that there is a risk of adopting a myopic conservation objective in which the preservation of existing adaptation is valued too highly relative to maintenance of the long-term evolutionary capability of a population. Loss of additive genetic variation in fitness is, in general, expected to reduce the rate of response to natural selection (Fisher's Fundamental Theorem). Furthermore, the loss of alleles which do not contribute substantially to additive fitness variation because of their rarity, but which do contribute to allele diversity, also represents a loss of evolutionary capacity when the environment changes. These points were made forcefully by Moritz (1999):

'I suggest that the conservation goal should be to conserve ecological and evolutionary processes, rather than to preserve specific phenotypic variants – the products of those processes. From this perspective, we should seek to conserve historically isolated, and thus independently evolving, sets of populations (i.e. evolutionary significant units, ESU). This can require manipulation of the component management units (MUs), some of which may be phenotypically distinct. . . . under some circumstances it is appropriate to mix individuals from different MUs within an ESU. These circumstances include augmentation of remnant populations that are showing signs of inbreeding depression or increased fragmentation and the use of mixed stocks for reintroductions into modified or changing environments or for introductions into novel environments. These actions are consistent with the goal of maintaining processes, but the extent to which differences in adaptation or coadaptation constrain the viability of populations subject to translocation needs further exploration.'

14.4 Use of cryopreserved sperm in the conservation of Atlantic salmon

Sperm that survive freezing to $-196°C$ (the temperature of liquid nitrogen) remain viable for an estimated 200–32 000 years (Stoss 1983). Not only do genes within individual frozen spermatazoa remain largely unchanged over this period, but so do the gene pools of cryopreserved gene banks; genetic variation is not lost from the sample and no directional changes in allele frequencies can occur. Cryopreserved sperm can thus be used to (1) gene bank wild populations for eventual restoration of natural runs, though with some limitations, (2) infuse future wild populations with additional genetic variation, and (3) minimise genetic change in LGB populations (discussed below).

14.4.1 Methods for the cryopreservation of milt

Procedures for successfully cryopreserving sperm (yielding > 80% post-thaw fertilisation success) vary among groups of fishes, even among the salmonids, and have been reviewed by Scott and Baynes (1980), Harvey (1993) and Lahnsteiner (2000). Procedures developed for Atlantic salmon, which have been used in the cryopreservation of milt from thousands of individuals (Gausen 1993), can be found in Stoss and Refstie (1983). The process involves the following steps: collecting and shipping of milt, evaluation of sperm viability, freezing of milt, long-term storage of milt, thawing of milt and post-thaw fertilisation. Salmon may be stripped in the field, and milt cooled on ice to $0-4°C$ for shipping to facilities elsewhere for cryopreservation; sperm properly stored on ice in oxygen-filled bags can remain viable for several days (Cloud and Osborne 1997).

Prior to cryopreservation, sperm viability should be tested by conducting fertilisation assays, evaluations of sperm motility, or one of several other methods discussed by Lahnsteiner (2000). Freezing of milt appears to be a critical step in the cryopreservation process – salmon milt should cool at a rate of ($20-30°C/min$) (Stoss 1983). One commonly used method of achieving a more controlled rate of cooling is to first place 50–100 μl aliquots of milt into 50–100 μl depressions made in dry ice ($-78.5°C$); pellets are then transferred to plastic cryotubes prior to immediate submersion in liquid nitrogen (Piironen 1993; Cloud and Osborne 1997; Lahnsteiner 2000). Alternatively, sperm may be frozen in individual 0.25 or

0.5 ml straws (hollow tubes tens of centimetres in length, and a few millimetres in diameter developed for sperm cryopreservation of domestic cattle), by placing into liquid nitrogen vapour for 15 min, followed by immediate submersion in liquid nitrogen (Cloud and Osborne 1997). The expense of supercooled storage space will likely limit the volume of sperm preserved from each male, particularly when one or more populations are being gene banked. Thawing of cryopreserved sperm is also critical and must be done very rapidly (Lahnsteiner 2000). Straws should be thawed in water baths at 30–40°C, and pellets in a temperature-controlled solution of 0.12 M $NaHCO_3$ (Stoss and Refstie 1983). Semen should be used to fertilise eggs as soon after thawing as possible.

The general consensus in the literature is that, while the fertilisation of a few eggs in the laboratory is certainly achievable by skilled technicians, the process is complicated and favourable results are less likely when untrained personnel are involved in the freezing and thawing of sperm. Additional advances are needed to make the process more robust and to increase fertilisation success of large numbers of eggs using small aliquots of thawed cryopreserved milt. Still, the scale of the Norwegian gene bank programme (Gausen 1993, see Box 14.5) clearly demonstrates that this technology is sufficiently developed to permit gene banking of dozens of individuals from multiple populations.

14.4.2 Use of cryopreserved milt in the restoration of wild salmon populations

Much of the remaining genetic variation in small, threatened populations can be conserved by sampling and cryopreserving sperm or milt from an adequate number of males. Genetic material stored in this way can be preserved for an almost unlimited period, but new individuals cannot, of course, be created using salmon sperm alone. Thorgaard and Cloud (1993) review two approaches to reconstituting original native populations from cryopreserved sperm. Firstly, cryopreserved sperm from an extirpated population can be used to fertilise eggs from a nearby extant population. The disadvantages of this approach are (1) considerable effort and time are required to reconstitute an approximation of the original native gene pool, (2) some introgression of genetic material from the host population will be unavoidable, (3) maternal genetic material (mtDNA and sex-linked nuclear DNA: nDNA) is lost, and (4) genetic change associated with multiple generations of captive rearing will be incurred in the production of the final generation of juveniles intended for release into wild river habitat.

The second approach discussed by Thorgaard and Cloud (1993) involves producing embryos with all paternal inheritance (androgenesis). Unfertilised eggs obtained from females from a nearby extant donor population would first be irradiated to inactivate the genetic material, and then fertilised using normal cryopreserved sperm from the original native population. Androgenic diploids consisting of DNA solely derived from the original native population would be produced by repressing the first cleavage division. Because such individuals would be homozygous at all loci (all pairs of alleles would be identical), survival and reproduction would likely be greatly reduced. Additional crosses would be necessary to restore heterozygosity and wild fitness. The disadvantages of this approach are (1) considerable effort and time are required to produce outbred populations of individuals, (2) genetic change associated with multiple generations of captive rearing will be incurred in the production of a final generation of heterozygous juveniles intended for release into wild river habitat, (3) maternal genetic material (mtDNA and sex-linked nDNA) is lost, and (4) the treatment used to block cleavage markedly reduces the survival of embryos.

Box 14.5 The Norwegian Atlantic salmon gene bank programme.

In 1988, the Advisory Committee of the Sea Ranching Programme recommended the establishment of LGBs for several stocks of Atlantic salmon in Norway (Gausen 1993; Walsø 1998). The main purpose of the programme was to create living reservoirs of genetic material for later use in re-establishing or enhancing threatened stocks. Only the most seriously threatened salmon stocks, those that were no longer capable of surviving in their native river habitat, were preserved in LGBs.

In Norway there are three LGB facilities for Atlantic salmon (Fig. B14.5.1), harbouring 28 anadromous stocks and one landlocked population. The Directorate for Nature Management operates two LGBs, while Statkraft (the Norwegian State Power Systems) operates the third. The LGBs are situated in Eidfjord (south-western Norway), Haukvik (central Norway) and Bjerka (northern Norway) (Fig. B14.5.1).

Live gene bank operations

Atlantic salmon are captured from rivers as adults, and kept in nearby tanks for a short period, until milt and eggs can be stripped. A proportion of milt is cryopreserved and transferred to a central storage facility, and fresh milt and unfertilised eggs to one of the three LGB centres (Fig. B14.5.2). Eggs from individual females are fertilised using milt from individual males, disinfected, and transferred to hatching cylinders; fertilised eggs from each male and female pair are kept separate. After hatching, each family group is reared separately until fish are approximately 7 cm in length, at which time individuals are fin clipped so as to permit identification to family, and combined into stock-specific tanks. During the second year of their life, about 30 smolts from each family group are tagged to permit individual identification. For the next two years, the different generations are kept isolated in separate rooms, thereby minimising the impact of possible fish disease outbreaks, while simplifying the tracking and administration of families, year classes and stocks. When it is time for the rebuilding of river populations, eyed eggs are transferred to local hatcheries where individuals are reared for release back into the river as fry, parr or smolts.

At all times, a great deal of effort is taken to keep fish free of disease. The health of all breeding fish is checked regularly, and strict protocols have been established for the handling and transport of fish, eggs and milt. All eggs are sterilised before being transferred from hatcheries to LGB facilities. Fresh water is used in all LGB centres, because seawater is frequently infected with *Aeromonas salmonicida* proteobacteria, the causative agent of furunculosis, as well as various other fish pathogens. To minimise the risk of spreading diseases to rivers, only disinfected eggs are transferred from LGBs to local hatcheries.

Cryopreservation of salmon milt

The freezing of salmon sperm at $-196°C$ (the temperature of liquid nitrogen) enables the preservation of genetic material for a virtually unlimited period. Milt (2 ml), enough to fertilise about 3000 eggs, is taken from each male. Emphasis is placed on collecting milt from geographically disparate locations within Norway, and from different environments. Stocks are prioritised on the basis of degree of risk of extirpation, scientific value and importance to fisheries. The milt bank now contains material from about 6500 individuals, representing 169 stocks (salmon from a single river, or from a particularly large tributary). Attempts are made to collect milt from at least 50 individuals from each stock. Milt is collected over a period of at least 3 years in order to reduce overrepresentation of siblings or closely related individuals in certain spawning runs.

The history and future of live gene bank salmon

All stocks currently maintained in Bjerka and Haukvik LGB facilities have been obtained from rivers that harbour, or which have harboured in the recent past, the pathogen *Gyrodactylus salaris*. *G. salaris* was unintentionally introduced into many Norwegian rivers in the 1970s and early 1980s through the importation of infected salmon from Baltic Sea hatcheries (Johnson *et al.* 1999). Introduction of *G. salaris* to a particular Norwegian river places the native population at imminent risk of extinction; the average short-term decline in parr densities in 14 infected Norwegian rivers has been 86% (Johnson *et al.* 1999), and there is no evidence that Norwegian salmon have developed resistance to the pathogen. For the high production Vefsna and Driva rivers, the long-term effects

Box 14.5 (cont'd)

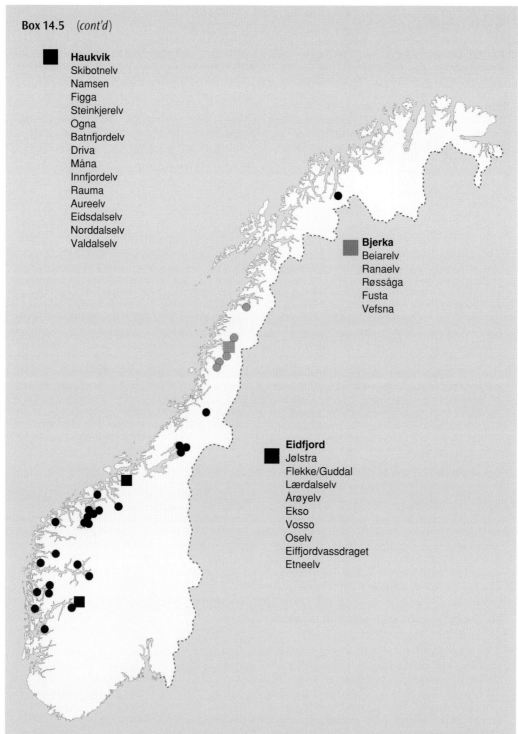

Fig. B14.5.1 Locations of threatened Norwegian stocks of Atlantic salmon (coloured circles), and the three LGBs (squares) in which they are maintained. (Image credit: Norwegian Directorate for Nature Management.)

Box 14.5 (cont'd)

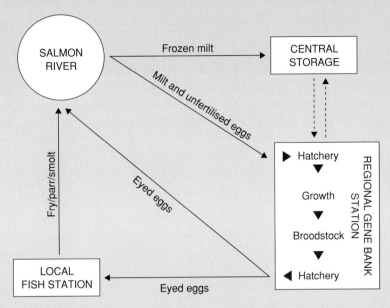

Fig. B14.5.2 Schematic of operations for the Norwegian gene bank programme. (Image credit: Norwegian Directorate for Nature Management.)

Table B14.5 History of live gene bank supported river stocks of Atlantic salmon in Norway, including principal threat(s), period of broodstock collection, and details on recent or planned restoration.

River	Threat(s)	Period of broodstock collection	Restoration of stocks		Comments
			Period	Life stage used	
Beiarelva	GS	1991–94	1998–2001	Eyed eggs	Chemical treatment 1994. Removed from LGB in 2001
Ranaelva	GS	1992–95, 2001–02	2005–10	Smolt, parr, eyed eggs	Chemical treatment/barrier 2003/2004
Røssåga	GS	1992, 1994, 2002–03	2005–10	Smolt, parr, eyed eggs	Chemical treatment/barrier 2003/2004
Fusta	GS	1994–96	2010–15	Unknown	Chemical treatment/barrier 2008/2009
Vefsna	GS	1994–96, 1998	2010–15	Unknown	Chemical treatment/barrier 2008/2009
Skibotnelva	GS	1995, 1998, 2001, 2004	Unknown	Unknown	Chemical treatment/barrier – time not decided
Namsen	GS	1990, 1995	Unknown	Unknown	Landlocked salmon. Stock in LGB for safety reasons
Figga	GS	1990–92	2003–12	Fry, eyed eggs	Chemical treatment/barrier 2002. Fry downstream barrier 2007/eyed eggs upstream 2008–12
Steinkjerelva	GS	1989–90, 1992, 1997–98	2003–08	Parr, eyed eggs	Chemical treatment 2002

Box 14.5 (cont'd)

Table B14.5 (cont'd)

River	Threat(s)	Period of broodstock collection	Restoration of stocks Period	Restoration of stocks Life stage used	Comments
Ogna	GS	1989–92, 1997, 1999	2003–08	Parr, eyed eggs	Chemical treatment 2002
Batnfjordselva	GS	1989–90, 1992–93	1995–2001	Fry	Chemical treatment 1994, 2003/2004. Still maintained in GB for safety reasons
Driva	GS	1990–92		Unknown	Chemical treatment/barrier – time not decided
Måna	GS	1990–92	1995–2004	Fry	Chemical treatment 1993. Rivers infected with GS in same fjord system. Stock in LGB for safety reasons
Innfjordelva	GS	1993–94, 1997–2001	2007–11	Unknown	Chemical treatment 2006/2007
Rauma	GS	1989–93	2007–11	Smolt, parr, eyed eggs	Chemical treatment 2006/2007
Aureelva	GS	1990	1995–98	Fry	Chemical treatment/barrier 1988. Removed from LGB in 1999
Eidsdalselva	GS	1989–92	1995–2004	Fry	Chemical treatment 1990. Removed from LGB in 2004
Norddalselva	GS	1991–93	1996–2004	Fry	Chemical treatment 1990
Valdalselv	GS	1989–92	1995–2004	Fry	Chemical treatment 1990. Removed from LGB in 2004
Jølstra	HD, AC, ES	1992–96, 1997	1999 →	Smolt, parr, eyed eggs	Liming in some tributaries
Flekke/Guddal	AC	1995, 1997–99	2002 →	Eyed eggs	River is perpetual limed from 1997
Lærdalselva	GS, HD	1992–93, 1995–98, 2000–04	2006–10	Eyed eggs	Chemical treatment 2005/2006
Årøyelva	HD, ES	1992–96, 2003–04	1998 →	Smolt, parr	
Ekso	ES, HD, AC	1994–99, 2001–04	2000 →	Eyed eggs	
Vosso	HD, AC, ES	1991–98, 2001	1999 →	Unknown	Lower part perpetual limed from 1994
Loneelva	PL, ES, FD	1991–98	1999–2004	Fry, eyed eggs	Removed from LGB in 2004
Oselva	ES	1992–95	2001–2002	Eyed eggs	Removed from LGB in 2004
Eidfjordvassdraget	SL, ES, HD	1994–96, 2000–04	2002 →	Parr, fry, eyed eggs	
Etneelva	ES	1991–96	1999 →	Eyed eggs	

HD: Hydropower development, AC: Acidification, ES: Escaped farmed salmon, PL: Pollution, FD: Fish diseases, SL: Salmon lice, GS: *Gyrodactylus salaris*.
Box contributed by Øyvind Walsø, Directorate for Nature Management, oyvind.walso@dirnat.no

Box 14.5 *(cont'd)*

of *G. salaris* infection have been a 97% reduction in parr density (Johnson *et al.* 1999). Treatment for *G. salaris* includes a combination of barrier construction and chemical treatment, using either the plant poison Rotenone or acidified aluminium. After the parasite is eradicated from a particular river, the original salmon stock is reintroduced from the LGB.

The stocks maintained in the Eidfjord facility face a number of threats, including barriers to fish passage and changes in hydrology associated with hydroelectric dam development, acidification, interaction with large numbers of escaped farmed salmon, salmon lice, pollution and *G. salaris* (Table B14.5).

Nine of the 29 salmon stocks that have been maintained in LGBs have been reintroduced into their river of origin; seven are no longer retained in captivity but two are being kept as a hedge against future catastrophes (Table B14.5). Twelve additional stocks are under restoration, while the seven remaining stocks await eradication of *G. salaris* from their native rivers. One stock of landlocked salmon is maintained in an LGB for safety reasons.

The ability to preserve salmon embryos for long periods of time would greatly simplify the future recovery of populations but, more importantly, could also mitigate some of the potential risks associated with the restoration of wild salmon runs from cryopreserved sperm or from multigenerational LGB populations. Unfortunately, no reports of successful cryopreservation of salmon embryos (or eggs) were found. Although salmon embryos do not appear to survive cryopreservation, isolated cells from blastoderms often do; viable offspring could be reconstituted by transplanting such cells into host embryos obtained from an extant salmon population (of course, the transplanted cells would have to contribute to the germ-cell lineage), or by transplanting diploid nuclei to enucleated host eggs (Thorgaard and Cloud 1993). While promising, these techniques require further development before being of practical use.

14.4.3 Addition of genetic variation to impoverished future populations

Salmonids have generally low estimated ratios of effective to census population (0.1–0.2) (Hedrick *et al.* 1995). Therefore, isolated salmon populations consisting of as many as 50–100 mature individuals are expected to accumulate inbreeding and to lose genetic variation in surprisingly few generations. Controlled hatchery releases of juveniles produced using cryopreserved sperm derived from earlier genetically diverse populations, and eggs from females obtained recently from the wild, could increase variability in subsequent wild generations. Very few immigrants are required to maintain genetic variability and to mitigate inbreeding depression (see above). However, given the potential for hatchery-released salmon to lower the wild fitness of the natural population (Chapter 11; Reisenbichler *et al.* 2003), such measures should only be taken if wild populations appear to be suffering from small-population effects such as inbreeding depression, or if loss of long-term evolutionary potential is a concern.

14.4.4 Minimising genetic change between founder and subsequent generations of live gene bank populations

Expected genetic changes in long-term LGB populations from drift and adaptation to captive conditions can accumulate over time and negatively impact the wild fitness of the future generation intended for transfer into natural river habitat. Cumulative changes such as these could be markedly reduced by utilising cryopreserved sperm obtained in an early generation to fertilise eggs of females in subsequent generations, as proposed by Smith (1977, cited in Sonesson *et al.* 2002). By utilising sperm from the founder males and from the first generation of LGB males, a large proportion of the genetic variation in the founder generation can be conserved, because alleles from the founder females are represented in the semen of the first generation of LGB males (Sonesson *et al.* 2002).

In the final LGB generation intended for release into wild river habitat, a large number of early-stage juveniles (hundreds of thousands to millions) would need to be produced to ensure the return of a modest number of spawning adults, assuming reasonable levels of freshwater and marine mortality. Given the small volume of cryopreserved milt available, and its reduced viability relative to fresh milt, it is unlikely that cryopreserved sperm could be used in the production of the last LGB generation intended for release into wild river habitat. However, cryopreserved sperm could be used to produce the third and second-last captive LGB generations. This procedure could greatly minimise inbreeding and the loss of genetic variation and domestication selection, as half the gametes that contribute to these latter generations would be obtained from individuals collected from the wild, or that have experienced a single generation of captive rearing. It may also be important to use cryopreserved sperm in intermediate generations to mitigate potential outbreeding-related problems in the event that significant divergence occurs between the founder and pre-release generations.

14.5 Research

14.5.1 Monitoring the loss of genetic variation and accumulation of inbreeding

Once the last founder broodstock has been incorporated into the programme, genetic variation will be lost due to drift at a rate inversely proportional to the effective population size. The broodstock management programme described will only minimise the rate of loss although, in theory at least, such programmes can be very effective in this regard. In practice, the maintenance of genetic variation will depend on a number of factors, including (1) the accuracy of the original kinship assessment among individuals in the founder generation, (2) the accuracy of assignment of offspring to parents in subsequent generations, (3) the ability to recover offspring from specific crosses either from the wild or captivity (due to, for example, among-family variation in mortality), and (4) divergence between the original programme design and that which is actually achieved given unforeseen problems and logistic constraints. Therefore, monitoring genetic variation will be important to assess the efficiency of the programme so that, if losses are unacceptable, modifications in captive breeding and rearing programmes can be made.

Several measures of genetic variation, including allelic richness, gene diversity and founder equivalent genomes, will be monitored. Allelic diversity, the mean number of unique alleles observed, is a very sensitive indicator of population bottlenecks (Spencer *et al.* 2000) and likely reflects the capacity of the final LGB generation, to be released into the natural river environment, to adapt to future challenges (Nevo 1978). Gene diversity, the probability that two alleles drawn at random from a population are different (Chapter 4; Lacy 1995), is more relevant to the maintenance of additive genetic variation and rapid microevolution (Moritz 1999; Doyle *et al.* 2001), both important for the final LGB generation to adapt to wild conditions.

Founder genome equivalent (FGE) is a concept introduced by Lacy (1989). This is the 'number of equally contributing founders with no random loss of founder alleles in descendants that would be expected to produce the same genetic diversity as in the population under study'. FGE is expressed relative to the baseline and estimates the loss of gene diversity (Lacy 1995). Here, comparisons will be made with the original founder broodstock collection obtained from the wild. Individual inbreeding coefficients will be estimated from microsatellite data following the 'method of moments estimator' for two-gene relationships developed by Ritland (1996), a measure of identity versus non-identity of alleles. Levels of inbreeding will also be estimated using the d^2 approach of Coulson et al. (1998). This measure incorporates information on likely relatedness of different microsatellite alleles, which the authors suggest, in some circumstances, may better reflect recent inbreeding than measures of heterozygosity.

Once one complete generation of pedigree information becomes available, inbreeding and relatedness coefficients will be calculated from pedigree information using path analysis or the additive relationship matrix approach (Lynch and Walsh 1998). Wright's population inbreeding coefficient (F) will also be estimated in subsequent generations and compared with values from the original founding population. Possible effects of inbreeding on growth, survival and reproduction (number of eggs and number of eggs successfully fertilised) will be evaluated by comparing all three in subsequent generations with values from the original founding generation. Since other effects (environment, domestication, accumulation of deleterious alleles, and so on) may also influence these parameters, growth, survival and reproductive output will also be compared among LGB individuals that have experienced different levels of individual inbreeding.

14.5.2 Identification of individuals, and evaluation of the relative efficacy of alternate management strategies

The Stewiacke (iBoF) River is a comparatively large and complex river system and no facilities currently exist for enumerating returning adults. Using previous fence counts, electrofishing surveys to determine juvenile densities, and multiple index models, Gibson and Amiro (2003) estimated that fewer than ten salmon returned to spawn in 2000–02. In the Big Salmon River, information from swim-through surveys of adults and redd counts suggest that a few dozen salmon returned to spawn in 2000–02 (Gibson et al. 2003b). One of the objectives of the proposed iBoF research programme is to determine whether remaining juveniles and returning adults were produced by wild salmon spawning in natural river habitat or by LGB salmon spawning in captivity or in the wild.

Examination of molecular genetic markers in recovered juveniles, and subsequent parentage (Box 14.2) and grandparentage analyses (Letcher and King 2001), will be used to determine the origin (wild, LGB or wild × LGB) of recovered salmon. This information, and data from several mtDNA single nucleotide polymorphisms (SNPs) currently being surveyed in all iBoF salmon, will be used to help refine above estimates of numbers of returning adults.

A number of management practices, each with advantages and disadvantages of unknown significance, have been recommended to mitigate negative environmental (behavioural and physiological conditioning) and genetic effects of captive breeding and rearing on the survival and breeding success of salmon in the wild (Table 14.6). Some of these recommendations have been incorporated into management strategies proposed or in use for iBoF salmon (Fig. 14.11). Offspring from predetermined parental crosses could be reared and released following one

Table 14.6 Alternate management strategies in use or proposed for inner Bay of Fundy salmon.

Management strategy	Likely effects on long-term fitness (genetic costs and benefits)	Short- and medium-term management implications
1 Release of adults, captively reared in fresh water or in sea pens, to spawn in the wild; in-river incubation of eggs in natural redds; development in wild river environment; recapture as late parr or smolts; captive rearing to maturity	Possible reduced accumulation of deleterious mutations due to benefits of mate choice (sexual selection) and early exposure to natural selection; reduced adaptation to captive environment due to exposure of fry to natural conditions only. Relaxed selection for ocean survival and return behaviour	Increased census population of returning adults in the rivers. Hatchery not necessarily required
2 Self-selection of mates but captive spawning of individuals; incubation of eggs in captivity; release of juveniles as unfed fry; rearing in wild river environment; recapture as late parr or smolts; captive rearing to maturity	Possible reduced accumulation of deleterious mutations due to benefits of mate choice (sexual selection) and early exposure to natural selection; reduced adaptation to captive environment due to exposure of early life-history stage salmon to natural conditions; increased effective population size. Possible negative effect is relaxed selection for natural spawning behaviour, redd-cleaning, etc. Relaxed selection for ocean survival and return behaviour	Hatchery is required. Increased census population of returning adults in the rivers
3 Minimisation of mean kinship breeding programme; river incubation of eggs in egg baskets or streamside incubators; volitional emergence of fry and development in wild river environment; recapture as late parr or smolts; captive rearing to maturity	Reduced accumulation of inbreeding; reduced loss of genetic variation; reduced within-family selection for captive rearing conditions due to broodstock management programme; possible reduced accumulation of deleterious mutations due to early exposure to natural selection; reduced adaptation to captive environment due to exposure of fry to natural conditions only. Possible negative effect is relaxed selection for natural spawning behaviour, redd-cleaning, etc. Possible accumulation of deleterious mutations because self-selection of mates is inhibited. Relaxed selection for ocean survival and return behaviour	Increased census population of returning adults in the rivers. Hatchery required. Considerable genetic expertise required on a continuing basis. Reduced animal husbandry costs. Uncertain success of egg basket release and limited deployment of streamside incubators
4 Minimisation of mean kinship breeding programme; incubation of eggs in captivity; release as unfed or early feeding fry; development in wild river environment; recapture as late parr or smolts; captive rearing to maturity	Reduced accumulation of inbreeding; reduced loss of genetic variation; reduced within-family selection for captive rearing conditions due to broodstock management programme; reduced accumulation of deleterious mutations and adaptation to captive environment due to exposure of early life-history stage to natural conditions. Possible negative effect is relaxed selection for natural spawning behaviour, redd-cleaning, etc. Possible accumulation of deleterious mutations because self-selection of mates is inhibited. Relaxed selection for ocean survival and return behaviour	Increased census population of returning adults in the rivers. Hatchery is required. Considerable genetic expertise required on a continuing basis

Table 14.6 (cont'd)

Management strategy	Likely effects on long-term fitness (genetic costs and benefits)	Short- and medium-term management implications
5 Minimisation of mean kinship breeding programme; incubation of eggs in captivity; release as autumn fingerlings; some development in wild river environment; recapture as late parr or smolts; captive rearing to maturity	Reduced accumulation of inbreeding; reduced loss of genetic variation; reduced within-family selection for captive rearing conditions due to broodstock management programme. Possible negative effect is relaxed selection for natural spawning behaviour, redd-cleaning, etc. Possible accumulation of deleterious mutations because self-selection of mates is inhibited. Relaxed selection for ocean survival and return behaviour	Increased census population of returning adults in the rivers. Hatchery is required. Considerable genetic expertise required on a continuing basis
6 Minimisation of mean kinship breeding programme; incubation of eggs in captivity; release as smolts; recovery of returning adults for introgression into captive broodstock	Reduced accumulation of inbreeding; reduced loss of genetic variation; reduced within-family selection for captive rearing conditions due to broodstock management programme. Possible accumulation of deleterious mutations because self-selection of mates is inhibited. Possible negative effect is relaxed selection for natural spawning behaviour, redd-cleaning, etc. Relaxation of selection for natural survival to smolt	Increased census population of returning adults in the rivers. Hatchery is required. Considerable genetic expertise required on a continuing basis. Recapture of breeders is required
7 Recovery of returning adults; minimisation of mean kinship breeding programme; introgression back into the captive broodstock	Reduced accumulation of inbreeding; reduced loss of genetic variation; reduced within-family selection for captive rearing conditions due to broodstock management programme. Possible accumulation of deleterious mutations because self-selection of mates is inhibited. Possible negative effect is relaxed selection for natural spawning behaviour, redd-cleaning, etc. Relaxation of selection for natural survival to smolt	Increased census population of returning adults in the rivers. Hatchery is required. Considerable genetic expertise required on a continuing basis. Recapture of breeders is required
Current iBoF gene bank programme Combination of several of the above strategies including natural and captive spawning, natural captive development of juveniles, and transfers between wild and natural environments at various stages as in strategies 1–7. All animals are linked in a common pedigree and database	Inbreeding and random loss of genetic diversity are minimized. Genetic effects of relaxed selection on components of natural fitness (e.g. natural spawning behaviour, return success) are monitored through the linked pedigrees and performance database. The effects of relaxed selection on these traits will be compensated for in the mating design. In later generations when all the founder diversity has been captured and distributed over the population, a programme for self-selection of mates in the hatchery may be introduced	Increased census population of returning adults in the rivers. Hatchery is required. Considerable genetic expertise required on a continuing basis. Recapture of breeders is required

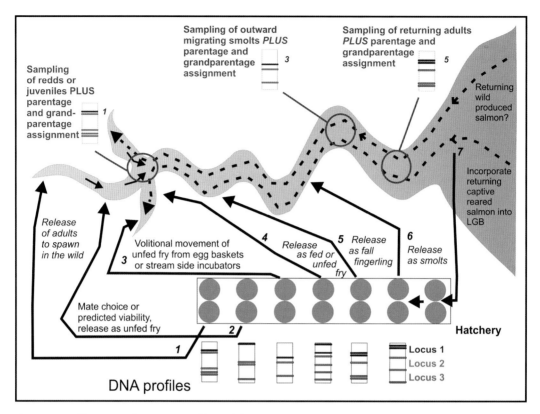

Fig. 14.11 Different sets of parents reared in captivity may be used in the production of the next generation according to various strategies (1–7 in this figure; see also Table 14.6) in use or proposed for iBoF salmon. Parentage and grandparentage analysis of parr, outward migrating smolts, or returning adults may be used to determine whether individuals at various life-history stages were produced by (1) wild salmon spawning in river habitat, (2) LGB parents directly, in any of the above 1–7 management strategies, or (3) offspring of LGB parents spawning in the wild. In addition to determining whether observed production is due to remnant wild salmon spawning in river habitat versus LGB salmon spawning in captivity or in the wild, this research will also help evaluate the relative efficacy of the different management programmes in terms of survival to various stages, and ultimately reproductive success.

management strategy, and other sets of families reared and released under other management regimes. Genotype analysis, parentage and grandparentage determinations of wild-exposed late-stage parr or smolts could be used to trace individuals back to their specific families, and hence the specific management strategies under which they were spawned, reared and released. This will provide information on the relative survival of salmon managed under different conditions in captivity and in the wild, and help in adapting future management practices to best achieve programme objectives.

Along with the maintenance of pedigree information for each LGB individual, records will be kept on numbers of months (or years) of wild exposure accrued by individuals along particular lineages, and the management strategy under which they were spawned, reared and released. Analysis of the accumulation of genetic effects over multiple generations may provide greater resolution in detecting the impact of different aspects of captive breeding and rearing on survival and reproductive success of LGB salmon (M. Kinnison, pers. comm.).

14.6 Summary and conclusions

- Captive breeding and rearing, despite the best of intentions, will bring about physiological, behavioural and genetic changes that will very likely lower the fitness of released individuals in the wild.
- An LGB programme should only be initiated if there is high certainty that the probability of persistence of a population or species in the wild at a given time is very low. Furthermore, it should be borne in mind that most reintroduction programmes fail, though common contributing factors include the lack of resources, lack of research into programme efficacy and premature cessation of activities.
- The genetic changes expected in a captive breeding and rearing programme include (1) inbreeding, (2) loss of genetic variation, (3) adaptation to captive rearing conditions, and (4) the accumulation of deleterious alleles.
- In addition to reducing the ability of salmon to survive and reproduce successfully in the wild, the loss of variation may also limit the ability of released populations to adapt to changing conditions.
- Loss of genetic variation and accumulation of inbreeding can be minimised by following prescribed breeding programmes discussed above, including those that minimise global co-ancestry in subsequent generations.
- Adaptation to captive conditions is likely a more important risk than previously thought, and in the iBoF LGB programme is being minimised by maximising exposure of early juveniles to natural selection.
- Ongoing limited, prescribed use of cryopreserved milt obtained in the first and second generation may also help minimise most of the above genetic changes, and increase the likelihood of populations persisting upon their release back into natural habitat.

14.7 Management recommendations

If the decision is made to protect or shelter populations until conditions in the wild improve, then the following steps should be taken so as to minimise loss of genetic variation and fitness through time:

- Attempt to elucidate the degree and pattern of relatedness (kinship) in the founding generation using molecular genetic markers and procedures that identify full and half-sib groupings in the absence of parental genotype information.
- Consider tracking genealogies in subsequent generations by rearing families separately until large enough to tag so as to permit identification of individuals, or by using molecular genetic markers and parentage analysis.
- Consider use of pedigree information and an appropriate mating programme to minimise inbreeding and loss of genetic variation through time.
- Where possible, maximise exposure to natural selection, especially in early life-history stages.
- During the captive phase, attempts should be made to simulate the natural environment as much as possible.
- Cryopreserve milt from males in the founding and second generation for prescribed, limited use in later generations so as to minimise negative genetic changes through time.

- Evaluate the rates of loss of genetic variation, using both pedigree information and molecular genetic analyses.
- Evaluate survival of different family lineages with varying amounts of cumulative time spent exposed to natural selection, both under captive and wild conditions.
- Experiment with mate selection and predicted viability measures, and evaluate their impact on fitness in captivity and in the wild.
- Experiment with different management practices, and develop genotype-based methods to track survival of offspring spawned and reared under different conditions.

Acknowledgements

The broodstock management programme implemented here would not have been possible without the availability of molecular genetic markers kindly provided by Eric Verspoor and Tim King. Microsatellite genotype information was contributed by Eric Verspoor and David Knox (FRS Marine Laboratory, Aberdeen, Scotland, UK) and by Andrea Cox (CMB Laboratory, Bedford Institute of Oceanography, Dartmouth, Nova Scotia, Canada). Christophe Herbinger provided software for the reconstruction of kinship and assisted with earlier assessments of relatedness. Carolyn Harvie, Roger Souloudre and Anthony Joyce helped with programming and database developments. Ellen Kenchington, Eric Verspoor and Christophe Herbinger provided valuable advice on the genetic management of iBoF broodstock. Trevor Goff, Shane O'Neil, and staff at Mersey, Coldbrook and Mactaquac Biodiversity facilities carried out the captive breeding and rearing of most iBoF LGB salmon. Charles Ayer and staff at the New Brunswick Minto Hatchery also assisted with the breeding and rearing of iBoF salmon. Peter Amiro, Jamie Gibson, Ross Jones, Leroy Anderson, Derek Knox and Alain Caissie provided tissue samples of salmon collected in the field. Additional thanks go to Carolyn Harvie, Trevor Goff, Eric Verspoor, Lee Stradmeyer and three anonymous reviewers for comments that helped improve later versions of this manuscript.

Further reading

For a discussion on how programmes based on pedigree information may be more effective at conserving genetic variation than approaches that maximise the number of breeders and equalise family size, see Ballou and Lacy (1995). Different types of pedigree-based management programme are compared in Fernandez and Caballero (2001) and practical considerations dealt with in Fernandez *et al.* (2001). Miller and Kapuscinski (2003) discuss many of the genetic risks associated with captive rearing and captive breeding of salmonids, and how to mitigate them.

References

Allendorf, F.W. and Ryman, N. (1987) Genetic management of hatchery stocks. In: N. Ryman and F. Utter (Ed.) *Population Genetics and Fishery Management*, pp. 141–159. University of Washington Press, Seattle, WA.

Amiro, P.G. (1987) Similarities in annual recruitment of Atlantic salmon to sport fisheries of inner Bay of Fundy rivers and stock forecasts for 1987. *DFO CAFSAC Research Document* 87/58.

Amiro, P.G. and Jefferson, E.M. (1996) Status of Atlantic salmon in Salmon Fishing Area 22 and 23 for 1995, with emphasis on inner Bay of Fundy stocks. *DFO Atlantic Fisheries Research Document* 96/134.

Armbruster, P., Hutchinson, R. and Linvell, T. (2000) Equivalent inbreeding depression under laboratory and field conditions in a tree-hole-breeding mosquito. *Proceedings of the Royal Society of London, Series B*, **267**: 1939–1945.

Ballou, J.D. and Lacy, R.C. (1995) Identifying genetically important individuals for management of genetic variation in pedigreed populations. In: J.D. Ballou, M. Gilpin and T.J. Foose (Ed.) *Population Management for Survival and Recovery*, pp. 76–111. Columbia University Press, New York.

Bijlsma, R., Bundgaard, J. and Boerema, A.C. (2000) Does inbreeding affect the extinction risk of small populations? Predictions from *Drosophila*. *Journal of Evolutionary Biology*, **13**: 502–514.

Bryant, E.H., Backus, V.I., Clark, M.E. and Reed, D.H. (1999) Experimental tests of captive breeding for endangered species. *Conservation Biology*, **13**: 1487–1496.

Caballero, A.C. and Toro, M.A. (2000) Interrelations between effective population size and other pedigree tools for the management of conserved populations. *Genetic Research*, **75**: 331–343.

Caballero, A.C. and Toro, M.A. (2002) Analysis of genetic diversity for the management of conserved subdivided populations. *Conservation Genetics*, **3**: 289–299.

Campton, D.E., Allendorf, F.W., Behnke, R.J., Utter, F.M., Chilcote, M.W., Leider, S.A. and Loch, J.J. (1991) Reproductive success of hatchery and wild steelhead. *Transactions of the American Fisheries Society*, **120**: 816–827.

Chilcote, M.W., Leider, S.A. and Loch, J.J. (1986) Differential reproductive success of hatchery and wild summer-run steelhead under natural conditions. *Transactions of the American Fisheries Society*, **115**: 726–735.

Cloud, J.G. and Osborne, C. (1997) *Cryopreservation of salmonid sperm*. Department of Biological Sciences, University of Idaho, Moscow, ID.

Coulson, T., Pemberton, J., Albon, S., Beaumont, M., Marshall, T., Slate, J., Guinness, F. and Clutton-Brock, T.H. (1998) Microsatellites reveal heterosis in red deer. *Proceedings of the Royal Society of London, Series B*, **265**: 489–495.

Crandall, K.A., Bininda-Emonds, O.R.P., Mace, G.M. and Wayne, R.K. (2000) Considering evolutionary processes in conservation biology. *Trends in Ecology and Evolution*, **15**: 290–295.

Crow, J.F. and Kimura, M. (1970) *An Introduction to Population Genetics Theory*. Harper & Row, New York.

Cuenco, M.L., Backman, T.W.H. and Mundy, P.R. (1993) The use of supplementation to aid in natural stock restoration. In: J.G. Cloud and G.H. Thorgaard (Ed.) *Genetic Conservation of Salmonid Fishes*, pp. 269–294. Plenum Press, New York.

Doyle, R.W., Perez-Enriquez, R., Takagi, M. and Taniguchi, N. (2001) Selective recovery of founder genetic diversity in aquaculture broodstock and captive, endangered fish populations. *Genetica*, **111**: 291–304.

Ducharme, L.J.A. (1969) Atlantic salmon returning for their fifth and sixth consecutive spawning trips. *Journal of the Fisheries Research Board of Canada*, **26**: 1661–1664.

Falconer, D.S. and Mackay, T.F.C. (1996) *Introduction to Quantitative Genetics*. Longman, Harlow, UK.

Fernandez, J. and Caballero, A. (2001) Accumulation of deleterious mutations and equalization of parental contributions in the conservation of genetic resources. *Heredity*, **86**: 1–9.

Fernandez, J. and Toro, M.A. (1999) The use of mathematical programming to control inbreeding in selection schemes. *Journal of Animal Breeding and Genetics*, **116**: 447–466.

Fernandez, J., Toro, M.A. and Caballero, A. (2001) Practical implementation of optimal management strategies in conservation programmes: a mate selection method. *Animal Biodiversity and Conservation*, **24**: 17–24.

Fleming, I.A. and Gross, M.R. (1993) Breeding success of hatchery and wild coho salmon (*Oncorhynchus kisutch*) in competition. *Ecological Applications*, 3: 230–245.

Fleming, I.A. and Petersson, E. (2001) The ability of released, hatchery salmonids to breed and contribute to the natural productivity of wild populations. *Nordic Journal of Freshwater Research*, 75: 71–98.

Foose, T.J., Lande, R., Flesness, N.R., Rabb, G. and Read, B. (1986) Propagation plans. *Zoo Biology*, 5: 139–146.

Gall, G.A.E. (1987) Inbreeding. In: N. Ryman and F. Utter (Ed.) *Population Genetics and Fishery Management*, pp. 47–88. University of Washington Press, Seattle, WA.

Gausen, D. (1993) The Norwegian Gene Bank Programme for Atlantic salmon (*Salmo salar*). In: J.G. Cloud and G.H. Thorgaard (Ed.) *Genetic Conservation of Salmonid Fishes*, pp. 181–187. Plenum Press, New York.

Gibson, A.J.F. and Amiro, P.G. (2003) *Abundance of Atlantic salmon* (Salmo salar) *in the Stewiacke River, NS, from 1965 to 2002*. CSAS Research Document 2003/108.

Gibson, A.J.F., Jones, R.A., Amiro, P.G. and Flanagan, J.L. (2003a) *Abundance of Atlantic salmon* (Salmo salar) *in the Big Salmon River, NB, from 1951 to 2002*. CSAS Research Document 2003/1119.

Gibson, A.J.F., Amiro, P.G. and Robichaud-LeBlanc, K.A. (2003b) *Densities of juvenile Atlantic salmon* (Salmo salar) *in inner Bay of Fundy rivers during 2000 and 2002 with reference to past abundance inferred from catch statistics and electrofishing surveys*. CSAS Research Document 2003/121.

Gjerde, B., Gunnes, K. and Gjedrem, T. (1983) Effect of inbreeding on survival and growth in rainbow trout. *Aquaculture*, 34: 327–332.

Hallerman, E.M. (2003a) Inbreeding. In: E.M. Hallerman (Ed.) *Population Genetics: Principles and applications for fisheries scientists*, pp. 215–237. American Fisheries Society, Bethesda, MD.

Hallerman, E.M. (2003b) Coadaptation and outbreeding depression. In: E.M. Hallerman (Ed.) *Population Genetics: Principles and applications for fisheries scientists*, pp. 239–260. American Fisheries Society, Bethesda, MD.

Hard, J. (1995) Genetic monitoring of life-history characters in salmon supplementation: problems and opportunities. In: H.L. Schramm and R.G. Piper (Ed.) *Uses and Effects of Cultured Fishes in Aquatic Ecosystems*, pp. 307–314. American Fisheries Society, Bethesda, MD.

Hartl, D. and Clark, A.G. (1989) *Principles of Population Genetics*. Sinauer Associates, Sunderland, MA.

Harvey, B. (1993) Cryopreservation of fish spermatozoa. In: J.G. Cloud and G.H. Thorgaard (Ed.) *Genetic Conservation of Salmonid Fishes*, pp. 175–178. Plenum Press, New York.

Heath, D.D., Heath, J.W., Bryden, C.A., Johnson, R. and Fox, C.W. (2003) Rapid evolution of egg size in captive salmon. *Science*, 299: 1738–1740.

Hedrick, P.W., Hedgecock, D. and Hamelberg, S. (1995) Effective population size in winter-run chinook salmon. *Conservation Biology*, 9: 615–624.

Hedrick, P.W., Hedgecock, D., Hamelberg, S. and Croci, J. (2000) The impact of supplementation in winter-run chinook salmon on effective population size. *Journal of Heredity*, 91: 112–116.

Henderson, J.N. and Letcher, B.H. (2003) Predation on stocked Atlantic salmon (*Salmo salar*) fry. *Canadian Journal of Fisheries and Aquatic Sciences*, 60: 32–42.

Herbinger, C.H., O'Reilly, P.T. and Verspoor, E. (2006) Unravelling first generation pedigrees in wild endangered salmon populations using molecular genetic markers. *Molecular Ecology*, 15: 2261–2275.

Jennions, M.D. and Petrie, M. (1997) Variation in mate choice and mating preferences: a review of causes and consequences. *Biological Review*, 72: 283–327.

Jessop, B.M. (1976) Distribution and timing of tag recoveries from native and nonnative Atlantic salmon (*Salmo salar*) released into Big Salmon River, New Brunswick. *Journal of the Fisheries Research Board of Canada*, 33: 829–833.

Johnson, B.O., Møkkelgjerd, P.I. and Jensen, A.J. (1999) The parasite *Gyrodactylus salaris* on Atlantic salmon in Norwegian rivers: status report when approaching year 2000. *NINA Oppdragsmelding*, **617**: 1–129. (in Norwegian with English abstract)

Kincaid, H.L. (1976a) Effects of inbreeding on rainbow trout populations. *Transactions of the American Fisheries Society*, **105**: 273–280.

Kincaid, H.L. (1976b) Inbreeding in rainbow trout (*Salmo gairdneri*). *Journal of the Fisheries Research Board of Canada*, **33**: 2420–2426.

Kincaid, H.L. (1977) Rotational line crossing: an approach to the reduction of inbreeding accumulation in trout brood stocks. *Progressive Fish Culturist*, **39**: 179–181.

Kincaid, H.L. (1983) Inbreeding in fish populations used in aquaculture. *Aquaculture*, **33**: 215–227.

Lacy, R.C. (1989) Analysis of founder representation in pedigrees: founder equivalents and founder genome equivalents. *Zoo Biology*, **8**: 111–123.

Lacy, R.C. (1995) Clarification of genetic terms and their use in the management of captive populations. *Zoo Biology*, **14**: 565–578.

Lahnsteiner, F. (2000) Semen cryopreservation in the Salmonidae and in the Northern pike. *Aquaculture Research*, **31**: 245–258.

Landry, C., Garand, D., Duchesne, P. and Bernatchez, L. (2001) 'Good genes as heterozygosity': the major histocompatibility complex and mate choice in Atlantic salmon (*Salmo salar*). *Proceedings of the Royal Society of London, Series B*, **268**: 1279–1285.

Langefors, A., Lohm, J., Grahn, M., Anderson, O. and von Schantz, T. (2001) Association between major histocompatibility complex class IIB alleles and resistance to *Aeromonas salmonicida* in Atlantic salmon. *Proceedings of the Royal Society of London, Series B*, **268**: 479–485.

Letcher, B. and King, T. (2001) Parentage and grandparentage assignment with known and unknown matings: application to Connecticut River Atlantic salmon restoration. *Canadian Journal of Fisheries and Aquatic Sciences*, **58**: 1812–1821.

Li, W.H. (1978) Maintenance of genetic variation under the joint effect of mutation, selection and random drift. *Genetics*, **90**: 349–382.

Lohm, J., Grahn, M., Langefors, A., Andersen, O., Storset, A. and von Schantz Torbjorn (2002) Experimental evidence for major histocompatibility complex-allele-specific resistance to a bacterial infection. *Proceedings of the Royal Society of London, Series B*, **269**: 2029–2033.

Lynch, M. and Walsh, B. (1998) *Genetic Analysis of Quantitative Traits*. Sinauer, Sunderland, MA.

Lynch, M., Conery, J.R. and Burger, R. (1995) Mutational meltdowns in sexual populations. *Evolution*, **49**: 1067–1080.

Madsen, T., Shine, R., Olsson, M. and Wittzell, H. (1999) Conservation biology: restoration of an inbred adder population. *Nature*, **402**: 34–35.

Maynard, D.J., Flagg, T.A. and Mahnken, C.V.W. (1995) A review of seminatural culture strategies for enhancing the post release survival of anadromous salmonids. In: H.L. Schramm and R.G. Piper (Ed.) *Uses and Effects of Cultured Fishes in Aquatic Ecosystems*, pp. 307–314. American Fisheries Society, Bethesda, MD.

Maynard, D.J., Flagg, T.A., Mahnken, C.V.W. and Schroder, S.L. (1996) Natural rearing technologies for increased postrelease survival of hatchery-reared salmon. *Bulletin of National Research Institute of Aquaculture*, **2**: 71–77.

Meffert, L.M. (1999) How speciation experiments relate to conservation biology. *BioScience*, **49**: 701–715.

Miller, L.M. and Kapuscinski, A.R. (2003) Genetic guidelines for hatchery supplementation programmes. In: E.M. Hallerman (Ed.) *Population Genetics: Principles and applications for fisheries scientists*, pp. 329–355. American Fisheries Society, Bethesda, MD.

Mills, D. (1989) *Ecology and Management of Atlantic Salmon*. Chapman & Hall, London.

Moritz, C. (1999) Conservation and translocations: strategies for conserving evolutionary processes. *Hereditas*, **130**: 217–228.

Nei, M. (1973) Analysis of gene diversity in subdivided populations. *Proceedings of the National Academy of Sciences of the USA*, **70**: 3321–3323.

Nevo, E. (1978) Genetic variation in natural populations: patterns and theory. *Theoretical Population Biology*, **13**: 121–177.

Petit, R.J., Mousadik, A. and Pons, O. (1998) Identifying populations for conservation on the basis of genetic markers. *Conservation Biology*, **12**: 844–855.

Piironen, J. (1993) Cryopreservation of sperm from the brown trout (*Salmo trutta M. lacustris* L.) and Arctic charr (*Salvelinus alpinus* L.). *Aquaculture*, **116**: 275–285.

Pray, L.A. and Goodnight, C.J. (1995) Genetic variation in inbreeding depression in the red flour beetle *Tribolium castaneum*. *Evolution*, **49**: 176–188.

Pray, L.A., Schwartz, J.M., Goodnight, C.J. and Stevens, L. (1994) Environmental dependency of inbreeding depression: implications for conservation biology. *Conservation Biology*, **8**: 562–568.

Reisenbichler, R.R. (1988) Relation between distance transferred from natal stream and recovery rate for hatchery coho salmon. *North American Journal of Fisheries Management*, **8**: 172–174.

Reisenbichler, R.R. and Brown, G. (1995) Is genetic change from hatchery rearing of anadromous fish really a problem? In: H.L. Schram and R.G. Piper (Ed.) *Uses and Effects of Cultured Fishes in Aquatic Ecosystems*, pp. 578–579. American Fisheries Society, Bethesda, MD.

Reisenbichler, R.R., Utter, F.M. and Krueger, C.C. (2003) Genetic concepts and uncertainties in restoring fish populations and species. In: R.C. Wissmar and P.A. Bisson (Ed.) *Strategies for Restoring River Ecosystems*, pp. 149–183. American Fisheries Society, Bethesda, MD.

Ritland, K. (1996) Estimators for pairwise relatedness and individual inbreeding coefficients. *Genetic Research*, **67**: 175–185.

Ryman, N. (1970) A genetic analysis of recapture frequencies of released young of salmon (*Salmo salar* L.). *Hereditas*, **65**: 159–160.

Ryman, N. and Laikre, L. (1991) Effects of supportive breeding on the genetically effective population size. *Conservation Biology*, **5**: 325–329.

Saccheri, I., Kuussaari, M., Kankare, M., Vikman, P., Fortelius, W. and Hanski, I. (1998) Inbreeding and extinction in a butterfly metapopulation. *Nature*, **392**: 491–494.

Scott, A.P. and Baynes, S.M. (1980) A review of the biology, handling and storage of salmonid spermatozoa. *Journal of Fish Biology*, **17**: 707–739.

Shaklee, J.B., Smith, C., Young, S., Marlowe, C., Johnson, C. and Sele, B.B. (1995) A captive broodstock approach to rebuilding a depleted Chinook salmon stock. In: H.L. Schramm and R. G. Piper (Ed.) *Uses and Effects of Cultured Fishes in Aquatic Ecosystems*, p. 567. American Fisheries Society, Bethesda, MD.

Smith, B., Herbinger, C.M. and Merry, H.R. (2001) Accurate partition of individuals into full-sib families from genetic data without parental information. *Genetics*, **158**: 1329–1338.

Sonesson, A.K., Goddard, M.E. and Meuwissen, T.H. (2002) The use of frozen semen to minimize inbreeding in small populations. *Genetic Research*, **80**: 27–30.

Spencer, C.C., Neigel, J.E. and Leberg, P.L. (2000) Experimental evaluation of the usefulness of microsatellite DNA for detecting demographic bottlenecks. *Molecular Ecology*, **9**: 1517–1528.

Stoss, J. (1983) Fish gamete preservation and spermatozoan physiology. In: W.S. Hoar, D.J. Randall and E.M. Donaldson (Ed.) *Fish Physiology*, Vol. 9, Part B, pp. 305–350. Academic Press, New York.

Stoss, J. and Refstie, T. (1983) Short-term storage and cryopreservation of milt from Atlantic salmon and sea trout. *Aquaculture*, **30**: 229–236.

Tave, D. (1993) *Genetics for Fish Hatchery Managers*. Van Nostrand Reinhold, New York.

Thorgaard, G.H. and Cloud, J.G. (1993) Reconstitution of genetic strains of salmonids using biotechnical approaches. In: J.G. Cloud and G.H. Thorgaard (Ed.) *Genetic Conservation of Salmonid Fishes*, pp. 189–196. Plenum Press, New York.

Utter, F. (2003) Genetic impacts of fish introductions. In: E.M. Hallerman (Ed.) *Population Genetics: Principles and applications for fisheries scientists*, pp. 357–378. American Fisheries Society, Bethesda, MD.

Utter, F., Seeb, J.E. and Seeb, L.W. (1993) Complementary uses of ecological and biochemical genetic data in identifying and conserving salmon populations. *Fisheries Research*, **18**: 59–76.

Verspoor, E., O'Sullivan, M., Arnold, A.L., Knox, D. and Amiro, P.G. (2002) Restricted matrilineal gene flow and regional differentiation among Atlantic salmon (*Salmo salar* L.) populations within the Bay of Fundy, Eastern Canada. *Heredity*, **89**: 465–472.

Vila, C., Sundqvist, A., Flagstad, O., Seddon, J., Bjornerfeldt, S., Kojola, I., Casulli, A., Sand, H., Wabakken, P. and Ellegren, H. (2003) Rescue of a severely bottlenecked wolf (*Canis lupus*) population by a single immigrant. *Proceedings of the Royal Society of London, Series B*, **270**: 91–97.

Walsø, Ø. (1998) The Norwegian Gene Bank Program for Atlantic Salmon (*Salmo salar*). *Proceedings of Action before extinction: an international conference on conservation of fish genetic diversity*, 16–18 February in Vancouver, BC.

Wang, J. (1997) More efficient breeding systems for controlling inbreeding and effective populations. *Heredity*, **79**: 591–599.

Wang, S., Hard, J. and Utter, F. (2002) Salmon inbreeding: a review. *Reviews in Fish Biology and Fisheries*, **11**: 301–319.

Wedekind, C. and Müller, R. (2004) Parental characteristics versus egg survival: towards an improved genetic management in the supportive breeding of lake whitefish. *Annales Zoologici Fennici*, **41**: 105–115.

Weitzman, M.L. (1992) On diversity. *Quarterly Journal of Economics*, **107**: 363–405.

Westemeier, R.L., Brawn, J.D., Simpson, S.A., Esker, T.L., Jansen, R.W., Walk, J.W., Kershner, E.L., Bouzat, J.L. and Paige, K.N. (1998) Tracking the long-term decline and recovery of an isolated population. *Science*, **282**: 1695–1698.

Whitlock, M.C. (2000) Fixation of new alleles and the extinction of small populations: drift load, beneficial alleles, and sexual selection. *Evolution*, **54**: 1855–1861.

Woodworth, L.M., Montgomery, M.E., Briscoe, D.A. and Frankham, R. (2002) Rapid genetic deterioration in captive populations: causes and conservation implications. *Conservation Genetics*, **3**: 277–288.

15 Atlantic Salmon Genetics: Past, Present and What's in the Future?

J. L. Nielsen

Upper: an angler playing a fish on the River Dee, Scotland, renowned for salmon returning to the river from the sea every month of the year. (Photo credit: D. Hay.) Lower: a salmon on the River Spey, Scotland, being released after being caught. (Photo credit: J. Webb.)

Atlantic salmon, *Salmo salar*, have played a significant role in human culture and economy for thousands of years, both as a commodity for consumption and as a symbol of the vast power and abundance of nature. They have been incorporated into the fabric of society through art and science over the course of western civilisation and stand as an icon of human ability to coexist with other species on which he depends.

15.1 Past

Long before humans were ever involved in their distribution and abundance, Atlantic salmon, *Salmo salar*, evolved from members of the relatively primitive higher teleostean (bony) fishes (Fig. 15.1); fossil records include the Eocene *Eosalmo*, an archaic trout found in North America and considered the oldest salmonine record (Wilson and Williams 1992; Nelson 1994). Sytchevskaya (1986) described Eocene (approximately 58 million years ago) and Oligocene (approximately 38 million years ago) salmonids from East Asia, but preservation of these samples was insufficient for confirmation.

Historical biogeography of this group is clouded by the fact that northern areas now inhabited by *Salmo* were repeatedly glaciated throughout the Pleistocene, the last ice age covering most of northern Europe and North America. Regan (1920) first suggested that one ancestral diadromous species (fish making regular or seasonal movements between freshwater and marine environments) differentiated into two independent salmonid lineages, the Atlantic basin *Salmo* and the Pacific basin *Oncorhynchus*, but he gave no evidence for the temporal scale or directionality of this divergence. The lack of native freshwater brown trout, *Salmo trutta*, in North America has been used to support a European origin for *Salmo* (Power 1958; Stearley 1992). Diversity of the genus *Salmo*, centring in Eastern Europe, suggests a possible point of origin (Behnke 1968). However, the current intergenic relationships in Salmoninae are still vague (Philips and Oakley 1997; Crespi and Fulton 2004).

Early life-history patterns in the progenitors of diadromous salmon, evolving from marine or freshwater forms, also remain controversial (Thorpe 1982, 1988; McDowall 1988, 2002). Indeed, several authors have suggested, but for different reasons, that variation in migratory and reproductive behaviour exhibited by salmon and trout makes the assignation of an original character state ambiguous (Gross 1987; Smith and Stearley 1989; Stearley 1992; McDowall 2002). Regardless of their evolutionary origins and life-history relationships, Atlantic salmon have survived in Europe and eastern North America over long periods of time. Considering their enduring history alone, they deserve our best efforts at conservation, protection and preservation.

Atlantic salmon are phenomenal in their adaptation to life on earth. Highly migratory anadromous Atlantic salmon cross extreme gradients of environmental condition ranging from freshwater to marine habitats in migrations that span most of the North Atlantic Ocean. They range from the frigid waters of the European and Canadian Arctic to temperate Atlantic waters off the northern coast of Spain. For centuries these fish swam freely with complete disregard for abstract boundaries or political divides set by humans within and outside their place of origin. It is only in recent history that their existence has been compromised by the increasing biological fragility and isolation of aquatic systems imposed by human activity on their critical habitats. Today Atlantic salmon have to face the negative effects of freshwater and marine pollution, loss of habitat due to forestry, urban development and agriculture,

Fig. 15.1 The past: *Gaudryella* (Patterson 1970) a small 10–15 cm fish from Upper Cretaceous (100 million years BP) limestone deposits in Lebanon belonging to the same taxonomic Order as salmon and trout (Salmoniformes). (Photo credit: D. Hay.)

climate change, over-harvest at sea, and genetic manipulation during artificial husbandry and propagation (Dodson *et al.* 1998; Friedland *et al.* 2000).

15.2 Present

The Atlantic salmon has been extensively studied by the scientific community for over 200 years with thousands of research papers, books, proceedings of symposia and popular articles.

Chapters in this book have demonstrated that early genetic studies of Atlantic salmon were primarily dedicated to the use of molecular markers to identify stocks in open ocean fisheries for allocation of harvest and later to identify country or river of origin for migrating adult salmon. Conservation genetics of Atlantic salmon has evolved as a field of study only recently as stocks and populations declined precipitously or disappeared. More recently genetic markers have been used to analyse the impacts of escaped farmed or cultivated fish on wild populations.

Understanding the genetics of this species is inherently difficult due to the diversity and complexity of habitats the animal uses throughout different life stages and the broad diversity of temporal and spatial scales found in its life history. Enormous effort, coupled with large amounts of dollars and euros, has been dedicated to propagation, conservation and recovery research and activity for Atlantic salmon. In western society, scientific study is an ongoing dynamic without clear end-points. One question always leads to another and seemingly definitive answers are never without contradiction. The recent evolution of ideas and information in the field of genetics is a clear case in point. This book reviews many aspects of ecology and genetics required for a better understanding of the conservation requirements for Atlantic salmon in today's environments.

Common characteristics of disputes surrounding the management and conservation of Atlantic salmon are no different than any other resource dispute involving human use and nature. These conflicts primarily involve four elements: (1) competing demands for a limited resource; (2) conflict based on confusion and divergent interpretations of complex scientific information (especially true in contemporary genetics); (3) overlapping jurisdictions and decision-making authorities; (4) competition among often incompatible strategies developed by different ownerships, each with their own historical or cultural perspective (see McKinney and Harmon 2004). The very nature and complexity of Atlantic salmon life history predicates a clash of values, competing interests and complicated relationships among divergent interest groups. Multiple jurisdictions, both political and geographic, naturally induce conflicts among different mandates, laws and policies pertaining to this species.

Add to this mix the dynamic evolution of the genetic sciences with new techniques and methods frequently coming to the table (Fig. 15.2) and you have the makings of deep conflict and confusion. Fundamentally it is a question of who gets to decide how Atlantic salmon move forward throughout this century. Can we retain viable commercial and recreational economies, a sustainable presence of wild Atlantic salmon, and the healthy aquatic ecosystems on which they (and we) depend? Can we have it all or just one or more of the pieces? Can genetics really answer these questions? This is a conundrum of contemporary culture not just limited to the discussion of Atlantic salmon where we try to decide what is worth preserving in the natural environments while retaining and improving the quality of human life on earth.

15.3 Future

Recent dramatic technological advances in genetics have marked quantum leaps in our understanding of Atlantic salmon as an organism and as a species. Over 25 years ago genetic technology began to find its place in fisheries research and management. Multiple landmark discoveries have followed, including recent efforts to sequence the entire genome of several fish species. Due to these technological advances, conservation genetics for Atlantic salmon is

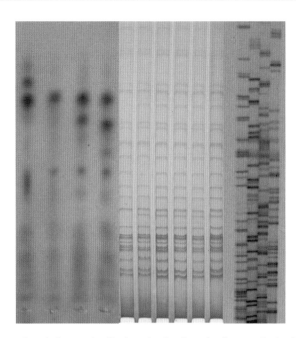

Fig. 15.2 The present: genetic variation resolved in the Atlantic salmon by three methods of electrophoretic analysis contributing to current understanding of genetic variation at the molecular level. Left: heritable variation in two closely related enzymes, malate dehydrogenase and NAD-dependent malic enzyme (each vertical track represents a different individual). Centre: base sequence variants detected by restriction enzyme HaeIII digestion of total mitochondrial DNA. Right: a DNA sequencing gel for a single individual, each track showing the fragments generated using a different radioactively labelled base C, G, A or T. These methods are described in detail in Chapters 3 and 4. (Photo credit: E.Verspoor.)

in a state of dynamic evolution. Lessons learned from the study of the human genome are starting to be translated into a demand for new technological and scientific applications in the study of this ecologically and economically important species (Thorsen *et al.* 2005).

Currently several major initiatives are under way in efforts to sequence the Atlantic salmon genome. The Genomics Research on Atlantic Salmon Project (GRASP) is mapping the chromosomes of salmon and plotting genes of known function on specific chromosomes in order to learn more about how genes work in specific races of Atlantic salmon. GRASP is part of Genome Canada, considered to have strategic significance for the management of Atlantic salmon in the North Atlantic Ocean. The GRASP approach includes BAC contig mapping, large-scale sequencing, genotyping, approaches in functional genomics, proteomics and bioinformatics (Rise *et al.* 2004a). BAC (bacterial artificial chromosome) contig mapping involves breaking up the salmon genome into overlapping pieces and inserting these into a bacterial DNA and then identifying a set of cloned bacterial cultures which provide a contiguous set of salmon DNA fragments which encompass the whole salmon genome. These cultures can then be used to study specific DNA and chromosomal regions and to understand DNA structure and function (i.e. genomics), the proteins produced by genes (proteomics), which involves the complex statistical analysis of DNA and protein variation (i.e. bioinformatics).

In addition to the chromosome maps, GRASP will assist research on how different stocks of Atlantic salmon respond to unique river conditions, including temperature, pollutants, parasites, viruses and bacteria (Rise *et al.* 2004b; Ng *et al.* 2005). GRASP brings together universities, government agencies, research institutes and industry from throughout Canada and the US and has fostered participation in major international genomics research programmes in Ireland, Scotland, France and Norway dedicated to the Atlantic salmon genome.

GRASP has developed useful microarrays from known genes stringently selected from Atlantic salmon and rainbow trout, *Oncorhynchus mykiss*, expressed sequence tag (EST) databases. These databases presently contain over 300 000 sequences from over 175 salmonid cDNA libraries derived from a wide variety of tissues and different developmental stages of fish (von Schalburg *et al.* 2005a). GRASP has recently made its Atlantic salmon cDNA microarrays available for a cost of $US70.00 for the 3700 gene chip and $120.00 for the 16 000 gene chip (http://web.uvic.ca/cbr/gasp.html). This makes the new genomic technology available to any research agency studying Atlantic salmon or other related species. Shared EST databases and cDNA microarrays also provide the platform for interspecific investigations using data mining and DNA hybridisation techniques (Rise *et al.* 2004a).

The Salmon Genome Project (SGP) funded by the Norwegian Research Council has recently posted new sequence data for Atlantic salmon in the public domain with large-sequence data sets submitted to GenBank and contig sequence searches available through the SGP database (www.salmongenome.no/cig-bin/sgp.cig). These efforts will benefit researchers, policy makers, fishers and salmon farmers in both aquaculture activities and wild salmon conservation applications through accelerated development of activities directed at salmonid health, vaccine development, and factors involving local adaptation and population viability.

Two genetic manipulation techniques – chromosome set manipulation (Thorgaard 1983) and recombinant DNA methods (Donaldson and Devlin 1996), which have been the focus of considerable research and controversy since the 1980s in the development of favourable traits in aquaculture production (Kapuscinski 2005) – will also benefit from a fuller knowledge of the Atlantic salmon genome by providing new targets for genetic manipulation in the development of desired traits. Genetic manipulation techniques are also proposed to be useful for biological control of invasive fish species with potential benefits associated with sterile releases (Cotter *et al.* 2000), avoidance of deleterious gene spread (Thresher *et al.* 2002), engineered overdominance (Davis *et al.* 2001), conditional lethality (Grewe 1996), female-specific lethal constructs (Gould and Schliekelman 2004) and engineered fitness disadvantages with insertion of selfish or intentional Trojan genes (Muir and Howard 1999; Burt 2003; Howard *et al.* 2004). All of the new genetic manipulation approaches are experimental and must be seriously considered in light of alternatives and associated risks (Burgman 2005; Kapuscinski and Patronski 2005).

New genomic tools and analyses also offer unique views into the past. The ancestors of all salmonids underwent one or more genome duplications at some point in their history (Allendorf and Thorgaard 1984). Therefore, the species is considered pseudo-tetraploid with identical gene sequences still found on different duplicated chromosomes. Comparisons of BAC sequences representing duplicated regions and associated DNA sequences are revealing important aspects of the nature of these duplications and the dynamics of how gene–gene interactions evolve over evolutionary time scales (Vasemägi *et al.* 2005). The implications of gene duplication in evolution and adaptation in Atlantic salmon have been discussed for a long time, but recent findings using genome sequence and functional genomics have contributed significantly to many unresolved questions involved in the management and conservation of Atlantic salmon. For instance, the mechanisms of sex determination in salmonids have long been a mystery. New information about genes involved in sex control aid our understanding of this important and critical life stage (Artieri *et al.* 2006). Genes involved in maturation and development will also be important components of the future in the development of monosex aquaculture stocks and enhanced reproduction in wild salmon stocks (von Schalburg *et al.* 2005b).

The enormous data sets generated by modern genetic techniques also demand new analytical approaches. This includes the development of networks among different laboratories to co-ordinate data collection and realise increased benefits from our ability to analyse effectively extensive data sets spanning the species' complete geographic range, giving insight into subtle differences in population structure. Standardisation of hypervariable genetic markers across many laboratories interested in the conservation of Atlantic salmon can be used to monitor levels of biodiversity across time and space and to assess changes in effective population size (i.e. number of adults effectively contributing genes to the next generation; N_e) in critical areas. One wonders if new genetic monitoring techniques will eventually replace more traditional census methods.

Electronic integration of genetic information using geographic information systems (GIS) also provides fine-scale information on genetic variation and biodiversity across geographic and/or political barriers contributing to broad-scale conservation efforts. GIS analyses of genetic variation can also provide new insights into regional evolutionary history for the species in relationship to historic patterns of glaciation, volcanism or other geologic activity leading to vicariant or discontinuous biogeographic distributions. These diverse analytical scales help managers define conservation priorities and make decisions on what scale of biodiversity is most critical to restoration goals.

The new post-genomic discipline of bioinformatics develops new strategies for the integration of large genomic data sets developed from different biological organisms based on studies of transcriptomics, proteomics, structure–function genomic organisation and systematic biology (gene interactions, signal transduction pathways and gene network structure). Understanding the structure and regulation of a species' genome and the processes that regulate gene expression (Fig. 15.3) and local adaptation demand new informational and computational technologies. Many recent international scientific forums have been dedicated to the problems of bioinformatics and integrated data analyses (see Kolchanov and Hofestaedt 2004). While we are still developing genome information for different salmonid species, these new computational tools developed from other genera can be used to direct research addressing issues involving DNA organisation, patterns of mutagenesis and propensity for disease, systematics of genomic structure, conservation of gene function and translation across time and space, structural biology of multigene complexes, genetic models of metabolic and endocrine pathways, and new statistical analyses of microarray data. One critical element of bioinformatics will be the ability to compare information across genomes of different species and organisms by digestion of immense masses of data with new analytical strategies. This will revolutionise our concepts of population structure and evolutionary adaptation.

Dramatic advances in genetics are also leading us down a critical path of understanding individual phenotypes and patterns of local adaptation through in-depth studies of functional genomics (Hofmann 2003). The long-term impact of functional genomics, both biologically and economically, will be enormous. But this impact will depend to a great extent on standardisation and simplification of protocols, robust error accounting and the development of new analytical techniques. Two approaches in functional genomics are currently in common use: (1) analysis of specific genes thought to have functional relevance to the question at hand;

Fig. 15.3 The future: the observed gene expression in tissue from Atlantic salmon challenged and unchallenged by exposure to sea lice. Each of the 16 000 dots on the chip has DNA from a different gene to which the DNA being tested binds; the differences in colour and intensity reflect different patterns of gene expression. (Photo credit: C.S. Jones, L.R. Noble and S. Milne, produced as part of the UK LINK Aquaculture SAL031 Project.)

(2) a broader-based approach using microarrays and classification of data into functional categories based on phenotype (see Brownstein and Khodursky 2003). Both approaches have developed profound insights into gene function at the level of individual organisms. Current strategies for the identification of functional genetic variation underlying phenotypic traits of ecological and evolutionary importance represent just one more step in the development of research strategies focused on important genetic variation found in Atlantic salmon (Renn *et al.* 2004; Aubin-Horth *et al.* 2005a,b; see also review in Vasemägi and Primmer 2005).

Studies of gene expression using microarray technology are uncovering the molecular basis of socially controlled plastic phenotypes in different animal species, including Atlantic salmon. Aubin-Horth *et al.* (2005b) have contrasted brain gene expression profiles of different reproductive Atlantic salmon phenotypes (sneaker males and those that go to sea before maturation) which develop under different abiotic and biotic influences. Roughly 15% of the 3000 genes found to express in the brains of fish in this study differed between the two reproductive phenotypes. This cutting-edge research suggests that the genes involved in early maturation in male Atlantic salmon parr include genes from the expected range of biochemical processes, including well-known growth and reproduction pathways. An interesting additional finding is that other genes expressed in early reproductive phenotypes include elements of neural plasticity, including genes associated with memory, which could affect aspects of homing. Cluster analyses of gene expressions showed that sneaker males were more similar to immature females than immature males who will mature at sea (Aubin-Horth *et al.* 2005b). This study reached the conclusion that delayed maturation at sea may result from active inhibition of early maturation in fresh water, supporting what has been argued by some for many years, on physiological grounds (e.g. see Thorpe *et al.* 1998).

Recent scientific discoveries in salmon ecology, aquaculture, physiological adaptation, life history and genomics provide tools that help us understand current issues and conflicts; however, they lack comprehensive answers on the future of the Atlantic salmon, taking into consideration the collision of interests involved. The diversity of interests, views and opportunities surrounding the management of Atlantic salmon can create a quagmire of bureaucratic doctrine with no viable outcome. Diverse ownerships must realise the power of inclusive, informed, deliberate decisions made in the best interest of the species on multiple scales. Sound adjudication must consider all of the factors in a public process. For that to happen, the best information, the best possible science, and the implications of different policies and decisions must be readily available and easily understood. This book represents the first comprehensive collection of materials and information based on genetics that has been dedicated to resource managers and biologists concerned with conservation and survival of the Atlantic salmon as a species in its natural environments. It is our hope that it will contribute to a successful integration of science, policy and management leading to increasing viability of this dynamic species.

References

Allendorf, F.W. and Thorgaard, G. (1984) Tetraploidy and the evolution of salmonid fishes. In: B.J. Turner (Ed.) *Evolutionary Genetics of Fishes*, pp. 1–53. Plenum Press, London.

Artieri, C.G., Mitchell, L.A. and Ng, S.H.S. (2006) Identification of the sex-determining locus of Atlantic salmon (*Salmo salar*) on chromosome 2. *Cytogenetics Genome Research*, **112**: 152–159.

Aubin-Horth, N., Letcher, B.H. and Hofmann, H.A. (2005a) Interaction of rearing environment and reproductive tactic on gene expression profiles in Atlantic salmon. *Journal of Heredity*, **96**: 261–278.

Aubin-Horth, N., Landry, C.R., Letcher, B.H. and Hofmann, H.A. (2005b) Alternative life histories shape brain gene expression profiles in males of the same population. *Proceedings of the Royal Society of London, Series B*, **272**: 1655–1662.

Behnke, R.J. (1968) A new subgenus and species of trout, *Salmo* (*Platysalmo*) *platycephalus*, from south-central Turkey, with comments on the classification of the subfamily Salmonidae. *Mitteilungen Hamburgischen Zoologischen Museum und Institut*, **66**: 1–55.

Brownstein, M.J. and Khodursky, A.B. (Ed.) (2003) *Functional Genomics: Methods and protocols*. Humana Press, Totowa, NJ.

Burgman, M. (2005) *Risks and Decisions for Conservation and Environmental Management*. Cambridge University Press, Cambridge.

Burt, A. (2003) Site-specific genes as tools for the control and genetic engineering of natural populations. *Proceedings of the Royal Society of London, Series B*, **270**: 921–928.

Cotter, D., Donovan, V.O., Ó Maoleidigh, N., Rogan, G., Roche, N. and Wilkins, N.P. (2000) An evaluation of the use of triploid Atlantic salmon (*Salmo salar*) in minimizing the impact of escaped farmed salmon on wild populations. *Aquaculture*, **186**: 61–75.

Crespi, B.J. and Fulton, M.J. (2004) Molecular systematics of Salmonidae: combined nuclear data yields a robust phylogeny. *Molecular Phylogenetics and Evolution*, **31**: 656–679.

Davis, S., Bax, N. and Grewe, P. (2001) Engineered overdominance allows efficient and economical introgression of traits into pest populations. *Journal of Theoretical Biology*, **212**: 83–98.

Dodson, J.J., Gibson, R.J., Cunjak, R.A., Friedland, K.D., García de Leániz, C., Gross, M.R., Newbury, R., Nielsen, J.L., Power, M.E. and Roy, S. (1998) Elements in the development of conservation plans for Atlantic salmon (*Salmo salar*). *Canadian Journal of Fisheries and Aquatic Sciences*, **55** (Supplement 1): 312–323.

Donaldson, E.M. and Devlin, R.H. (1996) Uses of biotechnology to enhance production. In: W. Pennell and B.A. Barton (Ed.) *Principles of Salmon Culture, Developments in Aquaculture and Fisheries Science*, no. 29, pp. 969–1020. Elsevier Publishers, Amsterdam.

Friedland, K.D., Hansen, L.P., Dunkley, D.A. and MacLean, J.C. (2000) Linkages between ocean climate, post-smolt growth, and survival of Atlantic salmon (*Salmo salar* L.) in the North Sea. *ICES Journal of Marine Science*, **57**: 419–429.

Gould, F. and Schliekelman, P. (2004) Population genetics of autocidal control and strain replacement. *Annual Review of Entomology*, **49**: 193–217.

Grewe, P. (1996) *Review and evaluation of the potential of molecular approaches for the environmentally benign management of common carp (Cyprinus carpo) in Australian waters*. Commonwealth Scientific and Industrial Research Organization. CSIRO CRIMP Technical Report 10, Melbourne, Australia.

Gross, M.R. (1987) Evolution of diadromy in fishes. In: M.J. Dadswell, R.J. Klauda, C.M. Moffit, R.L. Saunders, R.A. Rulifson and J.E. Cooper (Ed.) *Common Strategies of Anadromous and Catadomous Fishes*, pp. 14–25. American Fisheries Society Symposium 1. Bethesda, MD.

Hofmann, H.A. (2003) Functional genomics of neural and behavioral plasticity. *Journal of Neurobiology*, **54**: 272–282.

Howard, R.D., DeWoody, J.A. and Muir, W.M. (2004) Transgenic male mating advantages provides opportunity for Trojan gene effect in a fish. *Proceedings of the National Academy of Sciences*, **101**: 2934–2938.

Kapuscinski, A.R. (2005) Current scientific understanding of the environmental biosafety of transgenic fish and shellfish. *Scientific and Technical Review Office International des Epizooties*, **24**: 309–322.

Kapuscinski, A.R. and Patronski, T.J. (2005) *Genetic methods for biological control of non-native fish in the Gila River Basin*. Minnesota Sea Grant Publication F20. University of Minnesota, St Paul, MN.

Kolchanov, N. and Hofestaedt, R. (Ed.) (2004) *Bioinfomatics of Genome Regulation and Structure*. Kluwer Academic Press, Dordrecht.

McDowall, R.M. (1988) *Diadromy in Fishes*. Croom Helm Publishers, London.

McDowall, R.M. (2002) The origin of the salmonid fishes: marine, freshwater . . . or neither? *Reviews in Fish Biology and Fisheries*, **11**: 171–179.

McKinney, M. and Harmon, W. (2004) *The Western Confluence: A guide to governing natural resources*. Island Press, Washington, DC.

Muir, W. and Howard, R.D. (1999) Possible ecological risks of transgenic organism release when transgenes affect mating success: sexual selection and the Trojan gene hypothesis. *Proceedings of the National Academy of Sciences*, **96**: 13853–13856.

Nelson, J.S. (1994) *Fishes of the World* (3rd edn). John Wiley & Sons, New York.

Ng, S.H.S., Chang, A., Brown, G.D., Koop, B.F. and Davidson, W.S. (2005) Type 1 microsatellite markers from Atlantic salmon (*Salmo salar*) expressed sequence tags. *Molecular Ecology Notes*. (Published online, DOI: 10.111/j.1471-8286.2005.01056)

Philips, R.B. and Oakley, T.H. (1997) Phylogenetic relationships among Salmoninae based on nuclear and mitochondrial DNA sequences. In: T.D. Kocher and C.A. Stepien (Ed.) *Molecular Systematics of Fishes*, pp. 145–162. Academic Press, San Diego, CA.

Power, G. (1958) The evolution of the freshwater races of the Atlantic salmon (*Salmo salar* L.) in eastern North America. *Arctic*, **11**: 86–92.

Regan, C.T. (1920) The geographical distribution of salmon and trout. *Salmon and Trout Magazine*, **22**: 25–35.

Renn, S.C.P., Aubin-Horth, N. and Hofmann, H.A. (2004) Biologically meaningful expression profiling across species using heterologous hybridization to a cDNA microarray. *BMG Genomics*, **5**: 42.

Rise, M.L., von Schalburg, K., Brown, G.D., Devlin, R.H., Mawer, M.A., Kuipers, N., Busby, M., Beetz-Sargent, M., Alberto, R., Gibbs, A.R., Hunt, P., Shukin, R., Zeznik, J.A., Nelson, C., Jones, S.R.M., Smailus, D., Jones, S.J.M., Schein, J.E., Marra, M.A., Butterfield, Y.S.N., Stott, J.M., Ng, S., Davidson, W.S. and Koop, B.F. (2004a) Development and application of a salmonid EST database and cDNA microarray: data mining and interspecific hybridization characteristics. *Genome Research*, **14**: 478–490.

Rise, M.L., Jones, S.R.M., Brown, G.D., von Schalburg, K.R., Davidson, W.S. and Koop, B.F. (2004b) Microarray analyses identify molecular biomarkers of Atlantic salmon macrophage and hematopoietic kidney response to *Piscirickettsia salmonis* infection. *Physiological Genomics*, **20**: 21–35.

von Schalburg, K.R., Rise, M.L., Cooper, G.A., Brown, G.D., Gibbs, A.R., Nelson, C.C., Davidson, W.S. and Koop, B.F. (2005a) Fish and chips: various methodologies demonstrate utility of a 16,006-gene salmonid microarray. *BMG Genomics*, **6**: 126.

von Schalburg, K.R., Rise, M.L., Brown, G.D., Davidson, W.S. and Koop, B.F. (2005b) A comprehensive survey of the genes involved in maturation and development of the rainbow trout ovary. *Biological Reproduction*, **72**: 687–699.

Smith, G.R. and Stearley, R.F. (1989) The classification and scientific names of rainbow and cutthroat trouts. *Fisheries*, **14**: 4–10.

Stearley, R.F. (1992) Historical ecology of Salmoninae, with special reference to *Oncorhynchus*. In: R.L. Mayden (Ed.) *Systematics, Historical Ecology, and North American Freshwater Fishes*, pp. 622–658. Stanford University Press, Stanford, CA.

Sytchevskaya, E.C. (1986) Paleocene freshwater fish fauna of the USSR and Mongolia. *Trudy Sovmestnaya Sovetsko-Mongol'skaya Paleontologicheskaya Ekspeditsiya*, **29**: 1–157.

Thorgaard, G.H. (1983) Chromosome set manipulation and sex control in fish. In: W.S. Hoar, D.J. Randall and E.M. Donaldson (Ed.) *Fish Physiology*, Vol. 9, pp. 405–434. Academic Press, New York.

Thorpe, J.E. (1982) Migration in salmonids, with special reference to juvenile movements in freshwater. In: E.L. Brannon and E.O. Salo (Ed.) *Proceedings of the Salmon and Trout Migratory Behavior Symposium*, pp. 86–97. School of Fisheries, University of Washington, Seattle, WA.

Thorpe, J.E. (1988) Salmon migration. *Science Progress*, **72**: 345–370.

Thorpe, J.E., Mangel, M., Metcalfe, N.B. and Huntingford, F.A. (1998) Modelling the proximate basis of life history variation, with application to Atlantic salmon, *Salmo salar* L. *Evolutionary Ecology*, **121**: 581–600.

Thorsen, J., Zhu, B., Frengen, E., Osoegawa, K., de Jong, P.J., Davidson, W.S. and Hoyheim, B. (2005) A highly redundant BAC library of Atlantic salmon (*Salmo salar*): an important tool for salmon projects. *BMC Genomics*, **6**: 50.

Thresher, R.E., Hinds, L., Grewe, P. and Patil, J. (2002) *Genetic control of sex ratio in animal populations*. International Publication No. WO 02/30183 A1. World International Property Organization (WIPO). United Nations, Geneva.

Vasemägi, A. and Primmer, C.R. (2005) Invited Review: Challenges for identifying functional genetic variation: the promise of combining complementary research strategies. *Molecular Ecology*, **14**: 3623–3642.

Vasemägi, A., Nilsson, J. and Primmer, C.R. (2005) Expressed sequence tag-linked microsatellites as a source of gene-associated polymorphisms for detecting signatures of divergent selection in Atlantic salmon (*Salmo salar* L.). *Molecular Biology and Evolution*, **22**: 1067–1076.

Wilson, M.V.H. and Williams, R.R.G. (1992) Phylogenetic, biogeographic, and ecological significance of early fossil records of North American freshwater fishes. In: R.L. Mayden (Ed.) *Systematics, Historical Ecology, and North American Freshwater Fishes*, pp. 224–244. Stanford University Press, Stanford, CA.

Glossary of Terms

Words highlighted within definitions are themselves defined in the glossary.

1SW salmon   an Atlantic salmon which has spent one winter at sea before returning to fresh water to spawn; see also **grilse**

2SW salmon   an Atlantic salmon which has spent two winters at sea before returning to fresh water to spawn

Adaptability   the degree to which an organism or species is able to adjust to a wide range of environments by **physiological** or **genetic** means

Adaptation   1. the process of **genetic** adjustment of the character of a **population**, or the species as a whole, to its environment which results in increased survival and reproductive success; 2. a character of an organism or **population** that increases its **fitness**

Adaptedness   the degree to which an organism is able to live and reproduce in a given set of environments; the state of being adapted

Adaptive   conferring an increased chance of survival or of reproductive success under a particular set of environmental conditions

Additive genetic variation   heritable variation seen in a trait or character which is the result of the combined effects of **alleles** at different **genes** and is equal to the simple sum of their individual effects

Adfluvial   referring to fish which live in lakes and migrate into rivers or streams to spawn

Adult   any mature salmon

Allele   one of two or more **DNA** sequence variants of a particular gene or **locus** occurring at the same chromosomal location

Allele richness   allele diversity at a **locus** or across **loci**, standardised for differences in sample size

Allelic diversity   a measure of number of different **alleles** at a **locus**, or the average number of **alleles** across **loci**

Allozyme   **heritable** variants of an enzymatic protein, corresponding to different **alleles** at an enzyme coding **locus**, usually identified by differences in mobility when subject to **electrophoresis**

Anadromous   having a life history which involves a migration to salt water and a return migration to fresh water to reproduce; as opposed to **non-anadromous**

Aquaculture   the culture and rearing of aquatic species

Aquaculture escapees   individuals that have escaped from aquaculture facilities, e.g. freshwater hatcheries, marine farm cages

Artificial selection   differential breeding among genetic types within a cultured **population**; includes **domestication** and **selective breeding** leading to genetic and **phenotypic** change of the trait(s) in the progeny generation; as opposed to **natural selection**

Backcross 1. to cross a first-generation hybrid individual (F_1) between two different genetic types or populations with either parental type or population; 2. an individual resulting from backcrossing

BC_1 the generation of individuals derived from a particular mating or breeding programme produced by backcrossing

Biological diversity the variability among living organisms from all sources – including terrestrial, marine, and other aquatic ecosystems – and the ecological complexes of which they are a part. This includes diversity within species, between species and of ecosystems; at its most fundamental, **genetic diversity**

Biological reference point (BRP) a measure of the **stock** size (biomass or numbers) or fishing mortality defined to set management or monitoring objectives

Bottleneck a short, temporary decrease in the size of a **population** usually lasting one or a few generations

Bottleneck effect the reduction in genetic variation resulting from a **bottleneck**

Broodstock mature fish captured or reared to provide gametes (i.e. eggs and sperm) for artificial propagation

Captive breeding breeding of fish maintained in captivity

Census population size the total number of individuals in a genetic population or, in some cases, in the breeding component of the population

Chromatin the threadlike molecular complex of **DNA** and proteins found in the cell nucleus

Chromosome chromatin complex in a cell nucleus containing a single linear **DNA** strand and carrying genetic information

Cline gradual spatial variation of a character, often paralleling variation in a climatic or other environmental gradient

Co-adaptation the harmonious **adaptation** of **alleles** at different **loci** by simultaneous **natural selection**

Common garden experiment an experiment in which the performance of two different groups of organisms are compared in the same environment to control for the effect of environment on monitored differences

Conservation limit (CL) demarcation of undesirable **stock** levels, or levels of fishing activity; the ultimate objective when managing **stocks** and regulating fisheries will be to ensure that there is a high probability that the undesirable levels are avoided

Conservation unit a group of one or more local populations that share a common evolutionary lineage or genetically determined characteristics, which justify a common treatment from the point of view of conservation

Conspecific belonging to the same species

Crossing over the process in which homologous **chromosomes** undergo **recombination** and exchange **DNA**

Cryopreservation the keeping of cells live for prolonged periods of time at subzero temperatures, typically in liquid nitrogen

Cultured salmon a salmon which has been born and retained in culture for part or all of its life cycle; as opposed to **wild salmon**, **native salmon** and **non-native salmon**

Degree days the product of the daily temperature times the number of days summed over a given period of time

Deleterious recessive allele an allele that, when in a **homozygous** state, has a negative effect on individual survival or reproductive **fitness**

Deme a local **population** of potentially interbreeding individuals = **genetic population**

Diploid a genetic state where two copies of a **gene** or **chromosome**, or sets of **chromosomes** are present; also see **haploid**

DNA the polymeric molecule, <u>d</u>eoxyribo<u>n</u>ucleic <u>a</u>cid, composed of desoxy ribonucleotide bases; see Chapter 3

DNA polymerase an enzyme that constructs **DNA** from its constituent nucleotides using single-strand **DNA** as a template

DNA profiling the characterisation of an individual by variation in its **DNA**

DO dissolved oxygen

Domestication 1. the process whereby inadvertent selection in culture changes the genetic character of a **population** so as to increase **fitness**; 2. inadvertent adaptation to the culture environment

Dominant allele an allele whose effects mask those of its paired **allele** when in a **heterozygous** state

Effective population size the size of an ideal **genetic population** with a 1 : 1 sex ratio, and no variance in family contributions to subsequent generations, with the same amount of **inbreeding** or **genetic drift** as a given real **population**; as sex ratios and family contributions are not usually equal, the effective size is usually smaller than the number of breeders; see N_e

Endemic native to a particular area

Environmental variance the variation in the **phenotype** that is attributable to the environment (i.e. non-genetic)

Epistasis the masking of the action of **alleles** at a gene **locus** by **allelic** variation at other gene **loci**

Eukaryote A single- or multi-celled organism containing distinct cellular nucleus containing DNA

Evolution in a **population** context, refers to any change in **allele** frequencies over time; also known as microevolution

Evolutionary potential the capacity of a **population** to change genetically, usually in the context of **adaptation** to environmental change

F_1 generation the first filial (hence F) generation derived from a particular mating or breeding programme

F_2 generation the second filial (hence F) generation derived from a particular mating or breeding programme using **F_1 generation** fish

Fecundity the potential reproductive capacity of a female salmon measured by the potential number of viable eggs it produces

Fitness the ability of an individual, or **population**, in a given environment to survive and produce offspring so as to pass **genes** on to the next generation(s). **Fitness** is an estimate of how good an individual is at passing; see also **relative fitness**

Fixation the state of a **locus** where all **alleles** but one are lost from a **population** either by **selection** or **genetic drift**

Fluvial relating to a river

Founder effect changes in **allele** frequencies that occur when a new **population** is established by migrants from one or more established populations

Founder population the original **population**, usually small, established when a species colonises a new location

Fry   young salmon from when they cease to be dependent on the yolk sac as primary source of nutrition until they have dispersed and become territorial

F_{ST} a statistical measure of the proportion, from 0 to 1, of genetic variance within or among populations

Gamete   a mature sperm or egg containing a single, **haploid chromosome** set

Gametic phase disequilibrium   the condition where the **alleles** at two different **loci** are not associated randomly in an individual

Gene   linear stretches of **DNA** that constitute the functional units of inheritance and, through **transcription** and **translation**, control cellular processes

Gene bank   a collection of **gametes** or individuals, maintained to conserve **genetic diversity** for subsequent exploitation in **selective breeding** or as part of a species or **population** conservation programme

Gene diversity   the variance in **allele** frequencies in a **population** at a given **locus** and equal to the expected **heterozygosity** under Hardy–Weinberg equilibrium

Gene flow   the incorporation of **genes** from one **population** or species into another through **migration** and interbreeding

Gene locus   the physical location of a **DNA** sequence defining a gene; see **gene**

Gene pool   the collective allelic **variation** at all gene **loci** possessed by the potential breeders in a sexually reproducing **genetic population**

Genetic distance   a measure of the differences in **allele** types and frequencies between two populations, or species, at one or a number of **loci**

Genetic diversity   differences among **genetic populations** within species and among **species** in the types and frequencies of **alleles** they possess at **DNA loci**

Genetic drift   random changes in the frequency of **alleles** due to chance differences in survival or reproductive success among different **genotypes**

Genetic sampling   the process of selecting a proportion of all **loci** for genetic analysis to provide information from which to draw inferences about the genome as a whole

Genetic population   a group of sexually reproducing individuals and their relatives, within which mating is more or less random but among which interbreeding is constrained, so that they constitute a distinct **gene pool**; see also **deme, genetic stock, stock**

Genetic stock   used by some as equivalent to a **genetic population**, i.e. a **stock** defined in population genetic terms

Genetic variance   the variation in **phenotype** attributable to the genetic variation; see also **genotype**

Genotype   the hereditary or genetic constitution of an individual defined by its particular combination of **allelic** variants possessed at one or more **loci**, or, in its most general sense, at all **loci**

Genotype × environment (G × E) interaction   when the **phenotypic** response by two or more **genotypes** to environmental change differs

Grilse   a small adult salmon after one sea-winter found in a river or coastal marine area; in practice often defined by size (e.g. < 3 kg)

Haploid   a genetic state where only one copy of a **gene** or **chromosome**, or set of **chromosomes**, is present; also see **diploid**

Hardy–Weinberg equilibrium   the distribution of genotypes at a locus which results from random mating when there is no mutation, migration, natural selection, or random drift

Hemizygous the state of having only one copy of a particular gene, e.g. in animals where males are defined by having a distinct Y chromosome, males are hemizygous for genes found on that chromosome

Heritability in general, the extent to which observed variation in a trait is determined by an individual's **genes**; in a specific sense, the proportion of observed variation in a trait in a **population** which is determined by **additive genetic variation**

Heritable a trait possessed by an individual by virtue of the **genes** derived from its parents

Heterochromatin chromatin that remains tightly coiled throughout the cell cycle

Heteroplasmy genetic heterogeneity within an organism with regard to its mitochondrial **DNA**

Heterozygosity the proportion of **heterozygote**s, either across **loci** within an individual or across individuals in a **population**; or see also **expected heterozygosity, observed heterozygosity, Hardy–Weinberg equilibrium**

Heterozygote an individual that carries two different **alleles** at a given **locus**

Heterozygote advantage when selection favours **heterozygote** individuals over the two respective **homozygotes**

Heterozygous the state of being a **heterozygote**

Homozygote individual with two copies of the same allele in one **locus**

Homozygous the state of being a **homozygote**

Hybridisation crossing of individuals from different **genetic populations** (intraspecific hybridisation) or species (interspecific hybridisation) which results in offspring

Hybrids offspring derived from **hybridisation**

Hydrochemistry the chemical composition of water which determines its physical properties and character

Hydrology the distribution, behaviour and character of water on the earth: the study thereof

Identical by descent (IBD) identical, as in the case of **alleles**, by virtue of derivation from a common ancestor

Immigration in a genetic sense, the movement of individuals into a **population** and subsequent successful reproduction; see also **gene flow**

Inadvertent selection changes in **allele** frequencies resulting from unintended non-random **selection** of **broodstock**, usually in populations in culture

Inbreeding 1. in general sense, the successful mating of closely related individuals (e.g. siblings or first cousins) or of individuals more closely related to each other than the average within a **population**; 2. when the average relatedness of individuals breeding in a **population** increases over time

Inbreeding depression a decline in the **fitness** of the individuals, with regard to either survival or reproductive success, due to **inbreeding**

Interphase the metabolically active, nondividing stage of the cell cycle

Introgression the movement of new **genes** into a wild **population** by **backcrossing** of **hybrids** arising from **hybridisation** with individuals from a second **population**

Island model with respect to **metapopulation** dynamics, a scenario where all component populations within a **metapopulation** exchange migrants equally

Isolation by distance gene flow is a function of the physical distance separating populations and gives rise to a positive association of physical separation and genetic differentiation between individuals or populations

Iteroparity spawning on more than one reproductive cycle

Kelt an anadromous salmon that has completed spawning but has not yet returned to the sea
Kinship coefficient a measure of the probability that **alleles** drawn randomly from two individuals are **identical by descent**
Kinship value an adjustment of **mean kinship** based on breeding potential of each animal in the present generation
Lacustrine relating to a lake environment
Life-history trait a trait which relates to the way that an organism lives its life
Lifetime reproductive success the number of viable offspring left by an individual in the next generation over the course of life
Lineage a genetically distinct grouping of individuals or populations by virtue of having evolved independently from such groups
Linkage the physical connection of **loci** on the same **chromosome**
Live Gene Bank a collection of live individuals selected to be genetically representative of particular **stocks** or **genetic populations**
Local adaptation the evolutionary adjustment of the genetic character of a **population** which increases **fitness** in its local environment
Local population the **genetic population** native to a particular locale
Loci plural of **locus**
Locus a specific stretch on the **DNA** of a **chromosome** including those defining **genes**
Maternal effect a non-heritable influence of the **phenotype** or **genotype** of the female parent on the development or growth of its offspring
Maturation the developmental process leading to individuals becoming capable of breeding
Mature parr usually a sexually mature male parr in an anadromous **population**; mature female parr are very rare
Mean kinship the average **kinship coefficient** of an individual relative to all other individuals in a **population**, including itself
Meiosis the two successive nuclear divisions of a diploid nucleus resulting in the formation of **haploid** gametes with half the genetic material of the original cell
Meristic characters body traits that can be counted (e.g. number of vertebrae, fin rays)
Metapopulation a regional group of **genetic populations** or **demes** with similar but independent risks of extinction where extinctions are countered by **migration** and recolonisation from the extant populations within the group
Microsatellite a region of **DNA** containing variable numbers of simple, tandemly repeated units of two to eight **DNA** bases, e.g. CTCTCT or CTGGCTGGCTGG
Migration the directed movement of an organism from one habitat or location to another; often also equated in genetics texts with **gene flow**
Milt the male seminal fluid of fish containing its sperm
Minisatellite a region of **DNA** containing variable numbers of tandemly repeated units of 10+ **DNA** bases; the character of the repeated sequence is often complex
Mitochondrial DNA (mtDNA) DNA found in the mitochondria, energy-producing organelles in the cell (see Chapter 2)
Mitosis the process of nuclear division in **haploid** or **diploid** cells producing identical daughter cells
Monophyletic derived from two or more distinct evolutionary lineages; see also **polyphyletic**
Morphometric characters body traits that can be measured (e.g. body length, fin size)
Multi-sea-winter (MSW) a salmon which has spent two or more winters at sea

Mutation   permanent change in the **DNA** that can be inherited

N_c the shorthand term used to refer to the census size of a population

N_e the shorthand term used to refer to the effective genetic size of a population; effective population size

Native salmon   a salmon whose ancestors are associated with a given location; see also **wild salmon, non-native salmon**

Natural selection   the natural process by which the **genotypes** in a **population** best suited to their environment survive better and leave more descendants than those less well suited; see also **fitness, adaptation**

Neutral genetic variation   genetic variation believed to be unaffected by natural selection

Non-anadromous   having a life history which does not involve a migration to saltwater but completion of the entire life cycle in freshwater; as opposed to **anadromous**

Non-native salmon   having ancestors not associated with a given location; see also **native salmon**

Non-neutral genetic variation   genetic variation believed to be affected by natural selection

Nucleotides   a class of basic molecules joined together to form **DNA** and **RNA**

Null allele   an allele that for a number of reasons produces no functional product, and therefore is not detected in standard analyses

Outbreeding   the mating of genetically different organisms, either from different populations of the same species or from different species

Outbreeding depression   a reduction in **fitness** arising from **outbreeding**

Pacific salmon   seven species in the genus *Oncorhynchus* native to the rivers draining into the Pacific Ocean

Panmixis   random mating; individuals within a **population** have equal opportunity and probability for mating with each other

Parr   juvenile salmon after the fry stage, named for the characteristic black 'parr' marks which develop at this time on the sides of their bodies

PCR   short for Polymerase Chain Reaction, a molecular method to selectively amplify and produce large quantities of a targeted stretch of DNA sequence which can then be analysed by RFLP analysis or sequencing. Theoretically, PCR can take one molecule and produce measurable amounts of identical DNA in a short period of time. It is used in DNA fingerprinting and DNA sequencing

Performance trait   a trait which affects the physiology, morphology or behaviour of an individual

Phenotype   the state of an individual with respect to a given trait, e.g. blue with respect to eye colour, arising from the expression of its **genotype** in its particular environment; in the general sense, the overall character of an individual

Phenotypic correlation   the degree of association between the **phenotype** values of two traits

Phenotypic expression   the observed character of an individual for a given **genotype** in a particular environment

Phenotypic plasticity   the capacity to alter the **phenotype** arising from a given **genotype** in response to variation in environmental conditions

Phenotypic trait   an observable feature of an individual that results from the interaction between its **genotype** and the environment

Phenotypic variance   the total amount of variation observed in the **phenotype**; can be partitioned into genetic and environmental variances

Phylogeography the geographical distribution of evolutionary lineages and the determining historical processes; the study thereof

Polyandry the mating of one female with many males

Polygyny the mating of one male with many females

Polyphyletic derived from two or more distinct evolutionary lineages; see also **monophyletic**

Polytypic a taxon such as a species which is composed of two or more subtaxa; as opposed to **monotypic**

Population see **genetic population**

Post-smolt young anadromous salmon in the sea, at stage from leaving river until middle of first sea-winter

Precautionary approach a concept enshrined in Principle 15 of the Rio Declaration of the UN Conference on Environment and Development which states: 'In order to protect the environment, the precautionary approach shall be widely applied by States according to their capabilities. Where there are threats of serious or irreversible damage, lack of full scientific certainty shall not be used as a reason for postponing cost-effective measures to prevent environmental degradation'

Quantitative trait a character which shows measurable continuous variation

Quantitative trait loci (QTL) locations on **chromosomes** with **genes** contributing to the inherited variation observed in a **quantitative trait**

Random genetic drift see **genetic drift**

Reaction norm a property of the **genotype** which relates to the nature and extent of the **phenotypic plasticity** it displays

Recessive allele an **allele** whose effect is expressed in an individual only when it occurs in a homozygous state

Recombination the process whereby strands of DNA during **meiosis** break and recombine, exchanging complementary stretches of **nucleotide** bases generating new **allelic** variants

Recruitment the addition to a **stock** or **population** of new individuals as a result of reproduction

Recruits the new individuals of a year class entering a **stock** or **population**

Redd the physical structure on a stream bed made of gravel and created by a salmon for the deposition of its eggs

Reference point a value derived from an agreed scientific procedure which corresponds to a state of the resource and/or of the fishery used as a guide for management

Relative fitness the relative ability of an individual, **genotype** or **phenotype**, to survive and contribute genetically to the subsequent generation(s); see also **fitness**

Reproductive isolation the inability to interbreed, usually used in the context of species but also at a conceptual level applicable to isolated populations

Reproductive segregation the spatial or temporal separation of the reproduction of different groups of individuals

Restriction enzymes enzymes, known as endonucleases, which cut **DNA** when they encounter particular sequences of four to six bases

RFLP Restriction Fragment Length Polymorphism is variation detected in the base sequence of **DNA** using restriction enzymes; see **restriction enzymes**

RNA (ribonucleic acid) a usually single-stranded polymeric molecule consisting of ribonucleotide building blocks

Sea age the number of winters that salmon has remained at sea before returning to spawn

Secondary sexual character   a **phenotypic trait** which is associated with the sex of an individual but not part of the reproductive organs

Segregation   the two **alleles** at a **locus** in a parental organism will segregate into different gametes at **meiosis**

Selection   see **natural selection, artificial selection, inadvertent selection, domestication**

Selective breeding   the deliberate choosing of individuals with desired **phenotypic** or genotypic traits to establish or perpetuate a **population** in culture

Selectively neutral   an **allele** in the **population** whose frequency is influenced more by **genetic drift** than by **natural selection**

Semelparous (semelparity)   spawning on a single reproductive cycle only

Sex chromosomes   the **chromosome** pair in eukaryotic organisms whose divergent **gene** make-up determines the sex of an individual; in many species, such as humans, these are visibly structurally divergent (see Chapter 3)

Smolt   fully silvered juvenile salmon migrating or about to migrate to sea

Smolt age   the number of winters post-hatch that a salmon remained in fresh water prior to emigration as a **smolt**

Spawner escapement   numbers of salmon that survive to spawn, usually calculated after the removal effects of fisheries, predation and disease have occurred

Spring salmon   multi-sea-winter salmon which return to fresh water early in the year relative to the overall distribution, generally before May

Stock   a group of individuals of a species defined on the basis of arbitrary management criteria, such as river of origin, area of capture, or time of capture, which may encompass part of, all of or more than one **genetic population**. This term is generally used to describe salmon either originating from or occurring in a particular area; a **stock** defined in genetic terms is a **genetic population**

Strain   a cultured **population** of individuals showing a particular **phenotype** as a result of its unique genetic character

Subpopulation   a term, confusingly, used interchangeably in the literature with **genetic population**, particularly in the theoretical literature, when **population** is used to refer to the species as a whole

Sustainable use   the use of components of biological diversity in a way and at a rate that maintains biological diversity in the long term so it can meet the needs and aspirations of present and future generations

Thermal tolerance   the temperature conditions within which a **genetic population** or species can exist, something which typically varies with life stage

Transgenic   an organism to which genetic material derived from a different species has been added by human intervention

Triploid   an organism with three sets of **chromosomes**, one more than the normal complement of two

VNTR   encompassing both **microsatellite** and **minisatellite loci**, an acronym used to refer to **loci** with Variable Number of Tandem Repeats of **DNA** bases

Wild salmon   having spent its entire life cycle in the wild

Index

1 sea-winter salmon (1SW), 37

Acantholingua ohridana, 120–2
acidity of sites, 31
adaptation theories, 196–8
    see also local adaptations
Aeromonas salmonicida salmonicida, 370
AFLP (amplified fragment length polymorphism) analysis, 102, 169
aggression, and domesticated stock, 209, 366–7, 369
all-female triploid salmon, 387
alleles, 69, 481
    frequency analysis and phenotypic divergence, 215–16
    and mutation, 71–2, 93–5
    and population size, 241–3, 250, 251, 256
    technical artefacts, 92–3
    see also genetic variation
allozymes, 77, 481
    variation analysis, 96–8
        effects of selective harvesting, 315
        in Europe, 131–7
        for non-anadromous populations, 144–6
        in Western Atlantic, 137–43
        for wild–domesticated salmon comparisons, 364–5
amplified fragment length polymorphism (AFLP) analysis, 102, 169
anadromous populations
    classification and taxonomy, 21
    distribution and geographical range, 21–2, 25–8
    habitat needs cf. non-anadromous populations, 404–6
    life cycle patterns
        egg survival, 30–2
        homing and return, 38–40
        marine life, 36–8
        parr sexual maturation, 34–5
        parr movements, 35
    reproduction, 28–30
    smolt migration, 36
    skin colouration's and markings, 19–20, 36
androgenic fish, 65
assignment tests, 103–4, 273–80
    application to Atlantic salmon, 274–7
    background to methodology, 277–80
    power of tests, 279–80
    range of methods, 278–9
assortative mating, 92
Atlantic salmon (*Salmo salar*)
    background and history, 471–2
    classification and taxonomy, 21
    current concerns, 472–3
    distribution and geographic range, 21–2
    future considerations, 473–8
    genetic factors
        amino acid sequences, 83
        amount of DNA material, 60
        characterisation of wild populations, 95–102
        gene map, 68
        number of genes, 68
    hybrids, 180–4
    life history variations, 22–5
    sex determination, 63
    survival rates, marine phases, 37–8
    see also anadromous populations; non-anadromous populations

BAC (bacterial artificial chromosome) contig mapping, 474
bag nets, 301
Baltic Sea, 26–7, 346–7
    consequences of habitat disturbances, 413
    mitochondrial DNA (mtDNA) divergence studies, 134–5
    phylogeographic diversity studies, 133–7
    mtDNA analysis, 134–5
basal transcription factors, 66
baseline sampling, 288

Bay of Fundy (Canada)
    distribution of salmon, 24, 239
    live gene banking programmes, 432–46
    management recommendations, 460–2
    phylogeographic diversity, 138–43
    rates of population decline, 432
Baysian methodologies, 274, 277–80
    cluster techniques, 279
    mixture modelling, 280, 285–7
bend nets, 301
Big Triangle Pond (Newfoundland), 43
biodiversity
    general overview, 158–61
    losses due to habitat disturbances, 411–12, 413–15
    phylogeographic diversity, 121–53
    regional and local structures, 153–8
bioinformatics, 476
body size of populations, and fishing, 314–16
'Boreal' race salmon, 131–3
'bottlenecks', 90, 240–1, 242
    study examples, 252–5
breeding sites, choice issues, 28–30
Bristol Cove River (Newfoundland), 44–5
brown trout (*Salmo trutta*), 21
    evolutionary relatedness, 120–2
    hybridisation, 30, 180–4
        with farm-escaped salmon, 374–5
    interspecific competition, 30
Burrishoole system studies, 210
    farm escapes, 377–82
    lifetime success and performance characteristics studies, 326
Bygglansfjord Lake (Norway), 41, 43

'C value paradox', 60
C values, 60
Candlestick Pond (Newfoundland), 45
captive broodstock programmes, 341
Carlin tagging, 271
catch methods, 301–3
catch and release fishing, 303
catch statistics, 303–4
cDNA microarrays, 475
cell replication, 60–3
'Celtic' race salmon, 131–3
census population sizes (N_c), 247
'central dogma of molecular biology', 66
centromeres, 63
choice of mate *see* mate choice
chromatin
    cell division processes, 60–3

nature and structure, 60, 61
    replication and growth, 60–3
chromosome manipulation, 65
chromosome maps, 474
chromosome set manipulation, 475
chromosomes
    cell division processes, 60–3
    detection, 76–7
    karyotypes for Eastern Atlantic salmon, 124
    karyotypes for Western Atlantic salmon, 125
    replication and growth, 60–3
    structural organisation, 60, 61
chum *see Oncorhynchus keta* (chum)
cichlid fish, 122
climate change, 412–18
closed cultures, 385–6
coastal fishing, 301–3
coastal protection zones, 386, 388
coded wire tagging, 271
codominant alleles, 69
codons, 62, 66
coefficient of heritability (h^2), 94–5
Columbia River, 281
'common garden' experiments, 210–11
competition
    among emerging fry, 33
    among emerging parr, 34
    among farm-escapees, 370
    among males, 29, 35
    among nesting females, 28
    among transferred stock, 208
conservation guidelines
    and effective population size, 301
    see also management and conservation
Cove Brook, 143
cryopreservation of sperm, 386, 451–8
cultured stock *see* domesticated stock; gene banks
cutthroat trout (*Oncoryhynchus clarkii*), 21
    evolutionary relatedness, 120–1

D-loop haplotype variations, 126–8
    see also mitochondrial DNA (mtDNA)
dams
    consequences of habitat reduction, 407–8
    mitigation and conservation programmes, 345–51
databases of genotypes, 275–6
deformities, amongst domesticated stock, 209
deletions, 89
deme *see* population
'designer strains', 338
diet and feeding patterns, marine phases, 37

'diploid' states, 58
directed genetic change *see* selection
disassortative mating, 92
diseases and parasites, 370–71
   local adaptive response studies, 211
distribution and geographical range, 21–2
   anadromous populations, 25–8
   historical origins, 147–53
   non-anadromous populations, 41–2
   see also phylogeographic diversity
divergence studies *see* phylogeographic diversity
DNA
   characteristics and molecular structure, 58–60, 61, 63–6
   measurements and analysis methods, 60, 74–83
   replication, 60–3
   see also mitochondrial DNA (mtDNA); nuclear DNA (nDNA)
DNA base sequences, 67–8, 79–80
   analysis methods, 78–9
DNA restriction enzyme analysis, 80, 82
domesticated stock
     breeding behaviours, 373–4
     founder effects, 362–3
     genetic differences, 362–8
     phenotypic traits, 365–8
   composition information needs, 271, 291–3
     genetic tagging methods, 271–91
     traditional tagging techniques, 271
   mixed stock vs. population-specific stock, 272–3
   morphological responses, 209
   origin of strains, 363
   performance comparisons, 209, 338–9
   phenotypic comparisons, 365–8
   see also farm-escaped salmon; fisheries exploitation
drift nets, 301, 304, 316
Drosophila melanogaster, 241

Eastern Atlantic salmon
   phylogeographic diversity, 131–7
     evidence for evolutionary divergence, 123–30
ecological correlates, for fitness-related traits, 204–8
Economy River (Inner Bay of Fundy), 239
effective population size (N_e), 89–90, 104, 243–51
   analysis using molecular markers, 170–1
   basic concepts, 243–5
   calculation methods, 249–51
   and census population size, 247
   determining minimum sizes, 245–7
   examples, 252–5
   impact of fishing, 313–14
   influencing factors, 247–9
   mathematical equations, 248
   in subdivided populations, 310
egg sizes, 30
egg survival characteristics
   anadromous populations, 30–2
   non-anadromous populations, 45
electrophoresis, 96–8
electrophoretic analysis, 474
energy production, cellular level, 67
environmental conditions
   and declining populations, 316
   general characteristics, 25–7
   and hybridisation, 182
   and local adaptation, 197–8, 201–3, 217–18
     clinal distributions, 208
     management implications, 219
   see also habitats
epistatic genetic variance, 75
estuary-only marine migration, 23
Europe, phylogeographic diversity, 131–7
evolutionary divergence, 121–53, 198
   evidence studies, 123–30
   glaciation impacts, 148
   see also phylogeographic diversity
evolutionary relatedness, 120–1
   see also relatedness patterns
exploitation *see* fisheries exploitation
Exploits River (Newfoundland), 305
expressed sequence tag (EST) databases, 475
extinction
   defined, 340
   evaluating risks, 340
   supportive breeding programmes, 341–2
extinction–recolinisation dynamics, 91, 309
   see also stocking programmes
extragenic DNA, 68–9

farm-escaped salmon, 328, 357–89
   Burrishole experiment, 377–82
   extent and magnitude, 360–2
   genetic differences from wild salmon, 362–8
   harm reduction measures, 385–7
   impact on wild population, 368–72
   Imsa experiment, 375–7
   introducing 'closed cultures', 385–6, 388
   long-term genetic implications, 382–7
   management recommendations, 388–9

reproduction impacts of farm salmon, 372–5
    studies and research, 375–82
farmed stock *see* domesticated stock
feeding areas, ocean, 37–8
feeding materials, marine diets, 37
FISH (fluorescent in situ hybridisation), 76
fisheries exploitation, 300–18
    background and history, 301–9
        catch methods, 301
        catch statistics, 303–4
        exploitation rates, 304, 305–9
        potential for selection, 304
    as ecological and evolutionary force, 309–13
        directed genetic change, 312–13
        undirected genetic erosion, 309–12
    and effective population size, 313–14
    impact on phenotypic and evolutionary
        characteristics, 314–17
    see also domesticated stock
Fisher's opposing forces of evolution, 198
fishless systems, 330–31
    reintroducing salmon, 330–39
        management considerations, 339
        timescales, 338
'fitness'
    and adaptation, 197–201
    and heterozygosity, 242–3, 244
founder effects, 241, 362–3
'founder-flush' models, 251
Fraser River, 281
freshwater–freshwater migration patterns, 35, 42–5
freshwater–marine water migration, 36–40
freshwater phases
    in anadromous populations, 25–36
    in non-anadromous populations, 42–5
F_{st} differentiation levels, 274–5, 279
functional genomics
    new techniques, 473–7
        key approaches, 476–7
    see also genetic identification techniques
furunculosis epidemics, 370

G × E interactions, 95, 200–201
Gadus morhua (Atlantic cod), 274, 301
gametes, 60
gametogenesis, 72–3
Gander River (Newfoundland), 42, 154–5, 305
Gaspereau River (Nova Scotia), 24
gene banks, 386, 425–64
    background and concerns, 426–32
        long-term genetic implications, 426–32

current programmes, 432–46
history of programmes, Norway, 455–7
management recommendations, 446–51, 463–4
research areas, 458–62
use for conservation of remnant populations, 446–51
use of cryopreserved sperm, 451–8
gene development, 69–71
gene duplication, 121
gene expression, 66, 70–1, 477–8
    cellular processes, 67
    regulation mechanisms, 66, 68–9
gene flow *see* migration (gene flow)
gene organisation, 66–9
    molecular nature and structure, 66
    number and distribution, 66–8
    occurrence of extragenic material, 68–9
gene promoters, 66, 71
gene silencers/repressors, 66
genetic 'architecture', 334–6
genetic change *see* genetic drift; selection
genetic differentiation levels (F_{st}), 92, 274–5, 279
genetic drift, 89, 217–18, 251
    and biodiversity, 122–3, 217–18
    effect on allelic frequencies, 250, 251, 256, 430
    and fisheries exploitation, 309–12
    long-term effects on captive populations, 431–2
    and non-anadromous populations, 146–7
genetic identification techniques, 271–92
    advantages over traditional tagging methods, 271–2
    databases of genotypes, 275–6, 476
    future possibilities, 473–8
    of individuals, 273–80
        application to Atlantic salmon, 274–7
        methodologies, 277–80
    of population contributions, 280–9
        application to Atlantic salmon fisheries, 282–9
        methodologies, 284–9
    limitations, 272
genetic 'mark–recapture' analysis, 105
genetic markers *see* molecular markers
genetic population *see* population
genetic variation, 71–83, 87–8
    basic origins, 71–2, 73
    characterisation using molecular markers, 88, 95–102, 119–20
    concepts, 89–90
    detection, 74–83

genetic variation (cont'd)
   and fitness-related traits, 199–201
   future directions, 107–9
   geographical distribution areas, 199
   impact of fisheries exploitation, 309–13
   impact of reduced population size, 240–2
      measuring loss, 242–3
      and minimum effective size of populations, 245–7
   importance of diversity, 240–2
   occurrence among populations, 88, 211–17
      and allele frequency analysis, 215–16
      differentiation analysis, 92–3
      relationships between populations, 90–1
   occurrence within populations, 88, 211–17
   quantitative trait analysis, 93–5
   range of environmental and phenotypic variations, 201
   scope, 72
   see also local adaptations; phylogeographic diversity
genetically modified salmon, 384–5
genome duplication, 65–6
Genomics Research on Atlantic Salmon Project (GRASP), 474–5
genotype by environment (G × E) interaction, 95, 200–201
'genotypes', 69
geographical boundaries, and local adaptations, 198–9
geological features
   general characteristics, 25, 41
   see also environmental conditions; glaciations and glacial cycles
Girnock Burn, 177–8
glaciations and glacial cycles, 122, 148–53, 159, 416
global climatic changes, 412–18
graylings, and habitat fragmentation studies, 408
grilses, 37
gross mutations, 72
growth hormone treatments, 366, 368
Gulf of Bothnia, 413
Gulf of St Lawrence, 37
Gullspangalven, 42, 45
gynogenetic fish, 65
Gyrodactylus salaris parasite, 211, 212–14, 313, 370, 455–7

habitats
   basic requirements, 400–403
   and genetic influences, 404–6
   degradation, 304, 316, 410–11
   expansion issues, 409–10
   fragmentation problems, 408–9
   and global climate change, 412–18
   loss of biodiversity, 411–12, 413
   reduction in availability, 407–8
   see also environmental conditions
'haploid' states, 58
Hardy–Weinberg equilibrium, 92–3, 249
hatchery-reared stock see domesticated stock
Haukvik gene bank (Norway), 425
heavy metal pollution, 411
heritability, 199
   and phenotype analysis, 75–6
   and quantitative genetic variation analysis, 94–5, 199
heterogametic sex, 63
heterozygosity, 69
   correlation with fitness, 242–3, 244
   measuring loss of genetic variation, 242–3
heterozygotes, 69
histones, 58–9
historical distribution patterns, 28
homing and return, cues and trigger factors, 38–9
homogametic sex, 63
homozygotes, 69
Hudson Bay, 42
Humber River (Newfoundland), 305
hybridisation, 30, 180–4, 276
   and farm escapes, 382–5
   with brown trout (*Salmo trutta*), 374–5
   interspecific, 90, 92–3, 107, 182–4
hydroelectric dams, mitigation and conservation programmes, 345–51

IA techniques see individual assignment (IA) methods
Iberian Peninsula, 414
iBOF salmon see Bay of Fundy (Canada)
Icelandic rivers, exploitation rates, 306
ICES see International Council for the Exploitation of the Sea
'idealised populations', 245, 248
identification techniques see genetic identification techniques
ideograms, 64, 76
Imsa experiment, 375–7
in-river fisheries, catch trends, 303
inbreeding, 90
   calculating coefficients (f), 260–2
   and kin recognition, 107, 185

measuring equilibrium deviancy, 92–3
and small populations, 256–9
    effects of accumulation over generations, 258–9
and supportive stocking programmes, 342
incubation periods, 31
Indian Bay Brook (Newfoundland), 305
individual assignment (IA) methods, 273–80
    application to Atlantic salmon, 274–7
    background methodology, 277–80
        different assignment methods, 278–9
        power of tests, 279–80
        principles of assignment tests, 277–8
infectious pancreatic necrosis (IPN), 370–1
infectious salmon anaemia (ISA), 370–1
inheritance mechanisms
    chromatin and chromosomes, 60–6
        nature and structure, 60, 61
        number and ploidy levels, 63–6
        replication and growth, 60–3
    DNA control, 58–60
    genes
        and development, 69–71
        and genome organisation, 66–9
    see also genetic variation
inherited resistance, local adaptive response studies, 211
inner BOF see Bay of Fundy (Canada)
insertions, 89
International Council for the Exploitation of the Sea (ICES), 302–3
introgression, 383
    see also hybridisation
IPN see infectious pancreatic necrosis
Ireland, phylogeographic diversity, 131–3
ISA see infectious salmon anaemia
'island model' of population structure (Wright), 90

'junk' DNA, 68
juveniles
    capture programmes, 341
    escapes, 362

Kapisidlit River (West Greenland), 195
Kelt reconditioning programmes, 341
Kielder hatcheries (Scotland), 334
kin recognition, 107, 185
kin-biased behaviours, 185–6
Klaralven River, 42, 45
Kogaluk River (N. America), 42
Kyrksaeterora strain, 363

Lake Imandra (Kola Peninsula), 41
Lake Jänisjärvi (Finland), 41
Lake Lagoda (Russia), 41, 42–3, 45, 147
Lake Ohrid (Macedonia), 120–1
Lake Onega (Russia), 41, 42–3, 147
Lake Ontario (North America), 42–3, 45, 331–2
Lake Saimaa (Finland), 41, 42
Lake Sandal (Finland), 41
Lake Vänern (Sweden), 41, 42–3, 45
Lake Victoria (Africa), 122
Landcatch strain, 276
'landlocked' forms, 23
    historical origins, 143, 152–3
    spatio-temporal distribution studies, 277
    see also non-anadromous populations
'large allele drop-out', 92
LeHave strain (Nova Scotia), 332
Lepeophtheirus salmonis (marine salmon lice), 370
LGB (live gene banking) see gene banks
lice infestations, 370, 455–6
life cycles
    general characteristics, 19–20
    key patterns, 23
    in anadromous populations, 28–40
    and local adaptation, 202–3
    in non-anadromous populations, 25, 42, 43–5
    physiological changes, 36
life span, effects of selective harvesting, 315
Liscomb River (Nova Scotia), 306
'listed' subspecies, 131
Little Codroy River (Newfoundland), 305
Little Gull Lake (Newfoundland), 44, 146–7, 154–5, 405
Little Moose Lake (New York), 45
local adaptations, 198–204, 406
    challenges to hypothesis, 211–18
    essential conditions required, 198–9
        evidence studies for conditions, 199–204
        evidence studies of adaptations, 204–11
            direct evidence, 210–11
            indirect and circumstantial, 204–9
        implications for management, 219
Loch Rannoch, 337
loci of DNA, 69
Lomond River (Newfoundland), 305
long-line fishing, 301, 304

M74 syndrome, 27
Machias River (Maine), 306
Mactaquac Dam, 348–50

Maine salmon
    exploitation rates, 306
    phylogeographic diversity, 138–43
        and distinctiveness characteristics, 143
        and non-anadromous populations, 146
major histocompatability complex (MHC) genes, 101, 108, 179–80
malic enzyme locus (*MEP-2**) allelic distributions, 96, 123, 208
management and conservation
    contentious issues, 473
    impact of population sizes, 240–65
        loss of genetic variation, 240–3
    preserving biodiversity, 130–1, 159–60
        use of metapopulation modelling, 263–4
    problems associated with captive breeding programmes, 427–9
    and stock composition information needs, 271, 291–3
marine life and distributions, 36–8
marine salmon lice, 370
'mark-recapture' analysis, 105
Markov chain Monte Carlo (MCMC) methods, 280
mate choice, 179–80
    and hybridisation, 180–4
    and kin recognition behaviours, 185–6
    under natural conditions, 179–80
mating systems
    analysis using genetic markers, 169–70
    definitions, 168
    and effective population size, 170–1
    reproductive success, 177–8
        and male alternative reproduction tactics, 171–7, 187
maximum likelihood assignment tests, 141, 275–6
maximum likelihood estimations (MLE), 280, 284–7
maximum sustainable yield (MSY), in subdivided populations, 311–12
meiosis, 62
Mellingselva (Norway), 43–4
Mendelian genes, 70
MEP-2* (malic enzyme locus), 96, 123, 208
messenger RNA (mRNA), 66–7
metapopulation theories
    and conservation of small populations, 263–4
    and reintroduction programmes, 338
    see also 'subdivided populations'
MHC see major histocompatability complex (MHC) genes

microarray development, 475
microsatellite DNA analysis, 69, 82–3, 99–101, 289–90
    and continental divergence studies, 129–31, 145–6, 275–6
    future directions, 108
    and population identification, 282
    and stock identification, 272–3, 275–7, 282, 289–92
    technical artefacts, 92
Middle Brook (Newfoundland), 305
migration (gene flow), 89
    and long-term phylogeographic divergence, 158
    measuring equilibrium deviancy, 92–3
    and metapopulations, 156–8
    models, 90–1
    and population reductions, 259–63
Miramichi River (New Brunswick), 143, 306
mitochondrial DNA (mtDNA), 58–60, 98–9
    and continental divergence studies, 126–9
    in Eastern Atlantic, 134–5
mitosis, 61–2
mixed-stock analysis (MSA), 103–4, 273, 280–9
    application to Atlantic salmon fisheries, 282–9
    application to Pacific salmon fisheries, 280–2
    background and methodologies, 284–9
        maximum likelihood estimates, 284–7
        mixed vs. baseline sampling, 288
        reliability estimates, 288–9
MLE see maximum likelihood estimations (MLE)
molecular markers, 88, 95–102
    allozyme electrophoresis, 96–8
    amplified fragment length polymorphism (AFLP), 102
    differences between wild and domesticated salmon, 364–5
    future directions, 107–9
    major histocompatability complex (MHC) genes, 101
    microsatellite DNA, 69, 82–3, 92, 99–101
    mitochondrial DNA, 98–9
    single nucleotide polymorphisms (SNPs), 101–2
Moray Firth, 307
morphological abnormalities, domesticated vs. wild stock, 209
mortality rates, 37–8
Mowi strain (Norway), 363
MSA see mixed-stock analysis (MSA)
multi-sea-winter salmon (MSW), 37
multidimensional scaling plots, 130–1, 142

multiple mating patterns, 29, 178–80, 184, 187–8
mutation, 71–2, 89
   see also genetic variation

Namsen River (Norway), 43–5
Narraguagus River (Maine), 306
NASCO *see* North Atlantic Salmon Conservation Organization
NE Placentia River (Newfoundland), 305
nest construction, 28–30
nets, recommendations to reduce escapes, 385
neural plasticity, 477–8
non-anadromous populations
   classification and taxonomy, 21
   distribution and geographical range, 21–2, 41–2
     phylogeographic diversity, 143–7
   habitat needs cf. anadromous populations, 404–6
   historical origins, 147–53
   see also 'landlocked' forms
non-random mating, 92
North Atlantic Salmon Conservation Organization (NASCO) 302–3
   guidelines
     for rebuilding programmes, 353
     for stocking programmes, 353
Norwegian Atlantic salmon gene bank programme, 453–7
nuclear DNA (nDNA)
   characteristics and structure, 58–60
   and continental divergence studies, 129–31
   organisation, 63–6
null alleles, 92

oceans
   feeding areas, 37
   fishing methods, 301–3
   marine distribution patterns, 37
odours and mate choice, 179
oligotrophic streams, 27
Oncoryhynchus clarkii (cutthroat trout), 21
   evolutionary relatedness, 120–1
Oncoryhynchus genus, 21, 120–1, 240
Oncorhynchus gorbuscha (pink salmon), 240, 280–1
   effects of fishing, 314
Oncorhynchus keta (chum), 280–1
Oncorhynchus kisutch (coho salmon), 281–2
Oncorhynchus nerka (sockeye salmon), 314
Oncorhynchus tshawytscha (Pacific salmon)
   evolutionary relatedness, 120
   identification of stock, 280–2
   and selection practices, 312–13
   stock identification techniques, 280–2
Ouananiche Beck, 44–5
outbreeding depression, 130, 262–3, 371
   mechanisms involved, 263
oxygen requirements, 31

Pacific salmon (*Oncorhynchus tshawytscha*)
   evolutionary relatedness, 120
   identification of stock, 280–2
   and selection practices, 312–13
Pacific Salmon Treaty, 281
Pagrus auratus (New Zealand snapper), 313
parasites, local adaptive response studies, 211, 212–14
parentage assignment, 104–6
   analysis using molecular markers, 169–71, 274
parrs
   growth trajectories, 35
   mature males
     and reproductive success studies, 173–7, 187
     negative consequences of reproductive success, 176
   migration movements, 35
   problems with farm escapes, 362
     recapture programmes, 341
   sexual maturation, 34–5
     and new gene expression techniques, 477–8
   size–reproductive success comparisons, 175
PCA plots *see* principle component analysis (PCA), plots
PCR *see* polymerase chain reactions (PCR)
Penobscot River (Maine), 140, 143, 156, 157, 277
   exploitation rates, 306
phenotypic variation, 74–6
   domesticated vs. wild salmon, 365–8
   and local adaptation, 197–219
   and selective harvesting, 314–17
photoperiod variations, 218
phylogeographic diversity, 121–53
   conclusions and overview, 158–9
   conservation and management implications, 159–60
   gross differences and range-width, 123–31
   historic origins, 147–53
   in Eastern Atlantic salmon, 131–7
   in resident (non-anadromous) salmon, 143–7
   in Western Atlantic salmon, 137–43
   see also genetic variation
pink salmon *see Oncorhynchus gorbuscha* (pink salmon)

Pleistocene ice sheets, 148–53
Poecitiopsis monacha, 244–5
point mutations, 71–2, 89
pollution effects, 411
polymerase chain reactions (PCR), 80, 81
polymorphism, 76
   and molecular markers, 101–2
polytypic species, 21
population
   concept and terminology, 87
   identification techniques, 280–9
   importance of diversity, 240–2
   size reductions
      and effective population size, 243–51
      effects of genetic drift and selection, 251–6
      effects of inbreeding, 256–9
      gene flow and local adaptation, 259–63
      loss of genetic variation, 240–3
   study limitations and issues, 102–7
   types of study, 102–3
      detecting declines, 104
      effective population size estimates, 104
      mixed stock analysis and assignment tests, 103–4
      parentage assignment, 104–6
      relatedness estimates, 106–7
   see also phylogeographic diversity
population genetics, 89–93
   basic concepts, 89–90
   differentiation, 92–3
   structural models, 90–1
   see also phylogeographic diversity
population size estimates *see* effective population size (Ne)
population viability analysis (PVA), 340
principle component analysis (PCA) plots, 139
protein electrophoresis, 77–9
protein synthesis, cellular processes, 67
purging effects, 251
PVA *see* population viability analysis

quantitative genetics, 93–5
   cf. population genetics, 93–4
   genotype by environment interaction, 95
   integration with molecular genetics, 95
quantitative trait loci (QTL), 82–3, 108, 199

rainbow trout, 21
   evolutionary relatedness, 120–1
ranching
   described, 327
   see also stocking programmes

recessive alleles, 69
recolonisation programmes
   deliberate restocking programmes, 330–9
   natural occurrences, 333
   see also stocking programmes
recombinant DNA techniques, 475
recombination, 72, 89
redd superimposition, 29
   see also multiple mating patterns
re-diploidisation, 65–6
regulation of fisheries, 302–3
   potential consequences, 317–18
relatedness patterns
   estimate analysis, 106–7
      using molecular markers, 169–70
   and fitness, 185–6
   see also evolutionary relatedness
replication errors (DNA content), 61, 71–2
reproduction patterns
   anadromous populations, 28–30
      male parrs vs. anadromous males, 173–7
   barriers, 87–8
   non-anadromous populations, 42–5
   sex ratios, 29
   see also mating systems
reproductive isolation, 203–4
reproductive success estimates, natural conditions, 177–8
resistance to parasites, and local adaptation, 211
restricted maximum likelihood (REML) variance estimates, 94
river entry, 40
river fishing *see* in-river fisheries
River Ason (Spain), 195, 308
River Bidasoa (Spain), 308
River Brandon (Ireland), 131
River Burrishoole (N. Ireland), 210, 307, 326, 377–8
River Bush (N. Ireland), 307
River Connon (Scotland), 325
River Dee (Scotland), 156, 177–80
River Dove (England), 330
River Drammenselev (Norway), 308
River Eira (Norway), 308
River Erne (N. Ireland), 307
River Esk (Scotland), 307
River Isma (Norway), 308
River Kem (Russia), 41
River Kennet (England), 332
River Laerdalselv (Norway), 308
River Mersey (England), 330
River Nansa (Spain), 308

River Neva (Russia/Baltic), 147
River Nidelva (Norway), 41
River Otra (Norway), 41
River Polla (Scotland), 156, 357
River Scaur (Scotland), 325
River Selja (Estonia), 330–31
River Shin (Scotland), 325
River Spey (Scotland), 308
River Tana (Finland), 308
River Tay (Scotland), 399
River Tummel (Scotland), 336–7, 399
River Tyne (England), 334, 335–6
River Vig (Russia), 41
River Wye (Wales), 308
RNA polymerase, 66
RNA (ribonucleic acid), 66
'run timing', 39–40
Ryman–Laikre effect, 341, 344

Saguenay River (Quebec), 42, 140, 154
Salmo salar (Atlantic salmon)
   background and history, 471–2
   classification and taxonomy, 21
   current concerns, 472–3
   distribution and geographic range, 21–2
   future considerations, 473–8
   genetic factors
      amino acid sequences, 83
      amount of DNA material, 60
      characterisation of wild populations, 95–102
      gene map, 68
      number of genes, 68
   hybrids, 180–4, 374–5
   life history variations, 22–5
   sex determination, 63
   survival rates, marine phases, 37–8
   see also anadromous populations; non-anadromous populations
Salmo salar salar subgroup, 132, 143, 160
Salmo salar sebago subgroup, 132, 143, 160
Salmo trutta (brown trout)
   evolutionary relatedness, 120–2
   hybridisation, 30, 180–4
      with farm-escaped salmon, 374–5
   interspecific competition, 30
Salmon River (New Brunswick), 306
Salmonidae subdivisions, 120–2
Salmothymus obtusirostris, 121–2
Salvrthymus genus, 121
Sand Hill River (Labrador), 305
satellite DNA, 69

scale analyses, 271
Scotland, fisheries exploitation rates, 307–8
sea cages
   escapes, 361–2
   recommendations to reduce escapes, 385–6
sea migration, 36–8
sea surface temperatures, 38
'Sebago salmon', 42
selection, 89, 123
   and fishery exploitation, 304, 309, 312–13
   measuring equilibrium deviancy, 92–3
   in small populations, and genetic drift, 251–6
   see also domesticated stock
Selja River (Estonia), 277
sequencing (DNA) *see* DNA base sequences
sex determination, 63–4
sex ratios, 29
   and effective population size, 248–9
sexual characteristics, anadromous populations, 29
sexual maturation, 34–5
   effects of selective harvesting, 315
   see also reproduction patterns; spawning behaviours
Shannon River, 131
single nucleotide polymorphisms (SNPs), 101–2, 169
size *see* body size of populations; size of populations
size of populations
   impact on genetic variation, 240–2
      measuring loss, 242–3
   see also effective population size (N_e); small populations
skin colourations and markings, 19–20, 36
Skjern River (Denmark), 277
small populations
   general characteristics, 240
   and effective population size, 243–51
   effects of genetic drift and selection, 251–6
   effects of inbreeding, 256–9, 260–2
   gene flow and local adaptation, 259–63
   loss of genetic variation, 240–3
Smallwood Reservoir (Laborador), 415
smolts, migration patterns, 36–8
'sneaking' behaviours, 29, 35, 477–8
social structures, 184–6
   definitions, 168
   kin recognition, 107, 185
   kin-biased behaviours, 185–6
   relatedness patterns and fitness, 185–6
SPAM software program, 285

spatial boundaries, 154–6
spawning behaviours
    female nest construction behaviours, 28–30
    mature parrs, 35, 173–7, 187
    multiple events, 29, 178–80, 184, 187–8
'spring fish', 338
St Croix River (New Brunswick), 143
St-Jean river (Quebec), 140, 143, 277
St John River (New Brunswick), 276, 347, 349–50
Ste-Marguerite River (Quebec), 177–8
'stepping stone model' of population structure (Kimura and Weiss), 90
sterile salmon, 384–6, 389
Sternopygus obtusirostris, 121
Stewiacke River (Canada), 459–62
stocking programmes, 326–53
    described, 327
    enhancement scenarios, 343–5
        management recommendations, 345
    impact on wild populations, 328
    implications of stock transfers, 130–1, 160
    mitigation scenarios, 345–51
        management recommendations, 351
    nature of strains used, 327–8
        and farm strains, 338–9
    rehabilitation scenarios, 339–42
        management recommendations, 342
    reintroduction where salmon are extinct, 330–9
        genetic considerations, 338
        management recommendations, 339
        time considerations, 338
    transplants between rivers, 143
        and local adaptation studies, 208
STRUCTURE method, 274
'subdivided populations', 310–12
subspecies designations, support evidence, 132, 143–7
Sunndalsora strain, 365
survival rates, marine migration phases, 37–8

tagging, 156, 255
    and spatial boundaries, 154–6
    traditional identification methods, 271
    use of genetic identification methods, 271–3
        individual assignments, 273–80
        population assignments, 280–9
        recommendations, 292–3
taxonomic classifications, 21
Terra Nova River (Newfoundland), 305
territory issues, establishing locations, 34
tetraploidisation, 65
TF see transferrin (*TF*) variation analysis studies
thermal requirements
    anadromous populations, 27, 31, 38
    and local adaptation, 200
Torrent River (Newfoundland), 305
traits and gene phenotypes, 70
    analysis, 74–6, 94–5
    effects of selective harvesting, 315
    and local adaptation, 197–219
    see also genetic variation
transcription factors, 66
transferrin (*TF*) variation analysis studies, 123
transgenic salmon, 384–5
translocated populations, and local adaptation studies, 208
triploidy, 65, 387, 389
Tummel River (Scotland), 336–7, 399

Ungava Bay (Quebec), 41, 416

variation amongst individuals *see* genetic variation; phylogeographic diversity
video surveillance measures, 386
VNTR loci, 365

Wahlund effect, 92
Walton, Izaak, 196
water temperatures
    and global warming, 412–18
    and phenotypic traits, 202, 218
    requirements for anadromous populations, 27, 31, 38
Western Arm Brook (Newfoundland), 305
Western Atlantic salmon
    phylogeographic diversity, 137–43
        evidence for evolutionary divergence, 123–30